STEP 7
Programming
MADE EASY
IN LAD, FBD, and STL

STEP 7

Programming

MADE EASY

IN LAD, FBD, and STL

A Practical Guide to
Programming S7300/S7-400
Programmable Logic Controllers

FIRST EDITION

C.T. Jones

This book is dedicated to my boys
Bryan and Isaac—both admirable young men

About This Book

STEP 7 Programming Made Easy, in LAD, FBD, and STL is a comprehensive STEP 7 programming book. The book is a follow-up to *STEP 7 in 7 Steps*, which is a comprehensive configuration guide for the various components of the S7-300 and S7-400 controllers—including the network configuration. These two books are independent of one another, yet they are also a complimentary set.

STEP 7 Programming Made Easy is organized into the following parts:

Part I—Introduction to STEP 7

Part I starts with a thorough discussion of S7 memory, including a description of each addressable memory area and the application uses of each area. The first chapter is rounded out with a discussion of the various S7 data types and the data formats and representations of each data type. Next covered, is a complete chapter on the S7 data block—a primary data storage resource for STEP 7 programs. Understanding these two chapters is essential for working with the book's remaining topics.

Part II—Ladder Diagrams (LAD) and Function Block Diagrams (FBD)

The first chapter of Part II introduces the LAD and FBD programming languages—their operations categories, main construction elements, basic operations, and editing guidelines. Each of the remaining chapters of Part II presents a LAD/FBD instruction set category—starting with Binary Logic Operations and ending with Program Flow Control and Block Operations. Each chapter begins with an overview and discussion of how this group of operations is generally applied. Then in a side-by-side fashion, descriptions and programming for each operation are presented for LAD and FBD.

Part III—The Statement List (STL) Language

The first chapter of Part III introduces the STL programming language—its operations categories, main construction elements, basic operation, and editing guidelines. Each of the remaining chapters of Part III presents an STL instruction set category—starting with Binary Logic Operations and ending with Program Flow Control and Block Operations. Each chapter begins with an overview and discussion of how this group of operations is generally applied. Following the STL operations category overview, the chapter presents a discussion of individual operations and the associated programming.

Part IV—The STEP 7 Program

The chapters of Part IV present a systematic approach to designing and implementing a STEP 7 program. In the first of these chapters you are introduced to a method of defining a machine or process as a grouping of logical or functional units that should lead to a definition of the STEP 7 program structure. Next you're introduced to basic STEP 7 program building blocks; you'll then revisit S7 I/O module addressing and how to assign symbolic addresses for convenient interpretation of your S7 program. You'll also learn key basics of organizing calls to the other blocks of your S7 program, using OB1—the main STEP 7 program block.

Part IV is rounded out with two final chapters that put it all together by respectively covering the general application and guidelines for developing your STEP 7 program, using Functions, Function Blocks, and Organization Blocks.

About the Author

C.T. Jones, a graduate of Howard University, has made his career in the Industrial Automation Industry for over 25 years. During his career he has served both as an automation user and in automation vendor roles including product marketing, applications engineering, and technical instructor. It was in the role of Electrical Systems Engineer, for Procter & Gamble, that a fascination with PLCs and their endless possibilities in automation drew him into the field that he has passionately pursued ever since.

Early on as an Electrical Systems Engineer, Mr. Jones became aware of the difficult position of end users given the tough balancing-act of keeping a plant operational, while staying abreast of automation technologies. As a result, he co-authored the premier book on PLC Technology—"Programmable Logic Controllers Concepts and Applications." His second book, "Programmable Logic Controllers—The Complete Guide to the Technology," was designed to give students and practicing engineers a ready guide to sound principles and practices for implementing all aspects of PLC technology.

While serving as Applications Engineer and Technical Instructor for Siemens, Mr. Jones devoted much of his time to developing step-by-step application notes for specific subjects and tasks of which users could immediately take advantage. This concept of providing task-related information, in a step-by-step approach, is reflected in Mr. Jones' most recent book "STEP 7 in 7 Steps—A Practical Guide to Implementing S7-300/S7-400 Programmable Logic Controllers." In this latest book, the goal was to present a structured approach to programming in LAD, FBD, and STL, and at the same time present STEP 7 programming concepts in an easily understood fashion.

Contents

Part I

Introduction to STEP 7

Chapter 1

Memory and Addressing

This chapter begins by introducing the different areas of which the S7 memory is comprised, along with a brief description of how each area is used. With this introduction you'll become familiar with the basic memory arrangement and usage concepts of bits, bytes, words, and double words. Becoming familiar with how these memory units are numbered and grouped is the first step in learning to access elements such as digital I/O, analog I/O, bit memory, temporary local memory, timers, and counters in your control program.

After learning about the S7 memory areas, we'll take a look at three methods of accessing these areas in your program. For example there's absolute addressing, the standard method in which you specify a memory area identifier and a specific location; there's symbolic addressing, in which you assign a name to each absolute address to aid in interpreting the I/O and data variables in your program; and finally we'll look at indirect addressing, a method in which the target address is referenced indirectly using address registers and pointers.

The chapter is rounded out with a discussion of the various data types and number formats employed by STEP 7. Here, you'll gain insight into the application of each elementary, complex, and parameter data type. You'll learn how each type is represented and interpreted by STEP 7. A sound grasp of S7 data characteristics and the various formats in which data is represented will be essential as you start to define variables in the code blocks and data blocks of your program.

Main Topics

- S7 Memory Areas
- Absolute Addressing
- Symbolic Addressing
- Indirect Addressing
- Immediate Addressing
- Data Types and Formats

S7 MEMORY AREAS

S7 controllers, like other PLCs, contain memory areas that you will reference in the user program. For example, you will need to examine the status of digital inputs, control the status of digital outputs, read data from analog inputs and vary the signal to analog outputs. There are also timers and counters to be evaluated, and many other operations that require understanding of S7 memory and addressing.

In the S7-300/S7-400, the areas you need to become acquainted with include the *input image (I), output image (Q), bit memory (M), peripheral input (PI), peripheral output (PQ), local (L) stack, timer (T),* and *counter (C)* memory. Take a few moments to review Table 1-1—particularly the ID (identifier) column. After a quick review of the basic memory units by which the locations of these areas are accessed, we'll examine how each area is used. The STEP 7 data block, a static resource for storing program and process data, is presented in Chapter 2.

Table 1-1. Overview of S7 Memory Areas

Area	ID	Brief Description
Input Image	I	This area is a snapshot image of the '0' or '1' status of each connected discrete input. This image is read at the start of each CPU cycle—the current states of the inputs are used to evaluate the control program.
Output Image	Q	This area is an image of the '0' or '1' status of each connected discrete output, as determined by the control logic during the execution of the program. At the end of each CPU cycle, the values in this table are transferred to the output modules thereby switching the digital outputs accordingly.
Bit Memory	M	This area provides storage for interim results in your program—equivalent to scratchpad memory or to control relays. Sections may be defined as retentive.
Peripheral Input	PI	This area allows direct I/O bus read access to devices external to the CPU—for example discrete inputs, analog inputs, and intelligent modules.
Peripheral Output	PQ	This area allows direct I/O bus write access to devices external to the CPU—for example discrete outputs, analog outputs, and intelligent modules.
Timer	T	This area provides for storage of S7 Timer words—the locations accessed by the timer instructions in your program.
Counter	C	This area provides for storage of S7 Counter words—the locations accessed by the counter instructions in your program.
Local Stack	L	This area is dynamically allocated to each FC, FB, and OB at runtime. After a block is terminated, the local stack is reallocated to the next called block.
Data Block (Shared)	DB	Shared data blocks provide storage for the process data requirements of your program. This resource is part of the S7 program—not of system memory.
Data Block (Instance)	DI	Instance data blocks provide memory for data requirements of the associated Function Block—a resource of the S7 program—not of system memory.

Bits, Bytes, Words and Double Words

As described in the following discussions, each S7 memory area is organized in bytes starting from byte-0. In physical memory, of course, this starting byte-0 is a relative start for each area—but for practical purposes byte-0 is always the starting point for each area. With the exception of the timer, counter, peripheral input, and peripheral output areas, each area supports access to bit, byte, word, and double word locations. Each of these basic units of memory is illustrated in the following figure.

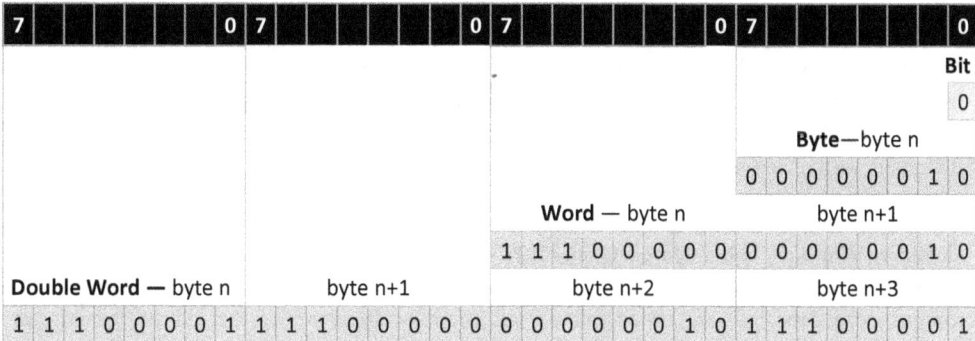

Figure 1-1. Illustration of S7 bit, byte, word, and double word memory locations

Bit Location

A bit location occupies the smallest unit of S7 memory—a single cell that contains a '1' or '0' status. These two binary values represent the two state conditions of a bit—for example On/Off, Open/Closed, Extended/Retracted, Running/Stopped, and True/False. Bit-7 is shaded below.

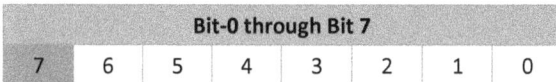

Bit-0 through Bit 7							
7	6	5	4	3	2	1	0

Figure 1-2. Memory bit illustration. Bit-7 is highlighted

Byte Location

A byte location is comprised of 8-bits that are numbered from 0-to-7 moving from right to left. The least-significant bit is bit-0; the most-significant bit is bit-7.

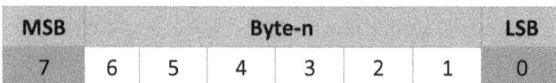

MSB		Byte-n				LSB	
7	6	5	4	3	2	1	0

Figure 1-3. Byte illustration. MSB and LSB are highlighted

Word Location

A word location is comprised of 16-bits that occupy two bytes in S7 memory—byte-n and byte n+1. Byte-n represents the starting byte of a word address, and the number that you specify when specifying the word address. Word-4 in a given area, for example, would be comprised of byte-4, and byte-5. Usually word addresses are assigned to start on even byte boundaries.

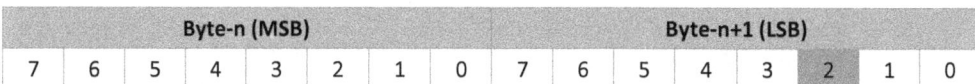

Byte-n (MSB)								Byte-n+1 (LSB)							
7	6	5	4	3	2	1	0	7	6	5	4	3	2	1	0

Figure 1-4. Word illustration. An S7 word consists of byte-n—the MSB, and byte-n+1—the LSB

Double Word Location

A double word location is comprised of four bytes—byte-n, byte n+1, byte n+2, and byte n+3. Byte-n is the starting byte of a double word address, and the number you specify when specifying the double word address. Double word-2 in a given area, for example, would be comprised of byte-2, 3, 4, and byte-5. Like word addresses, double word addresses usually are assigned on even byte boundaries.

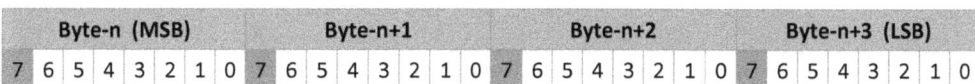

Byte-n (MSB)								Byte-n+1								Byte-n+2								Byte-n+3 (LSB)							
7	6	5	4	3	2	1	0	7	6	5	4	3	2	1	0	7	6	5	4	3	2	1	0	7	6	5	4	3	2	1	0

Figure 1-5. Double word illustration. Byte-n—the MSB, byte-n+1, byte-n+2, and byte-n+3—the LSB

ABSOLUTE ADDRESSING

The previous topic briefly introduced S7 memory areas and the addressable memory units that may be referenced in the STEP 7 program. In STEP 7, there are three methods of addressing: *absolute*, *indirect*, and *symbol* addressing. Since absolute addressing is the most commonly used method, that's where we'll start. Later we'll review how each area is addressed using indirect and symbol addressing.

In STEP 7, a unique *area identifier* indicates that a memory location is in a specific area. These area identifiers, shown in Table 1-1, are **I, Q, M, PI, PQ, L, T, C, DB**, and **DI**. In order, the identifiers are the input image, output image, bit memory, peripheral input, peripheral output, local memory, timer, counter, global data block, and instance data block. There are also identifiers that indicate the unit size of the location—'**B**', '**W**', or '**D**' to indicate byte, word, or double word respectively.

Absolute addressing, also referred to as *direct addressing*, is when an address is referenced by directly specifying an area and size identifier, followed by a 'start byte'. As each area is described, we'll see how the basic memory units of bit, byte, word, and double word are accessed with absolute addressing. In an absolute address, a bit is referenced by an area identifier followed by a byte and bit number separated by a period—for example **I 33.5** is bit 5 of input byte 33. A bit does not use a size identifier.

Input Image (I)

The S7 *input image* is also called the *process image of inputs (PII)*. Prior to each program cycle, the CPU reads the status of digital inputs and stores the results in the input image. The input circuits of each module are mapped onto the input image based on the starting byte address that is assigned to the module in the hardware configuration. Chapter 27 covers details of input module addressing.

You can address input image locations in the program using the area identifier '**I**'. An *input bit* is addressed **I byte.bit**—for example **I 9.7**; an *input byte* is addressed **IB n**—for example **IB8**; an *input word* is addressed **IW n**—for example **IW8**; and an *input double word* as **ID n**—for example **ID8**. The input image double word **ID8** and its mapping to an input module are illustrated below.

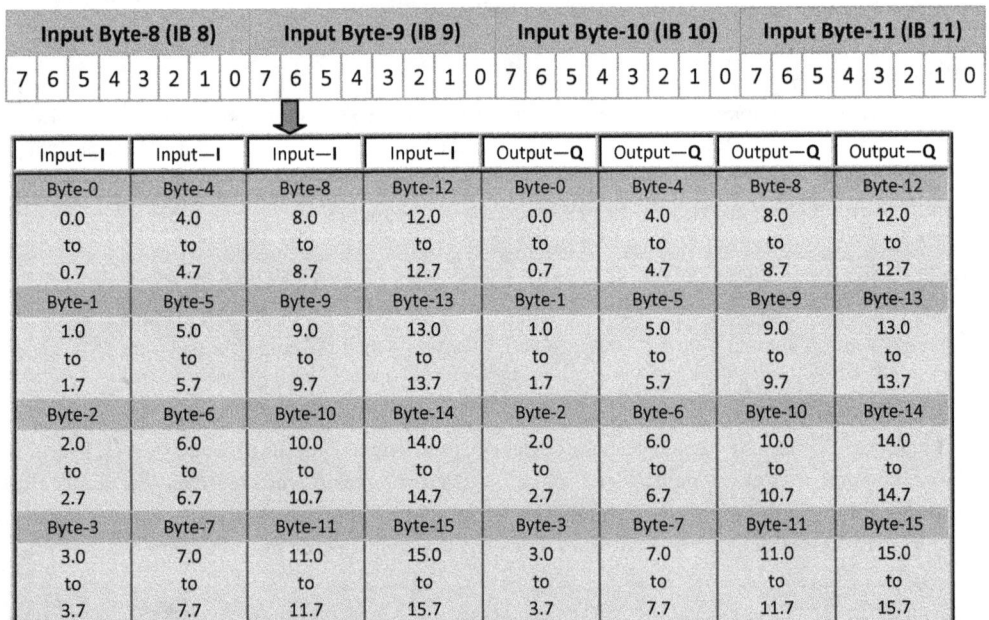

Input Byte-8 (IB 8)								Input Byte-9 (IB 9)								Input Byte-10 (IB 10)								Input Byte-11 (IB 11)							
7	6	5	4	3	2	1	0	7	6	5	4	3	2	1	0	7	6	5	4	3	2	1	0	7	6	5	4	3	2	1	0

Input—I	Input—I	Input—I	Input—I	Output—Q	Output—Q	Output—Q	Output—Q
Byte-0	Byte-4	Byte-8	Byte-12	Byte-0	Byte-4	Byte-8	Byte-12
0.0 to 0.7	4.0 to 4.7	8.0 to 8.7	12.0 to 12.7	0.0 to 0.7	4.0 to 4.7	8.0 to 8.7	12.0 to 12.7
Byte-1	Byte-5	Byte-9	Byte-13	Byte-1	Byte-5	Byte-9	Byte-13
1.0 to 1.7	5.0 to 5.7	9.0 to 9.7	13.0 to 13.7	1.0 to 1.7	5.0 to 5.7	9.0 to 9.7	13.0 to 13.7
Byte-2	Byte-6	Byte-10	Byte-14	Byte-2	Byte-6	Byte-10	Byte-14
2.0 to 2.7	6.0 to 6.7	10.0 to 10.7	14.0 to 14.7	2.0 to 2.7	6.0 to 6.7	10.0 to 10.7	14.0 to 14.7
Byte-3	Byte-7	Byte-11	Byte-15	Byte-3	Byte-7	Byte-11	Byte-15
3.0 to 3.7	7.0 to 7.7	11.0 to 11.7	15.0 to 15.7	3.0 to 3.7	7.0 to 7.7	11.0 to 11.7	15.0 to 15.7

Figure 1-6. On a 32-point digital input module that is assigned a start address of IB 8, as shown above, **IB 8** (input byte-8) would be the first of four groups of eight input circuits

Output Image (Q)

The S7 *output image* is also called the *process image of outputs (PIQ)*. The output circuits of each module are mapped to this area based on the starting byte address assigned to the module in the hardware configuration. During each CPU cycle the status of each output is determined by the program and written into the output image. At the end of each program cycle, the CPU transfers the new results of output image to the output modules to reflect the most recent program updates.

You can address output image locations in the program using the identifier 'Q'. An *output bit* is addressed **Q byte.bit**, for example **Q5.2**; an *output byte* is addressed **QB n**, for example **QB5**; an *output word* is addressed **QW n**, for example **QW4**; and an *output double word* as **QD n**, for example **QD4**. The *output double word* **QD 4** and its mapping to an output module are illustrated below.

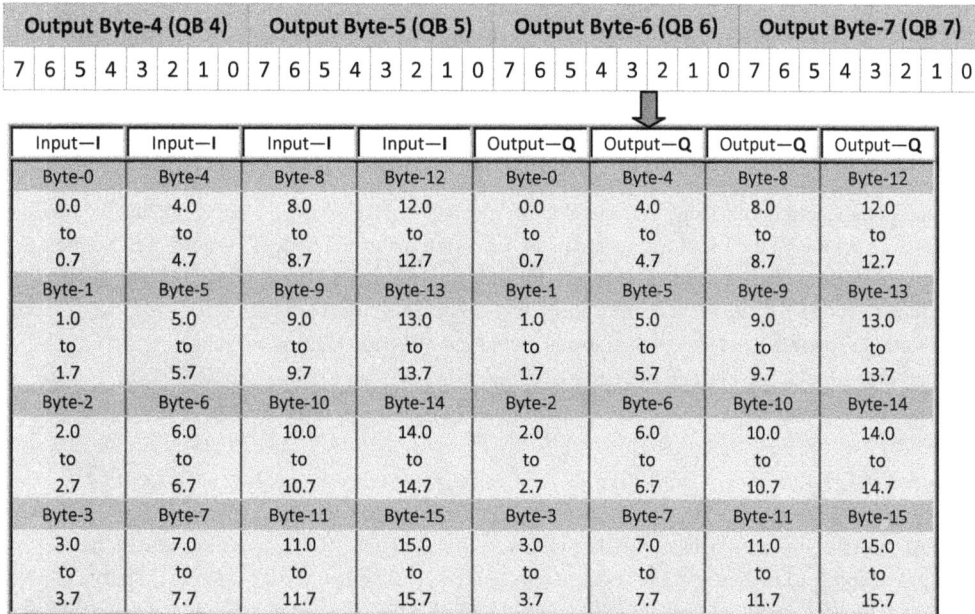

Output Byte-4 (QB 4)								Output Byte-5 (QB 5)								Output Byte-6 (QB 6)								Output Byte-7 (QB 7)							
7	6	5	4	3	2	1	0	7	6	5	4	3	2	1	0	7	6	5	4	3	2	1	0	7	6	5	4	3	2	1	0

Input—I	Input—I	Input—I	Input—I	Output—Q	Output—Q	Output—Q	Output—Q
Byte-0	Byte-4	Byte-8	Byte-12	Byte-0	Byte-4	Byte-8	Byte-12
0.0 to 0.7	4.0 to 4.7	8.0 to 8.7	12.0 to 12.7	0.0 to 0.7	4.0 to 4.7	8.0 to 8.7	12.0 to 12.7
Byte-1	Byte-5	Byte-9	Byte-13	Byte-1	Byte-5	Byte-9	Byte-13
1.0 to 1.7	5.0 to 5.7	9.0 to 9.7	13.0 to 13.7	1.0 to 1.7	5.0 to 5.7	9.0 to 9.7	13.0 to 13.7
Byte-2	Byte-6	Byte-10	Byte-14	Byte-2	Byte-6	Byte-10	Byte-14
2.0 to 2.7	6.0 to 6.7	10.0 to 10.7	14.0 to 14.7	2.0 to 2.7	6.0 to 6.7	10.0 to 10.7	14.0 to 14.7
Byte-3	Byte-7	Byte-11	Byte-15	Byte-3	Byte-7	Byte-11	Byte-15
3.0 to 3.7	7.0 to 7.7	11.0 to 11.7	15.0 to 15.7	3.0 to 3.7	7.0 to 7.7	11.0 to 11.7	15.0 to 15.7

Figure 1-7. On a 32-point digital output module that is assigned a start address of byte-4, as shown above, QB **4** (output byte 4) would be the first of four groups of eight output circuits

Bit Memory (M)

S7 *bit memory* is the same as the internal storage or scratch pad memory area of some PLCs. This area is typically used when a bit, byte, word, or double word location is needed for intermediate binary or digital operations. For example, internal outputs are generally needed for control logic interlocking. Intermediate results in arithmetic or other data operations might also use bit memory.

You can address bit memory locations in the program using the identifier '**M**'. A bit location is addressed **M byte.bit**, for example **M 101.2**; a byte is addressed **MB n**, for example **MB100**; a word is addressed **MW n**, for example **MW100**; and a double word as **MD n**, for example **MD100**. When configuring CPU parameters, contiguous areas of bit memory may be defined as retentive—otherwise, bit memory is non-retentive and is lost when power is removed.

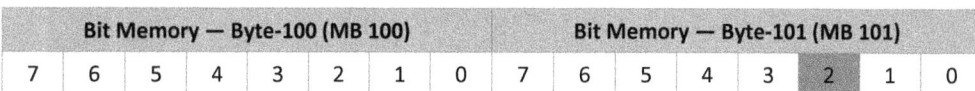

Bit Memory — Byte-100 (MB 100)								Bit Memory — Byte-101 (MB 101)							
7	6	5	4	3	2	1	0	7	6	5	4	3	2	1	0

Figure 1-8. Memory word-100 (MW100): A bit address is written M byte.bit—for example M101.2

Peripheral Input (PI)

Use of the *peripheral input memory* lets you access the data of an input module by allowing you to read directly from the module. Not only do the addresses of this area provide direct read access to an input module, but the read data reflects the instantaneous data at the point at which you access the module in your program. You'll see later, that analog input module data is always read using this memory area.

The address identifier for peripheral input is **PI**. When reading from peripheral input memory you may access a *peripheral input byte* (**PIB**), *peripheral input word* (**PIW**), or *peripheral input double word* (**PID**). Peripheral input bit addresses are not directly addressable—you may however move the peripheral location to bit-memory or another address area where bit addresses are accessible.

Direct Read Access to Digital Inputs

When connected discrete input modules are read on each scan, the CPU is actually accessing the Peripheral Input memory. The statuses of the inputs at that instance are mapped directly onto the Input Image memory. This is why it is said that the input image is a snapshot of the actual states of the inputs at the time of the input image update. The very nature of this process means that the Input data could be as much as one CPU cycle old.

By accessing digital inputs using peripheral input memory in your program, you eliminate the normal delay associated with the Input Image update at the end of each CPU cycle. This normal I/O update is what is reflected in the Input Image (**I**) memory and addresses like **I 37.5** or **IB 2**, or **IW 44**. In most situations, the Input Image data is quite adequate—especially since most mechanical sensors do not have a repeatability that is shorter than the normal cycle time. **PIW 22** is illustrated below.

PIW 22	Byte-22								Byte-23							
Digital Input Module	7	6	5	4	3	2	1	0	7	6	5	4	3	2	1	0

Figure 1-9. Peripheral input word-PIW 22—comprised of peripheral input bytes PIB 22/PIB 23

In LAD and FBD you can read an input module directly using the MOVE operation, where the desired peripheral input address is specified as the source address (**IN**) of the MOVE—you may specify any location you wish as the destination address (**OUT**). With a digital input module, as shown in the example below, you might read **PIB 22** and use the MOVE operation to update the corresponding Input Image memory byte **IB 22**. In the remainder of the program, you may now examine the Input Image addresses—especially if you need access to the bit addresses.

As shown in the STL part of the example, the Load (**L**) operation is used to access Peripheral Input memory and to read a digital input module directly.

LAD Example: Read Peripheral Input Byte-22	**FBD Example:** Read Peripheral Input Byte-22
Network 1 : Read Peripheral Input Byte 22 I33.7 MOVE ——| |——— EN ENO ——— PIB22 — IN OUT — IB22	Network 1 : Read Peripheral Input Byte 22 MOVE I33.7 — EN OUT — IB22 PIB22 — IN ENO

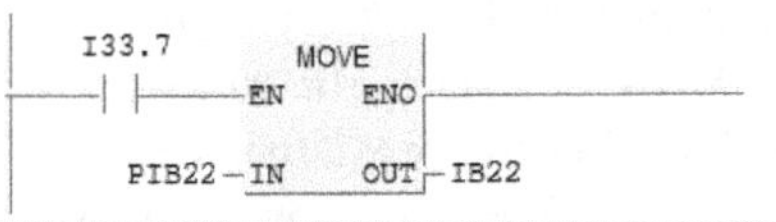

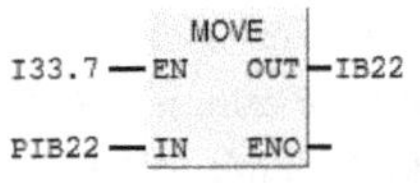

STL Example: Read Peripheral Input Byte-22
L PIB 22   // load peripheral input byte—PIB 22 to accumulator 1.
T IB 22   // transfer accumulator 1 contents (PIB 22) to input image byte IB 22.

Figure 1-10. Illustration of STEP 7 code for immediate read access of a digital input module

Reading Analog Inputs

Unlike digital inputs, analog inputs are not read cyclically to an input image that is subsequently used by the control program. In the S7 world, each analog input channel that is used must be read on demand in the control program. You must access the module directly, for each input channel, using the appropriate address in peripheral input memory.

Since analog input data is represented in a 16-bit word, each analog input address is prefixed with the identifier **PIW** for *peripheral input word*. Peripheral input address **PIW 256** is illustrated below.

PIW 256	Byte-256								Byte-257							
Analog Input Channel	7	6	5	4	3	2	1	0	7	6	5	4	3	2	1	0

Figure 1-11. Peripheral input word-PIW 256—comprised of peripheral input bytes PIB 256/PIB 257

In LAD/FBD the MOVE operation is used to read analog inputs. An analog input channel is read by specifying the associated peripheral input address as the source **IN** of the **MOVE** operation. Let's take for example you want to read the first analog input (channel 0) of the analog input module assigned byte-256 as the start address. You would specify **PIW 256** as the source **IN** of the **MOVE** operation. In the example below, the raw analog value is placed in **MW 44**—the **MOVE** destination.

In STL, as shown in the example, the Load (**L**) operation is used to access Peripheral Input memory and to read an analog input channel. The data read from analog input channel 0 (**PIW 256**) is placed in the Bit Memory location **MW 44**, using the Transfer (**T**) operation.

LAD Example: Read Analog Input Channel 0	**FBD Example:** Read Analog Input Channel 0
| Network 1 : Read Analog Input Channel 0<br><br>```
        MOVE
    ─EN      ENO─
PIW256 ─IN   OUT├─MW44
``` | Network 1 : Read Analog Input Channel 0<br><br>```
             MOVE
   ... ─EN      OUT├─MW44
   PIW256 ─IN   ENO├─
``` |

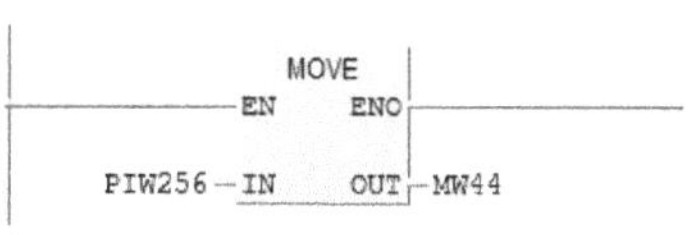

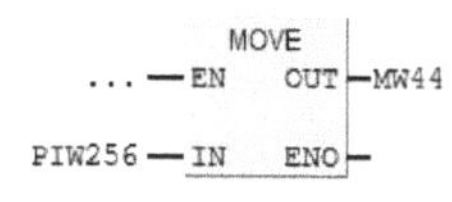

STL Example: Read Analog Input Channel 0
L PIW 256   // load peripheral input word—PIW 256 to accumulator 1 (AIN chan-0).
T MW 44   // transfer accumulator 1 contents to bit memory word—MW 44.

Figure 1-12. Illustration of STEP 7 code for reading input data from an analog input channel

Note: The STL *Load (L)* operation always moves the specified location or variable to accumulator 1. In the STL example shown above, the address of the analog input channel you wish to read is specified as the operand in the load operation. In this case the analog input channel address is PIW 256. Once the value is read into the accumulator, you may then move the value to a temporary or permanent storage location. The value is moved from accumulator 1 using the STL *Transfer (T)* operation. The *Transfer* operation always operates on the value in accumulator 1.

Peripheral Output (PQ)

Use of the *peripheral output memory* lets you control the data of an output module by allowing you to write directly to the module. Not only do the addresses of this area provide direct write access to an output module, but the data that you write to the module is reflected on the outputs at the point at which you access the module in your program. You'll see later that analog output module data is always written using this memory area.

The address identifier for peripheral output is **PQ**. When writing to peripheral output memory you may access a *peripheral output byte* (**PQB**), *peripheral output word* (**PQW**), or *peripheral output double word* (**PQD**). You may not access peripheral output bit addresses directly—you may however move a peripheral output location to bit-memory or another address area where bit addresses are accessible.

Direct Write Access to Digital Outputs

When connected discrete output modules are updated on each scan, the CPU is actually transferring or mapping the Output Image memory directly onto the Peripheral Output memory, which in turn controls the discrete output modules. The status of each output in the Output Image reflects the latest updates from the point at which it was modified in the program. In this normal process the values that are used to update the outputs reflect the status at the instance it changed in the control program. The very nature of this process means that the Output Image data may be as much as one cycle old.

By accessing digital outputs using Peripheral Output memory in your program, you are bypassing the Output Image memory, and writing directly to the addresses of an output module. With this type of access, you eliminate the normal delay associated with addresses like **Q17.3** or **QW 12** of the Output Image table that are updated at the end of each CPU cycle. In most situations, the Output Image data is quite adequate—especially since most mechanical actuators do not have a response that is shorter than the normal cycle time. Peripheral output address **PQW 12** is illustrated below.

PQW 12	Byte-12								Byte-13							
Digital Output Module	7	6	5	4	3	2	1	0	7	6	5	4	3	2	1	0

Figure 1-13. Peripheral output word PQW 12—comprised of peripheral output bytes PQB 12/PQB 13

In LAD/FBD you write to an output module directly using the **MOVE** operation, where the desired Peripheral Output address is specified as the destination (**OUT**). As the source address (**IN**), you may specify the location in which the data you are writing to the output is stored. With an output module, for example, you might write to the 16 bits of **PQW 34**. In doing so, you are immediately updating digital output channels that correspond to the Output Image address **QW 34**. In STL, the Transfer (**T**) operation is used to access Peripheral Output memory and to write to a digital output module directly.

LAD Example: Write Peripheral Output Byte-34	FBD Example: Write Peripheral Output Byte-34
Network 1 : Direct write to Peripheral Output Word 34	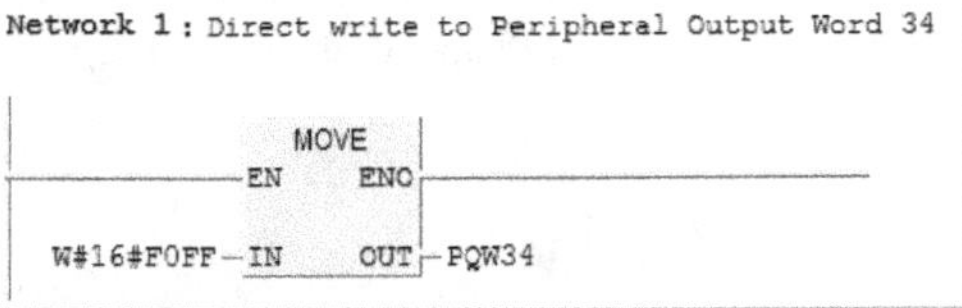 Network 1 : Direct write to Peripheral Output Word 34
| ```
          MOVE
      —EN      ENO—
 W#16#F0FF —IN   OUT —PQW34
``` | ```
             MOVE
     ... —EN      OUT —PQW34
 W#16#F0FF — IN   ENO—
``` |

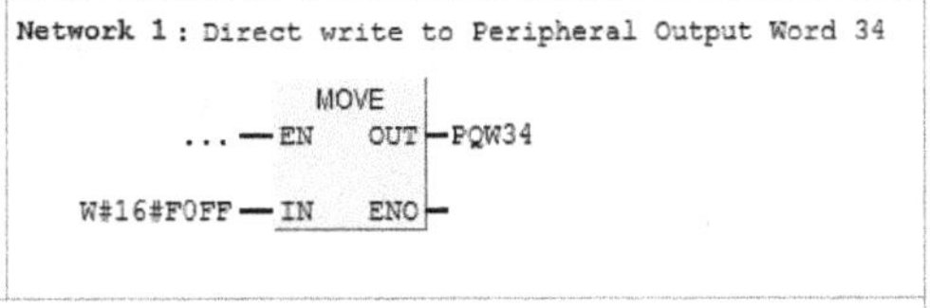

STL Example: Write Peripheral Output Byte-34
L W#16#F0FF   // load the constant value of F0FF (1111 0000 1111 1111) to accumulator 1.
T QW 34   // transfer accumulator 1 contents to output image word 34

Figure 1-14. Illustration of STEP 7 code for immediate write access of a digital output module

Controlling Analog Outputs

Unlike digital outputs that are cyclically updated from an output image, analog outputs are not updated cyclically by an analog output image that is updated by the CPU. Each analog output channel that is used must be written to on demand in the control program. You must access the module directly, for each output channel, using the appropriate address in the peripheral output memory.

Since analog output data is represented in a 16-bit word, each analog output address is prefixed with the identifier PQW for peripheral output word. Peripheral output **PQW 256** is illustrated below.

PQW 256	Byte-256								Byte-257							
Analog Output Channel	7	6	5	4	3	2	1	0	7	6	5	4	3	2	1	0

Figure 1-15. Peripheral output word-PQW 256—comprised of bytes PQB 256, PQB 257

In LAD/FBD the **MOVE** operation is also used to write (control) analog outputs. An analog output is controlled by specifying the desired analog output address as the destination (**OUT**) of the MOVE operation. You may specify the location in which the data you are sending to the output is stored as the source location (**IN**). Let's take for example you want to send the value in MW 44 to the third output channel (channel 2) of the analog output module. The module is assigned a start byte address of PIB 256. You would specify **PQW 260** as the destination (**OUT**) of the **MOVE** operation.

In STL, the Transfer (T) operation is used to access Peripheral Output memory and to write to an analog output channel. In the STL part of the example below, the data in the Bit Memory location MW 44 is written to analog output channel 2—**PQW 260**.

LAD Example: Write to Analog Output Channel 1	**FBD Example:** Write to Analog Output Channel 1

Network 1 : Control Analog Output channel 1
 	
``
        MOVE
    —EN   ENO—
 	
  MW44 —IN   OUT— PQW260
``
	Network 1 : Control Analog Output channel 1
	
	``
	           MOVE
	  ... —EN OUT— PQW260
	
	  MW44 — IN ENO—
	``

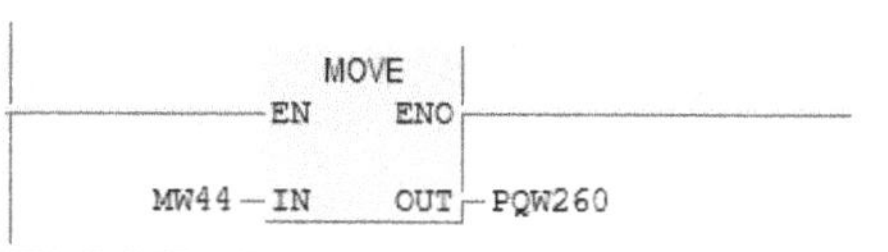

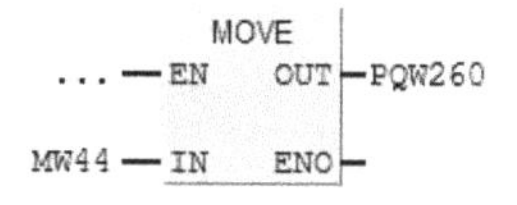

STL Example: Write to Analog Output Channel 1

L MW 44   // load analog output data (bit memory word MW 44) to accumulator 1.

T PQW 260   // transfer accumulator 1 contents to peripheral output word—PQW 260 (AOUT chan-1)

Figure 1-16. Illustration of STEP 7 code for writing data to an analog output channel

Note: The STL *Transfer (T)* operation always moves the contents of accumulator 1 to the specified location or variable. In the STL example shown above, peripheral output address PQW 260 is the address of the analog output channel. The value you wish to transfer to the analog output is first moved to accumulator 1 using the STL *Load* operation. In this case, analog value is in bit memory word MW 44. The *Load* operation always loads the specified value to accumulator 1.

Timer Memory (T)

Timer Memory holds the software timers that you enter in your program. Each timer is stored in one word that contains the preset value and the time base. The preset, a BCD value from 000-999, is stored in bits 0-11; the time base is encoded in bits 12-13 – 00 = 10ms, 01= 100ms, 10 = 1s, and 11 = 10s.

Each timer address, when programmed, is prefixed with the identifier 'T'. Timers are addressed **T0, T1, T2**, and so on, as shown below. The number of available timers is CPU-dependent. In the STEP 7 program, you may use timer addresses to designate a simple timer output or box operation. For further timer details, see the S5TIME data type later in this chapter, under *S7 Data Types and Formats.*

Timer Memory																
Time Base				Timer Preset Value in BCD (000-999)												
15	14	13	12	11	10	09	08	07	06	05	04	03	02	01	00	Timer Word
X	X	0	0	0	1	0	1	0	1	0	1	0	0	0	0	T0
X	X	0	1													T1
X	X	1	0													T2
X	X	1	1													T(n)

Figure 1-17. Illustration of Timer Word format. The preset value of T0 = 550 x .01, or 5.5 sec.

Counter Memory (C)

The *Counter Memory* holds the software counters that you program. Like S7 timers, each counter reserves a single word location. Each counter address, when programmed, is prefixed with the identifier 'C'. Starting with the first counter location, counters are addressed **C0, C1, C2**, and so on, as shown below. The number of available counters is CPU-dependent. In the STEP 7 program, you may use counter addresses to designate a simple counter output or box operation.

Counter Memory																
Bits 12-15 Unused				Counter Value in BCD (000-999)												
15	14	13	12	11	10	09	08	07	06	05	04	03	02	01	00	Timer Word
X	X	X	X	0	1	0	1	0	1	0	1	0	0	0	0	C0
X	X	X	X													C1
X	X	X	X													C2
X	X	X	X													C(n)

Figure 1-18. Illustration of Counter Word format

Local Memory (L)

Local memory (the L-stack) is memory allocated to an S7 block at runtime for handling of temporary variables declared in the block—for example an OB, FC, or FB. When the block terminates, the L-stack is made available to the next called block—and data from the previous block is overwritten, and thus why local memory is temporary.

Generally, temporary (TEMP) variables of local memory are addressed using the symbolic variable names assigned in the declaration section of the block. Local memory variables may also be addressed using their absolute addresses. The identifier 'L' prefixes local memory addresses. A local memory bit is addressed **L byte.bit**, for example **L 42.6**; a local byte is addressed **LB n**, for example **LB 2**; a local word is addressed **LW n**, for example **LW 2**; and a local double word is as **LD n**, for example **LD 4**.

The total size of the local memory is CPU dependent. The size of the local stack available to a block is actually the size allocated to the organization block from which the block is called directly or indirectly. By default, the total local memory of a CPU is equally divided among each OB priority class. The default allocation is 256 bytes. In the S7-400, and in some S7-300s, the allocation of local memory among OB priority classes may be modified.

Absolute Addressing Summary

A bit address is always referenced with the area identifier followed immediately by a byte and bit number separated by a period. **I 33.5**, for example, is bit 5 of input byte 33. For locations other than a bit, the area identifier is followed by a size identifier 'B', 'W', or 'D' to indicate a byte, word, or double word location. These width identifiers are not applicable to timer and counter memory.

A summary of how each S7 memory area is addressed is provided with examples in the following table. With the exception of local memory, each area may be referenced anywhere in the STEP 7 program. You may also reference these areas in the STEP 7 Monitor/Modify Variables utility using absolute or symbolic addresses. The input (**I**), output (**Q**), Local (**L**), and Bit Memory (**M**) memory areas all allow bit, byte, word, and double word access. Peripheral (**PI** or **PQ**) memory supports all address widths except for bit access. Counter and Timer addresses (**Cn** or **Tn**) may be used in bit operations to check the counter/timer status, and in word operations to check the current count or the remaining time.

Table 1-2. Absolute addressing summary for accessing S7 memory areas (continued on next page)

Area	Identifier	Access Units	Example	Description of Example
Input Image (PII)	I	I = Bit	I 30.7	Input Byte 30, bit 7
		IB = Byte	IB 30	Input Byte 30
		IW = Word	IW 30	Input Word 30; bytes 30-31
		ID = Double Word	ID 30	Input Double Word 30; bytes 30-33
Output Image (PIQ)	Q	Q = Bit	Q 44.7	Output Byte 44, bit 7
		QB = Byte	QB 44	Output Byte 44
		QW = Word	QW 44	Output Word 44; bytes 44-45
		QD = Double Word	QD 44	Output Double Word 44; bytes 44-47
Bit Memory	M	M = Bit	M 23.6	Memory Byte 23, bit 6
		MB = Byte	MB 23	Memory Byte 23
		MW = Word	MW 24	Memory Word 24; bytes 24-25
		MD = Double Word	MD 8	Memory Double Word 8; bytes 8-11
Counter	C	C = Counter	C 64	Counter 64
Timer	T	T = Timer	T 12	Timer 12
Local Stack	L	L = Bit	L 2.7	Local Memory Byte 2, bit 7
		LB = Byte	LB 2	Local Memory Byte 2
		LW = Word	LW 2	Local Memory Word 2; bytes 2-3
		LD = Double Word	LD 2	Local Memory Double Word 2

Area	Identifier	Access Units	Example	Description of Example
Peripheral Input & Peripheral Output	PI	PIB = Byte	PIB 44	Peripheral Input Byte 44
		PIW = Word	PIW 66	Peripheral Input Word 66
		PID = Double Word	PID 82	Peripheral Input Double Word 82
	PQ	PQB = Byte	PQB 44	Peripheral Output Byte 44
		PQW = Word	PQW 66	Peripheral Output Word 66
		PQD = Double Word	PQD 82	Peripheral Output Double Word 82

SYMBOLIC ADDRESSING

A symbol is an alphanumeric name or label that is assigned to an absolute memory address or block local variable. In STEP 7 you can actually assign a symbol name to memory locations, code blocks and data blocks, as well as other S7 objects. By assigning a meaningful name to each absolute address, you and others will later find the program easier to interpret, and troubleshoot.

Global Symbols

A *global symbol* is, as the name implies, a symbolic address that may be used anywhere in your program, as a substitute for the associated absolute address. Global symbols may even be used in S7 monitoring utilities. Global symbols are defined in the Symbol Table, using the Symbols Editor. Recall, that the **Symbol Table** is a STEP 7 object located under the **S7 Program** object. Each S7 Program has its own symbol table. In a multi-CPU configuration, each CPU has its own S7 Program, and each program has its own symbol table. As seen in the table below, you may assign global symbols to all S7 memory areas and blocks, as well as the other S7 objects.

Table 1-3. S7 memory areas, blocks, and objects to which global symbols may be assigned

Memory Areas	Block Objects	Other S7 Objects
Inputs (I)	FC	Variable Table (VAT)
Outputs (Q)	FB	User Defined Type (UDT)
Bit memory (M)	DB/SDB	
Timers (T),	SFC	
Counters (C)	SFB	
Peripheral inputs (PI)	OB	
Peripheral outputs (PQ)	DB	

Symbol addresses may be assigned to bit, byte, word, and double word locations for I/O and bit memory areas. For peripheral memory, only byte, word, and double word locations are supported. Table 1-4, shows some of the typical memory units and S7 objects to which you may assign symbolic names As for assigning a symbolic name to a block, you may enter a symbol in the properties dialog, when you create the block or directly in the symbols table at any time.

You'll learn more about global symbol assignments later in the Chapter 27 discussion of STEP 7 program development. There you'll also see that you can assign symbols to discrete and analog I/O module signals using either the Symbol Table Editor or the Hardware Configuration Tool.

Table 1-4. Examples of global symbol addresses assigned to S7 memory areas and blocks

Symbol	Address	Type	Comment
STRT_PMP	I 40.2	BOOL	EMS START PUMP
ES_PSINPUT	IB 30	BYTE	8- EMERGENCY PULL SWITCH INPUTS
ES_PBINPUT	IW 44	WORD	16- EMERGENCY STOP PB INPUTS
SPSET_D1	ID	DWORD	7-DIGIT TWS DRIVE 1 SPEED SETTING
ARM_FWD1	Q 108.5	BOOL	ARM TILT FORWARD 1
M1DOUTS	QW	WORD	SEQUENTIAL OUTPUT WORD OF 16-BITS
SPDSPLY_1	QD	DWORD	7-DIGIT BCD SPEED DISPLAY DRIVE 1
C_LVL	M 4.6	BOOL	COOLING LEVEL LOW
OP_ACK	MB	CHAR	TEMPORARY ACKNOWLEDGEMENT BYTE
CASC_SP1	MW 60	WORD	CASCADE MODE SPRAY 1
SHFT1T1	MD 64	DWORD	SHIFT 1 TEMPORARY TOTAL RUN1
PSIFAULT	PIB 30	BYTE	DIRECT READ DIGITAL INPUT BYTE 30
CWFI1	PIW 260	WORD	COOLING WATER FLOW INDICATOR 1—AIN
AIRVALVES	PQB 30	BYTE	DIRECT WRITE DIGITAL OUTPUT BYTE 30
D1SPC	PQW 486	WORD	DRIVE 1 SPEED CONTROL—AOUT
L1_CTRL	FC1	BLOCK_FC	CONTROL BLOCK LINE 1
SPC_FB1	FB 22	BLOCK_FB	SPECIAL CONTROL PROCEDURE 1
MAINBLK	OB 1	BLOCK_OB	MAIN CYCLICAL PROGRAM BLOCK
PRDDTA	DB 3	BLOCK_DB	PRODUCTION DATA BLOCK STORE
PARMSTR1	DB 22	BLOCK_DB	INSTANCE DATA BLOCK STORE FOR FB22
LOOPVAR	VAT10	VAT	CHART FOR MONITORING LOOP VARIABLES
LOOP1	UDT12	UDT	LOOP 1—PROCESS VARIABLES

Local Symbols

A *local symbol* refers to a variable name defined in the declaration section of a code block. As the name implies, use of the local symbol is restricted to the block in which it is defined. In the figure below, FB53, the two variables of type INT have the local symbol names of CH0_AVG and CH1_AVG; the single BOOL variable has the local symbol name DONE. You'll learn more about local symbol assignments later in the Chapter 27 discussion of STEP 7 program development.

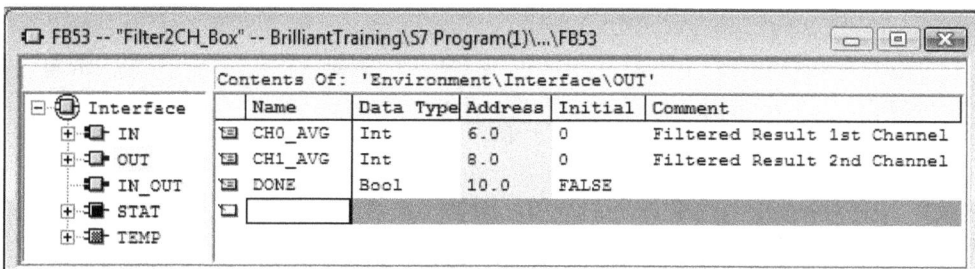

Figure 1-19. Illustration of local symbols

INDIRECT ADDRESSING

Indirect addressing is a method in which the address or value upon which an instruction operates is not specified directly in the statement. Instead another variable, referred to as a 'pointer', indirectly references or 'points to' the target address. By using a pointer, which generally can be manipulated, you can programmatically modify the target address during runtime. In STL, not available in LAD/FBD, there are two types of indirect addressing—*memory indirect* and *register indirect*.

Before getting started with indirect addressing, let's take a look at two pointer formats used by STEP 7 — the *area-internal pointer* and *area-crossing pointer*.

Area-Internal Pointer (P#byte.bit)

The area-internal pointer, as the name implies, points to an address within an S7 memory area. The format of area-internal pointer is represented in a 32-bit double word as shown below. A 16-bit byte address that has a theoretical range of 0-65,535 is stored in bits 3-18; a bit number with a range of 0-7 is encoded in bits 0-2; and bits 19-31 are unused. Together the byte and bit value represent the address you wish to access. In the case of a byte, word or double word address, the byte represents the byte or start byte of the address, and the bit number must be '0'—for example 32.0. Whether the address represents a bit, byte, word or double word address, a bit number must always be specified.

As seen later, with area-internal indirect addressing, the area is explicitly entered in the STL statement.

Area-Internal Pointer																						
31		19	18	17	16	15	14	13	12	11	10	9	8	7	6	5	4	3	2	1	0
0		X	0	0	0	0	0	0	0	0	0	0	0	0	0	0	1	0	0	0	0
bits 19—31 (unused)		byte number (byte range 0-65,535)																	bit no. 0-7		

Figure 1-20. The area-internal pointer format: **P#byte.bit**

Area-Crossing Pointer (P#Abyte.bit)

As with the area-internal pointer, the area-crossing pointer value is represented in a 32-bit double word format. In fact, the format is almost identical. A 16-bit byte address is stored in bits 3-18; and a bit number with a range of 0-7 is encoded in bits 0-2. The difference here is that a memory area ID is encoded in bits 24-26. When the area-crossing method is used, the pointer notation is P#Abyte.bit, where 'A' represents the memory area. This pointer supports addressing across memory area boundaries. Examples of pointer constants using this notation are P#M4.5, and P#Q2.0.

Area-Crossing Pointer																								
31		26	25	24	....	19	18	17	16	15	14	13	12	11	10	9	8	7	6	5	4	3	2	1	0
1		0	0	0	.....		0	0	0	0	0	0	0	0	0	0	0	0	0	0	1	0	0	0	0
area ID (Table 1-8)				byte number (byte range 0-65,535)																	bit no. 0-7			

Figure 1-21. Area-crossing pointer format: **P#Abyte.bit**

Unlike the area-internal pointer, which depends on the STL statement to reference the memory area, the memory area is directly encoded in the area-crossing pointer. The area of the address, for example **P, I, Q, M, DB, DI**, or **L** is encoded in bits 24, 25, and 26 of the pointer—as shown in Table 1-5.

Table 1-5. Binary encoding for S7 memory areas in the area-crossing pointer format

Area	Description	Bits 24-26	Area	Description	Bits 24-26
P	Peripheral I/O	000	DBX	Global Data	100
I	Input Image	001	DIX	Instance Data	101
Q	Output Image	010	L	Local Memory	110
M	Bit memory	011	L	Local Memory (Previous Block)	111

Memory-Indirect Addressing

Memory-indirect addressing is a method in which the target address is contained in a memory location that I will refer to as a *pointer location*. With memory-indirect addressing, there are two types of addresses that you may access—these two categories are distinguished as those accessed using a 16-bit word as the pointer location, and those that use a 32-bit double word as the pointer location.

Memory-Indirect Pointing to Timer, Counter, or Block

Using a word location, as shown in Figure 1-22, you may indirectly access a timer, counter, Function, Function Block, or Data Block (global or instance). In this method, the value in the location is a number that is equivalent to the number of a timer, counter, FC, FB, or DB. The pointer location may reside in bit memory (M), local memory (L), a global data block (DB), or an instance data block (DI). Remember that local memory is temporary, and is only available during block execution.

The word pointer location, shown below, is represented as an unsigned integer that uses bits 0-11. The range for this number, 0-65535, represents the theoretical address range when accessing a timer, counter, FC, FB, and DB—the actual range is based on the CPU limitations on these objects.

Word Pointer															
15	14	13	12	11	10	09	08	07	06	05	04	03	02	01	00
0	0	0	0	0	0	0	0	0	0	0	1	0	0	0	0
unsigned integer number for Timer, Counter, FC, or FB (0-65535 = address range)															

Figure 1-22. Word location to address areas T, C, FC, FC, DB, and DI in memory-indirect addressing

When using memory-indirect addressing, the instruction operand identifies the S7 address memory area and size (if applicable) identifier. Identifiers in this case include **C** for Counter, **T** for Timer, **FC** for Function, **FB** for Function Block, **DB** for global Data Block, and **DI** for Instance Data Block. Some example statements of memory-indirect addressing are shown in Table 1-6.

Table 1-6. Memory indirect addressing examples: Blocks (FC, FB, DB, and DI), Timers, and Counters

Example	Description
OPN DB[LW 12]	Open the global data block whose block number is contained in local word LW 12
CU C[MW 40]	Increment the UP counter whose address is the number in bit memory word MW 40
UC FC[DIW 10]	Make unconditional call to the FC whose number is contained in data word DIW 10
CC FB[DBW 2]	Make conditional call to the FB whose number is contained in data word DBW 2
A T[MW 14]	Check the status of the timer whose address is the number in bit memory word MW 14
OPN DI[LW 14]	Open the instance data block whose block number is in bit memory word MW 12

The examples below both illustrate memory indirect addressing as used to access a timer, counter, FC, FB, or DB. In each of these cases, as is always the case with memory indirect addressing, a number that represents the address is loaded to your pointer location—a word location is required in this case.

Example 1: Memory indirect method to indirectly open a DB

L	+75	// Load an integer value of 75 to accumulator 1
T	MW 4	// Transfer the contents of accumulator 1 (+75) to bit memory word MW 4
OPN	DB[MW 4]	// Open data block **DB75** (value contained in MW 4)

Example 2: Memory indirect method to indirectly start a PULSE timer

OPN	DB20	// Open data block **DB20**
L	+16	// Load the integer value 20 into accumulator 1
T	DBW 2	// Transfer the contents of accumulator 1 (+16) to data word DBW 2
L	S5T#500MS	// Load a timer preset value of 500 milliseconds into accumulator 1
SP	T[DBW 2]	// Start the Pulse timer **T16**, using the preset value of 500 ms

Memory-Indirect Pointing to a Memory Location

Memory-indirect addressing is also used to access addresses in S7 memory areas including input image (**I**), output image (**Q**), bit memory (**M**), peripheral input (**PI**), peripheral output (**PQ**), local (**L**) memory, and data elements (**DB** or **DI**). As with memory-indirect addressing of timers, counters, and blocks, here again a pointer location holds a number that represents the address you wish to access. The pointer location, in this case, is a 32-bit double word. Take a moment to look at that double word format, which was shown in Figure 1-20 (area pointer).

As previously described, the 32-bit pointer location may reside in bit memory (M), local memory (L), a global data block (DB), or an instance data block (DI). Remember that local memory is temporary, and is only available during block execution. Whereas the theoretical range of the byte value represented in bits 0-11 is 65,535, the actual range will be based on the limitations of each memory area in a specific CPU, as well as the start byte value, and the data unit you are attempting to access.

For example, byte address may have an even or odd address, so the largest byte address if supported is 65535. Since word addresses must be divisible by two, the largest address, if supported, is 65534. Finally, double word addresses must be divisible by four, so the largest address, if supported, is 65532. Furthermore, any rule that applies to specific area with absolute addressing is also true here. For example, bit addresses are not possible with peripheral areas PI and PQ.

When using memory-indirect addressing, the instruction operand identifies the S7 memory area and size (if applicable) identifier. Some examples of address identifiers include I (input bit), IB (input byte), IW (input word), ID (input double word), Q (output bit), QB (output byte), QW (output word), and QD (output double word). Notice that with bit address identifiers, like I and Q, no size identifiers are required. Some example statements of accessing S7 memory areas using memory-indirect addressing are shown in Table 1-7.

Table 1-7. Memory indirect addressing examples: S7 memory areas I, Q, M, L, PI, and PQ

Example	Description
A I [MD 4]	Perform AND operation on the input image bit whose address is in MD 4
S Q [DBD 8]	Set the output image bit whose address is in global data double data word DBD 8
L IB [DID 10]	Load to accumulator 1 the input image byte whose address is in instance data double data word DID 10
T PQW [LD 12]	Transfer data from accumulator 1 to the peripheral output word whose address is in local memory double word LD 12
L PIW [MD 8]	Load to accumulator 1 the peripheral input word whose address is in bit memory double word MD 8

In the following examples of memory indirect addressing, the address value is loaded to accumulator 1 as an area-internal pointer constant. The area-internal pointer constant notation of P#byte.bit was introduced earlier. The pointer is then moved from the accumulator to pointer location.

In the first example, the statuses of input word IW 256 are moved to output word QW 256. The pointer in this first example would be the same if you were indirectly accessing input bit address 256.0. The second example indirectly accesses an input bit address and an output address.

Example 1: Memory indirect method to access an input image word and an output image word

L	P#256.0	// Load the pointer value of 256.0 (2#0000 0000 0000 0000 0000 1000 0000 0000)
T	MD 8	// Store the value in the pointer location memory double word MD 8
L	IW [MD 8]	// Load the contents of input image word IW 256 to accumulator 1
T	QW [MD 8]	// Transfer the statuses of the input image word to the output image word QW 256

Example 2: Memory indirect method to access an input bit address and an output bit address

L	P#8.7	//Load the pointer value of 8.7 (2#0000 0000 0000 0000 0000 0000 0100 0111)
T	MD 2	// Store the pointer value 8.7 to the pointer location memory double word MD 2
A	I [MD 2]	// Check the status of the input bit I 8.7
=	Q [MD 2]	// Assign the RLO status to the output bit Q 8.7

Register-Indirect Addressing

Register-indirect addressing is implemented using S7 address registers AR1 and AR2, and supports indirect access to S7 memory areas. There are two reg-indirect methods—*area-internal*, and the *area-crossing* methods. These methods, although similar to memory indirect addressing, are considered more efficient since they use the CPU address registers. Each method, however, has its own merits.

The area-internal and area-crossing methods are differentiated in how the pointer is implemented. With area-internal, the area identifier and data size are explicitly entered in the STL statement that accesses the target address. Because the area identifier is explicitly stated, the area cannot be manipulated—hence the term area-internal. Area-crossing, on the other hand, encodes the area identifier within the pointer—therefore the identifier can be manipulated, and access to different areas is possible. Both the area-internal pointer (Figure 1-20) and area-crossing pointer (Figure 1-21), are represented in 32-bits.

Register Indirect—Area-Internal

Like with memory-indirect addressing, register-indirect area internal addressing uses a two part pointer. The first part is an S7 memory area and ++size identifier, explicitly stated with the instruction operand. Some examples of area-size identifiers include I (input bit), IB (input byte), IW (input word), ID (input double word), Q (output bit), QB (output byte), QW (output word), and QD (output double word). Notice that with bit address identifiers like I, Q, and M, only the area is required. The second part of the pointer is the S7 address register AR1 or AR2, and a byte.bit offset. The final pointer value is the result of adding the pointer value in the 32-bit address register to the specified offset.

Table 1-8 shows some example statements of accessing S7 memory areas using register indirect area-internal addressing.

Table 1-8. Register indirect area-internal addressing examples of areas I, Q, M, PI, PQ, and DB

Example	Description
A I [AR1, P#6.2]	Perform AND operation on the input image bit whose address is derived from the pointer in address register AR1 plus the offset pointer P# 6.2
= DIX [AR2, P#0.0]	Assign the RLO status to the instance data bit whose address is the result of adding to the pointer in address register AR2 the offset pointer P# 0.0
L IB [AR1, P#80.0]	Load to accumulator 1 the input image byte whose address is the result of adding to the pointer in address register AR1 plus the offset of 80 bytes
T PQW [AR2, P#40.0]	Transfer data from accumulator 1 to the peripheral output word whose address is derived from the pointer in address register AR2 plus the offset of 40 bytes

Once a pointer value is loaded in accumulator 1 it may then be loaded to an address register (AR1 or AR2) using the appropriate instruction. Since both the address register and the byte.bit offset have the same format, the final address is calculated by adding the byte value of the address register to the byte offset and the bit value of the address register to the bit value of the offset. In this addition the byte values are added in decimal, the bit values are added in octal (0-7). Since a byte is comprised of 8-bits, a carry over to the byte is required if the result of the bit addition is equal to or greater than 8.

The following examples illustrate how to work with address registers and the register indirect area-internal method for accessing S7 memory. Recall, that using the area-internal pointer notation of P#byte.bit, you can load to the accumulator a byte.bit address, where the byte designates a start byte address; the bit value specifies a bit number, or is zero for byte, word, or double word addresses.

In the first example, output bit Q10.0 is controlled by input image bit I8.7. In the second example, input image byte IB8 is indirectly accessed and then indirectly transferred to bit memory word MW106.

Example 1: Register indirect addressing using the area-internal method with pointer P#byte.bit

L	P#8.7	// Load the pointer value of 8.7 to the accumulator
LAR1		// Store the pointer value of accumulator 1 (P#8.7) in address register AR1
A	I [AR1, P#0.0]	// Complete AND operation on the input image bit whose address is derived by adding the pointer of AR1 (8.7) to the address offset of 0.0 (I 8.7)
=	Q [AR1, P#1.1]	// Assign the RLO status to the output image bit whose address is derived by adding the pointer of AR1 (8.7) plus the byte offset of 1.1. The address operated upon is **Q**10.0.

Example 2: Register indirect addressing using the area-internal method

L	P#6.0	// Load a double word pointer value of 6.0
LAR2		// Store the pointer of accumulator 1 (P#6.0) to address register AR2
L	IB [AR2, P#2.0]	// Load the input byte whose address is derived by adding the pointer of AR2 (6.0) plus a byte offset of 2.0. The address is **IB** 8.
T	MW [AR2, P#100.0]	// Transfer accumulator contents to the bit memory word whose address is derived by adding the pointer of AR2 (6.0) plus a byte offset of 100.0. The address operated upon is MW 106.

Register Indirect—Area-Crossing:

Register indirect area-crossing also uses S7 address registers AR1 and AR2. This method is similar to the register indirect area-internal method, but differs in that it is implemented to allow you to manipulate the address pointer to access addresses across memory areas. In this method the S7 memory area is encoded in the double word pointer, as shown in Figure 1-21. Areas I, Q, M, PI, PQ, L, DI, and DB, as presented in Table 1-5, are each identified by a unique binary code.

In an STL statement that uses the area-crossing method of register indirect addressing, a two part address points to the address that the instruction will process. The first part of the address is a data unit size identifier, like **B** for 'byte', **W** for word, or **D** for 'double word'. There is no indicator for a bit address. The area, again, is encoded in bits 25, 26, and 27 of the address register. The second part of the pointer is comprised of S7 address register AR1 or AR2 plus a specified byte.bit offset. The final pointer value is the result of adding the pointer value in the address register to the specified offset. Some example statements of accessing S7 memory areas using register indirect area-crossing addressing are shown in Table 1-9.

Table 1-9. Register indirect area-crossing addressing examples of areas I, Q, M, PI, PQ, and DB

Example	Description
A [AR1, P#6.2]	Perform AND operation on the bit whose memory area and location is derived from the pointer of address register AR1 plus the pointer offset of P#6.2.
= [AR2, P#0.0]	Assign the current RLO status to the bit whose memory area and location is contained in address register AR2.
L B [AR1, P#80.0]	Load to accumulator 1 the byte address whose memory area and location is calculated by the contents of address register AR1 plus 80 bytes.
T D [AR2, P#40.0]	Transfer data from accumulator 1 to the double word whose memory area and location is calculated by the contents of address register AR2 plus 40 bytes.

Note: In area-crossing addressing the area is encoded in bits 24, 25, and 26 of the pointer (Table 1-5).

With area-crossing the pointer is loaded to accumulator 1 using the pointer notation P#Abyte.bit, where the byte designates the start byte of the location; the bit specifies the bit number, and is '0' in the case of byte, word, or double word addresses; and 'A 'represents the memory area. Examples of this notation are P#M 4.5, and P#Q 2.0.

Once a pointer value is loaded in accumulator 1 it may then be moved to address register AR1 or AR2, using the appropriate instruction. Since both the address register and the byte.bit offset have the same format, the final address is calculated by adding the byte value of the address register to the byte offset and the bit value of the address register to the bit value of the offset. In this addition the byte values are added in decimal, the bit values are added in octal (0-7). Since a byte is comprised of 8-bits, a carry over to the byte is required if the result of the bit addition is equal to or greater than 8.

The examples illustrate working with address registers and the register indirect area-crossing method.

Example 1: Register indirect addressing using the area-crossing method

L	P#I 8.7	//Load the area-crossing pointer constant of I 8.7 to the accumulator.
LAR1		//Store the pointer value (I 8.7) to address register AR1.
L	P#Q 8.7	//Load the area-crossing pointer value of Q 8.7 to the accumulator 1.
LAR2		//Store the area-crossing pointer value (Q 8.7) to address register AR2.
A	[AR1, P#0.0]	//Perform AND operation on the bit whose address is the pointer of AR1 (I 8.7) plus the byte offset of 0.0. The new address operated on is I 8.7.
=	[AR2, P#1.1]	//Assign the RLO status to the bit whose address is the pointer value of AR2 (Q8.7) plus the offset of P#1.1. The new address is Q10.0.

Example 2: Register indirect addressing using the area-crossing method with pointer P#Abyte.bit

L	P#I 6.0	//Load the area-crossing pointer constant of I 6.0 to the accumulator .
LAR2		// Store the pointer value (I 6.0) to address register AR2.
L	P#M 6.0	// Load the area-crossing pointer value of M 6.0 to the accumulator.
LAR1		// Store the area-crossing pointer value (M 6.0) to address register AR1.
L	B [AR2, P#2.0]	// Load the byte whose address is the pointer of AR2 (M 6.0) plus the offset P#2.0 (2 bytes, plus 0 bits). The address operated upon is MB 8.
T	W [AR2, P#100.0]	// Transfer the accumulator 1 contents to the word whose address is the pointer value of AR2 (M 6.0) plus the offset P#100.0 (100 bytes, 0 bits. The new address is MW 100.

Further details for working with S7 address registers AR1 and AR2 are provided in Chapter 17, *Load and Transfer Operations*.

IMMEDIATE ADDRESSING

In STEP 7 *immediate addressing* refers to a method in which the actual value upon which an instruction operates is not accessed from memory, but is provided implicitly or explicitly in one of the following ways: the first method involves a constant that is entered as a parameter immediately following the statement; the second method includes instructions that implicitly operate on the result of logic operation (RLO) or some other value built into the instruction—example operations include SET or NOT, which affect the RLO; and the final method involves instructions that operate on a value currently in accumulator 1 or in some cases in accumulator 2. Where the accumulator is involved, the value would have been previously loaded.

Examples of each of these immediate addressing cases are illustrated in the following table.

Table 1-10. Examples of immediate addressing in STEP 7

	Example	Description
❶	L +75	Load the integer value of +75 to accumulator 1.
❷	L 2#1000_0001_1111_0101	Load the binary value 1000 0001 1111 0101 to accumulator 1.
❸	L 'BUSY'	Load the four CHAR characters 'BUSY' to accumulator 1.
❹	L B#(175, 255)	Load 2 unsigned byte values to accumulator 1.
❺	L C#0999	Load the BCD count value of 0999 to accumulator 1.
❻	AW W#0FFF	Combine hex value with accumulator 1 using AND Word.
❼	SET	Set the result of logic operation (RLO) to 1.
❽	INC 5	Increment value in accumulator 1 by +5 (range = 1-255).
❾	SLD	Shift 32-bits of accumulator 1 by **n**-bits (**n** is in accumulator 2).

DATA TYPES AND FORMATS

A data type is used to assign a defined set of characteristics to a data variable that you use in your program. A variable's type defines its width in bits, how the binary data is represented, and the range of valid values. By assigning a type to a variable, you indicate to STEP 7 how the data contained in the variable should be interpreted. How data is represented and interpreted is referred to as its format—for example, a signed integer, or a character string. This data interpretation affects the numeric range of the variable. Data type INT, for example, supports a range of -32768 to + 32767. These parameters, as you might guess, will dictate the data type that you will assign to each variable you define.

As you start to develop your program, you will need to define variables that are used locally by a code block (OB, FC, and FB) or that are contained in a data block. In either case, as you declare each variable, you will need to explicitly assign its data type. Also, as you begin to specify constant values or variables as instruction parameters, you will need an understanding of each data type and its usable formats. For example, as you enter a timer constant, or an integer constant in an arithmetic operation, you'll need to know how to specify an appropriate value.

The following discussions will introduce STEP 7 *Elementary*, *Complex*, and *Parameter* data types.

Elementary Data Types

Elementary Data Types are that group of data types that represent variables that involve single data elements, all of which are 32-bits or less. Variables of elementary data type may be declared in data blocks, organization blocks, functions, and function blocks. Elementary variables may also be supplied as parameter entries to blocks (FC, FB, SFB, and SFC) or to STEP 7 LAD/FBD or STL instructions. Elementary types include BOOL, BYTE, WORD, DWORD, INT, DINT, REAL, DATE, TIME, S5TIME, TIME-OF-DAY, and CHAR. These types and their basic uses are described in the following discussions.

Table 1-11. Overview of Elementary Data Types

Data Type	Bits	Brief Description
BOOL	1	Assign to bit variables, whose range is TRUE/FALSE, or '0' or '1'.
BYTE	8	Assign to byte (8-bit) variables. Raw binary $0\text{-}255_{10}$; $00\text{-}FF_{16}$.
WORD	16	Assign to word (16-bit) variables. Raw binary 0-65,535.
DWORD	32	Assign to double word (32-bit) variables. Raw binary 0-4,294,967,295.
INT	16	Assign to integer (16-bit) variables, whose range is -32,768 to +32,767.
DINT	32	Assign to integer (32-bit) variables, whose range is -2^{31} to $(2^{31}-1)$.
REAL	32	Assign to IEEE floating point (32-bit) variables.
CHAR	8	Assign to printable ASCII character variables.
S5TIME	32	Assign to timer preset variables, whose range is 0H_0M_0S_0MS to 2H_46M_30S_0MS.
DATE	32	Assign to variables, whose range is IEC Date in **1-day** increments.
TIME	32	Assign to variables, whose range is IEC Time in **1-ms** increments.
TIME_OF_DAY/TOD	32	Assign to variables, whose range is Time-of-Day in **1-ms** increments.

BOOL

Data type BOOL reserves a single bit, for a variable that takes on the Boolean values of **TRUE/FALSE**. In your code blocks for example, you might specify a BOOL variable instead of a bit memory address like M 33.3. The BOOL data type should be assigned to any variable that takes on the values of TRUE or FALSE and that is represented by binary '0' or '1'.

Bit	Bit	Bit	Bit	Bit	Bit	Bit	Bit
07	06	05	04	03	02	01	00
1	0	1	0	0	0	0	0
TRUE	FALSE	TRUE	FALSE	FALSE	FALSE	FALSE	FALSE

Figure 1-23. Eight bits—each bit is declared as a BOOL variable

BYTE

A variable of data type BYTE reserves 8-bits. These bits are not evaluated individually, but as an 8-bit hexadecimal number. Each hex digit is represented in four-bits; therefore a BYTE variable holds a two-digit Hex value. A constant of data type BYTE is specified using the identifier B#16#, where 'B' is for byte, and '16' is for hex. The range of hex values is B#16#00 to B#16#FF. Hex AF_{16} is shown below. The BYTE data type might be assigned to any variable that may be encoded in as a hex value.

8	4	2	1	8	4	2	1	Weight
07	06	05	04	03	02	01	00	Bit No.
1	0	1	0	1	1	1	1	
A				F				B#16#AF

Figure 1-24. A BYTE variable

WORD

Variables of data type WORD reserve 16-bits. The 16-bits of the data type WORD are unstructured and are not evaluated as a unit. This unstructured format allows a WORD variable to support several bit-oriented uses. A WORD variable, for example, may hold a 16-bit binary string of 1s and 0s, a 4-digit Hexadecimal number, a 3-digit count value in BCD, or two unsigned decimal values. Each of the WORD formats is described below.

WORD—As 16-bit String

A WORD variable is specified as a 16-bit binary string using the constant identifier **2#**. An example is 2#0000_1011_1111_1111. The range of this format is 2#0000_0000_0000_0000 to 2#1111_1111_1111_1111. This option might be used in a word logic operation to compare the 16-bits of an input module to an expected 16-bit pattern. Another example might be to manipulate sixteen digital outputs at once. You might use the MOVE operation for this purpose, where the binary pattern would be the source word and the word address of the output module the destination.

32768	-	-	-	-	-	1024	512	128	64	32	16	8	4	2	1	Weight
15	14	13	12	11	10	09	08	07	06	05	04	03	02	01	00	Bit No.
0	0	0	0	0	0	0	1	0	1	0	1	0	1	0		0
1	1	1	1	1	1	1	1	1	1	1	1	1	1	1	1	65535

Figure 1-25. Unsigned 16-bits

WORD—As 4-Digit Hexadecimal/3-Decade BCD

A variable of data type WORD is specified as a hexadecimal number using the identifier **W#16#**. 'W' is for word length, '16' is for hex. An example is **W#16#0FFF**. Each hex digit is represented using four-bits, allowing a 4-digit Hex number in a WORD variable. The permissible range of values for 4-digit Hex is from **W#16#0000** to **DW#16#FFFF**. You might use this format option when up to a 4-digit hex number is required, or when a shorter notation for specifying a binary string is needed—for example **W#16#FFFF** instead of 2#1111_1111_1111_1111. Hex values are often used in Word logic operations.

The 16-bit hex format may also support a 3-decade BCD value when you use only the hex digits 0-to-9 to avoid invalid BCD numbers. The valid range for BCD numbers in a WORD variable is ±999. In this format, bits 12-15 are used to indicate the sign, but only bit-15 is relevant—the remaining bits do not matter. Negative values are denoted by '1111' and positive values are denoted by '0000'. When specified using the 16-bit hex identifier, the 3-decade BCD range is **W#16#F999** to **W#16#0999**.

8	4	2	1	8	4	2	1	8	4	2	1	8	4	2	1	Weight
15	14	13	12	11	10	09	08	07	06	05	04	03	02	01	00	Bit No.
0	0	0	0	1	0	0	1	1	0	0	1	1	0	0	1	
0				9				9				9				W#16#0999

Figure 1-26. WORD variable interpreted as 16-bit hexadecimal. Example value is +999 BCD

WORD—As a Count Value

A variable of data type WORD may also be specified as a counter value, which might be loaded to a counter on command. Recall that each counter value reserves a single word and is interpreted as a 3-digit BCD value. When you specify a counter value as a constant input to an instruction, or as an initial value of a WORD variable declaration, the count value is preceded by the identifier **C#**. The permissible range for this format is **C#000** to **C#999**.

8	4	2	1	8	4	2	1	8	4	2	1	8	4	2	1	Weight
15	14	13	12	11	10	09	08	07	06	05	04	03	02	01	00	Bit No.
1	0	0	1	0	0	0	1	0	0	0	0	0	0	0	0	
0				1				0				0				C#100

Figure 1-27. WORD variable—16-bit hexadecimal interpreted as counter value

WORD—As 2-Bytes Unsigned Decimal

A variable of data type WORD may be specified as 2-bytes of unsigned decimal numbers, using the identifier **B#**. When entered as a constant, the two numbers are enclosed in parentheses and separated by a comma—for example **B# (25, 175)**. This format maximizes 16-bit data storage where positive values from 0-255 are sufficient. An example is a loop index where the maximum count is +10. The maximum range of each byte is from 0-255 or **B#** (0, 0) to **B#** (255,255).

128	64	32	16	8	4	2	1	128	64	32	16	8	4	2	1	Weight
15	14	13	12	11	10	09	08	07	06	05	04	03	02	01	00	Bit No.
1	1	1	1	1	1	1	1	1	1	1	1	1	1	1	1	
255								255								B#(255,255)

Figure 1-28. A WORD variable interpreted as 2-bytes unsigned decimal

DWORD

Variables of data type DWORD reserve 32-bits. The 32-bits of the data type DWORD are unstructured and are not evaluated as a unit. This unstructured format allows the DWORD variable to support several bit-oriented uses. A DWORD variable, for example, may be a 32-bit binary string of 1s and 0s, an 8-digit Hexadecimal number, or four unsigned decimal values. Because the binary equivalent of Hex 0-9 is the same for BCD, DWORD variables may also be used in BCD operations. Each of the DWORD formats is described below.

DWORD—As 32-bit String

A DWORD variable is specified as a 32-bit binary string, using the identifier **2#**. An example is **2#0000_1011_1111_1111_0000_1011_1111_1111**. The minimum value of this format is when all 32-bits are '0'; the maximum value is when all 32-bits are '1'. As an example, this option might be used in a Word logic operation to compare the 32-bits of an input module to an expected 32-bit pattern; or as a mask pattern in 32-bit word logic operation. Still another example might be to manipulate thirty-two digital outputs simultaneously. You might use the MOVE operation for this purpose, where the binary pattern would be the source and the double word address of the output module the destination.

DWORD—As 8-Digit Hexadecimal/7-Decade BCD

A variable of data type DWORD is specified as a hexadecimal number using the identifier **DW#16#**. 'DW' is for double word length, '16' is for hex. An example is **DW#16#0FFF_A0FF**. Each hex digit is represented using four-bits, allowing an 8-digit Hex number in a DWORD variable. The permissible range of values for 8-digit Hex is from **DW#16#0000_0000** to **DW#16#FFFF_FFFF**.

The 32-bit hex format may also support a 7-decade BCD value when you use only the hex digits 0-to-9 to avoid invalid BCD numbers. The valid range for BCD numbers in a DWORD variable is ±9,999,999. In this format, bits 28-31 are used to indicate the sign, but only bit-31 is relevant—the remaining bits do not matter. Negative values are denoted by '1111' and positive values by '0000'. When specified using the 32-bit hex identifier, the 7-decade BCD range is **DW#16#F9999999** to **DW#16#09999999**.

8 4 2 1	8 4 2 1	8 4 2 1	8 4 2 1	8 4 2 1	8 4 2 1	8 4 2 1	8 4 2 1	Weight
31—28	27—24	23—20	19—16	15—12	11—08	07—04	03—00	Bit No.
0000	1001	1001	1001	1001	1001	1001	1001	
0	9	9	9	9	9	9	9	DW#16#09999999

Figure 1-29. DWORD variable interpreted as 32-bit Hexadecimal. Example value is +9,999,999 BCD

DWORD—As 4-Bytes Unsigned Decimal

A variable of type DWORD may be specified as 4-bytes of unsigned decimal numbers using the identifier **B#**. When entered as a constant, the four values are separated by commas and enclosed in parentheses. An example is **B#** (255, 255, 255, 1). This format maximizes 32-bit data storage where positive values from 0-255 are sufficient. Examples might be preset count values. The maximum range of each byte is from 0-255 or **B#** (0, 0, 0, 0) to **B#** (255, 255, 255, 255).

128 64 32 16 8 4 2 1	128 64 32 16 8 4 2 1	128 64 32 16 8 4 2 1	128 64 32 16 8 4 2 1	Weight
7 0	7 0	7 0	7 0	Bit No.
1 1 1 1 1 1 1 1	1 1 1 1 1 1 1 1	1 1 1 1 1 1 1 1	1 1 1 1 1 1 1 1	
Byte-0 (0-255)	Byte-1 (0-255)	Byte-2 (0-255)	Byte-3 (0-255)	

Figure 1-30. DWORD variable interpreted as 4-bytes unsigned decimal

INT

Variables of data type INT represent signed decimal integers in a 16-bit word. The permissible range for variables of data type INT is -32,768 to +32,767. You may enter a constant of type INT in the declaration area of a code block, data block, or as an input value to an instruction, simply using the appropriate sign—for example +2575. No identifier is required when you use an integer constant.

S	-	-	-	-	-	1024	512	128	64	32	16	8	4	2	1	Weight
15	14	13	12	11	10	09	08	07	06	05	04	03	02	01	00	Bit No.
0	0	0	0	0	0	0	1	0	1	0	1	0	1	0	1	
							512	+	64	+	16	+	1	+	1	+597

Figure 1-31. Data type INT variable, interpreted as signed 16-bit integer

The two's complement method is used when representing DINT variables. The sign bit in the 16-bit representation is bit-15. A '0' in the sign bit indicates a positive value; a '1' indicates negative. The remaining 15 bits hold the double integer value. The remaining bits (0-14) hold the actual value.

The two's complement of any number is obtained by inverting each bit from right-to-left, starting with that position which follows the first '1' bit. This method is equivalent to inverting each bit, and then adding +1. The values of +597 and -597 are shown below.

S	-	-	-	-	-	1024	512	128	64	32	16	8	4	2	1	Weight
15	14	13	12	11	10	09	08	07	06	05	04	03	02	01	00	Bit No.
0	0	0	0	0	0	0	1	0	1	0	1	0	1	0	1	+597
1	1	1	1	1	1	1	0	1	0	1	0	1	0	1	1	-597

Figure 1-32. Two's Complement representation of data type INT negative numbers

DINT

Variables of data type DINT represent 32-bit signed decimal integers in an S7 double word. The permissible range for variables of data type DINT is -2,147,483,648 to +2,147,483,647. You will typically use DINT variables when an operation involving integer numbers is expected to exceed the range of data type INT. The *long integer* identifier 'L#' is used to enter an initial DINT constant in a code block, data block, or as an input value to an instruction—for example, L#500000.

The two's complement method is used when representing DINT variables. The sign bit in the 32-bit representation is bit-31. A '0' in the sign bit indicates a positive value; a '1' indicates negative. The remaining bits (0-30) hold the actual value. -2147483647 and +2147483647 are shown below.

S	32-bit DINT Value							
31—28	27—24	23—20	19—16	15—12	11—08	07—04	03—00	Bit No.
0111	1111	1111	1111	1111	1111	1111	1111	+2147483647
1000	0000	0000	0000	0000	0000	0000	0001	-2147483647

Figure 1-33. Data type DINT variable, interpreted as signed 32-bit integer

REAL

Variables of type REAL represent floating-point numbers—for example 25.375 or 0.31415. REAL variables are used to work with fractional numbers or to handle calculations or measured quantities that need floating-point accuracy. Internally, REAL variables are represented in the IEEE 32-bit floating-point format. STEP 7 represents REAL numbers in 32-bit or a double word memory location. In this format, bits 0-22 represent the *mantissa,* or the significant digits of the number; and bits 23-30 represent the *exponent,* a signed power of 10. Bit-31 is the sign bit; '0' indicates a positive value and '1' indicates a negative value.

S	exponent		mantissa	
31	30	23	22	... 0
	2^7	2^0	2^{-1} 2^{-2}...	...2^{-23}

Figure 1-34. DWORD variable interpreted as 4-bytes unsigned decimal

The permissible range for variables of data type REAL is -3.402833E+38 to -1.175494E-38 and +1.175494 E-38 to +3.402833 E+38. This range of REAL numbers, which supports full accuracy, represents what is referred to as normalized floating-point numbers; STEP 7 also supports so-called de-normalized floating-point numbers.

You may enter a REAL constant in the declaration area of a code block, data block, or as an input value to an instruction, by simply using the appropriate sign and format—for example with +2.575E3, no identifier is required.

Using the REAL notation, shown in Table 1-12, you may specify an initial value for a declared variable or a REAL value constant input to a STEP 7 instruction. This shorthand notation places the first of the significant (non-zero) digits to the left of the decimal point and up to six additional significant digits to the right of the decimal point. The significant digits are followed by the exponent. An uppercase 'E' or lowercase 'e' is valid. The exponent represents the number of places the decimal point must be moved to the left or right of the most-significant digit in order to reflect the original value. A positive exponent reflects a move to the right—a negative exponent reflects a move to the left.

Table 1-12. REAL (Floating-point) constants in scientific notation

Original Value	REAL Notation	Original Value	REAL Notation
12345.67	1.234567 e +04	2000.0	2.0 E +03
.1234567	1.234567 e -01	20003.0	2.0003 E +04
-12.34567	-1.234567 e +01	.01	1.0 E -2
-0.0000023	-2.3 e -06	.001	1.0 E -3
+0.000000000675	6.75 e -10	.0001	1.0 E -4
1,234,567,000	1.234567e+9	10000	1.0 E +4

CHAR

A variable of data type CHAR reserves one byte (8-bits) and represents a single ASCII character. Valid characters include printable characters—including upper and lower case alphabets, numbers 0-9, and excluding the characters 127 (DEL) and 34 ("). To specify a CHAR constant or an initial CHAR value, you simply enclose the character using single quote marks—for example 'N' or 'Y'. The quote marks, which are always required, are automatically inserted by STEP 7. CHAR (ASCII) characters are stored in hexadecimal as shown in the following figure.

Although the CHAR type represents a single ASCII character, when you are writing to a CHAR variable you may use the MOVE instruction to transfer two characters to a word location—for example 'NO'; or 4 characters—for example 'abcd', to a double word location.

128	64	32	16	8	4	2	1	Weight
07	06	05	04	03	02	01	00	Bit No.
0	1	0	1	1	0	0	1	B#16#59
			Y					'Y'

Figure 1-35. CHAR character represented in hexadecimal

In addition to printable ASCII characters, some special characters are also possible. When entered, special characters are prefixed with the dollar symbol as shown in the table below.

Table 1-13. Notation for Special CHAR Characters

ASCII Character	S7 CHAR Notation	ASCII Code (Hex)
Dollar Sign	'$$'	24
Apostrophe	'$' '	27
Line Feed Control (LF)	'$L' or '$l'	0A
New Page Control (FF)	'$P' or '$p'	0C
Carriage Return Control (CR)	'$R' or '$r'	0D
Tab Control	'$T' or '$t'	09

DATE

A variable of data type DATE reserves an S7 word to represent a date value as an unsigned fixed-point number. Internally the date value is interpreted as a positive number that represents the total days since January 01, 1990. This IEC format for specifying a date uses the prefix identifier **DATE#** or **D#**. The year, month and day are then separated by hyphens—for example **DATE#2002-08-22**. Data type DATE supports dates from January 1, 1990, through December 31, 2168, as shown below.

DATE#1990-01-01 through DATE#2168-12-31

TIME

Variables of data type TIME reserve an S7 double word. A constant of data type TIME is defined in units of days (D), hours (H), minutes (M), seconds (S), and milliseconds (MS) using the identifier **TIME#** or **T#**. Unit identifiers may be entered as upper or lowercase, and units may be omitted if not needed—for example **TIME#10H30M**. STEP 7 represents the time value as a signed 32-bit fixed-point number, which is interpreted as total milliseconds.

This IEC format for specifying a time value is intended for specifying a delay time—for example, a 'wait time' in your STL code. Both positive and negative values are supported, with the range as shown in the following example.

TIME#-24d20h31m23s647ms to TIME#+24D20H31M23S647MS

TIME_OF_DAY

A variable of data type TIME_OF_DAY reserve an S7 double word. The TIME_OF_DAY constant is specified using the identifier **TIME_OF_DAY#** or **TOD#** preceding the hours, minutes, and seconds, which are separated by colons—for example **TOD#**01:15:00. Milliseconds may be added to the time using a period as the separator, or omitting it altogether. An example is shown below.

TOD#-00:00:00 to TOD#23:59:59.999

STEP 7 represents the TIME_OF_DAY value as a signed 32-bit fixed-point number that is interpreted as total milliseconds.

S5TIME

A variable of data type S5TIME is specified to represent a preset value for a timer. This 16-bit format is the same as what was formerly used to store the STEP 5 timer preset. Using the identifier **S5TIME#** or **S5T#**, you may enter an initial S5TIME value in units of hours (H), minutes (M), seconds (S), and milliseconds (MS). An example is **S5T#**2H30M30S. You may specify the unit identifier in upper or lowercase, and units that are not required may be omitted. The maximum value that you may enter is 9,990 seconds (**S5T#**2H46M30S).

As shown in the following figure, the internal format of the S5TIME data type involves a single word encoded in two BCD parts—the time value and the time base. The time value, represented in the first 12 bits (00-11) of the timer word, allows a BCD value from 000-to-999. This value represents the total number of time base units for which the timer should time.

The time base units of 0.01 second, 0.1 second, 1.0 second, and 10 seconds are encoded in bits 12-13, and are represented by the BCD values of 0000, 0001, 0010, and 0011 respectively as shown below.

Time Base (00-11)				Time Value in 3-decades BCD (000-999)												Timer Word
15	14	13	12	11	10	09	08	07	06	05	04	03	02	01	00	Timer Word
0	0	0	1	0	1	0	1	0	1	0	1	0	1	0	1	T0
X	X	0.1 sec		5				5				5				55.5 sec.

00 = .01 sec
01 = 0.1 sec
10 = 1.0 sec
11 = 10 sec

Figure 1-36. Illustration of Timer Word format—the preset value of T0 = 555 x 0.1, or 55.5 sec

Since both components of the timer preset is in BCD, it is also possible to preload a timer using a WORD variable—for example in this case W#16#1555. This value being BCD might also be input from a BCD input device.

Elementary Data Types Summary

A summary of elementary data types is presented in the following table. Take a few moments to review these data types, and perhaps why they are referred to as elementary types. The usable formats of each data type are given—for example data type WORD may be used to represent a 16-bit binary string, a 4-digit hexadecimal value, a 3-digit BCD value, and two unsigned decimal values of 0-255. These values may all be referenced in the program as constants or variables. You may also reference these formats in the STEP 7 Monitor/Modify Variables utility as formats in which to view specific variables.

Table 1-14. Summary of S7 Elementary Data Types

Data Type	Bits	Format	Valid Constant Notation(s)/Range
BOOL	1	Boolean in Text	TRUE/FALSE
BYTE	8	Hexadecimal	B#16#0 to B#16#FF
WORD	16	Binary..........................	2#0 to 2#1111_1111_1111_1111
		Hexadecimal...............	W#16#0000 to W#16#FFFF
		BCD Count...................	C#000 to C#999
		Two Unsigned Bytes....	B#(0,0) to B#(255,255)
DWORD	32	Binary..........................	2#0 -to- 2#1111_1111_1111_1111_1111_1111_1111_1111
		Hexadecimal...............	DW#16#0000_0000 -to- DW#16#FFFF_FFFF
		Four Unsigned Bytes....	B#(0,0,0,0) -to- B#(255,255,255,255)
INT	16	Signed Decimal	-32768 to +32767
DINT	32	Signed Decimal	L#-2147483648 -to- L#2147483647
REAL	32	IEEE Floating Point	±1.175495e-38 -to- ±3.402823e+38
CHAR	8	ASCII Character	'A', 'B', 'C', 'D', and so on
S5TIME	16	S5 Time (10ms)	S5T#0H_0M_0S_10MS -to- S5T#2H_46M_30S_0MS
DATE	16	IEC Date (1-day)	D#1990-1-1 -to- D#2168-12-31
TIME	32	IEC Time (1ms units)	T#-24D_20H_31M_23S_648MS -to- T#24D_20H_31M_23S_647_MS
TIME_OF_DAY	32	Time of day (1ms)	TOD#0:0:0.0 -to- TOD#23:59:59.999

Complex Data Types

With *complex data types* you may define variables that exceed 32-bits. A complex variable may be the result of combining elementary types, for example DATE and TIME; or a variable that is comprised of multiple data elements of the same or different types, for example ARRAY and STRUCT. Complex data types may only be assigned to variables declared in global or instance data blocks, or as temporary local data in organization blocks, functions, and function blocks; and as parameters in FBs and FCs.

Complex data types ARRAY, DATE-AND-TIME, STRING, STRUCT, and UDT are described in the following discussions.

Table 1-15. Overview of Complex Data Types

Data Type	Brief Description
ARRAY	Supports grouping of multiple data elements of one type—treated as a unit.
DATE-AND-TIME	Supports variable that store the year, month, day, hours, minutes, seconds, milliseconds, and the day of week.
STRING	Supports string of up to 254 elements of data type CHAR, treated as a single unit.
STRUCT	Supports combining of data elements of different data types to form a single unit.
UDT	Support creation of data block or declared variable using a structure template.

ARRAY

An array variable represents a grouping of a fixed number of related data elements, all of which are of the same type—either elementary or complex variables. For example, if you were storing 10 timer preset values of a machine setup, you might create an array that contains 10 values of type S5TIME.

An array is declared as **ARRAY [x1..x2]**, where **x1** and **x2** are index values that define the array size in terms of a lower and upper index limits. The lower and upper index values are separated by two periods. The example array described above could be declared as **ARRAY**[1..10], and is a single dimension array of 10 elements. The integer index values may start at a negative, zero, or positive value, where the maximum range is -32768 to 32767. Some single dimension arrays are shown below.

Example 1: Single dimension arrays

a) ARRAY [1..8] = 1-dimensional array—8 elements.

b) ARRAY [0..3] = 1-dimensional array—4 elements.

c) ARRAY [-10..10] = 1-dimensional array—21 elements.

You may define an array variable of up to six dimensions—each dimension is separated by a comma. The ARRAY [**x1..x2,y1..y2,z1..z2**], for example, defines a 3-dimensional array where **x1**, **y1**, and **z1** are respectively the lower index limits of the first, second, and third dimensions; **x2**, **y2**, and **z2** are respectively the upper index limits of the first, second, and third dimensions. Some multi-dimensional arrays are shown below.

Example 2: Multi-dimensional arrays

a) ARRAY [1..4, 3..5] = 2-dimensional array—12 elements.

b) ARRAY [1..10, 1..4 , 1..3] = 3-dimenional array—120 elements.

Imagine that you wanted to collect 10 machine parameters, from 4 identical machines, and from 3 separate shifts. Using the 3-dimensional array as an example, the first dimension might represent the 10 machine parameters, the second dimension might represent the four machines, and the third dimension could represent the 3 shifts. The total number of elements in an array is the product of multiplying the number of elements in each dimension. The total number of elements in the array with the dimensions [1..10, 1..4, 1..3] is 10 x 4 x 3 = 120.

DATE_AND_TIME

A variable of data type DATE_AND_TIME is comprised of the date and time. The DT variable stores the year, month, day, hour, minute, second, millisecond, and day of the week. When specifying this variable in a data block, or as a variable in a function block (FB), an initial value is entered using the identifier **DATE_AND_TIME#** or **DT#**, followed by the units of yy-mm-dd:hh:mm:ss:ms. An example date and time entry is **DT#2009:12-31-23:59:59.999**, where milliseconds may be omitted.

Table 1-16. DATE AND TIME representation in memory

BYTE	CONTENTS	VALUE RANGE	BCD CODE	EXAMPLE
0	Year	1990-2089	00-99	B#16#09
1	Month	1—12	01-12	B#16#05
2	Day	1—31	01-31	B#16#25
3	Hour	00—23	00-23	B#16#08
4	Minute	00—59	00-59	B#16#00
5	Second	00—59	00-59	B#16#00
6	Two most significant digits of MSEC	00—99	00-99	B#16#23
7 (4 MSB)	Least significant digit of MSEC	0—9	0—9	B#16#6
7 (4 LSB)	Day of Week = DOW (1=SUN … 7=SAT)	1—7	1—7	B#16#5

Internally, this variable is stored in BCD format in 8-bytes. Starting with byte-n, the variable contains the Year (00-99), Month (1-12), Day (1-31), Hour (0-23), Minute (0-59), Second (0-59), Millisecond (0-999), and finally the Day of the Week (1-7), where Sunday is day 1 and Saturday is day 7.

Byte-0	Byte-1	Byte-2	Byte-3	Byte-4	Byte-5	Byte-6	Byte-7	
YEAR	MONTH	DAY	HOUR	MIN	SEC	MSEC		DOW
00001001	00000101	00100101	00001000	00000000	00000001	00100011	00000010	
09	05	25	08	00	00	23	0	2

Figure 1-37. Illustration of DATE AND TIME format. Example: Monday, 2009:05-25-08:00:00.23.

When it comes to handling DATE_AND_TIME variables, the following IEC functions are available.

Table 1-17. Standard IEC DATE and TIME Functions

Function	Name	Brief Description
FC 6	DT_DATE	Extract DATE from the DATE_AND_TIME (or DT) format
FC 7	DT_DAY	Extract the day of the week (DAY) from the DATE_AND_TIME format
FC 8	DT_TOD	Extract the time of day (TOD) from the DATE_AND_TIME (or DT) format
FC 3	D_TOD_DT	Combine DATE and TOD to DATE_AND_TIME (or DT) format
FC 33	S5TI_TIM	Convert S5TIME (or S5T) to TIME format
FC 40	TIM_S5TI	Convert TIME to S5TIME (or S5T) format
FC 1	AD_DT_TM	Add DT to TIME
FC 35	SB_DT_TM	Subtract TIME from DT (or DATE_AND_TIME)
FC 34	SB_DT_DT	Subtract DT (or DATE_AND_TIME) from DT (or DATE_AND_TIME)

STRING

A variable of type STRING lets you define an ASCII string of up to 254 characters. You can specify a string as STRING [**n**] where **n** defines the length, in characters, that the variable can accommodate. For example, STRING [5] will hold 5 characters. If the length is not explicitly specified, then the variable will default to 256 bytes, which includes a 2-byte header and up to 254 characters.

When you define a variable of data type STRING, you may enter an initial string enclosed with apostrophes. Special characters are also enclosed with apostrophes, but preceded by the dollar sign. The actual byte requirement of a string is minimized when you specify the exact number of characters—for example STRING [5], with initial value 'READY'. If the initial string is shorter than the specified length then the actual length is reserved. In the following illustration, a length of 6 was reserved, but the actual string length is 5.

Byte-0	Byte-1	Byte-2	Byte-3	Byte-4	Byte-5	Byte-6
MAX. LENGTH	ACT. LENGTH	CHAR	CHAR	CHAR	CHAR	CHAR
00000110	00000101	01010010	01000101	01000001	01000100	01011001
6	5	R	E	A	D	Y

Figure 1-38. Variable STRING[6] variable. Maximum length is 6, actual length is 5

When it comes to accessing and handling STRING variables, the following standard IEC functions are available and may be integrated in your program.

Table 1-18. Standard IEC STRING Functions

Function	Name	Brief Description
FC 2	CONCAT	Return a new string by concatenating specified string (IN1) with (IN2)
FC 17	INSERT	Insert specified sub-string (IN2) into specified string (IN1) at position (P)
FC 4	DELETE	Delete a number of characters (L) from a string, starting with character (P)
FC 31	REPLACE	Replace (L) characters of string (IN1), with string (IN2); start with character (P)
FC 11	FIND	Return start position of specified substring (IN2) within specified string (IN1)
FC 21	LEN	Return the length (valid number of characters) of a specified string (S)
FC 20	LEFT	Return string using (L) characters from the left side of specified string (IN)
FC 32	RIGHT	Return string using (L) characters from the right side of specified string (IN)
FC 26	MID	Return string using (L) characters of string (IN), starting with character (P)
FC 16	I_STRNG	Convert specified INT (16-bit integer) value (I) to a STRING
FC 5	DI_STRNG	Convert a DINT (32-bit integer) value (S) to a STRING
FC 30	R_STRNG	Convert a REAL value (IN) to a STRING
FC 38	STRNG_I	Convert specified string (S) to an INT (16-bit integer) value
FC 37	STRNG_DI	Convert specified string (S) to a DINT (32-bit integer) value
FC 39	STRNG_R	Convert specified string (S) to a REAL value

STRUCT

A variable of data type STRUCT represents a *structure,* comprised of a fixed number of elements that may involve any combination of data types. The data elements of a structure, like those of an array, are generally related. But, whereas the elements of the ARRAY must be of the same type, the elements of a structure may combine whatever elementary and complex data types that you choose. In the example below, the structure has three variables of type INT: Quantity, Speed_1, and Speed_2; a variable of type CHAR: Opcode; and two single variables of type BOOL, and an array of 10 BOOL elements.

Example 1: A Structure

Quantity   = 1000 cartons

Speed_1   = 750 rpm

Speed_2   = 1500 rpm

Opcode   = 'X'

Done   = FALSE

Sys_Ready = FALSE

In the declaration table, the STRUCT variable includes the data elements enclosed between the keywords STRUCT and END_STRUCT as shown in the following figure.

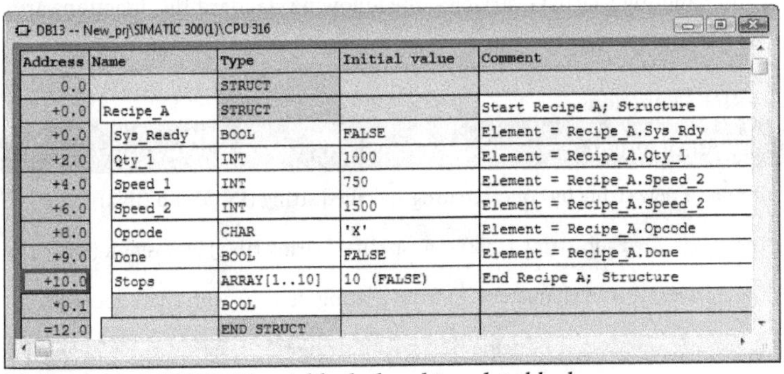

Address	Name	Type	Initial value	Comment
0.0		STRUCT		
+0.0	Recipe_A	STRUCT		Start Recipe A; Structure
+0.0	Sys_Ready	BOOL	FALSE	Element = Recipe_A.Sys_Rdy
+2.0	Qty_1	INT	1000	Element = Recipe_A.Qty_1
+4.0	Speed_1	INT	750	Element = Recipe_A.Speed_2
+6.0	Speed_2	INT	1500	Element = Recipe_A.Speed_2
+8.0	Opcode	CHAR	'X'	Element = Recipe_A.Opcode
+9.0	Done	BOOL	FALSE	Element = Recipe_A.Done
+10.0	Stops	ARRAY[1..10]	10 (FALSE)	End Recipe A; Structure
*0.1		BOOL		
=12.0		END_STRUCT		

Figure 1-39. A STRUCT variable declared in a data block

The individual elements of the structure are accessed using the structure name and the variable element name separated by a period—for example **Recipe_A.Speed_1**. Since each element in a structure represents a single data element, you may specify as a parameter in a box instruction, as well as in a user or system Function or Function Block. See Chapter 27 *Developing a STEP 7 Program* for declaring data elements of type STRUCT.

User Defined Type

A user defined type (UDT) is, as the name implies, a data type that you define as you wish. It is essentially a structure that may be comprised of variables of both elementary and complex data types. You can create a UDT in the Blocks folder of a program, using the SIMATIC Manager. This process is described later in Chapter 2 *Data Blocks*. The UDT, when created, is assigned a number (absolute address) from UDT 0 to UDT 65,535. You may also assign a symbolic address, for example **PIDLoop**, to a UDT. In the UDT named **PIDLoop**, shown in Figure 1-40, six variables have been defined. An individual variable named **PV**, contained in a UDT named **PIDLoop** is addressed as '**PIDLoop.PV**'.

Address	Name	Type	Initial value	Comment
0.0		STRUCT		
+0.0	Loop_No	WORD	W#16#1	Loop Number
+2.0	PV	REAL	0.000000e+000	Process Input Variable
+6.0	SP	REAL	6.500000e+002	Setpoint
+10.0	UL	REAL	6.750000e+002	Upper Limit Alarm
+14.0	LL	REAL	6.550000e+002	Lower Limit Alarm
+18.0	CTRL	WORD	W#16#0	Control Word
=20.0		END_STRUCT		

UDT1 -- "PID Loop" -- BrilliantTraining\S7 Program(1)\...\UDT1

Figure 1-40. A User-Defined Type is created from the SIMATIC Manager

Generally speaking, the UDT supports the creation of a data structure as a template, a unique set of variables, which can be used in two basic ways:

The first way in which the UDT template is used is as a predefined data type used when defining variables in code blocks or in data blocks. Remember that the UDT is a defined structure that is comprised of whatever variables you choose. If the UDT is already defined, you can select it from the data types in the **Type** field of the declaration table. UDT1, PIDLoop, shown in Figure 1-40, is a structure that contains six loop variables. In DB70 shown below, UDT1 is assigned as the data type to several process loops that share the same set of six process variables.

DB70 -- BrilliantTraining\S7 Program(1)

	Addr	Declaration	Name	Type	Initial value	Actual value	Comment
1	0.0	stat	loop1.Loop_No	WORD	W#16#1	W#16#1	Loop Number
2	2.0	stat	loop1.PV	REAL	0.000000e+000	0.000000e+000	Process Input Variable
3	6.0	stat	loop1.SP	REAL	6.500000e+002	6.500000e+002	Setpoint
4	10.0	stat	loop1.UL	REAL	6.750000e+002	6.750000e+002	Upper Limit Alarm
5	14.0	stat	loop1.LL	REAL	6.550000e+002	6.550000e+002	Lower Limit Alarm
6	18.0	stat	loop1.CTRL	WORD	W#16#0	W#16#0	Control Word
7	20.0	stat	loop2.Loop_No	WORD	W#16#1	W#16#1	Loop Number
8	22.0	stat	loop2.PV	REAL	0.000000e+000	0.000000e+000	Process Input Variable
9	26.0	stat	loop2.SP	REAL	6.500000e+002	6.500000e+002	Setpoint
10	30.0	stat	loop2.UL	REAL	6.750000e+002	6.750000e+002	Upper Limit Alarm
11	34.0	stat	loop2.LL	REAL	6.550000e+002	6.550000e+002	Lower Limit Alarm
12	38.0	stat	loop2.CTRL	WORD	W#16#0	W#16#0	Control Word
13	40.0	stat	loop3.Loop_No	WORD	W#16#1	W#16#1	Loop Number
14	42.0	stat	loop3.PV	REAL	0.000000e+000	0.000000e+000	Process Input Variable
15	46.0	stat	loop3.SP	REAL	6.500000e+002	6.500000e+002	Setpoint
16	50.0	stat	loop3.UL	REAL	6.750000e+002	6.750000e+002	Upper Limit Alarm

Figure 1-41. A UDT used as a data type in a data block. The UDT is also usable in logic blocks FC/FB

The second use of the UDT is as a template used to create multiple data blocks, all of which will have identical data arrangements. For example, the application may have several identical machines where each uses the same data structure; or perhaps in a batch process multiple data block must be created to manage different variations of the same recipe. The parameters are the same, but the values are different in each data block.

For more on user defined types, see Chapter 2 *Data Blocks* topics *Creating a User Defined Type*, *Creating a UDT-Based Data Block*, and *Entering UDT-Based Variables*.

Parameter Types

Parameter types are a special group of data types that allows timers, counters, function blocks (FBs), functions (FCs), data blocks (DBs), and pointers to be passed as block parameter in a code block.

Generally speaking, block parameters refer to those variables that may be defined in a FC or FB code block such that when the block is called, a different value may be passed to the block each time that the block is called. Development of these types of FCs and FBs is introduced and described in Chapter 28 *Developing Functions and Function Blocks*. Whereas data variables with the standard data types are generally passed via block parameters, parameter types are the applied data types when the unique group of variables listed in Table 1-19 are passed as block parameter variables.

You can learn more on block parameters and how they're used under the Chapter 28 *Developing Functions and Function Blocks* topic *Defining Block Parameter Variables*.

Table 1-19. Overview of Parameter Data Types

Data Type	Description	Bits	Actual Address Examples
TIMER	Timer	16	T42
COUNTER	Counter	16	C25
BLOCK_FC	Function	16	FC44
BLOCK_FB	Function Block	16	FB22
BLOCK_DB	Data Block	16	DB12
BLOCK_SDB	System Data Block	16	SDB50
POINTER	DB Pointer	48	P#DB10.DBX16.1 (DB Pointer); P#M14.0 or M14.0 (area - crossing)
ANY	ANY Pointer	80	Contiguous data area: P#DB5.DBX10.0 INT 16 Any variable of (elementary/complex/parameter) type

Note: The DB pointer format is shown in Figure 1-42; the ANY pointers format is in Figure 1-43

TIMER

A timer might be passed to an FC or FB that performs an operation in which the block is called repeatedly, while using a unique timer on each call. An example of such an application might be a conveyor system in which a series of AC motors are started sequentially, each after a time delay. Not only could such a block manage the passing of I/O data variables to start and stop each motor, but it could also pass a unique timer that delays the starting of each motor. Once the timer number is passed to the block, standard timer operations are possible within the block—for example, specify the timer preset, reset the timer, check the timer status, or check the remaining time. See Chapter 28 *Developing Functions and Function Blocks*.

COUNTER

Similar to a timer, a counter may also be passed to a code block as a parameter. Imagine a block that governs how often several identical large motors are started. If a counter tracks the number of times the motor is started, a single FB would suffice if a unique counter is passed to the block on each call of the block. Once the counter number has been passed to the block, normal counter operations are possible from that block. For example you may specify a counter preset, reset the counter, check the counter status, or check the current count. See Chapter 28 *Developing Functions and Function Blocks*.

BLOCK_FC and BLOCK_FB

It is also possible to develop an FC or FB in which a different FC (number) or FB (number) is passed to the code block. Imagine an application in which, based on various custom conditions, calling a different block on each call of the block, supports performing a configurable operation. Once the block number for an FC or FB has been passed to the block, the operations that are normally possible with respect to an FB or FC are all possible from within the block. For example you may call the FC and FB unconditionally (**UC** FB, **UC** FC); the FB may also be called conditionally (**CC** FB). In all cases where an FC or FB is passed as a parameter, it must not have any block parameters of its own.

BLOCK_DB and BLOCK_SDB

A data block (DB) number or system data block (SDB) number may also be passed to a code block when a formal parameter has been defined using the parameter type DB or SDB respectively. Recall that an SDB must be allocated as the instance DB for each call of an SFB (system function block). It is very likely that if an FB is passed as a parameter, then its associated instance SDB is also passed to the block. Imagine an application in which, based on various conditions, a unique FB or SFB performs an operation in which the block uses a unique DB on each call of the block. Once the DB number has been passed to the block, then normal DB operations are possible from within that block.

POINTER

By applying a parameter type of POINTER, you may pass a pointer location or pointer constant as the actual parameter to an FC or FB. The pointer parameter may be applied as an IN, IN/OUT parameter in FC or FB, and as an OUT parameter in an FC only. When supplying the actually parameter to the block, you may specify a DB pointer constant using the notation P#DBn.DBXByte.bit. The pointer constant may also be an area-crossing pointer—for example **P#M14.0**, or simply specified as **M14.0**. The DB Pointer format is shown on the following page.

ANY

By applying the parameter variable type ANY you may pass variables of all types of elementary or complex data types to an FC or FB. The programming of the FC or FB will dictate the actual variable address or data type that is applied to the parameter when the block is called. You may also specify a constant at the block parameter by using the ANY pointer notation of P#DBn.DBXy.0 DataType Length. In this notation, you may specify a contiguous data area to be passed to a block.

The ANY Pointer format is shown in Figure 1-42. There you'll see that in addition to containing the DB pointer, which itself contains an area pointer, the ANY pointer supports a memory area, start byte, data type, and length—by specifying a length a contiguous data area is defined. This notation is typically used in blocks such as SFC 20 (Copy) to specify a source data area, from which a contiguous data set can be accessed, and a destination data area, to which a contiguous data set can be copied.

P#DB5.DBX10.0 INT 16, for example points to the data area in DB 5, which starts at a byte offset of 10, and contains 16 contiguous variables of data type INT.

DB Pointer

The DB Pointer is an extension of the area pointer. In addition to containing the area pointer to S7 memory areas, I, Q, M, L, T, and C, the DB pointer supports specifying an INTEGER value that designates a global or an instance data block number. The DB pointer, represented in 6-bytes, is shown in Figure 1-42. If byte-n and byte-n+1 contains zero, the pointer simply contains an area pointer. Area pointer formats are given in Figure 1-20 and Figure 1-21.

BYTE	7	6	5	4	3	2	1	0
byte n								
byte n+1				Data Block Number (INT)				
byte n+2								
byte n+3				Area Pointer (32-bits)				
byte n+4								
byte n+5								

Figure 1-42. DB Pointer Data Structure

ANY Pointer

The ANY pointer is an extension of the DB pointer. In addition to containing a DB pointer, it supports the specification of a data type and a repetition factor. With the ANY pointer, you can specify an elementary or complex variable of any type, and specify a repetition value that lets you reference a contiguous block of data in a specific S7 memory area. This aspect of the ANY type is particularly useful when a contiguous block of memory locations must be handled, for example with the MOVE Function SFC 20, which copies a block of locations from one area to another area. The ANY pointer format, represented in 10-bytes, is shown in Figure 1-43.

BYTE	ANY Pointer—Data Types	ANY Pointer—Parameter Types
byte n	16#10	16#10
byte n+1	Type	Type
byte n+2	Quantity	Quantity
byte n+3	Quantity	Quantity
byte n+4	DB Number	16#0000
byte n+5	DB Number	16#0000
byte n+6	Area Pointer (32-bits)	Type
byte n+7	Area Pointer (32-bits)	16#00
byte n+8	Area Pointer (32-bits)	Number
byte n+9	Area Pointer (32-bits)	Number

Figure 1-43. ANY Pointer Data Structures

Chapter 2

Data Blocks

This chapter presents a discussion of S7 data blocks. In STEP 7 the data block (DB) resource supports the structuring and storing of data required in your program. In this chapter you'll learn how both Global and Instance DBs are applied, as well as how to address individual data elements of a data block using the absolute address or the symbolic address. You'll also learn how these data elements are accessed in LAD/FBD and in STL.

As you prepare to define, create, and enter initial data values for data blocks, you may wish to review the Chapter 1 *Memory and Addressing* discussion of S7 data types and the various data formats, if you have not done so already.

Main Topics

- Application Overview
- Creating Data Blocks
- Editing Shared Data Blocks
- Data Block Address Structure
- Accessing Data Elements

APPLICATION OVERVIEW

In S7 controllers, a data block (DB) is a program resource for organizing and storing both the constant and variable data required in your program. Typical constants might include presets for analog set points, PID loop data, or recipe data values in a batch process. Variables might include analog input and output data, various production accounting totals, or other values collected for SCADA purposes. You may create as many data blocks, as your storage needs dictate—within the limits of the CPU.

You may organize each DB as you wish, and the DB may contain variables of the same or of different data types. You determine the name, data type and order of each variable, and the number of variables stored in each DB. You may create elementary and complex variables. Whereas intermediate program results are typically stored in Bit Memory (M) or in the local memory (L) of a code block, data blocks are generally used to manage the bulk data requirements of your process and program. Data stored in a DB is retained even when the machine or process is stopped and when power is removed.

Shared and Instance Data Blocks

There are two data block types in STEP 7. When you create a DB you define the data block type based on how it will be used in conjunction with code blocks. A *shared DB* (also called *global DB*) is a common data storage container whose data may be read from or written to by any code block within the program. An *instance DB*, on the other hand, is a data block assigned exclusively to a specific Function Block as required memory for the FB. Only the assigned FB has read/write (R/W) access to the instance DB. When an instance DB is created, you must explicitly assign it to an FB.

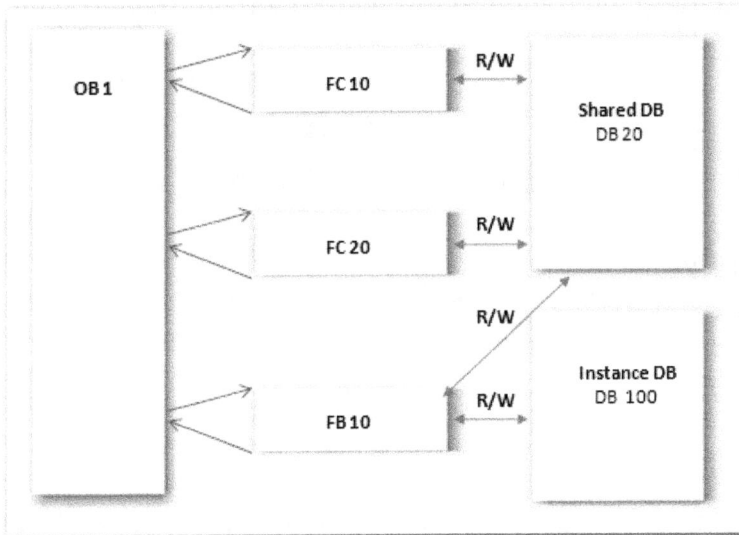

Figure 2-1. Illustration of Access to Instance and Shared (Global) Data Blocks

CREATING DATA BLOCKS

To create a new data block is to generate a new DB object. As described in the previous paragraph, there are two types of data blocks—shared DBs that may be accessed by any of the code blocks in your program; and instance DBs, each of which *is* associated with a Function Block (FB) as required memory. Each data block, regardless of its type, is comprised of a *block header* section, and a *variable declaration* section. When you generate a DB, the header dialog is where you define attributes of the data block, including the type. You may create a DB directly from the Simatic Manager or from within the LAD/FBD/STL program editor.

Creating a Shared DB

The DB properties dialog is shown below. To create a shared DB, first specify the **Name and Type** for the new DB. The name is actually the DB number—for example DB22 or DB34. The range of DB numbers is CPU dependent. In the field adjacent to the **Name and Type** field, you would select **Shared DB**. You can also assign a **Symbol Name**, which can be used in place of the absolute address (DB number) when referencing the DB; and a **Symbol Comment**, which may be up to 80 characters. Once a shared DB is generated, you may open the block to define and initialize the required data elements.

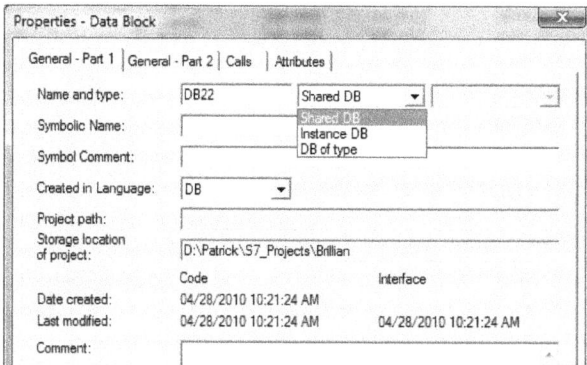

Figure 2-2. General DB Properties—creating a shared DB

Creating an Instance DB

Each FB you create will require an instance DB. You should create the FB and define its variables before creating the instance DB. The properties dialog is shown below. First, specify the **Name and Type** for the new DB. The name is actually the DB number—for example DB104. The range of DB numbers is CPU dependent. In the field adjacent to the **Name** and **Type** field, select **Instance DB** as the type of DB. In the immediately adjacent field, from the list of Function Blocks, select the FB to which this instance DB should be assigned. Finally, enter a **Symbol Name** that can be used in place of the absolute address (DB number) when referencing the DB; and finally, a **Symbol Comment**.

When you create the instance DB and assign it to an existing FB, the static (STAT) variables and variables that were declared as block parameters in the declaration area of the selected FB are automatically written to the instance DB by STEP 7. The data structure in the instance DB will be the same as in the declaration table of the associated FB. In this way, the static variables and the actual parameters of the FB are always stored in the instance DB and are available each time the FB is called. Remember, when you create an instance DB you will assign it to an FB that already exists.

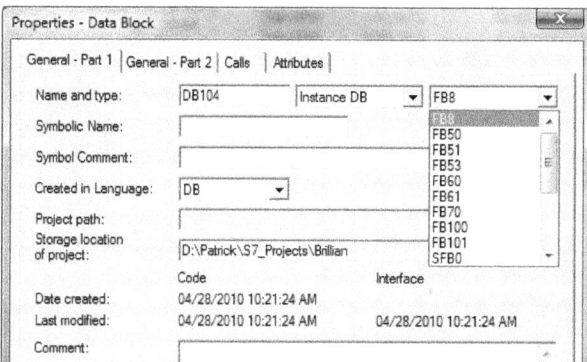

Figure 2-3. General DB Properties—creating an instance DB

Creating a User Defined Type

A STEP 7 user defined type (UDT) is a data type that you define as you wish—according to specific data requirements. The UDT, described in the Chapter 1 *Memory and Addressing* discussions of Data *Types and Formats*, is essentially a structure that may be comprised of variables of both elementary and complex data types.

Once defined the UDT may be applied as a DB template to create data blocks of an identical structure or as a variable template used to insert a canned variable of type structure. A variable created from a UDT is expected to be used throughout the program and can be declared in any code block or data block. When used as a data type, the variables of the UDT are automatically inserted.

The following steps describe how the UDT is applied as a template for creating data blocks. The steps for applying the UDT as a data type are presented in the topic *Entering UDT-Based Variables*.

To create a data block based on an existing UDT:

1. From the **S7Program** right-click on the **Blocks** folder and select **Insert New Object** ➤ **Type**.

2. Enter a **Name** (for example UDT1, UDT2, UDT3), **Symbol Name** (optional), and **Symbol Comment** up to 80 characters (optional). Click **OK** to generate the UDT.

3. From the **Blocks** folder, double-click on the UDT object to open the UDT (the declaration table with the keywords **STRUCT** and **END_STRUCT** are displayed).

4. In the **Name** field of the first row, enter the name of the new structure. In the example shown below, the name **Recipe** is used.

5. Click in the **Type** field and enter the keyword **STRUCT**; then advance through the remaining fields and to the next row using the **RETURN** key. A single blank declaration row and the last line of the structure with the keyword **END_STRUCT** will be inserted.

6. On the next row, enter the symbolic **Name**, data **Type, Initial value** and **Comment** for the first data element. In the example below, the name **Param_1** is used. Again, advance to the next row using the **RETURN** key.

7. Enter as many additional variables as required, using the previous step. In the following example 13 variables are entered in the structure.

8. Click **OK** to save the UDT.

Address	Name	Type	Initial value	Comment
0.0		STRUCT		
+0.0	Recipe	STRUCT		
+0.0	Param_1	INT	0	Machine Idle Speed
+2.0	Param_2	INT	0	Machine Low Speed
+4.0	Param_3	INT	0	Machine High Speed
+6.0	Param_4	DINT	L#0	Machine Shift 1 Total
+10.0	Param_5	DINT	L#0	Machine Shift 2 Total
+14.0	Param_6	BOOL	FALSE	Machine System Ready S
+14.1	Param_7	BOOL	FALSE	Machine Emergency Stop
+16.0	Param_8	S5TIME	S5T#0MS	Machine Initial Dwell
+18.0	Param_9	S5TIME	S5T#0MS	
+20.0	Param_10	S5TIME	S5T#0MS	
+22.0	Param_11	REAL	0.000000e+000	Machine Zone 1 Tempera
+26.0	Param_12	REAL	0.000000e+000	Machine Zone 2 Tempera
+30.0	Param_13	REAL	0.000000e+000	Machine Zone 3 Tempera
=34.0		END_STRUCT		
=34.0		END_STRUCT		

Figure 2-4. Example UDT 3

Creating a UDT-Based Data Block

The use of the UDT, described here, is as a template for creating data blocks that will all have the same data arrangement. This may be the case, for example, in an application where identical machines use the same data structure; or in a batch process where individual data blocks are used to store variations of the batch recipe. The machine or process parameters are the same, but the values representing a recipe is different in each data block. Use of the UDT eliminates the need to re-enter what could be hundreds of data values. The figure below illustrates using UDT3 to create data blocks DB201, DB202, and DB203.

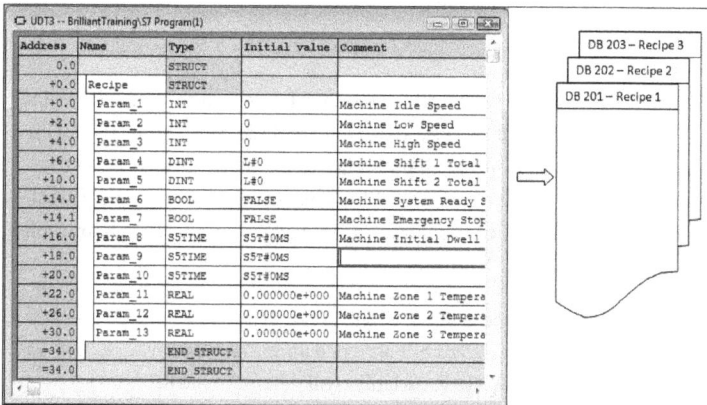

Figure 2-5. DB201, DB202, and DB203 are based on the structure of UDT 3

To create a data block based on an existing UDT

1. Open the **S7Program** in which you wish to create a UDT-based data block (the UDT must have already been created—See *Creating a User Defined Type*).

2. From the menu, select **Insert ➤ S7 Block ➤ Data Block**. The DB dialog opens.

3. In the first **Name and Type** field enter the DB number—absolute address of the DB; in the center **Name and Type** field, click the drop list and select **DB of Type**; then in the last **Name and Type** field select the UDT to use as a template in creating the new DB. **UDT3** is selected as the user defined type in this example.

4. Enter a **Symbol Name** and **Symbol Comment**—up to 80 characters; both are optional.

5. Click **OK** to save the DB (repeat for DB202, and DB203; or for as many DBs as required).

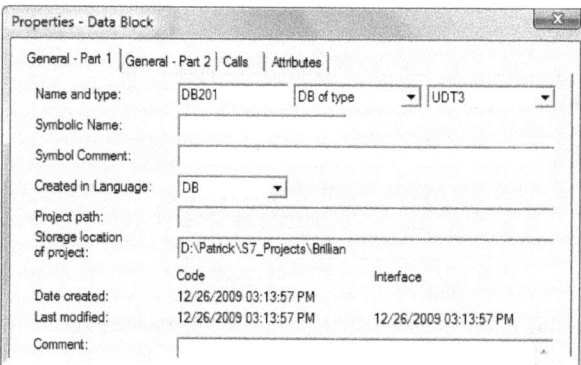

Figure 2-6. Data Block dialog—Creating DB201 using UDT 3

EDITING SHARED DATA BLOCKS

When you first generated data blocks in your program, you may have created both shared DBs and instance DBs. The initial data structure and content of each instance DB you created was initialized by the STEP 7 editor when you assigned the DB to a specific function block. On the other hand, shared DBs will contain process data that will be available to any of the blocks of your program (for reading or writing data). When you create a shared DB, you will determine the organization of the data and the initial values for each data element you insert. If the data blocks you wish to work with already exist, then you may insert the initial data, edit or add to the current data contents.

Entering Initial Data

A shared DB, like other blocks, can be edited offline or online, and may be opened from the SIMATIC Manager or from the LAD/STL/FBD editor. In either case, when you are done editing the DB you can determine whether it should be stored offline, online, or in both places. Variables are entered in the declaration table of a data block in a fashion similar to code blocks. This initial editing of a data block is done while in the *Declaration View* of the editor. For each data variable, you must assign a **Name** and data **Type**. The assignment of an **Initial value** and **Comment** is optional.

If an initial value is not entered, then a default value is taken as the initial value. This initial value will depend on the data type. For example, numeric variables not assigned an initial value are assigned a default of '0'; BOOL values default to FALSE; and CHAR variables will default to blank ' '.

DB10 -- "Zone Data" -- New_prj\SIMATIC 300(1)\CPU 316\...\DB10

Address	Name	Type	Initial value	Comment
0.0		STRUCT		
+0.0	TMP_PRESET_1	INT	350	Temperature Preset Zone1
+2.0	TMP_PRESET_2	INT	375	Temperature Preset Zone2
+4.0	TMP_PRESET_3	INT	400	Temperature Preset Zone3
+6.0	OVR_LMT_T1	BOOL	FALSE	ZONE 1 OVER Limit
+6.1	OVR_LMT_T2	BOOL	FALSE	ZONE 2 OVER Limit
+6.2	OVR_LMT_T3	BOOL	FALSE	ZONE 3 OVER Limit
+7.0	OPER_RESP_1	CHAR	'N'	OPERATOR RESPONSE 'YES'
+8.0	OPER_RESP_2	CHAR	'N'	OPERATOR RESPONSE 'NO'
+10.0	NEW_CNT	WORD	W#16#900	COUNTER OVERRIDE
=12.0		END_STRUCT		

Figure 2-7. Editing a Shared DB. Data variables are defined with initial values

To enter initial data in a data block:

1. From the SIMATIC Manager open the data block you wish to edit. The DB will open in the **Declaration View** in order to accept new variables.

2. Click in the **Name** field of the first data element, and enter a symbolic name.

3. Click in the **Type** field, right-click and select the data type for the data element.

4. Click in the **Initial value** field, and enter an initial value (optional).

5. Click in the **Comment** field and enter a comment (optional—up to 80 characters).

6. From the menu, select **Insert ➤ Declaration Line ➤ Before Selection** or **After Selection** to place the next variable either before or after an existing row, including the first row.

7. Enter as many additional variables as required, and click **File ➤ Save** to save the DB.

8. Click the **Download** button to save the DB online in the PLC (need online connection).

Modifying Actual Data

The data in a data block may include both process constants and variables. Some values, especially the variables, will be written to by your program while the system is running. If you need to edit initialized constant values you entered earlier, or the current values (actual values) of DB variables, then you will need to switch the DB editor to the *data view*. The data view will then show the **Initial value** column and the **Actual value** column—the **Actual value** column will be enabled to allow data entry.

When you open a DB in the data view each value in the **Actual value** column is the value that was last written by the active program. The data in this column is not automatically updated in this view, while the program is being processed. You may switch to an actual online update mode using the **F5** key—the values are then updated to reflect their current value (even as the values change).

With the DB displayed in the data view, you may also re-initialize the data elements of the data block. To re-initialize the values of a data block is to set the values in the **Initial value** column to match those values currently in the **Actual value** column.

\textbf{Addr}	\textbf{Name}	\textbf{Type}	\textbf{Initial value}	\textbf{Actual value}	\textbf{Comment}
0.0	TMP_PRESET_1	INT	350	0	Temperature Preset number 1
2.0	TMP_PRESET_2	INT	375	375	Temperature Preset number 2
4.0	TMP_PRESET_3	INT	400	400	Temperature Preset number 3
6.0	TMP_PRESET_4	INT	500	500	Temperature Preset number 4
8.0	OVR_LMT_T1	BOOL	FALSE	TRUE	ZONE 1 OVER Limit
8.1	OVR_LMT_T2	BOOL	FALSE	FALSE	ZONE 2 OVER Limit
8.2	OVR_LMT_T3	BOOL	FALSE	FALSE	ZONE 3 OVER Limit
8.3	OVR_LMT_T4	BOOL	FALSE	FALSE	ZONE 1 OVER Limit

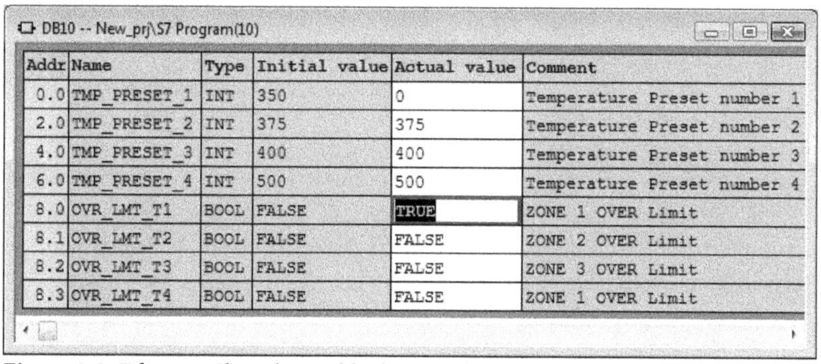

DB10 -- New_prj\S7 Program(10)

Figure 2-8. Editing a Shared Data block opened in the Data View

To modify initial data or re-initialize the data block:

1. From the SIMATIC Manager, open the data block you wish to edit.

2. From the menu, select **View ➤ Data View** in order to view the DB with an **Actual value** column enabled for editing the variables.

3. Click in an **Actual value** field to modify the value as required—or:

4. Modify as many values as required, then click **Save** to store the changes offline.

To re-initialize the data block:

1. Open the data block you wish to edit.

2. With the DB open in the data view.

3. From the menu select **Edit ➤ Initialize Data Block**. All of the DB **Initial values** are set to equal the **Actual values**.

Entering Elements of Type ARRAY

An array is declared as **ARRAY [x1..y1]**, where the dimension limits **x1** and **y1** define the array size. Up to six dimensions may be declared, where each dimension is separated by a comma. For example, ARRAY [1..2, 1..6], defines a two-dimension array with 12 elements. The start index value may start at a negative integer, zero, or a positive integer. For example, **ARRAY** [-10..10] is a one dimension array with 21 elements. The data type of an array, for example TIME_OF_DAY, is assigned on the declaration row immediately following the array name and size (See Figure below).

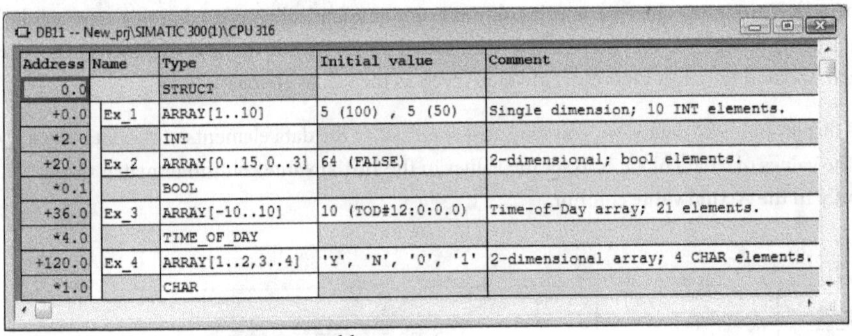

Address	Name	Type	Initial value	Comment
0.0		STRUCT		
+0.0	Ex_1	ARRAY[1..10]	5 (100) , 5 (50)	Single dimension; 10 INT elements.
*2.0		INT		
+20.0	Ex_2	ARRAY[0..15,0..3]	64 (FALSE)	2-dimensional; bool elements.
*0.1		BOOL		
+36.0	Ex_3	ARRAY[-10..10]	10 (TOD#12:0:0.0)	Time-of-Day array; 21 elements.
*4.0		TIME_OF_DAY		
+120.0	Ex_4	ARRAY[1..2,3..4]	'Y', 'N', '0', '1'	2-dimensional array; 4 CHAR elements.
*1.0		CHAR		

Title bar: DB11 -- New_prj\SIMATIC 300(1)\CPU 316

Figure 2-9. Declaring ARRAY variables

Initial values may be assigned to array elements by entering each value separated by a comma. The same value can be assigned to consecutive elements by enclosing the value in parentheses preceded by a repetition factor—for example 5(100) assigns a value of 100 to five consecutive values. Elements not assigned an initial value will default to a value of zero.

To enter array data elements:

1. Open a new data block or an existing DB you wish to edit.

2. Click in the **Name** field and enter the symbolic name of the new array variable.

3. Click in the **Type** field and enter the keyword **ARRAY**; then enter the array upper and lower index limits—for example [1..8] for a single dimension array, or [1..8, -10..10] for a 2-dimensional array. Recall that up to 6 dimensions may be defined.

4. In the **Initial value** field, assign each individual value using comma separation; use a repetition factor to assign the same value to consecutive elements; or let remaining unassigned data elements default to a value of zero. See the following examples.

   Example 1: Individual comma separated Data elements

   ARRAY [1..8]
   100, 100, 100, 100, 100, 550, 750, 1200

   Example 2: Repetition factor to assigns same value to consecutive elements

   ARRAY [1..8]
   5(100), 550, 750, 1200

   Example 3: After elements are initialized, unassigned data elements are zero filled

   ARRAY [1..15]
   5(100), 550, 750, 1200

Entering Elements of Type STRUCT

You may define a structure in the variable declaration table of a code block or in a data block. In the declaration table, you will insert the data elements of the STRUCT variable between the keywords STRUCT and END_STRUCT as shown in the figure below. The keyword END_STRUCT is inserted automatically by STEP 7. Remember, a structure may be comprised of any combination of variables of elementary and complex data types.

In your program, individual elements of the structure may be accessed using the structure name and the variable element name separated by a period—for example **Recipe_A.Sys_Rdy**. Since each structure element—for example **Recipe_A.Speed_1**, represents a single data element, it can be supplied as a parameter in a box instruction, as well as in a user FC, FB, or system block.

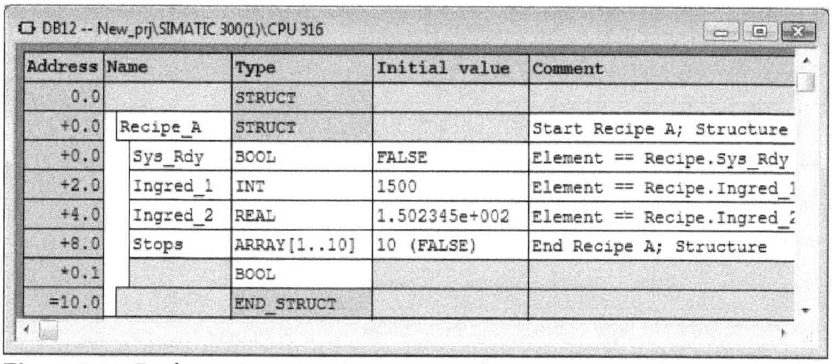

Address	Name	Type	Initial value	Comment
0.0		STRUCT		
+0.0	Recipe_A	STRUCT		Start Recipe A; Structure
+0.0	Sys_Rdy	BOOL	FALSE	Element == Recipe.Sys_Rdy
+2.0	Ingred_1	INT	1500	Element == Recipe.Ingred_1
+4.0	Ingred_2	REAL	1.502345e+002	Element == Recipe.Ingred_2
+8.0	Stops	ARRAY[1..10]	10 (FALSE)	End Recipe A; Structure
*0.1		BOOL		
=10.0		END_STRUCT		

Figure 2-10. Declaring DATE_AND_TIME and STRING, and STRUCT variables

To enter structure data elements:

1.   Open a new data block or code block you wish to edit.

2.   Click in the **Name** field and enter a symbolic name of the new structure—in this example the name is **Recipe_A**.

3.   Click in the data **Type** field and enter the keyword **STRUCT**; then advance to the next row using the **RETURN** key. A single blank declaration row and the last line of the structure— **END_STRUCT** are automatically inserted.

4.   On the next row, enter the symbolic **Name**, data **Type, Initial value** and **Comment** for the first data element. Again, advance to the next row using the **RETURN** key.

5.   Enter as many additional variables as required, using the previous step.

6.   To insert a new blank row, you may also select a declaration row and then from the menu select **Insert ➢ Declaration Line ➢ After Selection**.

7.   Click **File ➢ Save** to save the code block or data block.

Entering UDT-Based Variables

The use of the UDT, as described here, is like any other STEP 7 data type that may be applied to a variable that you declare in either a code block or data block. The UDT can be applied as often as required. The difference, however, is that once the variables that make up the structure are entered as the UDT, you do not have to enter them again. Each time you apply the UDT as the data type, the defined variables are implicitly inserted.

UDT1, shown in the figure below, is a structure containing six variables. Each of the variables is required in several process loops. Once the UDT is created, it can be applied as the data type of a single variable, thereby declaring all six variables of the UDT as a single entry.

UDT1 -- "PID Loop" -- BrilliantTraining\S7 Program(1)\...\UDT1

Address	Name	Type	Initial value	Comment
0.0		STRUCT		
+0.0	Loop_No	WORD	W#16#1	Loop Number
+2.0	PV	REAL	0.000000e+000	Process Input Variable
+6.0	SP	REAL	6.500000e+002	Setpoint
+10.0	UL	REAL	6.750000e+002	Upper Limit Alarm
+14.0	LL	REAL	6.550000e+002	Lower Limit Alarm
+18.0	CTRL	WORD	W#16#0	Control Word
=20.0		END_STRUCT		

Figure 2-11. UDT1—user defined data type contains six process loop variables

As illustrated in the figure below, UDT1 is applied as the data type to the variables named Loop1 through Loop8 in the declaration table of FB70.

FB70 -- "LoopMngr" -- BrilliantTraining\S7 Program(1)\...\FB70

Interface: IN, OUT, IN_OUT, STAT, TEMP

Contents Of: 'Environment\Interface\STAT'

Name	Data Type	Address	Initial Value	Comment
loop1	UDT 1	0.0		Loop 1 Variables
loop2	UDT 1	20.0		Loop 2 Variables
loop3	UDT 1	40.0		Loop 3 Variables
loop4	UDT 1	60.0		Loop 4 Variables
loop5	UDT 1	80.0		Loop 5 Variables
loop6	UDT 1	100.0		Loop 6 Variables
loop7	UDT 1	120.0		Loop 7 Variables
loop8	UDT 1	140.0		Loop 8 Variables

Figure 2-12. FB70 Loop1 to loop8 declared as variables of data type UDT1

To enter variables based on a UDT data type:

1. From the appropriate **S7Program**, open the code block or data block in which you will declare variables using a UDT data type (the UDT must have already been created).

2. Click in the **Name** field of the first data element, and enter a symbolic name (for example Loop1), then in the **Data Type** field enter UDT1 as the data type.

3. Click in the **Comment** field and enter a comment up to 80 characters (optional).

4. Enter the remaining variables as required (for example Loop2 through Loop3), and click **File ➤ Save** to save the block.

When the instance DB70 is created and associated with FB70, the data structure of FB70 is applied to DB70. You can see DB70 in Figure 2-41, where starting with Loop1 the six variables (of UDT1) of each loop are inserted in the data structure of the data block.

DATA BLOCK ADDRESS STRUCTURE

Once you have created the data blocks and entered the initial data, you can then access the data elements in your S7 program. When a data block is created, each data element is assigned a **Name**, data **Type**, **Initial value**, and **Comment**. The assigned **Name** is actually a symbolic name for the data element that may be used instead of the absolute **Address**. In the next discussion, we'll look at how STEP 7 assigns an absolute address to each data element and how it is accessed in the program.

Absolute Data Element Addresses

As data is entered into a data block, each element is stored in the order it is entered, starting from data byte zero. Each element reserves an area that corresponds to the length of the assigned type. The relative start byte address of each element is listed in the Address column, as shown below. As each element is entered, the next available byte is based on the storage requirement of the previous element.

Address	Name	Type	Initial value	Comment
0.0		STRUCT		
+0.0	TMP_PRESET_1	INT	350	Temperature Preset Zone1
+2.0	TMP_PRESET_2	INT	375	Temperature Preset Zone2
+4.0	TMP_PRESET_3	INT	400	Temperature Preset Zone3
+6.0	OVR_LMT_T1	BOOL	FALSE	ZONE 1 OVER Limit
+6.1	OVR_LMT_T2	BOOL	FALSE	ZONE 2 OVER Limit
+6.2	OVR_LMT_T3	BOOL	FALSE	ZONE 3 OVER Limit
+7.0	OPER_RESP_1	CHAR	'N'	OPERATOR RESPONSE 'YES'
+8.0	OPER_RESP_2	CHAR	'N'	OPERATOR RESPONSE 'NO'
+10.0	NEW_CNT	DWORD	DW#16#900	COUNTER OVERRIDE
=14.0		END_STRUCT		

Figure 2-13. Editing a Shared Data block. Data variables defined with initial values

When defining data elements you may use any valid data type, including elementary types BOOL, BYTE, INT, DINT, WORD, DWORD, REAL, CHAR, TIME, TIME_OF_DAY, and S5TIME. Complex data types may also be entered. Whether a data element uses a bit, byte, word, or double word, the start byte is reflected in the relative address of each element. As illustrated in Figure 2-13, and described in Table 2-1, word and double-word elements always start on the next available even-byte. Byte elements start on the next available byte. Bit elements always allocate the next available bit.

Table 2-1. Structure of absolute addresses in a data block

Address (Start Byte)	Element	Width	Reserved Area	Relative Address
0	TMP_PRESET_1	16-bits	Byte-0, Byte-1	DBW0
2	TMP_PRESET_2	16-bits	Byte-2, Byte-3	DBW2
4	TMP_PRESET_3	16-bits	Byte-4, Byte-5	DBW4
6	OVR_LMT_T1	bit	Bit-0 of Byte-6	DBX6.0
6	OVR_LMT_T2	bit	Bit-1 of Byte-6	DBX6.1
6	OVR_LMT_T3	bit	Bit-2 of Byte-6	DBX6.2
7	OPER_RESP_1	8-bits	Byte-7	DBB7
8	OPER_RESP_2	8-bits	Byte-8	DBB8
10	NEW_CNT	32-bits	Byte-10, 11, 12, and 13	DBD10

ADDRESSING DATA ELEMENTS

Once you have created the data blocks of your program and entered any initial data, you may then access the data elements from within your S7 program. When it comes to accessing data block variables, you may access data elements using *partial-addressing, full-addressing,* or *symbol-addressing.*

Data-Bits,—Bytes,—Words, and—Double Words

To access a data element, first open the DB using the appropriate instruction to reference the DB number—for example **OPN DB n** or **OPN DI n**. The parameters DI and DB represent the two registers that a CPU uses to hold the current open DB number. At any given time, two data blocks may be open. Although both global and instance DBs may be opened in either register, it is preferable that a shared DB is opened in the *DB-register* and an instance DB in the *DI-register.* Once open, any data element may be accessed using the correct data size identifier and start byte—for example DBW4. Examples of data bits, data bytes, data words, and data double words are shown in Table 2-2.

Table 2-2. DB Addresses identifiers and example addresses

Register	Data Unit	Identifier	Example	Description of Example
DB-Register (Shared DB)	Bit	DBX	DBX4.7	Bit 7 of Byte 4 of Shared DB (n)
	Byte	DBB	DBB4	Byte 4 of Shared DB (n)
	Word	DBW	DBW4	Word 4 of Shared DB (n)
	Double Word	DBD	DBD4	Double Word 4 of Shared DB (n)
DI-Register (Instance DB)	Bit	DIX	DIX12.4	Bit 4 of Byte 12 of Instance DB (n)
	Byte	DIB	DIB12	Byte 12 of Instance DB (n)
	Word	DIW	DIW12	Word 12 of Instance DB (n)
	Double Word	DID	DID12	Double Word 12 of Instance DB (n)

Note: Data elements are accessed with partial addresses using the DB or DI identifier based on which data block register (DB or DI) is used to open the data block.

In LAD/FBD, you can affect the DB or DI register when you use the **OPN** coil/output operation. After a DB is opened in the DB register, a data element may then be accessed in the following networks until a new DB is opened. In LAD/FBD, an instance DB is inserted above the associated FB when an FB is called—the DB is opened by default in the DI-register. In STL, the open statement is in the format OPN DB n—for example OPN DB22; or OPN DI n—for example OPN DI23. As shown in Table 2-2, a data word is addressed DBW if you use OPN DB; a word is addressed DIW if you use OPN DI.

DBX4.7 (data bit)

DBB4 (data byte—4)

DBW4 (data word-4 = byte-4 and byte-5)

DBD4 (data double word-4 = byte—4, byte—5, byte—6, and byte—7)

Figure 2-14. Illustration of a data bit, data byte, data word, and data double word location in a DB

Partial-Addressed Data Elements

With partial addressing, you are accessing a data element of an already opened global or instance data block. In STL, the open instruction is **OPN DB n** for a global DB and **OPN DI n** for an instance DB. In LAD/FBD the open instruction is in the form of a special coil/output, placed in a network without preceding logic. Above the coil/output you may reference the DB number.

Following the open DB instruction, any data element can be accessed in subsequent LAD/FBD networks or STL statements until a new data block is opened. This method is generally used if several elements are being accessed. Example of partial addresses after using OPN DB are DBX8.7, a data bit; DBB10, a data byte; DBW20, a data word; and DBD12, a double data word. If you had used OPN DI in STL, these locations would be addressed DIX8.7, DIB10, DIW20, and DID12 (See Table 2-2). Data element access using partial addressing is illustrated in the following examples.

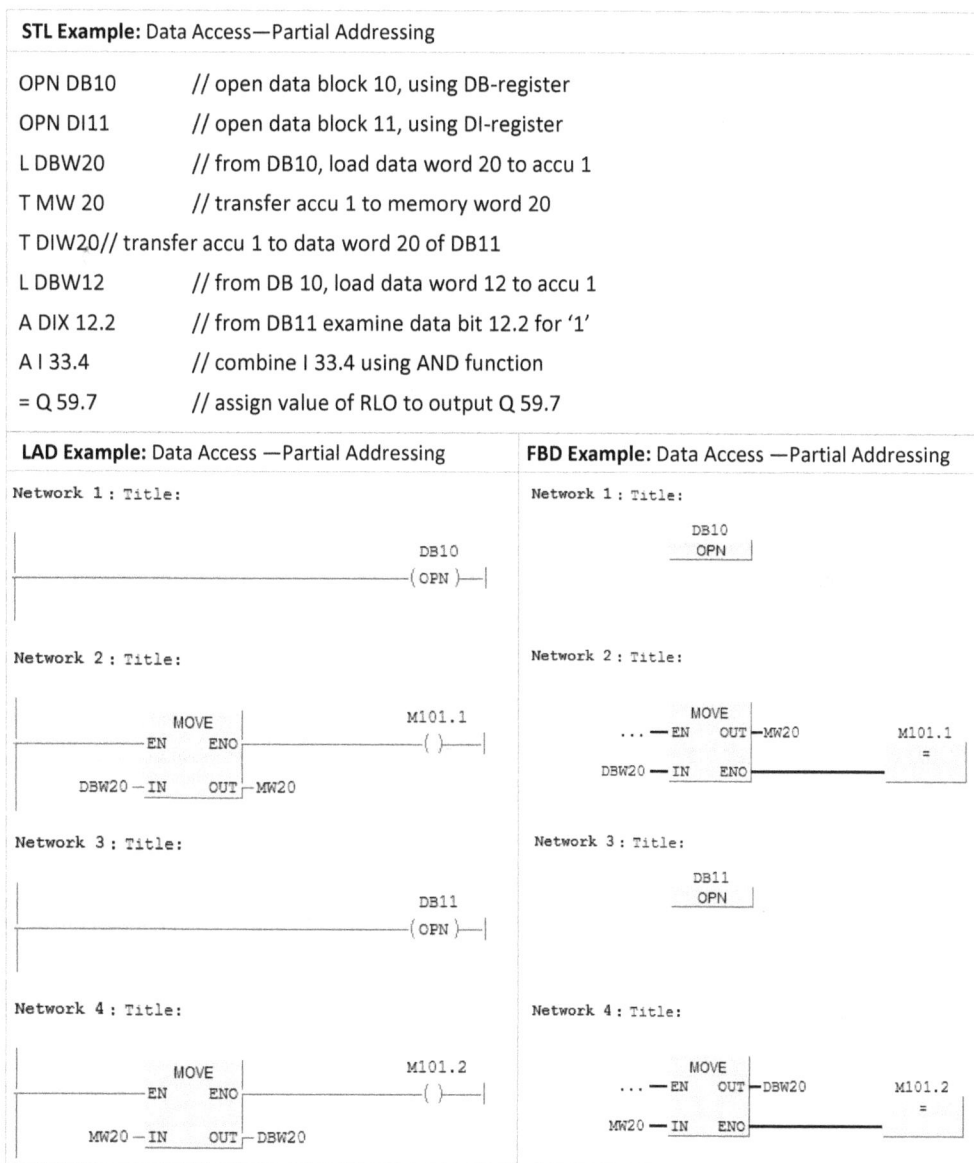

STL Example: Data Access—Partial Addressing

OPN DB10	// open data block 10, using DB-register
OPN DI11	// open data block 11, using DI-register
L DBW20	// from DB10, load data word 20 to accu 1
T MW 20	// transfer accu 1 to memory word 20
T DIW20// transfer accu 1 to data word 20 of DB11	
L DBW12	// from DB 10, load data word 12 to accu 1
A DIX 12.2	// from DB11 examine data bit 12.2 for '1'
A I 33.4	// combine I 33.4 using AND function
= Q 59.7	// assign value of RLO to output Q 59.7

LAD Example: Data Access —Partial Addressing

Network 1 : Title:
```
                          DB10
                       —( OPN )—|
```

Network 2 : Title:
```
        MOVE                M101.1
     —EN    ENO—          —( )—|
  DBW20 —IN   OUT—MW20
```

Network 3 : Title:
```
                          DB11
                       —( OPN )—|
```

Network 4 : Title:
```
        MOVE                M101.2
     —EN    ENO—          —( )—|
  MW20 —IN   OUT— DBW20
```

FBD Example: Data Access —Partial Addressing

Network 1 : Title:
```
              DB10
              OPN |
```

Network 2 : Title:
```
        MOVE
  ... —EN   OUT—MW20        M101.1
  DBW20 — IN  ENO            =
```

Network 3 : Title:
```
              DB11
              OPN |
```

Network 4 : Title:
```
        MOVE
  ... —EN   OUT—DBW20       M101.2
  MW20 — IN  ENO             =
```

Figure 2-15. Accessing data locations using partial-addressed operands

Full-Addressed Data Elements

Fully-addressed operands are referred to as such because you specify the data block number as part of the bit, byte, word, or double word address. With fully-addressed data elements you do not have to use an open data block operation first since the DB number is explicitly specified. Examples of this method include the bit address DB22.DBX8.7, the byte address DB22.DBB10, the word address DB22.DBW20, and the double word address DB22.DBD12.

Use of fully-addressed operands is typical when only a single data element is being accessed or when a data element is passed as a block parameter. In the example, a single word (DBW14) of DB11 is being moved to the same word address of DB21. While this method is straight-forward, a symbolic address, as shown in Figure 2-17, is also fully-addressed and is perhaps more easily interpreted.

LAD Example: Data Access—Full Addressing	**FBD Example:** Data Access—Full Addressing
Network 1 : Title: MOVE EN ENO DB11.DBW14 — IN OUT — DB21.DBW14	Network 1 : Title: MOVE ... — EN OUT — DB21.DBW14 DB11.DBW14 — IN ENO

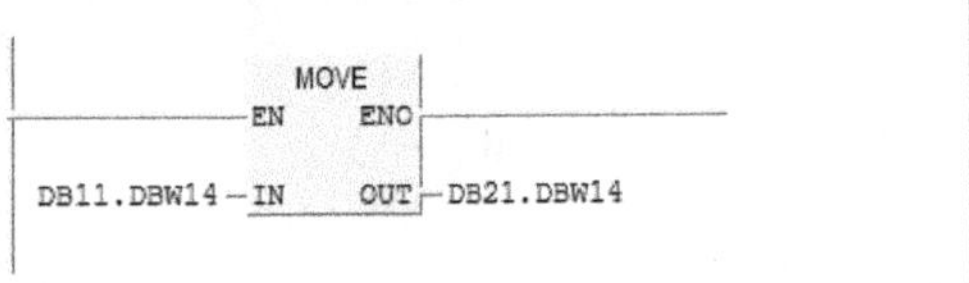

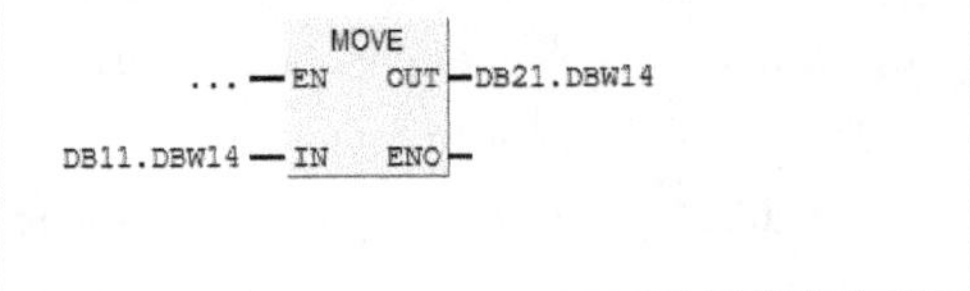

STL Example: Data Access with Full Addressing

Network 1 : Title:
 	
    L   DB11.DBW   14
 	
    T   DB21.DBW   14

Figure 2-16. Accessing data locations using fully-addressed operands

Symbol-Addressed Data Elements

Symbol addressing can also be applied with fully-addressed data elements. Symbol names are given to each data element as you enter the data elements to the data block. You also need to ensure that you assign a symbolic address to the data block. The symbolic addresses are then derived by combining the assigned symbolic name of the DB with the assigned name of each element in the data block.

In the example shown below, the data element **Zone_1.TMP_6** is in a data block whose symbol name is Zone_1. The specific data element is named TMP_6.

LAD Example: Data Access—Symbol Addressing	**FBD Example:** Data Access —Symbol Addressing
Network 1 : Title: MOVE EN ENO "Zone1". TMP_6 — IN OUT — "Zone2". Valu_6	Network 1 : Title: MOVE ... — EN OUT — "Zone2". "Zone1". Valu_6 Valu_6 — IN ENO

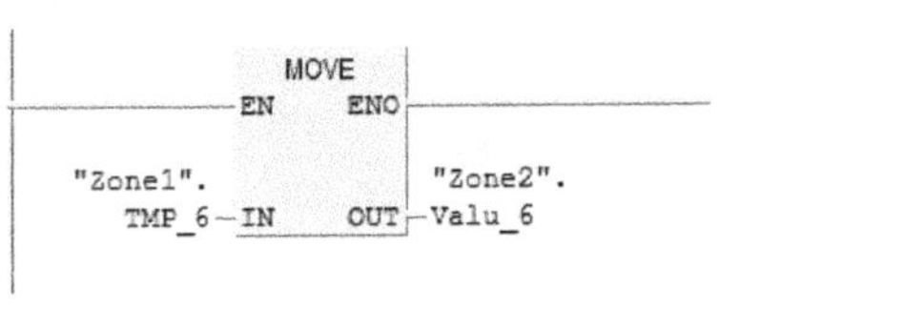

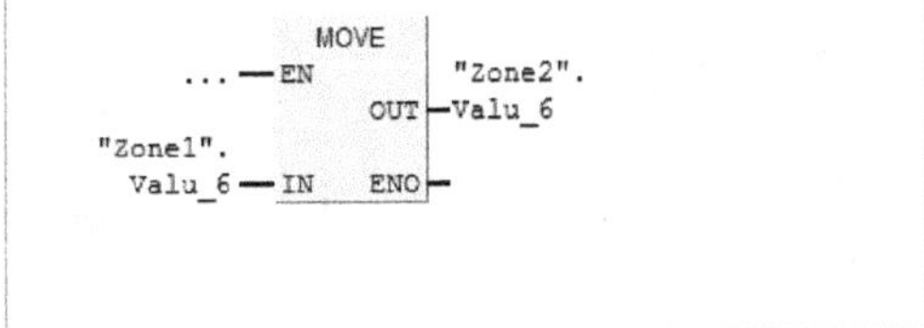

STL Example: Data Access—Full Addressing

Network 1 : Title:
 	
    L   "Zone1".TMP_6
 	
    T   "Zone2".Valu_6

Figure 2-17. Accessing data locations using fully-addressed operands

Part II

Ladder and
Function Block Diagrams

Chapter 3

Introduction to LAD and FBD

This chapter presents an overview of both LAD and FBD language representations. First we'll look at the basic instruction elements and how each is used as fundamental building blocks in developing a program. For example in LAD, basic elements include contacts, coils, and instruction blocks; in FBD there are positive and negated binary inputs, Boolean gates, standard and special output boxes, and instruction blocks. Next we examine how these basic program elements are combined to form a LAD or FBD network. The network is a basic unit of an S7 code block. To complete this part of each language overview, we examine how LAD and FBD networks are entered, edited, and evaluated.

In the eleven chapters following this chapter, each of the STEP 7 instruction categories, including Binary Logic, Move, Timer, Counter, Conversion, Compare, Integer and REAL Arithmetic, Shift-Rotate, Word Logic, Status Bits, Program Flow Control and Block Operations, is covered side-by-side for LAD and FBD.

Main Topics

- LAD Language—An Overview
- Guidelines for Editing in LAD
- FBD Language—An Overview
- Guidelines for Editing in FBD

LAD LANGUAGE—AN OVERVIEW

STEP 7 LAD is one of three basic languages used to develop the STEP 7 program. LAD corresponds to the IEC 1131-3 standard for *Ladder Diagram*—a graphic language based upon the representation of electromechanical relay circuit diagrams. STEP 7 LAD incorporates an extensive operations set, including contact, coil, and box instructions. The basic elements of the LAD language and the LAD/FBD programming editor are described in this section. If you are already familiar with ladder programming, you should easily adapt to programming in STEP 7 LAD.

LAD Instruction Set Categories

The LAD instruction set is comprised of the categories shown in Table 3-1. These categories are the same as those found in the FBD language, and for the most part in STL. In fact, LAD and FBD instructions are fundamentally based on the STL operations set, which represents the full STEP 7 operations set. For example, each of the box instructions of LAD and FBD are comprised of two or more STL operations. Starting in the following Chapter, *Binary Logic Operations*, the LAD and FBD instruction categories are described in subsequent chapters through Chapter 14 *LAD/FBD Program Flow Control and Block Operations*.

Table 3-1. S7 LAD Instruction Categories Summary

Instruction Categories		
Binary Logic Operations	Conversion Operations	Word Logic Operations
Move Operations	Comparison Operations	Status Bit Operations
Timer Operations	Integer and REAL Operations	Program Flow Control
Counter Operations	Shift-Rotate Operations	Block Operations

Basic LAD Elements

The basic elements of which the LAD instruction set is comprised are shown in Table 3-2. These basic elements, which include contacts, coils, box instructions, and logic construction elements, represent the fundamental building blocks of the LAD language—upon which the entire operations set is based. Using these elements, the LAD editor supports simple to complex logic construction of series and parallel circuits for developing binary control logic.

Table 3-2. Basic Program Elements in LAD

Contact Elements		Coil Elements		Box Instructions	
Standard	Special	Standard	Special	With EN/ENO	Without EN/ENO
-]/[-	-]NOT[-	-( )-	-(R)- -(S)-	INV_I —EN ENO—	CMP ==I —
-] [-			-(P)- -(N)-	—IN OUT—	—IN1
			-(#)- -SAVE)-		—IN2
LAD Logic Construction Elements					
Series Contacts (AND Logic)			**Parallel Contacts (OR Logic)**		
—| |————| |—			—| |— —| |—		

Contact Elements

Contact Elements allow you to examine the status of bit-addressable memory locations for '1' or '0' and to combine them to form series, parallel, or series/parallel logic combinations. These addresses may represent the signals of externally connected input devices, or any bit variable or address in S7 memory. From this control logic you can determine how internal or external outputs are controlled, whether or not a box instruction is executed, or if a specific code block should be called for processing.

Coil Elements

Using *coil elements* you can control the '1' or '0' (on/off) status of bit-addressable memory locations as required. These addresses may represent internal outputs, externally connected output devices, coils with special functions such as latching/unlatching (set/reset), or indicators for positive or negative edge transitions. The on/off state of these variables is determined by the control logic you insert to control their operation—a basic feature described later in this section as the LAD network.

Box Instructions

Box Instructions combine two or more S7 operations to form a single instruction in a box format. These box instructions move the LAD instruction set beyond binary operations to include operations involving digital data variables. Timers, counters, compares, conversions, arithmetic, shift-rotate, word logic, and move are all examples of box instructions. The greater part of the STEP 7 instruction set includes box operations.

As seen in Table 3-2, a box instruction usually has a number of inputs to the box and outputs from the box—these inputs and outputs may include both binary signals as well as digital data values. A binary input, for example will control some aspect of the box operation—like to count up in an up counter. A binary output provides some status indication, for example timer enabled. A digital value input is required, for example, when providing two values to be compared. In the illustration below a digital data value output from the timer box provides the remaining time in BCD.

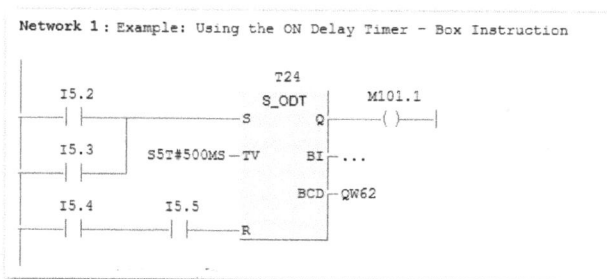

Figure 3-1. Illustration of a box instruction in a LAD network

Also shown in Table 3-2, is that box instructions may be distinguished as those with the EN/ENO (enable input/enable output) mechanism and those without this mechanism. Operations that use EN/ENO include digital data operations including integer and real arithmetic, conversion, shift-rotate, and word logic operations. The enable input line EN is where you will insert the control logic that determines if the operation should be executed. Box operations without the EN/ENO mechanism include timers, counters, and move operations. These latter box operations generally involve one or more other binary inputs to which control logic is inserted.

When EN is logic 1, the operation executes. The enable output ENO signals execution of the operation and whether an error occurred. If an error occurs while EN is at logic 1, ENO is set to logic 0; otherwise the status of ENO follows the status of EN. This operation is described in the following table.

Table 3-3. LAD Status Bit Operation and Evaluation.

If RLO	Then EN	Instruction Execution	Then ENO	
= 0	= 0	The instruction is not executed	= 0	
= 1	= 1	The instruction is executed without error	= 1	
= 1	= 1	The instruction is executed with error	= 0	

```
         INV_I
    —EN      ENO—
    —IN      OUT—
```

As you program box operations that incorporate the **EN/ENO** mechanism, you may connect the **ENO** line to drive a following contact or coil, or directly to the **EN** input line of another box operation. In this fashion you may place box operations in series, as shown in the figure below. A box operation in series with another box operation is only executed if the preceding operation processes without error.

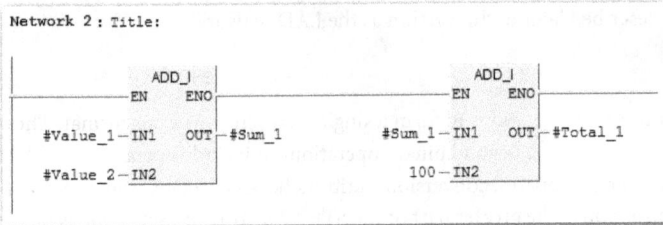

```
Network 2 : Title:

              ADD_I                          ADD_I
           EN      ENO                     EN      ENO

  #Value_1—IN1   OUT—#Sum_1      #Sum_1—IN1   OUT—#Total_1

  #Value_2—IN2                      100—IN2
```

Figure 3-2. Illustration of EN/ENO box instructions in series

LAD Logic Construction Elements

In LAD, the creation of simple series and parallel circuits allow binary input signals to be combined and evaluated according to requirements of the control logic. More complex circuits may be developed by combining series circuits with parallel circuits. As required by the control logic, you may evaluate each binary signal using either a normally-open or normally-closed contact element. Recall, that the normally-open contact examines for logic 1, and therefore yields a '1' result if the condition is '1'; and the normally-closed contact examines for logic 0, therefore yields a '1' result if the condition is '0'.

When you program a series circuit, the programmed contact elements are combined according to the logical AND function, since all of the series conditions must be TRUE in order to yield a logic 1 result. The parallel circuit allows contact elements to be combined according to the logical OR function, since any one of multiple binary conditions must be TRUE in order to yield a logic 1 result.

The LAD Network

A *LAD network* or *ladder rung* is the smallest part of a LAD program. The illustration below shows a basic LAD network—it consists of a set of *logic conditions* represented by contact elements arranged as a logic circuit. The network logic drives a coil instruction or represents the binary input to a box operation. A code block, written in LAD, generally consists of several networks. When the CPU is switched to Run Mode, and the block is called, each network is evaluated in the order it was entered.

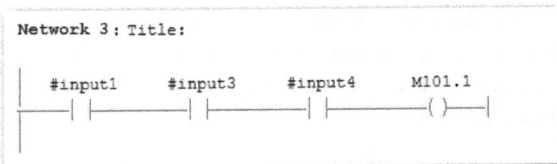

```
Network 3 : Title:

    #input1      #input3      #input4      M101.1
    —| |————————| |————————| |—————————( )—
```

Figure 3-3. Illustration of a basic LAD network

Evaluating Contacts and Coils in a LAD Network

By convention, a network coil will switch ON (1) if any left-to-right path of driving logic conditions is closed and thereby allowing logic power flow. In such a case, the network is said to be 'TRUE', meaning that the logic circuit is completed or has *logic continuity*. In STEP 7 this logic continuity is referred to as the *result of logic operations (RLO)*. If the circuit has complete logic continuity then it is said that the RLO = 1; otherwise, the RLO = 0. Furthermore, the logic state evaluated at any node along the circuit is also referred to as the result of logic operations. So, the RLO can be evaluated after any program element, after any combination of logic, or at the end of the logic circuit.

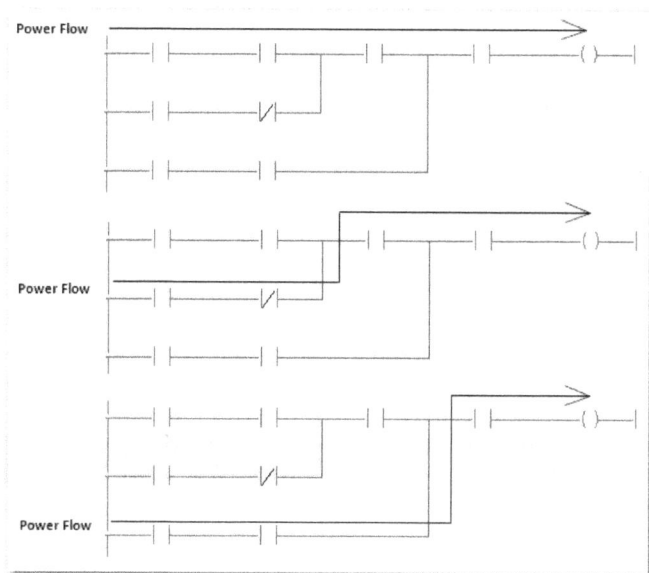

Figure 3-4. Evaluation of contact instructions in a LAD network

When you examine discrete input devices, remember that inputs are wired normally-open or normally-closed. A device that is wired normally-open presents a logic '1' input when the device is activated; and presents logic '0' when de-activated. This type of device is said to be *Active HIGH*. On the other hand, a device wired normally-closed presents a logic '1' when not activated, and presents logic '0' when activated. This type of device is said to be *Active LOW*.

How a device is wired and the effect it is expected to have in your control logic will determine which LAD contact element you should use to examine an input. In general, a normally-open contact is used when a logic '1' input is expected to cause power flow; a normally-closed contact is programmed when a logic '0' input is expected to cause power flow.

Evaluating Block Operations in a LAD Network

As with networks programmed with contact and coils, networks with box instruction are also affected by the result of logic operations. How box operations are affected by the RLO will differ depending on the specific instruction. In general, box operations that have the **EN/ENO** mechanism are only executed if the RLO = 1. Depending on the box operation, there may also be other input lines—for example the Reset line in timer and counter box instructions.

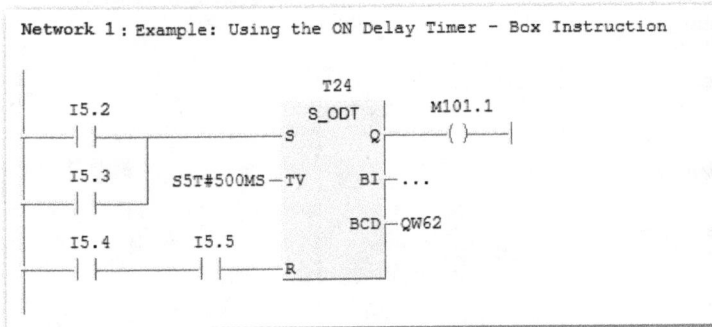

```
Network 1: Example: Using the ON Delay Timer - Box Instruction


                                T24
     I5.2                      S_ODT         M101.1
  ─┤ ├──────────────┬──────S          Q───────( )───┤
                     │
     I5.3            │   S5T#500MS─TV     BI ┌ ...
  ─┤ ├───────────────┘
                              BCD ┤QW62
     I5.4        I5.5
  ─┤ ├──────────┤ ├──────R
```

Figure 3-5. Evaluation of a block instruction in a LAD network

A summary of how box operations are evaluated in the LAD network is presented in the table below.

Table 3-4. Evaluation summary of Block Operations in LAD

EN/ENO	Operation Type	Brief Evaluation Description	Typical
NO	Timer	With the exception of the OFF-delay timer, all timers are started with a 0-to-1 transition of the RLO on the timer start line (S).	S_ODT ─S Q─ ─TV BI─ ─R BCD─
		The OFF-delay timer is started with a 1-to-0 transition of the RLO on the timer start line (S).	
	Counter	Counter Operation counts accordingly whenever the RLO on the count line transitions from 0-to-1.	S_CU ─CU Q─ ─S CV─ ─PV CV_BCD─ ─R
	Comparison	Comparison operations are executed irrespective of the RLO. Instead, the compare operation behaves like a normally-open contact that allows power flow when the outcome of the compare test is True (1) and interrupts power flow when the outcome is False (0).	CMP ==I ─ ─IN1 ─IN2
YES	Arithmetic	Executed when the RLO = 1 on the **EN** line	SHL_DW ─EN ENO─ ─IN OUT─ ─N
	Conversion	Executed when the RLO = 1 on the **EN** line	
	Word Logic	Executed when the RLO = 1 on the **EN** line	
	Shift-Rotate	Executed when the RLO = 1 on the **EN** line	
	Move	Executed when the RLO = 1 on the **EN** line	

GUIDELINES FOR EDITTING IN LAD

When it comes to entering and editing in LAD, you may be working with an existing block or with an entirely new block that you have created. In either case your first step will be to open the block from either the Simatic Manager or from the LAD/STL/FBD editor. With the block opened you will set the STEP 7 editor to work in LAD by using the menu and selecting **View ➤ LAD**.

General Editing in LAD

Before inserting any LAD program elements, you may want to ensure that the editor is in the *insert mode* as opposed to the *overwrite mode*. Like with most editors, the overwrite mode writes over a previously entered character or object based on how you place the cursor. Insert mode simply inserts a new item and pushes existing objects to the right based on the cursor position. Use the <**Insert**> key to toggle between these two edit modes. The current edit mode is displayed in the status bar in the next to the last field. Whereas insert mode is the default in normal editing mode, overwrite mode is generally used to overwrite an instruction when the wrong instruction has been entered. To overwrite an element, the new element must be of the same general type as the old element.

When a LAD instruction is inserted, red question mark characters '**???**' indicate that an address or parameter must be specified. Black characters '**...**' indicate that an address or parameter can be specified, but is optional.

Figure 3-6. LAD/STL/FBD Status bar, with the current status showing the editor in '**Insert** mode'

Inserting a LAD Network

A new network is inserted by a single click on the **New Network** toolbar icon, by double-clicking on the **New Network** object in the Program Elements window, or by dragging and dropping the New Network object from the Program Elements window to the desired position. Before any of these methods is used, you need to decide the position of the new network. If you have opened a new block for the first time, then the first network will appear by default.

To insert a network following an existing network, select that network and then use one of the previously described options. For example, select **Network 3** to insert the new network between **Network 3** and **Network 4**. To insert a network as the first network in an existing block, double-click on the Block Title line and then use one of the previously described New Network options.

Inserting LAD Contacts and Coils

To insert basic LAD elements, you may use the toolbar icons or Function Keys. **F2** = Normally-Open Contact; **F3** = Normally-Closed Contact; **F7** = Coil; **F8** = Branch Open; and **F9** = Branch Close. Other special contact and coil instructions not found on the toolbar are found in the Bit Logic subfolder of the Program Elements tab, or in one of the other subfolders—for example the Status Bits folder.

After inserting an element, simply address the element as required. Use the **View** menu and select **Display with ➤ Symbols Representation** if you wish to use symbol addresses.

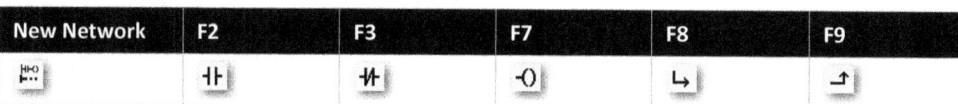

New Network	F2	F3	F7	F8	F9
┣┅┅	┤├	┤/├	‑( )	↳	⤒

Figure 3-7. Toolbar icons for basic LAD program elements

Element Placement

When editing in LAD, remember that contact elements are condition instructions and are generally placed to the left side of the network—preceding any output instructions. The various series and parallel arrangement that you create represents the control logic that will determine the RLO and whether or not a coil is activated or whether a box instruction is executed.

In LAD, coil instructions terminate the network and are placed in the right-most position as the last element of the network. Exceptions to this general rule for coil placement include the midline coil, the P-coil, and the N-coil. These coils may be placed anywhere along the network, excluding the left-most and right-most positions. Furthermore, these coils may not be placed in parallel branches. The table below provides a placement summary of LAD program elements.

Table 3-5. Summary of LAD binary output requirements and restrictions in a network

Coils that require preceding logic			
Output Coil	—( )—	Jump if NOT	—(JMPN)—
Set Coil	—(s)—	MCR Zone Start	—(MCR<)—
Reset Coil	—(R)—	Save RLO to BR	—(SAVE)—
Positive Edge Coil	—(P)—	Return	—(RET)—
Negative Edge Coil	—(N)—	Timer Coils—All	—(SE)—
Midline Coil	—(#)—	Counter Coils—All	—(CU)—
Coils that do not allow preceding logic			
Open DB Coil	—(OPN)—	MCR Activate	—(MCRA)—
MCR Zone End	—(MCR>)—	MCR De-Activate	—(MCRD)—
Coils that do not allow parallel outputs			
Jump	—(JMP)—	Jump if NOT	—(JMPN)—
CALL Coil	—(CALL)—	Return Coil	—(RET)—

Note: All coils not listed in this table may be inserted with or without preceding logic

Inserting LAD Elements in Series

You may insert LAD contacts, box instructions, and coils in series by dragging and dropping the object onto the network or using any of the previously described methods. By default, page settings for the LAD editor are set to 'A4 Portrait Layout'. With this setting you may place approximately 9 contact elements in series to drive a coil. If you insert more than the allowed number of elements, STEP 7 will issue a message that the page settings have been exceeded. You may modify the layout setting from the Options menu by selecting Customize ➤ **LAD/FBD** tab, and then modify the **Layout**.

As shown in column ❶ you may first insert as many series contacts as required; then, as shown in column ❷ you may go back and enter the appropriate addresses.

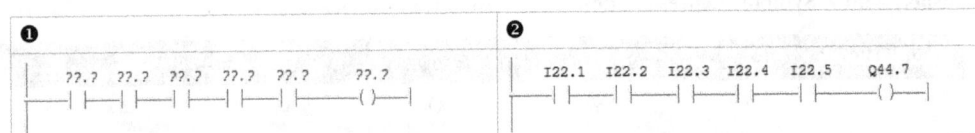

Figure 3-8. Inserting LAD elements in series

64

Inserting LAD Elements in Parallel

In LAD, clicking on the ladder rung between any two elements will cause the *node* that exists between two points to be highlighted. The point between the left rail and the first element in a rung is also a node, and is shown selected in Figure 3-9.

Contacts or other Instructions in Parallel

To inset a contact or other element in parallel with another, ❶ click on the node just in front of the element you want to parallel—the toolbar **open branch** icon will enable; ❷ click the **open branch** icon, and ❸ insert one or more instruction elements in the new parallel branch—the toolbar **close branch** icon will enable; and ❹ click the **close branch** element, to terminate the branch.

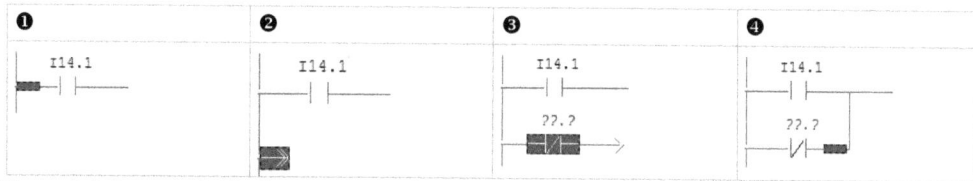

Figure 3-9. Inserting LAD elements in parallel—Part 1

Branching around Multiple Network Columns

The following figure illustrates how the branch is developed to close around more than one network element or column. In Figure 3-10, a new branch is inserted in column❶. After inserting a contact in the new branch, the toolbar **close branch** icon will enable—❷ the branch could be closed around **I 4.3** or **I 4.4**. To close the branch, ❸ click on the last element around which the branch should be closed; and ❹ click the **close branch** element, to terminate the branch.

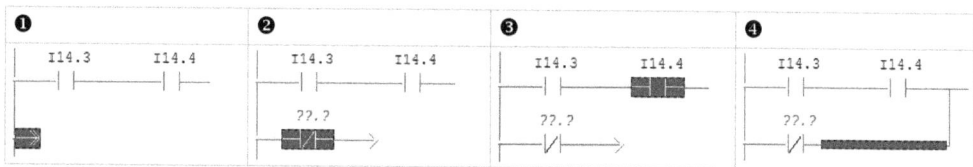

Figure 3-10. Inserting LAD elements in parallel—Part 1

Output Instructions in Parallel

Inserting output elements in parallel is accomplished in the same fashion as with contacts. While standard LAD outputs may be placed in parallel, parallel branches are not permitted with output coils **JMPN** (Jump if Not), **JMP** (Jump), **CALL** (call block), and **RET** (Return).

Inserting parallel outputs in LAD is shown below. ❶ Click on the node preceding the coil element around which you wish to create a parallel branch—the toolbar **open branch** icon enables. ❷ Click the **open branch** icon; ❸ click on the toolbar icon for coil to insert the new output element; and ❹ finally, enter the appropriate address. Additional parallel outputs are added by repeating this sequence.

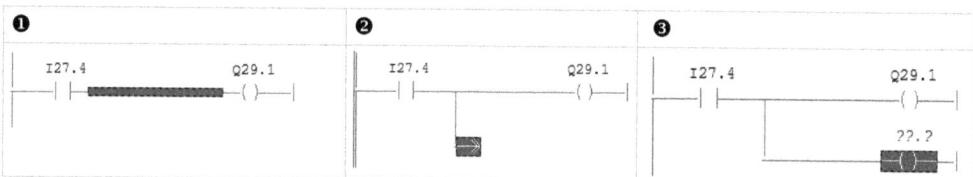

Figure 3-11. Inserting LAD elements in parallel—Part 2

Inserting LAD Program Comments

With each block you create in LAD you may enter 1) **Variable Comments** of up to 80-characters, for each variable you declare; 2) a **Block Title**, to briefly title the code block; 3) a **Block Comment**, to describe the purpose or operation of the block; 4) a **Network Title**, to title each logic network you enter; and 5) a **Network Comment** that describes the purpose or operation of each logic network. If the comment fields are not enabled (hidden), you may enable commenting from the editor menu.

From the menu, select **View** ➤ **Display with** ➤ **Comments**. This selection toggles between enabling and disabling comment entry and display. Steps for displaying, entering, and editing program comments are described in Chapter 27 *Designing and Implementing the STEP 7 Program.*

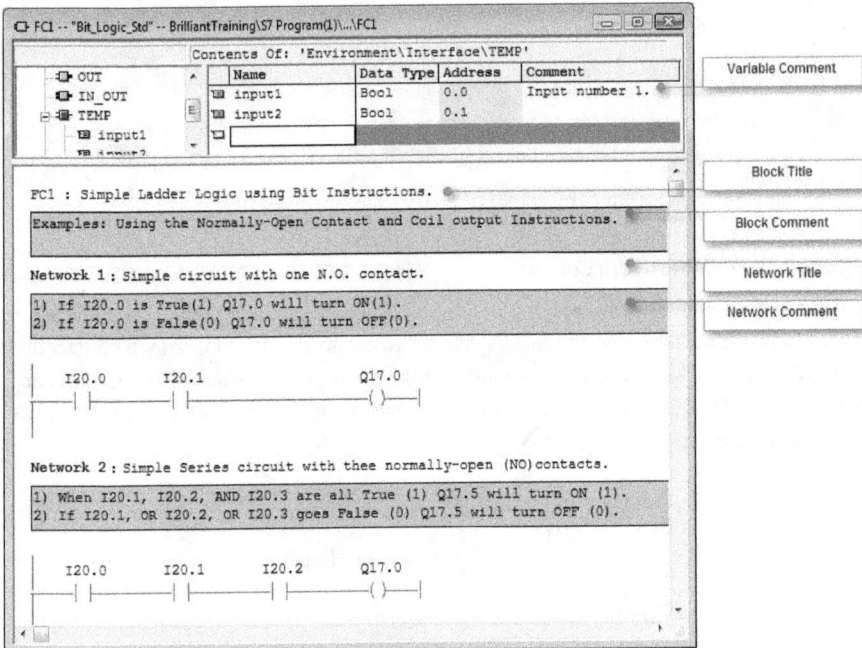

Figure 3-12. LAD Block window with Comments enabled and displayed

FBD LANGUAGE—AN OVERVIEW

STEP 7 FBD is one of three basic languages used to develop your STEP 7 program. FBD is short for *function block diagrams*, a language that is based on the graphic presentation of electronic gate logic diagrams. STEP 7 FBD incorporates an extensive operations set, including binary input, binary output, and box instructions. The basic elements of theFBD language and the LAD/FBD programming editor are described in this section. If you are already familiar with Boolean gate logic, you should easily adapt to programming in STEP 7 FBD.

FBD Instruction Set Categories

The instruction categories of which the FBD instruction set is comprised are shown in the list below. These categories are the same as those found in LAD, and for the most part in STL. In fact, FBD and LAD instructions are fundamentally based on the STL instruction set, which represents the full STEP 7 operations set. For example, each of the box instructions of FBD and LAD are comprised of one or more STL operations. Starting in the following Chapter, with Binary Logic Operations, the FBD and LAD instruction categories are described in subsequent chapters through Chapter 14 *LAD/FBD Program Flow Control and Block Operations*.

Table 3-6. S7 FBD Instruction Categories Summary

Instruction Categories		
Bit Logic Operations	Integer Arithmetic	Shift-Rotate
Move Operations	REAL Arithmetic	Status Bit Operations
Timers and Counters	Standard Math Functions	Program Flow Control
Conversions and Comparisons	Word Logic Operations	Block Operations

Basic FBD Elements

The basic elements of which the FBD instruction set is comprised are shown in the following table. These simple elements that include binary inputs, binary outputs, box instructions, and Boolean gate elements, represent the fundamental building blocks of the FBD language. The entire operation set is based upon these elements. Using these basic elements, the FBD editor supports simple to complex logic construction of AND (series) and OR (parallel) gates to develop binary control logic.

Table 3-7. Basic Program Instruction and Construction Elements in FBD

Binary Input Elements		Binary Output Elements		Box Instructions	
Standard	Special	Standard	Special	With EN/ENO	Without EN/ENO
Positive Input ⊣	Negated Input ⊸	=	S / R	INV_I — EN OUT — IN ENO	CMP ==I — IN1 — IN2
FBD Logic Construction Elements—Boolean Gates					
AND Gate		OR Gate		XOR Gate	
&		>=1		XOR	

Binary Input Elements

Binary Input instructions allow you to examine the status of any bit-addressable memory locations. Using positive and negated binary inputs you may check the status of connected discrete inputs and outputs or any S7 bit variable for a status of '0' or '1'. The status of these variables can be combined to form control logic that will determine how internal or external outputs are controlled, whether a box instruction is executed, or if a specific code block should be called for processing.

Binary Output Elements

Using *binary output* instructions you can control the '1' or '0' status of bit-addressable locations or variables. These binary outputs may represent internal outputs, externally connected outputs, or outputs with special functions such as latching/unlatching (set/reset), or indicators for positive or negative edge transitions. The on/off state of these variables is determined by the control logic you insert to control their operation—a basic feature described later in this section as the FBD network.

Box Instructions

Box Instructions combine two or more S7 operations to form a single instruction in a box format. These box instructions move the FBD instruction set beyond binary operations to include operations involving digital data variables. Timers, counters, compares, conversions, arithmetic, shift-rotate, word logic, and move are all examples of box instructions. The greater part of the STEP 7 instruction set includes box operations.

As seen in Table 3-7, a box instruction usually has a number of inputs to the box and outputs from the box—these inputs and outputs may include both binary signals as well as digital data values. A binary input, for example will control some aspect of the box operation—like to count up in an up counter. A binary output provides some status indication, for example timer enabled. A digital value input is required, for example, when providing two values to be compared. In the illustration below a digital data value output from the timer box provides the remaining time in BCD.

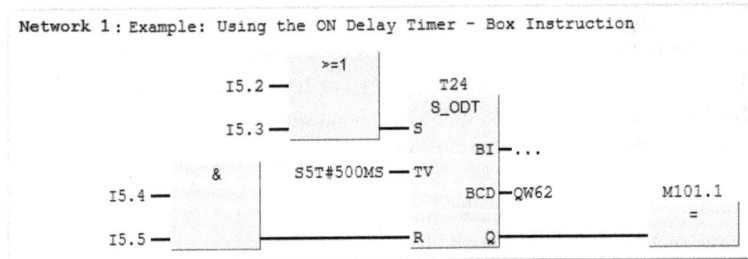

Figure 3-13. Illustration of a box instruction in an FBD network

Also shown in Table 3-7, is that box instructions may be distinguished as those with the EN/ENO (enable input/enable output) mechanism and those without this mechanism. Operations that use the EN/ ENO mechanism include digital data operations including integer and real arithmetic, conversion, shift-rotate, and word logic operations. The enable input line EN is where you will insert the control logic that determines if the operation should be executed. Box operations without the EN/ENO mechanism include timers, counters, and move operations. These latter box operations generally involve one or more other binary inputs to which control logic is inserted.

When EN is logic 1, the operation executes. The enable output line ENO signals execution of the operation and whether an error occurred. If an error occurs while EN is at logic 1, ENO is set to logic 0; otherwise the status of ENO follows the status of EN. This operation is described in Table 3-8.

Table 3-8. FBD Status Bit Operation and Evaluation

If RLO	Then EN	Instruction Execution	Then ENO	
= 0	= 0	The instruction is not executed	= 0	INV_I
= 1	= 1	The instruction is executed without error	= 1	— EN OUT —
= 1	= 1	The instruction is executed with error	= 0	— IN ENO —

As you program box operations that incorporate the **EN/ENO** mechanism, you may connect the **ENO** line to a following gate function, binary output, or directly to the **EN** input line of another box. In this fashion, you may place box operations in series, as shown in Network 2 below. A box operation in series with another box operation is only executed if the preceding operation processes without error.

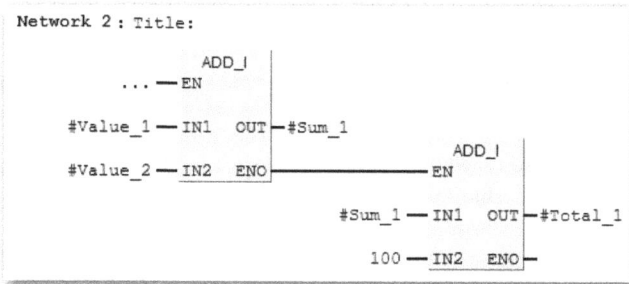

```
Network 2 : Title:
                     ADD_I
              ... — EN

    #Value_1 — IN1   OUT —#Sum_1
                                        ADD_I
    #Value_2 — IN2   ENO——————————— EN

                          #Sum_1 — IN1   OUT —#Total_1

                             100 — IN2   ENO —
```

Figure 3-14. Illustration of the EN/ENO box instruction in an FBD network

FBD Logic Construction Elements

In FBD, Boolean logic gates allow binary conditions to be combined and evaluated according to requirements of the control logic. Whereas LAD supports the development of series and parallel circuits using contact elements—development of control logic in FBD is supported using three basic logic gates—AND, OR, and XOR that accepts positive or negated binary input elements.

The AND gate function allows inputs to be combined in according to the AND truth table, since all inputs to the gate must be logic 1 in order to yield an output of logic 1. The OR gate allows binary inputs to be combined in parallel or according to the OR truth table, since any one of multiple binary inputs must be logic 1 in order to yield an output of logic 1. The XOR logic gate allows two binary input signals to be combined and evaluated according to the XOR truth table, since one of the inputs must be logic 1, but not both, in order to yield an output of logic 1.

Table 3-9. Boolean Logic Gates AND, OR, and XOR—each with the governing Truth Table

Boolean Logic Gates with Truth Tables		
AND	**OR**	**XOR**

AND Truth Table				OR Truth Table				XOR Truth Table		
IN1	IN2	OUT		IN1	IN2	OUT		IN1	IN2	OUT
0	0	0		0	0	0		0	0	0
0	1	0		0	1	1		0	1	1
1	0	0		1	0	1		1	0	1
1	1	1		1	1	1		1	1	0

(Gate symbols: AND = `&`, OR = `>=1`, XOR = `XOR`)

These three gates may be combined in simple to complex arrangements to form the logic that controls standard and special outputs such as *Set* and *Reset*, as well as FBD box operations like *move, timers, counters, compares, conversions, arithmetic, shift-rotate,* and *word logic.*

The FBD Network

An *FBD network* is the smallest part of an FBD program. Figure 3-15 shows a basic FBD network—it consists of *logic conditions*, represented by binary inputs to a logic gate, which in turn drives a binary output. The logic conditions of a network may also drive the binary input to a box operation. An FBD code block will generally consist of several networks, each of which is evaluated in the order it was programmed when the CPU is switched to Run mode, and the block is called for processing.

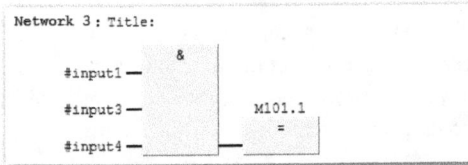

```
Network 3 : Title:

                    &
    #input1 ─

    #input3 ─                  M101.1
                                  =
    #input4 ─
```

Figure 3-15. Illustration of a basic FBD network

Evaluating Binary Inputs and Boolean Gates in an FBD Network

A key to understanding FBD is Boolean logic Gates AND, OR, and XOR. Each of these operations is governed by a rule that is often described by a so-called *truth table*. When logic conditions of a gate are met according to the logic function and its truth table, the gate is said to be 'satisfied' or 'TRUE'. A TRUE value is normally represented by logic '1'; a FALSE value is usually represented by logic '0'. This TRUE/FALSE value is assigned to the output of the gate as a result.

In STEP 7 this logic result is called the *result of logic operations (RLO)*. If a logic circuit has a TRUE result then it is said that the RLO = 1; otherwise, the RLO = 0. The logic state at any node in the logic network may also be referred to as the result of logic operation. The RLO may be evaluated after an input element, after any gate element, after any logic combination, or at the output of the logic circuit.

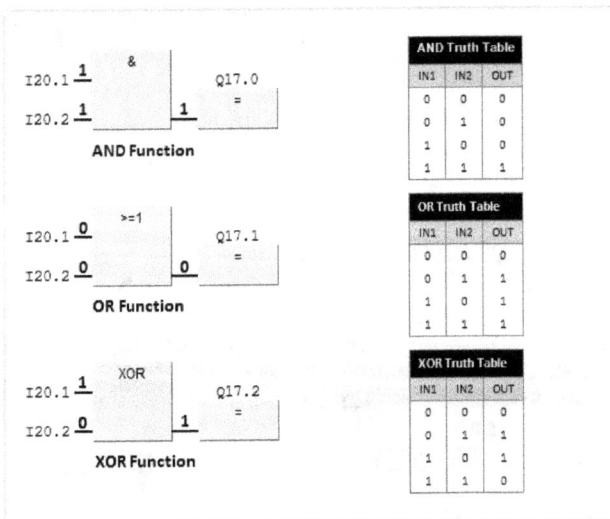

AND Truth Table		
IN1	IN2	OUT
0	0	0
0	1	0
1	0	0
1	1	1

OR Truth Table		
IN1	IN2	OUT
0	0	0
0	1	1
1	0	1
1	1	1

XOR Truth Table		
IN1	IN2	OUT
0	0	0
0	1	1
1	0	1
1	1	0

Figure 3-16. Basic evaluation of binary inputs to gate logic in an FBD network.

When you examine discrete input devices, remember that inputs are wired as normally-open or normally-closed. A device that is wired normally-open presents a logic '1' input to the gate when the device is activated, and presents a logic '0' to the gate when de-activated. This type of device is said to be *Active HIGH*. On the other hand, a device wired normally-closed presents a logic '1' when not activated, and presents logic '0' when activated. This type of device is said to be *Active LOW*.

How a device is wired and the effect it is expected to have in your control logic will determine which FBD binary input element you should use to examine an input. In general, a positive binary input is used when a logic '1' input is expected to produce a logic '1' at the gate element; a negated input is programmed when a logic '0' input is expected to produce a logic '1' at the gate element. In FBD, the negated binary input element serves the same purpose as does the LAD normally-closed input.

Evaluating Block Operations in an FBD Network

As with networks programmed with binary inputs, Boolean gate, and binary output elements, networks programmed with box instructions are also affected by the result of logic operations. How box operations are affected by the RLO will differ depending on the specific box instruction. In general, box operations that have the **EN/ENO** mechanism are only executed if the RLO = 1. Depending on the box operation, there may also be other input lines—for example the Reset line in timer and counter box instructions. A reset line is always processed if the RLO = 1.

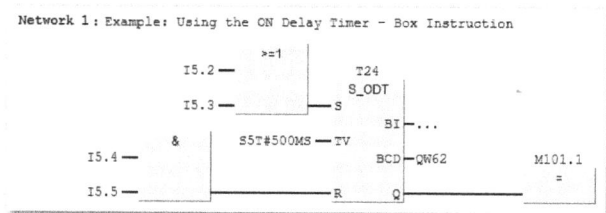

```
Network 1: Example: Using the ON Delay Timer - Box Instruction
                              >=1
               I5.2 —                    T24
                                         S_ODT
               I5.3 —             — S
                                      BI —...
                     &     S5T#500MS — TV
        I5.4 —                        BCD —QW62      M101.1
                                                        =
        I5.5 —                    — R    Q
```

Figure 3-17. Evaluation of box operations in an FBD network

A summary of how box operations are evaluated in an FBD network is shown in the following table.

Table 3-10. Evaluation summary of Block Operations in FBD

EN/ENO	Operation Type	Brief Evaluation Description	Example
NO	Timer	With the exception of the OFF-delay timer, all timers are started with a 0-to-1 transition of the RLO on the start line (S).	S_ODT / —S Q— / —TV BI— / —R BCD—
		The OFF-delay timer is started with a 1-to-0 transition of the RLO on the timer start line (S).	
	Counter	Counter Operation counts accordingly whenever the RLO on the count line transitions from 0-to-1.	S_CU / —CU Q— / —S CV— / —PV CV_BCD— / —R
	Comparison	Comparison operations are executed irrespective of the RLO. Instead the compare operation behaves like a normally-open contact that allows power flow when the outcome of the compare is True (1) and interrupts power flow when the outcome is False (0).	CMP ==I / —IN1 / —IN2
YES	Arithmetic	Executed when the RLO = 1 on the **EN** line.	SHL_DW / —EN / —IN OUT— / —N ENO—
	Conversion	Executed when the RLO = 1 on the **EN** line.	
	Word Logic	Executed when the RLO = 1 on the **EN** line.	
	Shift-Rotate	Executed when the RLO = 1 on the **EN** line.	
	Move	Executed when the RLO = 1 on the **EN** line.	

GUIDELINES FOR EDITING IN FBD

When it comes to entering and editing in FBD, you may be working with an existing block or with an entirely new block that you have created. In either case your first step will be to open the block from either the Simatic Manager or from the LAD/STL/FBD editor. With the block opened, you will set the STEP 7 editor to work in FBD by using the menu and selecting **View ➤ FBD**.

General Editing in FBD

Before inserting any FBD program elements, you may want to ensure that the editor is in the *insert mode* as opposed to the *overwrite mode*. Like with most editors, the overwrite mode writes over a previously entered character or object based on how you place the cursor. Insert mode simply inserts a new item and pushes existing objects to the right based on the cursor position. Using the <**Insert**> key, you may toggle between these two edit modes. The current edit mode is displayed in the status bar in the next to the last field. Whereas insert mode is the default in normal editing mode, overwrite mode is generally used to overwrite an instruction when the wrong instruction has been entered. To overwrite an element, the new element must be of the same general type as the old element.

When an FBD instruction is inserted, red characters "**???**"indicate that an address or parameter must be specified. Black characters "**...**" indicate that an address or parameter may be specified, but not required. This latter marker is more likely with box instructions like timers and counters.

Press F1 to get Help.	🖸 offline	Abs < 5.2	Nw 2	Insert Chg

Figure 3-18. LAD/STL/FBD editor Status bar. The current status shows the editor in **Insert** mode

Inserting an FBD Network

A new network is inserted by a single click on the **New Network** toolbar icon, by double-clicking on the **New Network** object in the Program Elements window, or by dragging and dropping the New Network object from the Program Elements window to the desired position. Before any of these methods is used, you need to decide the position of the new network. If you have opened a new block for the first time, then the first network will appear by default.

To insert a network following an existing network, select that network and then use one of the previously described options. For example, select **Network 3** to insert the new network between **Network 3** and **Network 4**. To insert a network as the first network in an existing block, double-click on the Block Title line, and then use one of the previously described New Network options.

Inserting Binary Inputs and Outputs

To insert basic FBD program elements, use the toolbar icons or keyboard Function Keys. **F2** = AND Gate; **F3** = OR Gate; **F7** = Assignment; **F8** = Binary Input; **F9** = Negated Binary Input; and **F11** = Insert Branch. Other special binary outputs not found on the toolbar—for example Set (S), Reset (R), and Midline outputs (#), are found in the Bit Logic subfolder of the Program Elements tab, or in one of the other subfolders—for example Status Bits, Jumps and Program Control.

After inserting an element, simply address the element as required. Use the **View** menu and select **Display with ➤ Symbols Representation** if you wish to use symbol addresses.

New Network	F2	F3	F7	F8	F9	F11
⊞	[&]	[≥1]	[=]	=[	⊣[	⊏

Figure 3-19. Toolbar icons for basic FBD program elements

Element Placement

When inserting gate elements and box operations, remember that together with the use of binary input elements, these operations support the development of the conditional logic that drives outputs or inputs to block operations. These operations are generally placed to the left side of the logic string—preceding any output instructions. The logic arrangement represents the control logic that will determine the RLO, and whether an output is activated, or if a box operation is executed.

In FBD, output instructions terminate the network and are placed in the right-most position as the last element in the network. Exceptions to this general rule for output placement include the midline coil, the positive edge output, and the negative-edge output. These outputs may be placed anywhere along the logic string excluding the left-most and right-most positions. Furthermore, these outputs may not be placed in parallel. The table below provides a placement summary of FBD program elements.

Table 3-11. Summary of FBD binary output requirements and restrictions in a network

Outputs that Require Preceding Logic Conditions			
Assign Output	=	Jump if NOT	JMPN
Set Output	S	MCR Zone Start	MCR<
Reset Output	R	Save RLO to BR	SAVE
Positive Edge	P	Return	RET
Negative Edge	N	Timer Outputs—All	—
Midline Output	#	Counter Outputs—All	—
Outputs for which Preceding Logic Conditions are not Permitted			
Open DB Output	OPN	MCR Activate	MCRA
MCR Zone End	MCR>	MCR De-Activate	MCRD
Outputs that must be placed in the right-most position of the logic string			
Set Count Value	SC —PV	Simple Timer—SF	SF —TV
Simple Up Count	CU	Simple Timer—SP	SP —TV
Simple Down Count	CD	Simple Timer—SE	SE —TV
Simple Timer—SD	SD —TV	Simple Timer—SS	SS —TV

Inserting Boolean Gates

As you start to build FBD networks, the following guidelines should be helpful when it comes to using Boolean gates AND, OR, and XOR. As previously stated, gate logic may be combined with other FBD instructions to complete the required network logic.

Guideline 1: Insert additional binary inputs

As shown below, ❶ an inserted gate by default has two positive binary input lines. To insert additional input lines, ❷ click and select the gate ❸ and with the gate selected, click on the binary input toolbar icon for each input you wish to add. As shown in ❹, additional input lines are inserted at the top of the gate above any existing inputs. You may insert as many inputs as the gate requires.

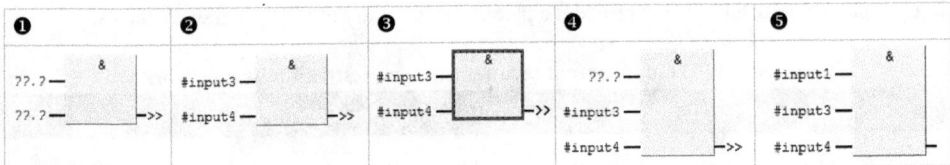

Figure 3-20. Inserting binary inputs in FBD

Guideline 2: Insert negated binary inputs

As previously stated, inserted binary inputs are positive binary input lines by default. To insert a negated binary input, ❶ select an existing or newly inserted binary input line—the negate binary input toolbar icon will enable, then ❷ click on the negate binary input icon to invert the selected input line.

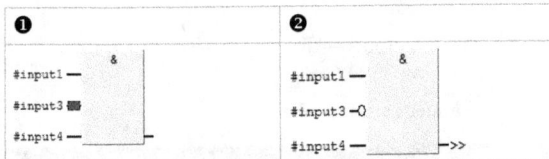

Figure 3-21. Inserting negated binary inputs in FBD

Guideline 3: Insert binary input between existing inputs

To insert a new binary input between existing input lines—for example between input1 and input2, as shown in ❷, select the input line after which you want to insert the new input line; then click on the binary input toolbar icon—❸ by default a positive binary input is inserted. ❹ Address the new input.

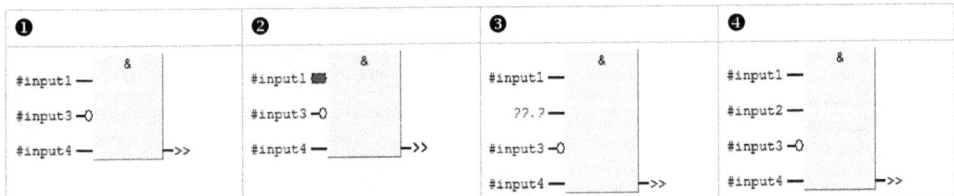

Figure 3-22. Inserting binary inputs between existing inputs in FBD

Guideline 4: Insert logic branches in a network

Only a single logic string may be developed in an FBD network. That is to say that in a network you may not develop two independent strings of logic, each of which has a separate output. The requirement is that all elements in the network be connected. Nevertheless it is possible to develop one or more logic branches off of a logic string, with each branch having a unique output. In FBD, a logic branch is developed using a *T branch* element.

To insert a branch off of a logic string, ❶ click on the node just in front of the element at the point you wish the branch to be placed—the toolbar **branch** icon will enable; ❷ click the **branch** icon to insert the new branch element; ❸ insert the desired element or branch logic and terminate the branch with an Assign output, or perhaps another binary output element. ❹ Enter the appropriate addresses.

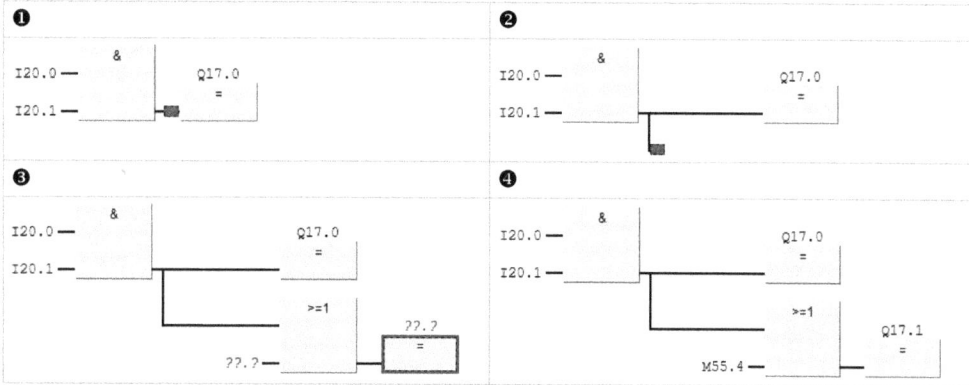

Figure 3-23. Inserting logic branches in an FBD network

Combining Box Operations with Boolean Gates

In FBD, you may combine Boolean gates and box operations as needed to develop the control logic. As you enter the FBD network, remember that only a single logic string may be developed—and that all elements must be connected. Although it is not required, inserting the FBD network starting with the right-most logic gate or box operation simplifies program entry. It may also be beneficial to sketch the logic diagram first—this practice is especially helpful with complex logic circuits.

In FBD, entry of a new operation or logic gate into an existing gate (or operation) begins by clicking on a node point at which you wish to insert a new gate or FBD operation. In FBD, like in LAD, a node always exists between two elements—for example between a binary input and a gate; between a gate and an output element. Three examples of node selection are shown below—❶ between a gate and its output; ❷ between a binary input and a gate; and ❸ between a binary output line and a T-branch element. The node that you select should reflect the point at which you would like to insert a new logic gate or instruction element—for example you may want to add an OR gate, or a counter instruction.

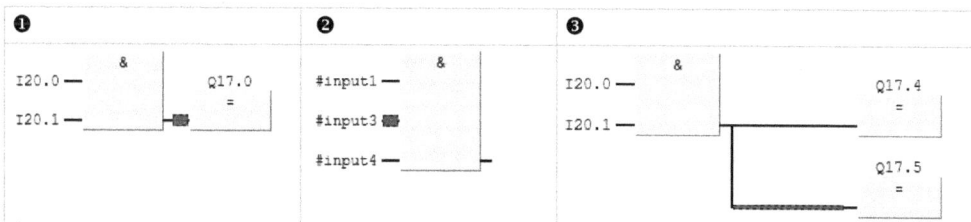

Figure 3-24. Combining Boolean gates in an FBD network

Figure 3-25 illustrates how an FBD diagram is expanded by inserting additional logic gates or STEP 7 operations.

As shown below, the AND gate ❶ will be expanded to form a circuit whose output is the result of ANDing two OR gates❻. First, as shown in ❶, click on a node at one of the AND gate inputs, where the first OR gate will be inserted; ❷ click **F3**, the OR gate function key; ❸ click on the node at the other AND gate input, where the second OR gate will be inserted; ❹ again, click **F3**, the OR gate function key. In this fashion, you complete your FBD network entry by inserting elements in either a forward or backward progression. As noted earlier, a sketch of your diagram will simplify your entry.

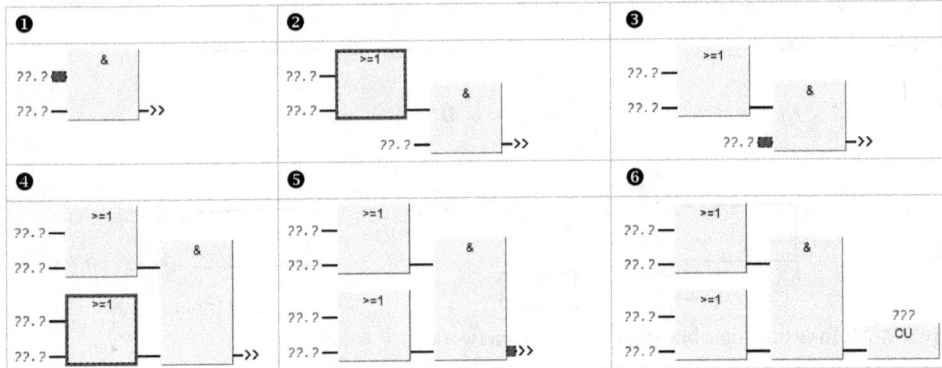

Figure 3-25. Inserting additional binary inputs in an FBD network

Inserting FBD Program Comments

With each block you create in FBD you may enter 1) **Variable Comments** of up to 80-characters, for each variable you declare; 2) a **Block Title**, to briefly title the code block; 3) a **Block Comment**, to describe the purpose or operation of the block; 4) a **Network Title**, to title each logic network you enter; and 5) a **Network Comment** that describes the purpose or operation of each logic network. If the FBD comment fields are not displayed, select **View** ➤ **Display with** ➤ **Comments**. This selection toggles between enabling and disabling comment entry and display.

Figure 3-26. FBD Block with Comments enabled and displayed

Chapter 4

LAD/FBD
Binary Logic Operations

This chapter introduces LAD/FBD Binary Logic Operations. In your S7 program development, these instructions are used primarily to create relay-equivalent control logic for on/off switching of externally connected devices. Binary logic operations also serve in the development of internal decision logic that in turn drives other STEP 7 operations—for example whether an arithmetic operation should be carried out, or whether a code block should be called. Decisions such as these are generally based upon control logic constructed of series and parallel circuits in LAD, or using Boolean operators AND, OR, and NOT combinations in FBD. In this chapter, Binary Logic instructions have been divided into Basic Logic Operations and Memory Operations.

Before setting out on your review of this and subsequent chapters, if you have not already done so, you may find it beneficial to first review the material of Chapter 1 *Absolute Addressing of S7 Memory Areas*. There you will learn important concepts of STEP 7 data types, memory areas, and addressing. These topics will aid you when it comes to assigning the address parameters with each instruction.

Main Topics

- Binary Logic Operations Overview

- Basic Logic Operations Descriptions

- Programming Basic Logic Operations in LAD

- Programming Basic Logic Operations in FBD

- Binary Logic—Memory Operations Overview

- Set/Reset Operations Descriptions

- Edge Detection Operations Descriptions

- Programming Binary Logic Memory Operations in LAD

- Programming Binary Logic Memory Operations in FBD

BASIC LOGIC OPERATIONS OVERVIEW

This first group of Binary Logic instructions, *basic logic operations*, is used in developing relay equivalent control logic and internal decision logic. With LAD instructions normally-open and normally-closed contacts, and the standard output, or FBD instruction that include positive and negated binary inputs, Boolean gates AND, OR, and XOR, and the assign output, bit logic instructions will support development of series, parallel, and series-parallel logic combinations.

With basic logic operations, you are able to examine and control the ON/OFF status of the S7 memory areas, including discrete I/O signals, presented in the process image of inputs and outputs, as well as declared variables of type BOOL.

Table 4-1. LAD/FBD Binary Logic—Basic Logic Operations Summary

LAD	FBD	Brief Description
Output Coil ??.? —( )—	Assign Output ??.? =	Assign the binary '0' or '1' result of logic operation (RLO) to the referenced coil/output address.
Normally-Open Contact ??.? —| |—	Binary Input —|	Check the referenced address for logic 1 status. A status of logic 1 yields a RLO = 1 to promote logic continuity through the contact/gate; logic 0 yields a RLO = 0 to interrupt logic continuity.
Normally-Closed Contact ??.? —|/|—	Negate Input —o|	Check the referenced address for logic 0 status. A status of logic 0 yields a RLO = 1 to promote logic continuity through the contact/gate; logic 1 yields a RLO = 0 to interrupt logic continuity.
NOT Contact —| NOT |—	See Note	Invert the RLO at the point at which the operation is inserted. In FBD 'invert RLO' is achieved with the 'Negate Binary Input'.
Midline Coil ??.? —(#)—	Midline Output ??.? #	Store intermediate RLO for the logic circuit up to and preceding the point of insertion.

BASIC LOGIC OPERATIONS DESCRIPTIONS

With basic logic operations you can examine and control the on/off status of bit addressable memory areas I, Q, M, or D, or of block variables of type BOOL. Simply stated, these operations allow you to develop logic to examine or influence the result of logic operations (RLO). Basic Logic Operations are described in the following discussions.

Normally-Open Contact/Binary Input

When the *normally-open (NO) contact* is evaluated, the bit address referenced above the contact is examined for a status of '1'. If a status of '1' is detected, at the referenced address, the contact closes and allows logic continuity through the contact; if '0' is detected, the contact assumes its normally-open state and continuity is interrupted.

In FBD, the *binary input* performs the same function as the NO contact in LAD—it examines a bit address for a status of '1'. If the address is '1', then '1' is passed to the input of the gate or box instruction. If the referenced address is '0', then '0' is passed to the input of the gate or box instruction.

In the following network, the input I20.0 must have a status of '1' to energize the output Q17.0.

LAD Example: Examine for '1' at I 20.0	**FBD Example:** Examine for '1' at I 20.0
I20.0 Q17.0 ├──┤ ├─────────────────────────( )──┤	Q17.0 = I20.0 ─ &

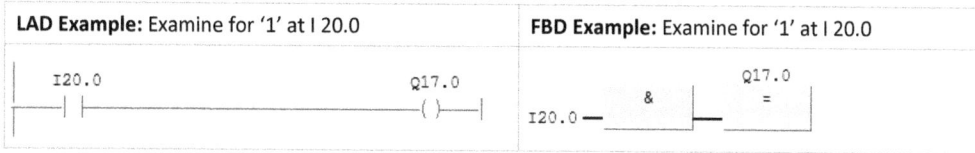

Figure 4-1. Illustration of LAD normally-open contact and FBD binary input instruction

Normally-Closed Contact/Negate Input

When the *normally-closed (NC) contact* is evaluated, the bit address referenced above the contact is examined for a status of '0'. If '0' is detected, the contact remains closed, and allows logic continuity through the contact; if '1' is detected, the contact opens and interrupts logic continuity.

In FBD, the *negate input* performs the same function as the NC contact in LAD—it examines the referenced address for a status of '0'. If the status is '0', then '1' is passed to the input of the gate or box instruction. If the referenced address is '1', then '0' is passed to the input of the gate or box instruction.

In the network below, the input I 20.1 must have a status of '0' to energize the output Q 17.1.

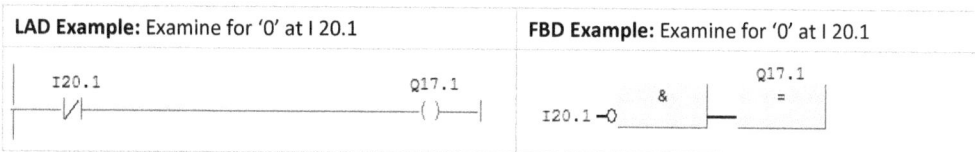

LAD Example: Examine for '0' at I 20.1	**FBD Example:** Examine for '0' at I 20.1
I20.1 Q17.1 ├──┤/├─────────────────────────( )──┤	Q17.1 = I20.1 ─O &&

Figure 4-2. Illustration of the LAD normally-closed contact and FBD negate input instruction

Output Coil/Assign Output

When the *output coil* is evaluated, the bit address referenced above the output is determined by the preceding logic conditions. If the network has logic continuity (RLO = 1), the referenced bit address assumes a status of '1'; if logic continuity is interrupted (RLO = 0), then the referenced address will assume a status of '0'.

When the referenced output is '1', any normally-open contacts of the same address will close throughout the program; all normally-closed contacts of the same address will open. Conversely, when the referenced output is '0', any normally-open contacts of the same address will open throughout the program; all normally-closed contacts of the same address will close.

The *assign output* in FBD performs the same function as the output coil in LAD. The result of logic operations (RLO) of the preceding logic is assigned to the address referenced by the assign output. If the RLO = 1, then the output is assigned '1'; otherwise the output is assigned '0'.

In the network below, output Q 43.7 is energized if I 20.0 **AND** I 20.1 **AND** I 20.3 have a status of '1'.

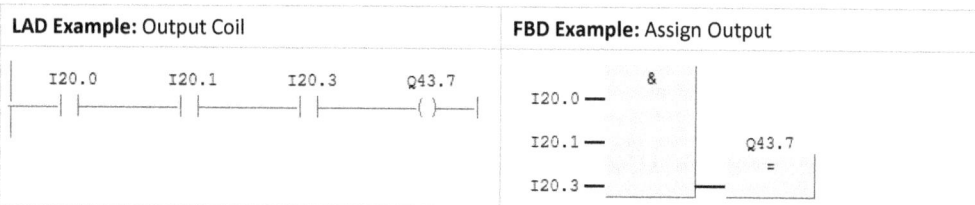

LAD Example: Output Coil	**FBD Example:** Assign Output
I20.0 I20.1 I20.3 Q43.7 ├─┤ ├──────┤ ├──────┤ ├──────( )──┤	& I20.0 ─ I20.1 ─ Q43.7 I20.3 ─ =

Figure 4-3. Illustration of LAD output coil and FBD assign output operation

NOT Contact/Negate Input

When the *NOT contact* is evaluated, the RLO is inverted at the point of insertion. If the RLO = 1 preceding the contact, then following the contact the power flow is inverted (RLO = 0). If the RLO = 0 preceding the contact, then following the contact the power flow is inverted (RLO = 1). The NOT operation does not reference any bit address, and may be placed between any two network nodes.

In FBD, the *negate input* performs the same function as the invert power flow (NOT) contact in LAD—it inverts the result of logic operation at the point at which the negate input is inserted. If the RLO = 1 prior to the negate input, then the RLO = 0 after the negate input; if the RLO = 0 prior to the negate input, then the RLO = 1 after the negate input.

In the following circuit, if I 23.2 AND I 23.4 both have a status of '1', then the RLO = 1 prior to the NOT contact—after the RLO is inverted the output Q 17.5 will de-energize. Output Q 17.5 will energize when I 23.2 OR I 23.4 .1 has a status of '0', or both have a status of '0'.

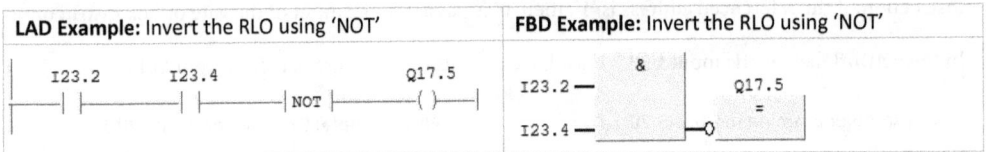

LAD Example: Invert the RLO using 'NOT'	FBD Example: Invert the RLO using 'NOT'

I23.2   I23.4   Q17.5
├──┤ ├────┤ ├──┤ NOT ├────( )──┤
	
	123.2 —┐ &
	        │ Q17.5
	123.4 —┘ =
	            —○

Figure 4-4. Illustration of LAD invert power flow (NOT) operation and FBD negate input instruction

Midline Output

In the following circuit, the *midline output* M 7.1 is activated when I 2.1 AND I 2.2 both have a status of '1'. The midline output M 7.2 is activated when M 7.1 AND I 2.3 AND I2.4 have a status of '1'. The midline output M 7.3 is activated when M 7.1 AND M 7.2 AND I2.5 have a status of '1'. The midline outputs M 7.1, M 7.2, and M 7.3 may be used throughout the program.

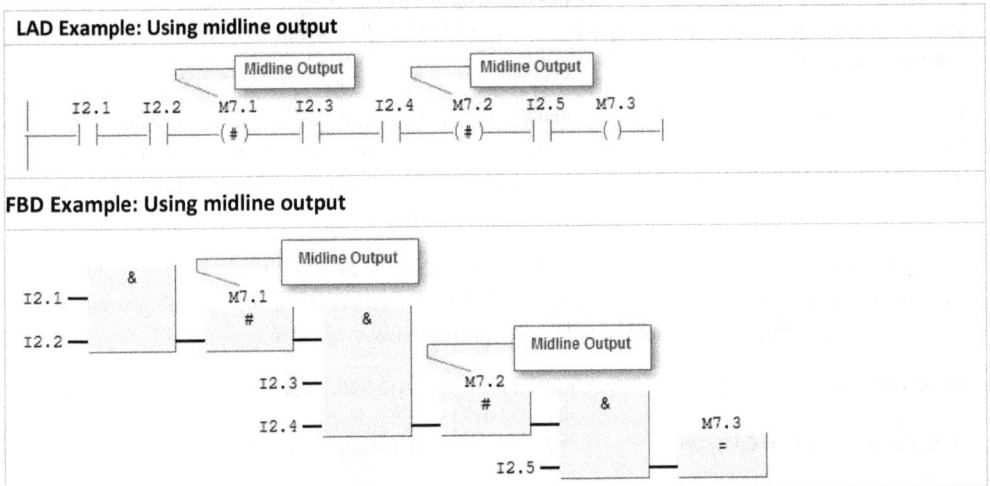

LAD Example: Using midline output

I2.1 I2.2 M7.1 [Midline Output] I2.3 I2.4 M7.2 [Midline Output] I2.5 M7.3
├──┤ ├──┤ ├──(#)──┤ ├──┤ ├──(#)──┤ ├──( )──┤

FBD Example: Using midline output

Figure 4-5. Illustration of LAD midline coil and FBD midline output instruction

PROGRAMMING BASIC LOGIC OPERATIONS IN LAD

Basic logic operations provide the basis elements for developing conditional logic that controls externally connected ON/OFF signals, to devices like motors and valves, as well internal logic signals used in program interlocking and other such binary signals. Such signals are used to enable or disable operations like timers, counters, arithmetic, comparisons, program jumps, and other parts of your program. Programming these basic logic operations in LAD is described in the following discussions.

Using the Normally-Open Contact

In LAD, you would use the normally-open (NO) contact instruction to examine an address for a status of '1'. If a signal status of '1' is detected, then the NO contact closes and allows continuity through the contact. If a signal status of '0' is detected, then the NO contact opens and interrupts logic continuity. In the case of physical inputs, what is actually being checked is the status presented at an input terminal. This status may reflect a binary '1' presented by a sensor whose NO contacts are activated, or by a sensor whose NC contacts are not-activated—yet presents a '1' to the input.

So, the NO contact checks the status of an input for '1' and then closes if the status is '1'—regardless of whether the signal is presented from a device wired NO or NC. Given this operation, how a device is wired (normally-open or normally-closed) is irrelevant. You must determine whether use of a NO or NC contact instruction suits the logic requirements. As a general rule, in a logic circuit where a signal state of '1' is expected to allow continuity and a signal state of '0' is expected to interrupt continuity, you should use a NO contact instruction.

In the example below, the NO contact closes and allows logic continuity (power flow) if I 27.4 is '1'. The output coil Q 29.1 is then activated.

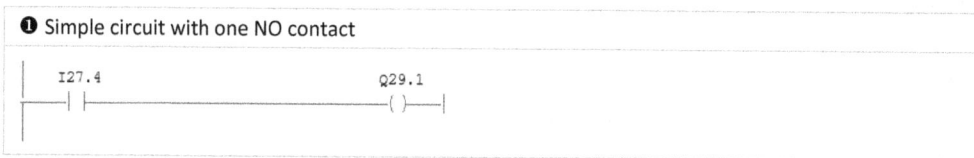

❶ Simple circuit with one NO contact

```
    I27.4                        Q29.1
   ─┤ ├─────────────────────────( )──┤
```

Figure 4-6. Using the normally-open contact to require a status of '1' to allow power flow

Using the Normally-Closed Contact

In a LAD, the normally-closed (NC) contact is used to examine the referenced address for a status of '0'. If a signal status of '0' is detected, then the NC contact closes (or remains closed) and allows continuity through the contact. If a signal status of '1' is detected, then the NC contact opens and interrupts logic continuity. In the case of physical inputs, what is actually being checked is the status presented at an input terminal. This status may reflect a binary '0' presented by a sensor whose NC contacts are activated, or by a sensor whose NO contacts are not-activated.

So, the NC contact checks the status of an input for '0' and then closes if the status is '0', regardless of whether the signal is presented from a device wired NO or NC. Given this operation, how the device is wired (normally-closed or normally-open) is irrelevant. You must determine whether use of a NC or NO contact instruction suits the logic requirements. As a general rule, in a logic circuit where a signal state of '0' is expected to allow logic continuity and a signal state of '1' is expected to interrupt logic continuity, you should use a NC contact instruction.

In the example below, the NC contact closes and allows power flow if I 27.5 is '0'. The output coil Q 29.2 is then activated.

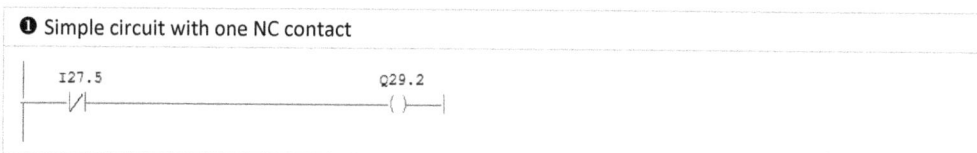

❶ Simple circuit with one NC contact

```
    I27.5                        Q29.2
   ─┤/├─────────────────────────( )──┤
```

Figure 4-7. Using the normally-closed contact to require a status of '0' to allow power flow

Using the Standard Output Coil

In LAD, the standard output coil is likely the most used binary output instruction. The coil is activated when the preceding RLO = 1, and is de-activated when the preceding RLO = 0. In short, the output coil is always assigned the state of the result of logic operations from any series or parallel path of logic that has complete logic continuity. If the RLO =1, the bit address referenced above the coil is assigned a status of '1'; if the RLO = 0, the bit address referenced above the coil is assigned a status of '0'.

In the first example below, if the output image address Q 30.3 is '1', the NO contact closes and the output image address Q 17.5 is activated and assigned a status of '1'. If the address Q 30.3 is '0', then the NO contact opens and the output image address Q 17.5 is de-activated and assigned a status of '0'. In the second example, if the bit address Q 30.3 is '1', the parallel outputs Q 17.5 and Q 17.6 are both activated and assigned a status of '1'. If the bit address Q 30.3 is '0', both outputs are de-activated and assigned a status of '0'.

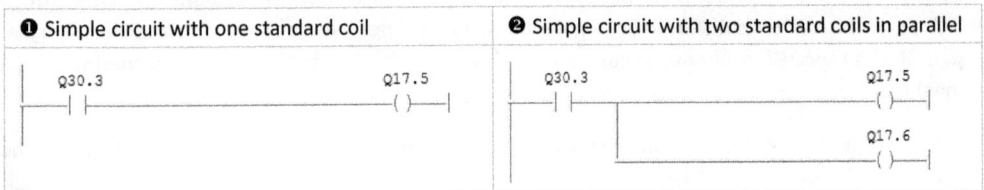

Figure 4-8. Using the standard output coil

Developing AND (Series) Circuits

When developing control logic, a series logic circuit is used when an output should be activated only when a group of two or more contacts all are closed. In a series circuit, contacts are connected inline one after another—when all of the contacts are closed the circuit has logic continuity or an RLO = 1.

In the first example below, the output address Q 17.0 is activated (logic 1) when the two NO contacts I 20.0 **AND** I 20.1 have a signal state of '1'. The output is de-activated whenever either of the two inputs has a signal state of '0'. In the second example, output address Q 17.0 is activated when all three NO contacts have a signal state of '1'.

In the third example, Q 17.1 is activated if the NO contact I 20.0 has a signal state of '1' **AND** the NC contact I 20.1 has a signal state of '0' (not activated).

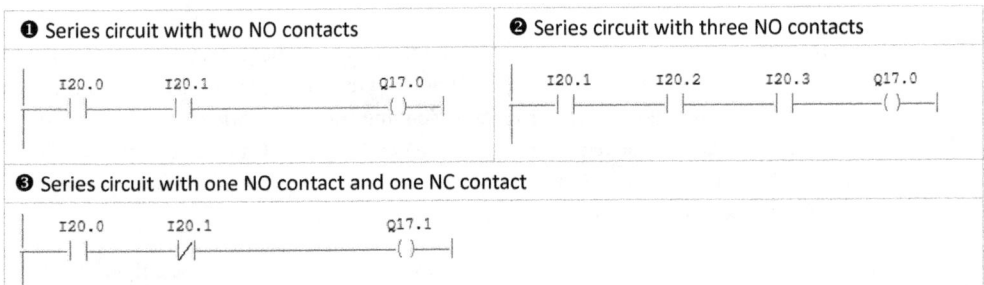

Figure 4-9. Creating a series logic circuit

Developing OR (Parallel) Circuits

When developing control logic, a parallel logic circuit is used when an output should be activated when any one of two or more contacts closes. In a parallel circuit, contacts are connected as branch circuits one underneath another—when contact closes, the circuit has logic continuity or an RLO = 1.

In the first example, the output Q 17.7 is activated (logic 1) when either NO contact I 20.2 **OR** I 20.3 has a signal state of '1'. The output is de-activated (logic 0) when all of the inputs is logic '0'. In the second example, three NO contacts are in parallel—you may parallel as many contacts as required. Output address Q 17.7 is activated when any one of the NO contacts has a signal state of '1'.

In the third example, the output Q 17.7 is activated (logic 1) if the NO contact I 20.3 has a signal state of '1' **OR** the NC contact I 20.4 has a signal state of '0'.

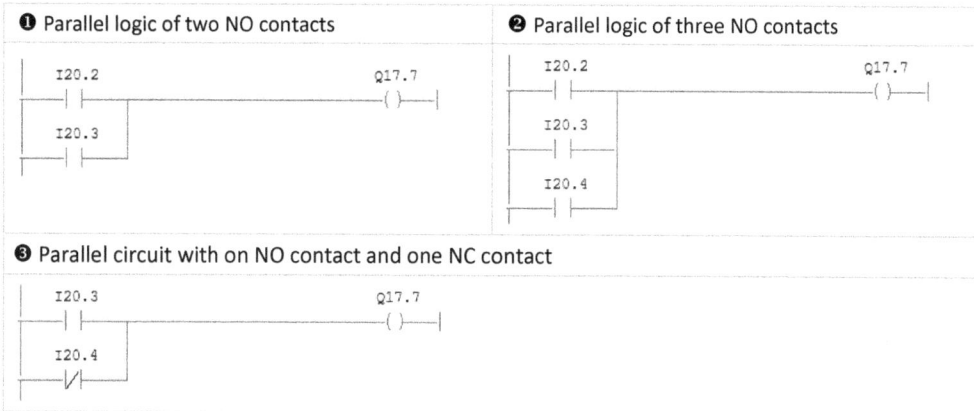

Figure 4-10. Creating a parallel logic circuit

Developing Exclusive OR Logic Circuits

When developing control logic, an exclusive-OR logic circuit is used when an output should be activated only when one of two inputs is closed, but not both. The exclusive OR logic function holds that a binary output signal **Y** has a signal state of '1' only when one of two binary signals (**A** or B) is '1'. If both signals (A and B) have a signal state of '1', then the binary output **Y** will have a signal state of '0'. This function is implemented in LAD using normally-open contacts combined as shown below.

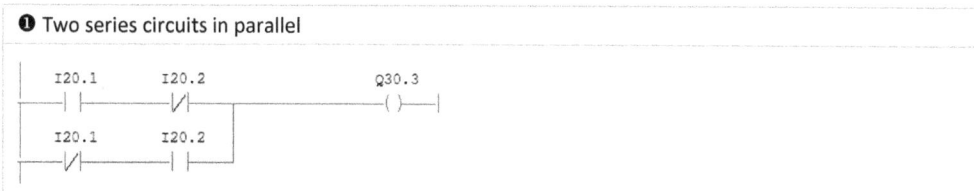

Figure 4-11. Creating an exclusive-OR logic circuit

In the example above, the output Q 30.3 is activated by the first branch if the signal status of I 20.1 is '1' **AND** the signal status of I 20.2 is '0', **OR** by the second branch if the signal status of I 20.1 is '0' **AND** the signal status of I 20.2 is '1'. The output Q30.3 is de-activated when the signal status for I 20.1 and I 20.2 are '1'. In the two parallel circuits above, apply logic '1' to both operands in either branch to test the result. Recall that a signal status of '1' causes the NC contact to open and interrupt continuity.

AND-before-OR Logic Combinations

The *and-before-or* combination places two or more series circuits in parallel—that is two or more series circuits are combined one above another. In such a combination a parallel branch has logic continuity (RLO = 1) if all of the series contacts in that branch are closed. The network has logic continuity if the RLO = 1 for the series circuit in either of the branches.

In the first of the following examples, the output address Q 30.2 is activated (logic 1), when the two NO contacts I 20.1 **AND** I 20.2 have a signal status of '1' **OR** when the two NO contacts I 20.3 **AND** I 20.4 have a signal status of '1'. The output is de-activated whenever the RLO = 0 in both series circuits.

In the second frame below, three series circuits are combined in parallel to drive the output image address Q 30.2. You may add as many series circuits in parallel as required.

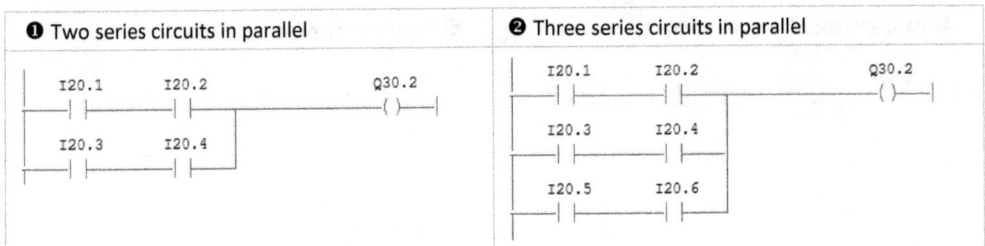

❶ Two series circuits in parallel	❷ Three series circuits in parallel
``` I20.1 I20.2 Q30.2 ─┤├──────┤├────────────( )──┤ I20.3 I20.4 ─┤├──────┤├─ ```	``` I20.1 I20.2 Q30.2 ─┤├──────┤├────────────( )──┤ I20.3 I20.4 ─┤├──────┤├─ I20.5 I20.6 ─┤├──────┤├─ ```

Figure 4-12. Creating series circuits in parallel

OR-before-AND Logic Combinations

The *or-before-and* combination places two or more parallel circuits in series—that is two or more parallel circuits are combined one after another. In such a combination each parallel circuit has logic continuity (RLO = 1) if either of the parallel contacts is closed. The network has complete logic continuity if the RLO = 1 for all of the parallel circuits that are in series.

In the first example below, the output address Q 30.6 is activated (logic 1) when I 22.1 **OR** I 22.3 is closed **AND** I 22.2 **OR** I 22.4 is closed. For example, Q 30.6 is activated if I 22.3 is closed **AND** I 22.2 is closed. The output is de-activated whenever the RLO = 0 in any one of the parallel circuits.

In the second frame below, three parallel circuits are combined in series to drive the output image address Q 30.6. All three parallel circuits must have an RLO = 1 to activate the output Q 30.6.

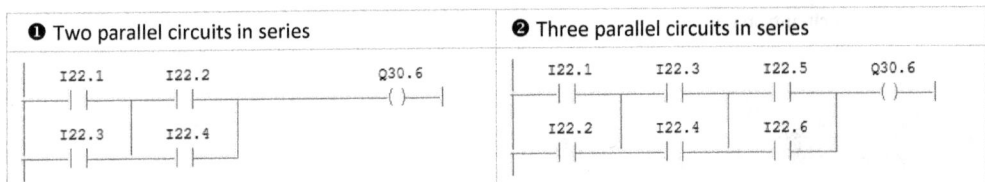

❶ Two parallel circuits in series	❷ Three parallel circuits in series
``` I22.1 I22.2 Q30.6 ─┤├──────┤├────────────( )──┤ I22.3 I22.4 ─┤├──────┤├─ ```	``` I22.1 I22.3 I22.5 Q30.6 ─┤├───────┤├───────┤├────( )──┤ I22.2 I22.4 I22.6 ─┤├───────┤├───────┤├─ ```

Figure 4-13. Creating parallel circuits in series

Negated Logic Circuits

In some cases, you will need to invert the result of logic operation to achieve the desired control logic. In LAD, the NOT contact may be placed between any two network nodes, for example between any two contacts. If the RLO = 1 prior to the NOT contact, then following the operation the RLO = 0. If the RLO = 0 prior to the NOT contact, then following the operation the RLO = 1. The NOT operation does not reference any bit address, and may be placed at any point in the logic network prior to the coil/output.

In the first example below, the output Q27.0 is activated if one or both of the contacts I 44.5 and I 44.6 is open—that is when RLO = 0 is presented to the NOT contact. When both contacts are closed RLO = 1 is presented to the NOT contact, and Q27.0 is de-activated after the RLO is inverted. This operation is the negated AND function or NAND.

In the second example below, the output Q27.0 is activated only when contacts M37.4 and M37.5 are both open—that is when RLO = 0 is presented to the NOT contact. If either or both of the contacts are closed, then RLO = 1 is presented to the NOT contact; after the inversion Q27.1 is de-activated. This operation is the negated OR function or NOR.

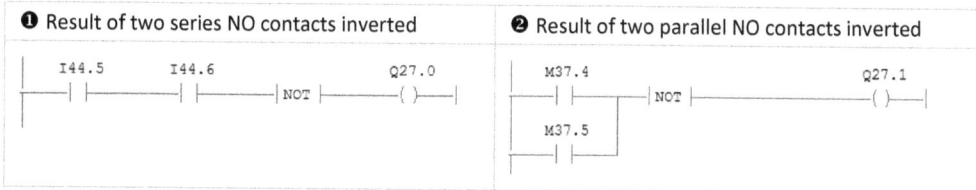

Figure 4-14. Negated logic combinations **NAND** circuit and **NOR** circuit

Using the Mid-Line Output

Use the midline Output to buffer the RLO at intermediate points in a LAD network. By using the midline coil, you can eliminate repeat combinations of logic that might appear in several networks as part of the total logic that controls the outputs. The midline output behaves like the standard coil in that it assumes the logic state of the preceding RLO. The difference is that the midline coil need not be placed at the end of a network; it can be placed in series with other contacts, but not connected directly to either ladder rail, directly after a branch connection, or at the end of a branch.

After examining the three networks in the first frame below, see how use of the midline output has resulted in the single network in the second frame.

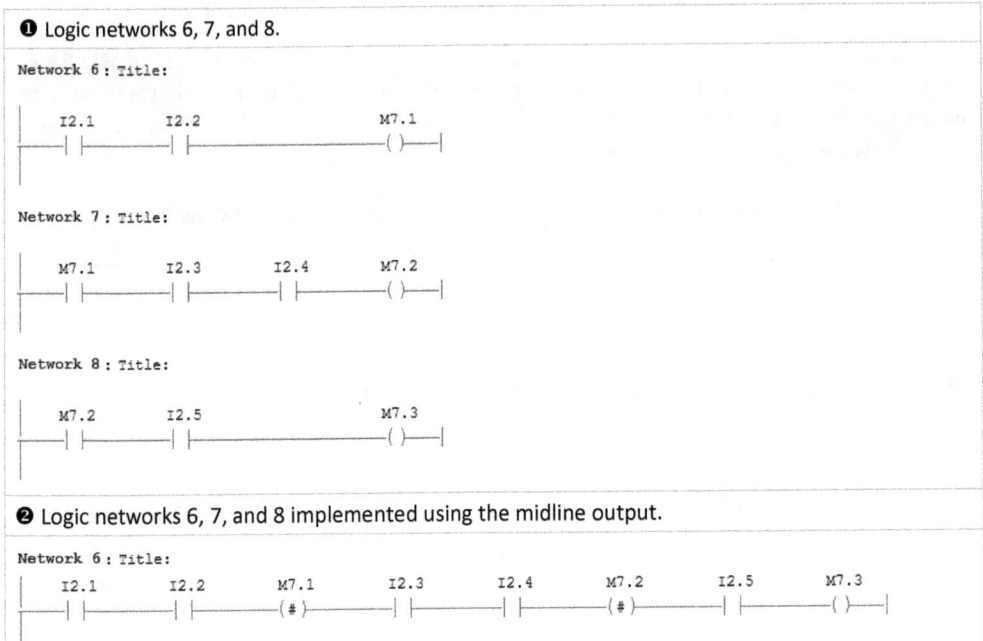

Figure 4-15. Use of the midline output to minimize logic duplication

PROGRAMMING BASIC LOGIC OPERATIONS IN FBD

Basic logic operations provide the basic elements for developing conditional logic to control externally connected ON/OFF signals to devices like motors and valves, as well as internal logic signals used for program interlocking and other such binary signals. Such signals are used to enable or disable operations like timers, counters, arithmetic, comparisons, program jumps, and other program parts. Programming for these basic logic operations in FBD is described in the following discussions.

Using the Binary Input

In FBD the binary input is used to examine a referenced address for a status of '1'. If a signal status of '1' is detected, then the binary input passes logic '1' to the gate function or other FBD operation. If a signal status of '0' is detected, then the binary input passes logic '0' to the gate function or other FBD operation. In the case of physical inputs, what is actually being checked is the status presented at an input terminal. This status may reflect a binary '1' presented by a sensor whose NO contacts are activated, or by a sensor whose NC contacts are not-activated—both present a '1' to the input.

86

So, the binary input checks the referenced input for a status of '1' and if '1' is detected, then logic 1 is passed to the gate, regardless of whether the signal is presented from a device wired NO or NC. Given this operation, whether a device is wired normally-open or normally-closed is irrelevant. You must determine whether use of the binary input or the negated binary input suits the logic requirements. A rule is, if you are developing a logic circuit where a signal state of '1' should allow continuity and a signal state of '0' should interrupt continuity, then you should use the binary input.

In the example, if I 27.4 is '1', then logic '1' is passed to the gate function and the output Q 29.1 is assigned a status of logic '1', which energizes the output.

❶ Simple gate with one binary input

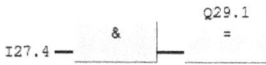

```
                        Q29.1
                &         =
    I27.4 ──        ──
```

Figure 4-16. Using the binary input to require a status of '1' to allow power flow

Using the Negate Input

In a FBD, the negate input is used to examine a referenced address for a status of '0'. If a status of '0' is detected, then the negate binary input passes logic '1' to the gate function or other FBD operation. If a signal status of '1' is detected, then the negate input passes a logic '0' to the gate function or other FBD operation. In the case of physical inputs, what is actually being checked is the status presented at an input terminal. This status may reflect a binary '0' presented by a sensor whose NC contacts are activated, or by a sensor whose NO contacts are not-activated—both present a status of '0'.

So, the negated input checks the referenced input for status of '0' and if '0' is detected then logic 0 is passed to the gate, regardless of whether the signal is presented from a device wired NO or NC. Given this operation, whether a device is wired normally-closed or normally-open is irrelevant. You must determine whether use of the binary input or the negated binary input suits the logic requirements. A rule is, if you are developing a logic circuit where a signal state of '0' should allow logic continuity and a signal state of '1' should interrupt logic continuity, then you should use the negated binary input.

In the following example, if the status of I 27.5 is '0', then logic 1 is passed to the gate function and the output Q 29.2 is assigned a logic '1' state, which energizes the output.

❶ Simple gate with one negated binary input

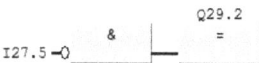

```
                        Q29.2
                &         =
    I27.5 ──○       ──
```

Figure 4-17. Using the negated binary input to require a status of '0' to allow power flow

Using the Assign Output

In FBD, the assign output is likely the most commonly used binary output instruction. The output operand is assigned a state of '1' when the preceding RLO = 1, and is assigned a state of '0' when the preceding RLO = 0. In short, the assign output is always assigned the state of the result of logic operations from the preceding logic string.

In the first example below, if the binary input address Q 30.3 is '1', then logic '1' is passed to the gate and the output image address Q 17.5 is assigned a status of '1'. If the binary input Q 30.3 is '0', then '0' is passed to the gate and the output image address Q 17.5 is assigned a status of '0'. In the second example, if the binary input address Q 30.3 is '1', the parallel outputs Q 17.5 and Q 17.6 are both assigned a status of '1'. If the bit address Q 30.3 is '0', both outputs are assigned a status of '0'.

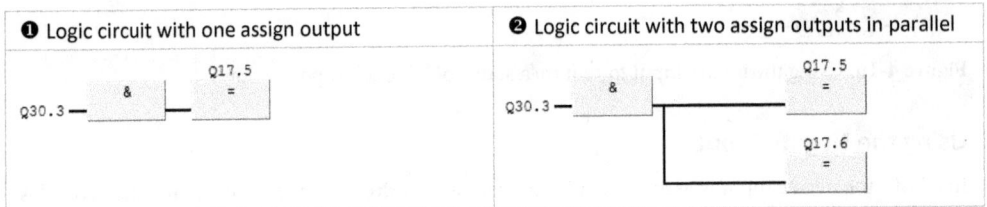

❶ Logic circuit with one assign output	❷ Logic circuit with two assign outputs in parallel
Q30.3 — & — Q17.5 =	Q30.3 — & — Q17.5 = / Q17.6 =

Figure 4-18. Using the assign output

Developing AND Gate Logic

When developing control logic, an AND gate is used when an output should be activated if all of the binary inputs to the gate pass logic '1' to the gate. The AND gate is equivalent to a series circuit—when all of the binary inputs to the gate pass logic '1', the gate has logic continuity or an RLO = 1.

In the first example below, the output Q 17.0 is assigned a status of logic '1' when the binary inputs I 20.0 AND I 20.1 have a signal state of '1'. The output is assigned a status of logic '0' whenever either of the two binary inputs has a signal state of '0'. In the second example, output image address Q 17.0 is assigned a status of logic '1' when all three binary inputs have a signal state of '1'.

In the third example, Q 17.1 is assigned a status of logic '1' if the binary input I 20.0 has a signal state of '1' AND the negated binary input I 20.1 has a signal state of '0'.

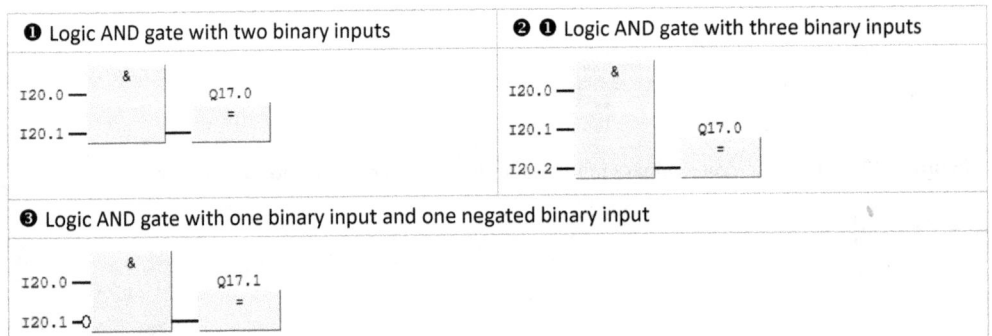

❶ Logic AND gate with two binary inputs	❷ ❶ Logic AND gate with three binary inputs
I20.0 — I20.1 — & — Q17.0 =	I20.0 — I20.1 — I20.2 — & — Q17.0 =
❸ Logic AND gate with one binary input and one negated binary input	
I20.0 — I20.1 —○ & — Q17.1 =	

Figure 4-19. Using the AND logic function to create a series logic circuit

Developing OR Gate Logic

When developing control logic, an OR gate is used when an output is to be activated when any one of the binary inputs to the gate pass logic '1' to the gate. The OR gate is equivalent to a parallel circuit—when any one of the binary inputs to the gate pass logic '1' to the gate, the gate has an RLO = 1.

In the first example below, the output address Q 17.7 is assigned a status of logic 1 when either NO contact I 20.2 OR I 20.3 has a signal state of '1'. The output is assigned a status of logic '0' when all of the inputs are logic '0'. In the second example, three binary inputs are combined using an OR gate—you may insert as many gate inputs as required. Output address Q 17.7 is assigned a status of logic '1' when any one of the binary inputs has a signal state of '1'.

In the third example, **Q 17.7** is assigned a status of logic '1' if the binary input **I 20.3** has a signal state of '1' **OR** the negated binary input **I 20.4** has a signal state of '0'.

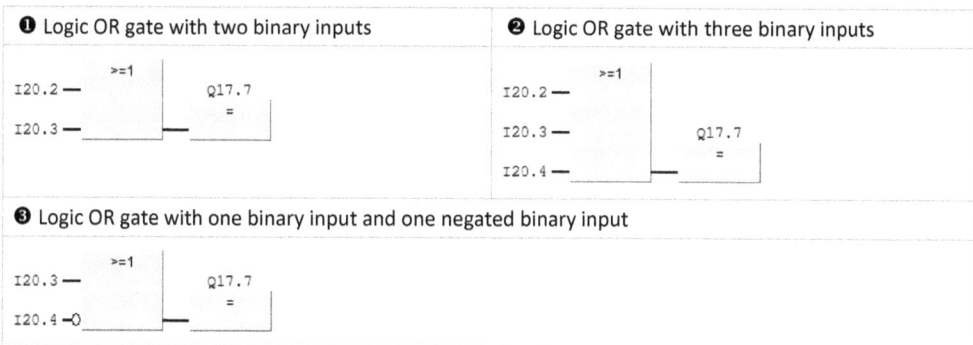

❶ Logic OR gate with two binary inputs	❷ Logic OR gate with three binary inputs
| ```
      >=1
I20.2 ─          Q17.7
I20.3 ─┘   └─      =
``` | ```
          >=1
I20.2 ─
I20.3 ─          Q17.7
I20.4 ─┘   └─      =
``` |
| ❸ Logic OR gate with one binary input and one negated binary input | |
| ```
      >=1
I20.3 ─          Q17.7
I20.4 ─0┘   └─     =
``` | |

Figure 4-20. Using the OR logic function to create a parallel circuit

Developing Exclusive OR Gate Logic

When developing control logic, the exclusive-OR logic gate is used should an output be activated only if one of two binary inputs presents a signal state of '1' to the gate, but not both. The exclusive OR logic function holds that a binary output signal **Y** has a signal state of '1' only when one of two binary signals (**A** or B) is '1'. If both signals (A and B) have a signal state of '1', then the binary output **Y** will have a signal state of '0'. This function is implemented in FBD as shown below.

❶ Logic OR circuit of two AND gates
| ```
       &
I33.1 ─
I33.2 ─0┘   └─   >=1
       &
I33.1 ─0          Q30.3
I33.2 ─┘   └─┘   └─  =
``` |

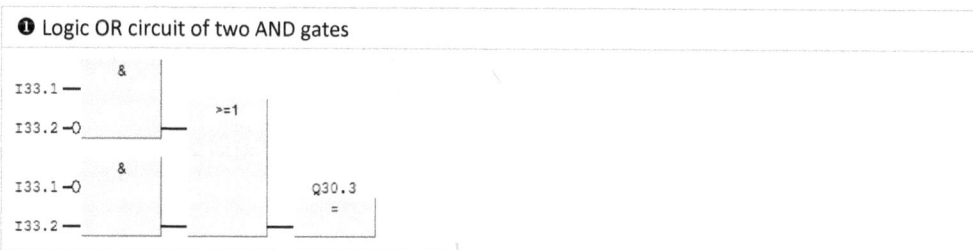

Figure 4-21. Using the Exclusive-OR logic function

In the example above, output address Q 30.3 is assigned a staus of '1' by the first AND gate if the signal state of I 33.1 is '1' AND the signal state of I 33.2 is '0' OR by the second AND gate if the signal state of I 33.1 is '0' AND the signal state of I 33.2 is '1'. The output is assigned a status of '0' whenever the signal state for I 33.1 and I 33.2 are '1'. In the two AND gates above, apply logic '1' to both operands to test the result. Recall, that a signal state of '1' causes the negated binary input to present logic '0' to the gate, thereby interrupting logic continuity in both AND gates. Q 30.3 is assigned a state of '0'.

AND-before-OR Logic Combinations

The *and-before-or* combines two or more AND gates as inputs to an OR gate—the equivalent of two or more series circuits in parallel. In such a combination a logic '1' is presented at the input of the OR gate when all of the inputs (positive or negated) to either of the AND gates present a logic '1'to the gate. The output of the OR gate is logic '1' if the RLO = 1 at the output of any one of the AND gates.

In the first example below, the output address Q 30.2 is assigned a status of logic 1 when a status of '1' is at the output of any one of the AND gates (I 20.1 AND I 20.2 have a signal state of '1' OR if I 20.3 AND I 20.4 have a signal state of '1'). The output Q 30.2 is assigned a status of logic 0 whenever the RLO = 0 at the output of all of the AND gates. In the second frame below, three AND gates are combined using the OR gate. You may insert as many AND gate inputs to the OR gate as required.

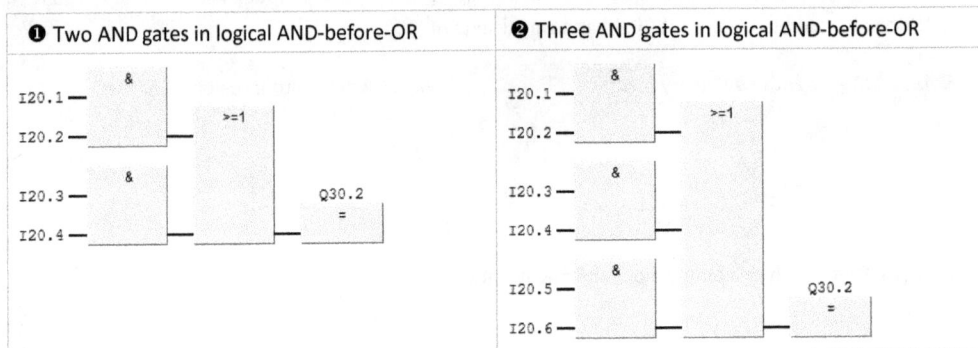

❶ Two AND gates in logical AND-before-OR	❷ Three AND gates in logical AND-before-OR

Figure 4-22. Using AND-before-OR to combine series circuits in parallel

OR-before-AND Logic Combinations

The *or-before-and* combines two or more OR gates as inputs to an AND gate—the equivalent of two or more parallel circuits in series. In such a combination, if either of the OR gate inputs (positive or negated) presents a logic '1', then the output of the OR gate has an RLO = 1. The output of the AND gate is assigned a status of '1' if the RLO = 1 at the output of all of the OR gates.

In the first example, the output Q 30.6 is assigned a status of logic 1, when the combination of I 22.1 OR I 22.3 is '1' AND I 22.2 OR I 22.4 is '1'. For example, Q 30.6 is activated if I 22.3 is '1' AND I 22.2 is '1'. The AND gate output is assigned as status of '0' when the RLO = 0 in any one of the OR gates.

In the second frame below, three OR gates are combined using the AND gate to drive the output image address Q 30.6. You may insert as many OR gate inputs to the AND gate as required.

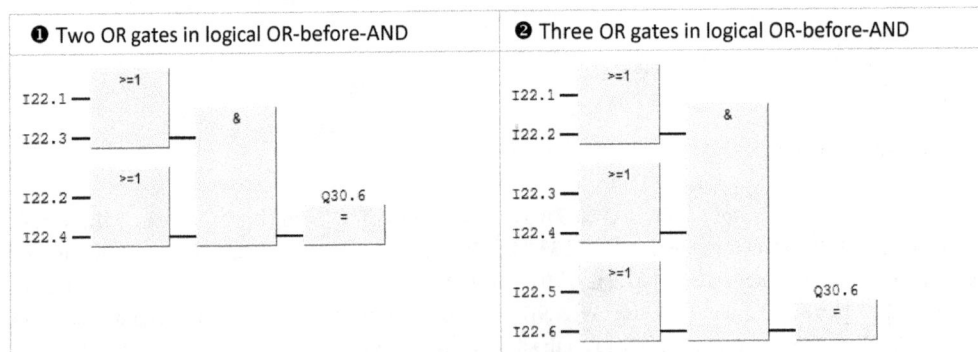

❶ Two OR gates in logical OR-before-AND	❷ Three OR gates in logical OR-before-AND

Figure 4-23. Using OR-before-AND to combine parallel circuits in series

Negated Logic Combinations

In FBD, the negate input is not only used to invert an input to a gate or box operation, but also to invert the result of logic operation. By inserting the negate input at the output of a logic gate or box operation you thereby invert the RLO. The negate input may be placed between any two nodes in a logic string. In the examples below, the negate input is placed between the logic gate and the assign output. This placement inverts the RLO after the logic gate has been evaluated. An output RLO = 1 becomes an RLO = 0; and an output RLO = 0 becomes an RLO =1.

In the first example, if both binary inputs I 44.5 AND I 44.6 are '1', a '1' is presented to the negate input—after inversion the RLO = 0 and output Q 27.0 is assigned a status of '0'. When either one or both inputs to the AND gate are '0' then '0' is presented to the negate input—after inversion the RLO = 1 and Q 27.0 is assigned as status of '1'. This operation is the negated AND function or NAND.

In the second example, only if both binary inputs M 37.4 and M 37.5 are '0' is the output Q 27.1 assigned a status of '1'—that is when a '0' is presented to the negate input. In all other cases a '1' is presented to the negate input, and Q 27.1 is assigned a status of '0'. This operation is the negated OR function or NOR.

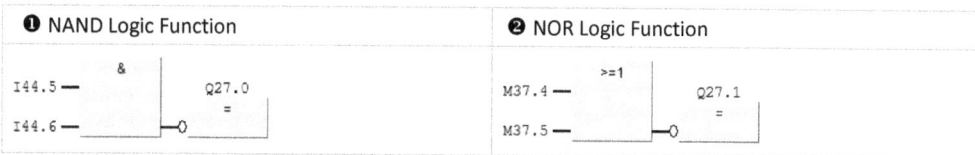

❶ NAND Logic Function	❷ NOR Logic Function

Figure 4-24. Negated logic combinations **NAND** and **NOR**

Using the Mid-Line Output

Using the mid-line output instruction, you can implement the function of the assign output at a point in the logic string other that at the end of the network. In this way, you can store the result of logic operation at intermediate points of a logic string in order to minimize logic duplication.

After examining the three networks in the first frame below, see how use of the midline output has resulted in the single network in the second frame.

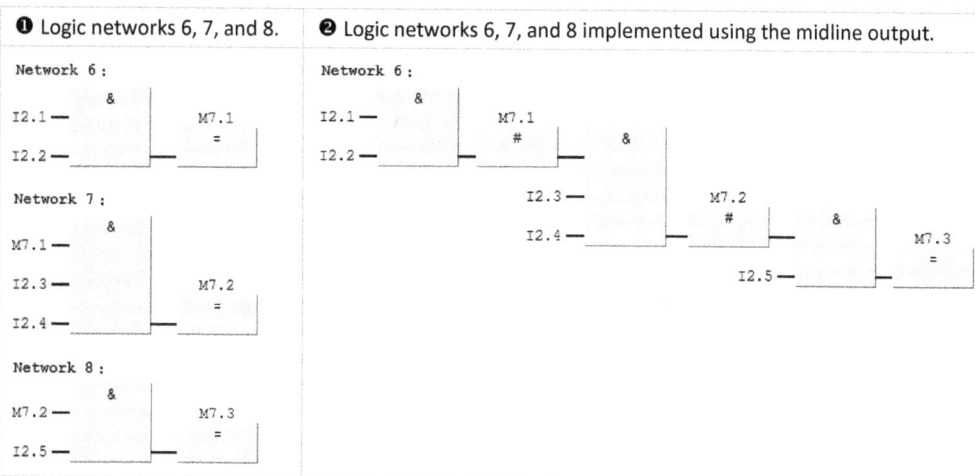

❶ Logic networks 6, 7, and 8.	❷ Logic networks 6, 7, and 8 implemented using the midline output.

Figure 4-25. Use of the midline output to minimize logic duplication

BINARY LOGIC –MEMORY OPERATIONS OVERVIEW

Binary Logic Memory operations, summarized in the table below, include edge detection and Set/Reset operations. Whereas the edge detection operations support the detection of positive-edge (0-to-1) and negative edge (1-to-0) transitions in your program, Set/Reset operations support latching and unlatching of bit variables or bit addressable memory.

Table 4-2. LAD/FBD Binary Logic—Memory Operations

LAD	FBD	Brief Description
??.? ——(P)——	??.? P ??.?—	**RLO Positive-Edge Detect.** Detect a 0-to-1 transition resulting from the logic preceding the **P**-coil. Signal edge detection with the output going high (1) for a single CPU scan.
??.? ——(N)——	??.? N ??.?—	**RLO Negative-Edge Detect.** Detect a 1-to-0 transition resulting from the logic preceding the **N**-coil. Signal edge detection with the output going high (1) for a single CPU scan.
??.? NEG Q ??.?—M_BIT	??.? NEG ??.?—M_BIT Q—	**Address Negative-Edge Detect.** Detect a 1-to-0 transition of a specific bit address. Signal edge detection with the output **Q** going high (1) for a single CPU program scan.
??.? POS Q ??.?—M_BIT	??.? POS ??.?—M_BIT Q—	**Address Positive-Edge Detect.** Detect a 0-to-1 transition of a specific bit address. Signal edge detection with the output **Q** going high (1) for a single CPU program scan (one shot).
??.? ——(S)——	??.? S ??.?—	**Set Coil/Output (Latch).** If preceding result of logic operations (RLO) is '1', then set the referenced address to '1'; remain at logic 1 until reset.
??.? ——(R)——	??.? R ??.?—	**Reset Coil/Output (Unlatch).** If preceding result of logic operations (RLO) is '1', then set the referenced address to '0'.
??.? SR …—R	??.? SR ??.?—S …—R Q—	**Set/Reset Flip-Flop (Latch-Unlatch).** Implement Set/Reset (Latch/Unlatch) function, giving priority to the Reset (**R**) if the RLO = 1 at both the **S** and **R** input lines during the same scan.
??.? RS R Q …—S	??.? RS ??.?—R …—S Q—	**Reset/Set Flip-Flop (Unlatch-Latch).** Implement Reset/Set (Unlatch/Latch) function, giving priority to the Set (**S**) if the RLO = 1 at both the **R** and **S** input lines during the same scan.

PROGRAMMING MEMORY OPERATIONS IN LAD/FBD

As with basic binary logic operations, code developed using binary logic memory operations is used to form the driving logic for externally connected discrete devices, as well as for triggering various internal operations. Programming these basic logic memory operations in LAD/FBD is described in the following discussions.

Set-Reset Operations

The Set and Reset coils/outputs, like the standard output coil operation, are final control elements that are driven by some combination of series/parallel logic. Also, as the standard output, these operations, when used to terminate a binary logic string in LAD and in FBD, may be inserted as parallel outputs. Set and Reset outputs may be used individually or as a pair. Used individually, for example, and as paralleled outputs, it is possible to set several bit addresses to '1' or reset several bit addresses to '0'.

When used as a pair the Set (**S**)/Reset (**R**) outputs, like the **SR** and **RS** box operations, support the latching and unlatching of bit variables or bit addressable memory locations. What the Set/Reset coil/output operations provide when programmed as a pair is provided as a single box function with the SR flip-flop and the RS flip-flop.

Similar to **S** and **R** coils/outputs, the **S** and **R** logic input lines of the flip-flop box operations are also controlled by some combination of series-parallel logic. With these operations, you may also insert contacts or other logic elements on the output line as well.

Set/Reset Dominance

When it comes to programming set and reset operations, *dominance* is an important characteristic. Dominance determines which of the **S** or **R** operation overrides the other if the RLO=1 for both operations during the same processing cycle. With set/reset output instructions, dominance is determined by which of the Set or Reset networks is programmed last. With the flip-flop operations, dominance is based on which of the two input lines is processed last in the box. In the **RS** flip-flop, the **S** input line is processed last, and therefore dominant. In the **SR** flip-flop, the **R** input line is processed last, and is therefore the dominant of the two. It is up to you to determine which should have priority.

Set Coil/Set Output: S

The **S** coil/output is used to set and maintain a specific bit address to logic 1, even if logic continuity is lost. When evaluated, if any path of logic conditions has continuity (RLO=1), the bit address referenced above the set operation will be set to '1'. The referenced bit address will remain latched at logic 1 until reset (unlatched) by a reset operation that references the same address.

In the following network, if the NO contact operands of input I 3.1 AND I 3.2 are closed, the output address Q 4.6 is set (latched) to '1'. Q4.6 remains at '1' even if logic continuity is lost and the RLO = 0.

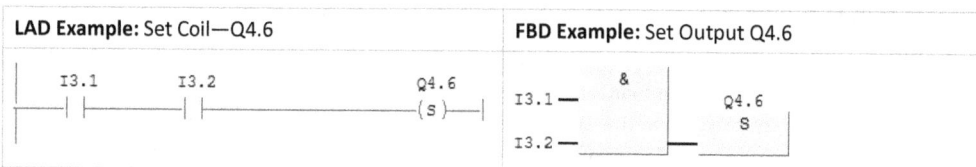

LAD Example: Set Coil—Q4.6	FBD Example: Set Output Q4.6

Figure 4-26. LAD/FBD illustration of Set operation

Reset Coil/Reset Output: R

The **R** coil/output is used to reset or unlatch a set operation that references the same bit address. When evaluated, if any path of logic conditions has continuity (RLO=1), the bit address referenced above the reset operation is reset to '0'.

In the following network, if the NO contact operands of input I 3.3 AND I 3.4 are closed, the output address Q 4.6 is reset (unlatched) to '0'.

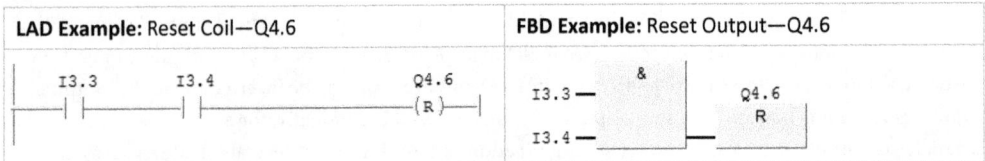

LAD Example: Reset Coil—Q4.6	FBD Example: Reset Output—Q4.6
I3.3 I3.4 Q4.6 ─(R)─	I3.3 — & Q4.6 R I3.4 —

Figure 4-27. LAD/FBD illustration of Reset operation

Set/Reset Flip-Flop: SR

The **SR** flip flop performs both the Set (latch) and Reset (unlatch) functions, and is used when the reset function must have priority (dominance). Both the Set (**S**) and the Reset (**R**) functions are executed only if the logic of their respective input line is satisfied (RLO=1). A loss of logic continuity (RLO=0) at either input has no effect on the operation or on the current status of the affected bit variable.

If logic driving the **S** line is satisfied (RLO=1), the address referenced above the SR flip-flop is set to logic 1 and will remain set until reset. If logic driving the **R** line is satisfied (RLO=1), the referenced bit address is reset to logic 0. The output **Q** will always reflect the state of the address referenced above the instruction. If both the **S** and **R** logic inputs are satisfied simultaneously (RLO =1 in same scan), the SR flip-flop would first be set and then subsequently reset, upon evaluation of the **R** line.

In the following network, M101.5 is set if I 3.5 is closed (1). The address Q 4.7 follows the bit memory address M101.5. The bit address M 101.5 is reset if the combination I 3.6 AND I 3.7 has an RLO =1. The address above the flip-flop must be defined as retentive if Q is to retain its state upon power loss.

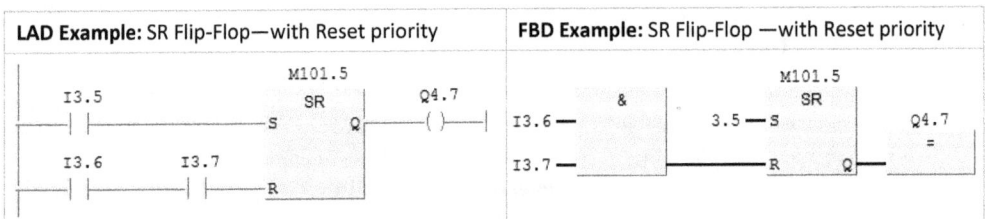

LAD Example: SR Flip-Flop—with Reset priority	FBD Example: SR Flip-Flop —with Reset priority
M101.5 I3.5 SR Q4.7 ─S Q─( )─ I3.6 I3.7 ─R	M101.5 & SR Q4.7 I3.6 — 3.5 — S = I3.7 — R Q─

Figure 4-28. LAD/FBD illustration of Set/Reset Flip-Flop operation

Reset/Set Flip-Flop: RS

The **RS** flip flop performs both the Reset (unlatch) and the Set (latch) functions, but is used when the set function must have priority (dominance). Both the Reset (**R**) and the Set (**S**) functions are executed only if the logic of their respective input line is satisfied (RLO=1). A loss of logic continuity (RLO=0) at either input has no effect on the operation or on the current status of the affected bit variable.

If RLO = 1 at the **R** line, the address referenced above the RS flip-flop is reset to logic 0 and will remain reset until set again. If logic driving the **S** line is satisfied (RLO=1), the referenced bit address will be Set to logic 1. The output **Q** will always reflect the state of the address referenced above the instruction. If both the **R** and **S** inputs are satisfied simultaneously (RLO =1 in same scan), the RS flip-flop would first be reset and then subsequently set, upon evaluation of the **S** line.

In the following network, bit address M101.3 is reset if input I 4.1 is closed. The address Q 4.5 follows the bit memory address M101.3. The bit address M101.3 is set if the combination of I 4.2 AND I 4.3 has an RLO = 1. An important note here is that the address above the flip-flop must be defined as retentive if **Q** is to retain its state upon power loss.

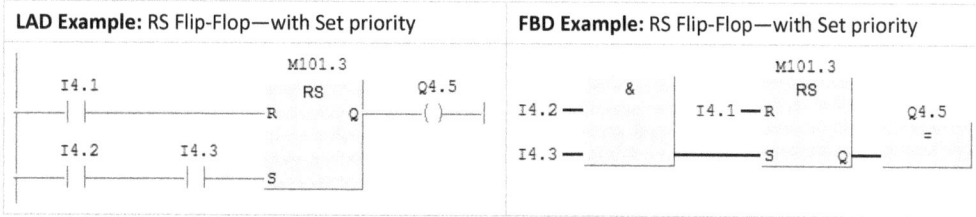

LAD Example: RS Flip-Flop—with Set priority	FBD Example: RS Flip-Flop—with Set priority

Figure 4-29. LAD/FBD illustration of Reset/Set Flip-Flop operation

Edge Detection Operations

An edge transition is a discrete signal change from 0-to-1 or from 1-to-0. A change from 0-to-1 is referred to as a positive-edge or rising-edge transition; and a signal change from 1-to-0 is referred to as a negative-edge or falling-edge transition.

Edge detection operations are useful when it comes to detecting when a motor is switched on or off, when a photo eye is first blocked, or when a button is first pushed. Because these operations indicate the detection of a rising or falling edge using a one-shot pulse, they are also useful for ensuring that certain operations like an arithmetic calculation, a block transfer, a bit shift, or other such operation are triggered once and only once on an edge transition.

Edge detection operations are especially useful in that they supplement the standard binary examination operations, which simply check whether a discrete signal is ON or OFF—these operations detect and indicate both positive and negative edge transitions. Like binary contact operations, edge evaluation instructions may be inserted anywhere in the logic string.

In STEP 7 there are two types of *edge detection operations*—both of which support detection of positive and negative edge transitions. The first type of edge evaluation detects an edge transition in a specific bit address or variable—these instructions are the *Address Positive-Edge Detect* and the *Address Negative-Edge Detect*. The second type of edge evaluation detects a transition in the RLO (result of logic operation) at a point in a logic network—these instructions are the *RLO Positive-Edge Detect* and the *RLO Negative-Edge Detect*.

RLO Positive-Edge Detect: P

The **RLO positive-edge detect** instruction detects a rising edge (0-to-1) transition in the result of logic operations that precede the instruction. The address you specify above the instruction saves the previous state of the RLO at the point of insertion, and compares it to the new RLO.

When the RLO preceding the P-coil makes a 0-to-1 transition, the RLO after the P-coil is set to logic 1 for one program scan (one shot pulse) and then returns to logic 0. The RLO must return to logic 0 and back to logic 1 again to retrigger the one-shot pulse. You may insert the P-coil, as required, at nodes between any two program elements—never at the last position of the network.

In the following network, the output Q 3.1 is reset (unlatched) when the combination of I 4.2 OR I 4.3 AND I 4.4 makes a 0-to-1 transition.

LAD Example: P-Coil—M 100.2	FBD Example: P-Output—M 100.2

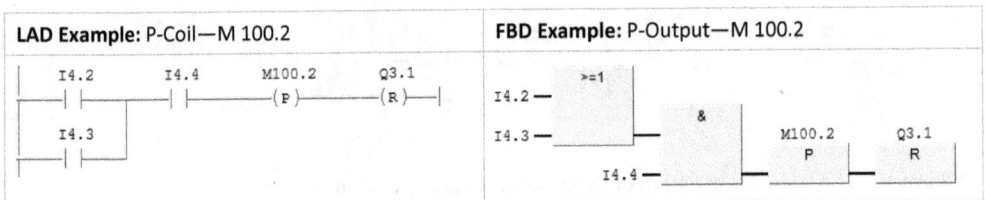

Figure 4-30. LAD/FBD illustration of Positive RLO Edge Detection

RLO Negative-Edge Detect: N

The **RLO negative-edge detect** instruction detects a falling-edge (1-to-0) transition in the result of logic operations (RLO) that precede the instruction. The address you specify above the instruction remembers the previous state of the RLO at the point of insertion, and compares it to the new RLO.

When the RLO preceding the N-coil makes a 1-to-0 transition, the RLO after the N-coil is set to logic 1 for one program scan (one shot pulse) and then returns to logic 0. You may insert the N-coil, as required, at nodes between any two program elements—never at the last position of the network.

In the following network, the output Q 3.0 is set (latched) when the logic combination of (I 4.2 OR I 4.3) AND I 4.4 makes a 1-to-0 transition.

LAD Example: N-Coil—M 100.1	FBD Example: N-Output—M 100.1

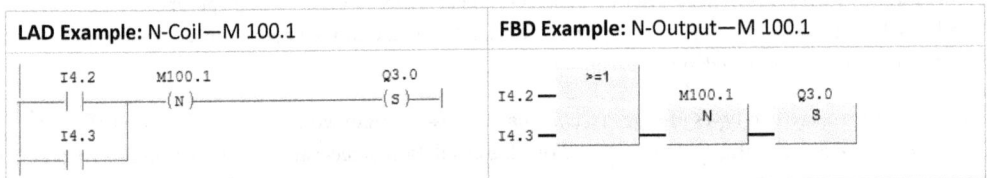

Figure 4-31. LAD/FBD illustration Negative RLO Edge Detection

Address Positive-Edge Detect: POS

The **address positive-edge detect** instruction detects a rising-edge (0-to-1) transition of the signal referenced by the bit address specified above the instruction. While the enabling logic is True (RLO=1) at the unlabeled input line to the box, a 0-to-1 transition of the edge trigger signal will cause a one-shot output (logic 1 for one program scan) to be produced at the **Q** output line.

The bit address you specify at the parameter **M_BIT** saves the signal state of the edge signal from the previous check, thereby enabling detection of the 0-to-1 transition.

In the LAD/FBD examples shown in the top cells, if I 4.2 is True (RLO=1) a one scan pulse is output at **Q** whenever I 4.1 makes a 0-to-1 transition—as a result, Q3.7 is set. In the examples shown in the bottom cells, there are no conditions combined with **Q** (edge evaluation is always enabled). A one scan pulse is output at **Q** whenever the input I 4.1 makes a 0-to-1 transition—as a result, Q3.7 is set.

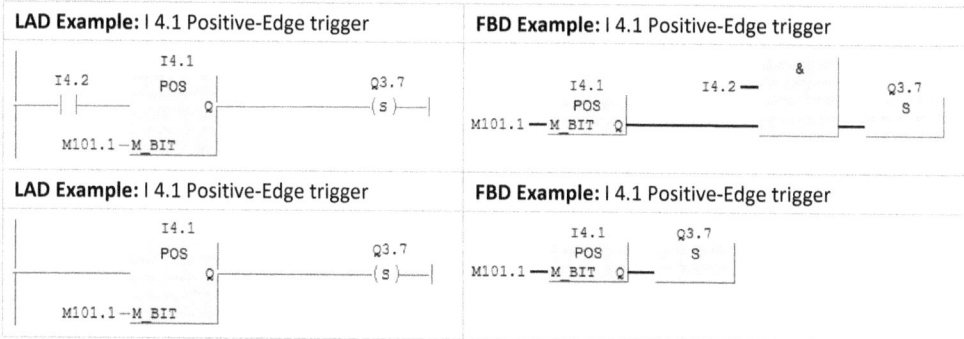

LAD Example: I 4.1 Positive-Edge trigger	FBD Example: I 4.1 Positive-Edge trigger
LAD Example: I 4.1 Positive-Edge trigger	**FBD Example:** I 4.1 Positive-Edge trigger

Figure 4-32. LAD/FBD illustration of Address Positive-Edge Detection

Address Negative-Edge Detect: NEG

The **address negative-edge detect** instruction detects a falling-edge (1-to-0) transition of the signal referenced by the bit address specified above the instruction. While the enabling logic is True (RLO=1) at the unlabeled input line to the box, a 1-to-0 transition of the edge trigger signal will cause a one-shot output (logic 1 for one program scan) to be produced at the **Q** output line.

The bit address you specify at the parameter **M_BIT** saves the signal state of the edge signal from the previous state check, thereby enabling detection of the 1-to-0 transition.

In the LAD/FBD examples shown in the top cells, if I 4.0 is True (RLO=1), a one scan pulse is output at **Q** whenever I 4.3 makes a 1-to-0 transition—as a result, Q3.6 is set. In the examples shown in the bottom cells, there are no conditions combined with **Q** (edge evaluation is always enabled). A one scan pulse is output at **Q** whenever the input I 4.3 makes a 1-to-0 transition—as a result, Q3.6 is set.

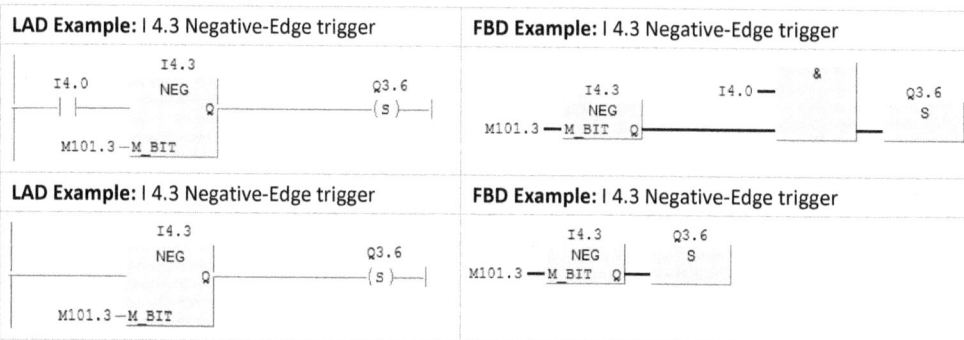

LAD Example: I 4.3 Negative-Edge trigger	FBD Example: I 4.3 Negative-Edge trigger
LAD Example: I 4.3 Negative-Edge trigger	**FBD Example:** I 4.3 Negative-Edge trigger

Figure 4-33. LAD/FBD illustration of Address Negative-Edge Detection

Chapter 5

LAD/FBD
Move Operations

In this chapter you'll be introduced to LAD/FBD Move operations. In S7 programs, the Move operation is used to copy data from one memory location or variable to another for various purposes—sometimes simply to store or protect data from being overwritten, and other times to prepare data for subsequent processing and use in another operation.

In addition to the STEP 7 Move operation, which handles only one value at a time, standard System Functions that support the simultaneous handling of multiple locations or variables as a contiguous block are covered in Chapter 28 *Developing Functions and Function Blocks*. These callable Functions, which include SFC 20 (BLKMOV—Copy Data Area), SFC 81 (UBLKMOV—Copy Data Area without Gaps), and SFC 21 (FILL—Pre-Fill Data Area Values), are available in the STEP 7 library, and are integrated in some S7 CPUs.

Main Topics

- Move Operations
- Move Operations Descriptions
- Programming Move Operations in LAD and FBD

MOVE OPERATIONS

The *move* instruction copies data from one location to another, while leaving the source location unchanged. Often data is moved to a new location prior to performing an operation that uses the data, or just after an operation has been performed, in order to preserve the data from being overwritten. Programmatically changing a timer preset, a process set point, or saving and stacking data for transfer to another device all are cases where move operations might be used.

A summary of sources from which data may be moved and destinations to which data can be copied is provided in the following table. Any memory area shown under Source (IN) may be moved to any memory area shown under Destination (OUT). In addition to moving memory locations and variables, you may also move constant values where valid constant notations may be entered as the Source of the move operation.

In STEP 7 LAD and FBD, the Move operation is used to copy the contents of a single location at a time. To transfer a block of data using a single operation, you will need to use the appropriate function (SFC) available in the STEP 7 standard library. Systems Functions SFC 20, SFC 21, and SFC 81 all deal with transferring multiple locations.

Table 5-1. Summary of valid Source/Destination areas in LAD/FBD MOVE operation

Memory Area	Source (IN)	Destination (OUT)	Area
Input Image	IB = Byte	IB = Byte	**Input Image**
	IW = Word	IW = Word	
	ID = Double Word	ID = Double Word	
Output Image	QB = Byte	QB = Byte	**Output Image**
	QW = Word	QW = Word	
	QD = Double Word	QD = Double Word	
Bit Memory	MB = Byte	MB = Byte	**Bit Memory**
	MW = Word	MW = Word	
	MD = Double Word	MD = Double Word	
Counter	C = Counter	**NOT A VALID DESTINATION**	Counter
Timer	T = Timer		Timer
Local Stack	LB = Byte	LB = Byte	**Local Stack**
	LW = Word	LW = Word	
	LD = Double Word	LD = Double Word	
Peripheral Inputs	PIB = Byte	**NOT A VALID DESTINATION**	**Peripheral Inputs**
	PIW = Word		
	PID = Double Word		
Peripheral Outputs	**NOT A VALID SOURCE**	PQB = Byte	**Peripheral Outputs**
		PQW = Word	
		PQD = Double Word	

Note: Peripheral Output addresses are not valid as a MOVE source location. Counter, Timer, and Peripheral Input addresses are not valid as MOVE destination locations.

MOVE OPERATIONS DESCRIPTIONS

Moving data in a PLC program generally involves taking the contents from one or more memory locations and copying it to one or more new locations. The location from which data is copied is called the *source;* the place to which it is copied is the *destination.*

MOVE Instruction Parameters

Parameters for the LAD Move operation are shown below. You may insert logic at the enable input line EN, to control when the operation is executed. You must then specify the parameter IN as the source value of Move operation and parameter OUT as the destination or the location to which the data is copied. When EN is at logic 1, the contents of the location or variable specified at IN are copied to the location specified at OUT.

Table 5-2. LAD/FBD MOVE Parameters: Elementary data types except BOOL, valid at IN and OUT

Parameter	Data Type	Description	Required?	MOVE Operation	
				LAD	**FBD**
EN		Enable Input Line		MOVE	MOVE
IN	Elementary	Move Source	YES	—EN ENO—	—EN OUT—
ENO		Enable Output Line	NO	—IN OUT—	—IN ENO—
OUT	Elementary	Move Destination	YES		

Valid entries for **IN** include REAL and integer constants, variables of all elementary data types except BOOL, and all valid memory operands. Examples include IB10, IW22, ID15, QB12, QW34, QD44, MB14, MW60, MD54, DB22.DW24, and DB22.DD12. The location you specify at OUT does not have to be the same size as IN; however, data is always moved right-justified. That is to say if the source is shorter than the destination, the data is copied to the right-most bits and the remaining bits are zero-filled. If the source is larger than the destination, then the excess bits are truncated from the left.

Move: MOVE

The *move word* instruction essentially performs a copy function. The parameters of the move instruction are shown in the table above. When the control line driving the box is TRUE, the contents of the location specified as the source are copied right-justified into the location specified as the destination location. The source location is left unchanged.

Table 5-3. Illustration of MOVE with a Word operand

Example: MOVE (Word)	
1010 0101 1111 1100	**IN** = (Source)
1010 0101 1111 1100	**OUT** = (Destination)

PROGRAMMING LAD/FBD MOVE OPERATIONS

You will use the MOVE operation to copy the contents of a memory location or variable to a new location or variable. Programming a MOVE operation is a matter of placing the move box into the LAD or FBD network, specifying an input parameter as the value to be moved, and an output parameter as the location to which the value is copied. In LAD and FBD the MOVE instruction operates on a single location, which must be an elementary data element of 8, 16, or 32 bits.

For data transfers involving multiple locations or complex data types such as arrays, you may use system blocks SFC 20 (BLCKMOV) and SFC21 (FILL).

In a MOVE operation, the source location is always moved to the destination right-justified. The basic move scenarios are illustrated in the following figures.

If the source and destination are of equal width, for example word-to-word, byte-to-byte, or double word-to-double word, then the contents are moved bit for bit to the destination.

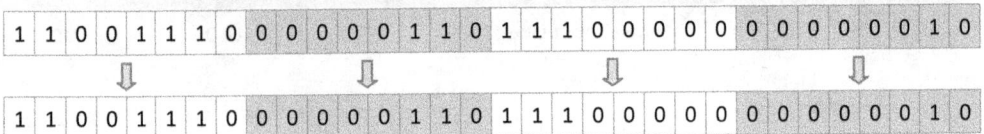

1	1	0	0	1	1	1	0	0	0	0	0	0	1	1	0	1	1	1	0	0	0	0	0	0	0	0	0	0	0	1	0

1	1	0	0	1	1	1	0	0	0	0	0	0	1	1	0	1	1	1	0	0	0	0	0	0	0	0	0	0	0	1	0

Figure 5-1. MOVE with source and destination of equal length

If the source is larger than the destination, then the excess bits are truncated from the left.

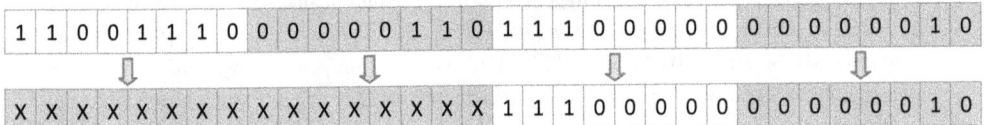

1	1	0	0	1	1	1	0	0	0	0	0	0	1	1	0	1	1	1	0	0	0	0	0	0	0	0	0	0	0	1	0

X	X	X	X	X	X	X	X	X	X	X	X	X	X	X	X	1	1	1	0	0	0	0	0	0	0	0	0	0	0	1	0

Figure 5-2. MOVE with source length greater than destination length

If the source is shorter than the destination, the data is copied to the right-most bits of the destination and the remaining bits are zero-filled.

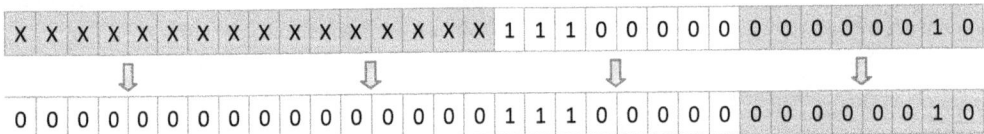

X	X	X	X	X	X	X	X	X	X	X	X	X	X	X	X	1	1	1	0	0	0	0	0	0	0	0	0	0	0	1	0

0	0	0	0	0	0	0	0	0	0	0	0	0	0	0	0	1	1	1	0	0	0	0	0	0	0	0	0	0	0	1	0

Figure 5-3. MOVE with source length less than destination length

Valid memory areas at the parameter IN, as shown in Table 5-1, include M, I, Q, PI, L, T, C, and **DB** memory; valid memory areas at OUT include M, I, Q, PQ, L, and DB memory. In the S7-300, the addresses of the entire input/output image areas (I/Q) may be specified at the IN and OUT parameters of the MOVE operation, regardless of whether the address is assigned to an installed module. In the S7-400, however, you should only access an address if it is assigned to an installed module. You should also be aware of the fact that data moved to Input Image memory will be overwritten on the next update of the I/O image areas.

Moving Constants

A constant of elementary data type may be moved from the source IN to the destination OUT. In the context of the Move operation, specifying a constant at OUT is invalid. Typical formats include hex constants—for example W#16#80FF; a BCD counter preset—for example C#100; and two unsigned bytes—for example B#(128,255). Moving a constant value—for example the value zero, may be needed to reset a variable. Examples of constant value notations are provided in Table 2-12.

In the example below, the hexadecimal word constant 80FF is always moved to word DBW20 of DB30.

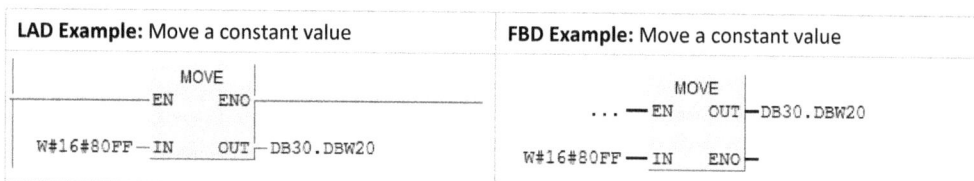

LAD Example: Move a constant value	FBD Example: Move a constant value
MOVE EN ENO W#16#80FF—IN OUT—DB30.DBW20	MOVE ...—EN OUT—DB30.DBW20 W#16#80FF — IN ENO

Figure 5-4. MOVE operation to copy a constant value

Moving Bit Memory (M)

Bit memory locations may be specified as the source IN or the destination OUT. You may specify byte (MB), word (MW), or double word (MD) locations. Moving a memory location—for example MW 34, may be required to save an intermediate calculation value, or transfer a new preset value to a setpoint.

In the example below, when the RLO = 1 at EN, the contents of bit memory word MW44 is moved to the output image word QW44.

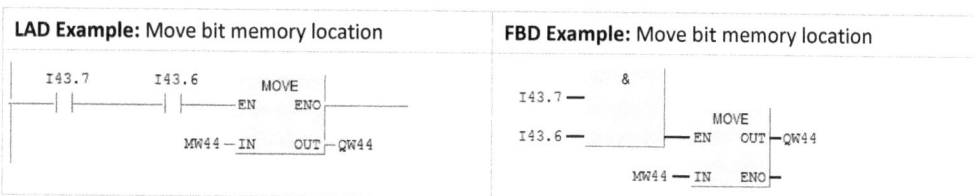

LAD Example: Move bit memory location	FBD Example: Move bit memory location
I43.7 I43.6 MOVE ─┤ ├──────┤ ├───EN ENO MW44 — IN OUT—QW44	& I43.7 — I43.6 — MOVE EN OUT—QW44 MW44 — IN ENO

Figure 5-5. MOVE operation to copy Bit Memory

Moving Input Image Memory (I)

Input Image locations may be specified as the source IN or destination OUT. Remember that data moved to input memory (OUT) is overwritten on the next update of the input image. Furthermore, in the S7-400, an input address should only be specified if a corresponding module is installed. The S7-300 is not restricted in this way. You may specify byte (IB), word (IW), or double word (ID) locations.

Moving an Input (I) image location—for example IW44, might be used to transfer the status of 16 digital inputs directly to 16 digital outputs that are connected to indicators; or perhaps to move the inputs of a 3-decade BCD device to a 3-decade BCD output device.

In the example, the value in input image word IW44 is always moved to output image word QW44.

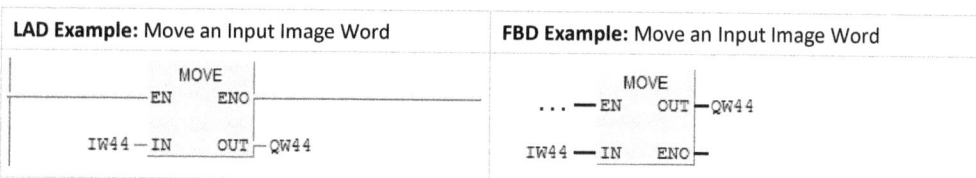

LAD Example: Move an Input Image Word	FBD Example: Move an Input Image Word
MOVE EN ENO IW44 — IN OUT—QW44	MOVE ...—EN OUT—QW44 IW44 — IN ENO

Figure 5-6. MOVE operation to copy Input Image Memory

Moving Output Image Memory (Q)

Output Image memory may be specified as the source IN or as the destination OUT. Recall that data moved to output memory affects the connected devices at the next output image update. In the S7-400, an output address should only be specified if a corresponding output module is installed. The S7-300 is not restricted in this way. You may specify byte (QB), word (QW), or double word (QD) locations.

Moving an Output Image address—for example QW18, might be used at the start of the CPU scan to move the status of the 16-points of an output module to a bit memory location that represents the previous status of the outputs. The current and previous states may then be compared at the end of the current CPU scan. In the figure below, the 16 outputs of QW18 are always moved to DB10.DBW22.

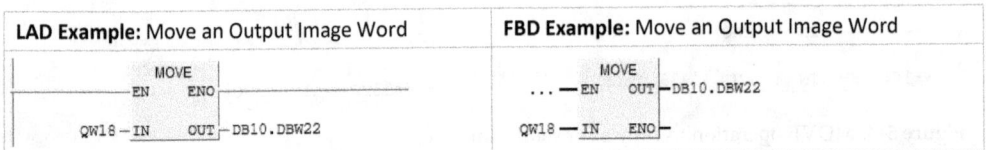

LAD Example: Move an Output Image Word	FBD Example: Move an Output Image Word
MOVE EN ENO QW18 — IN OUT — DB10.DBW22	MOVE ... — EN OUT — DB10.DBW22 QW18 — IN ENO

Figure 5-7. MOVE operation to copy Output Image Memory

Moving Peripheral Input Memory (PI)

Peripheral input memory (PI) is used with the Move operation to read data from an analog input module or from a digital input module, directly via the I/O bus. As discussed in more detail in Chapter 1 *Absolute Addressing of S7 Memory Areas*, peripheral memory input is always required to read data from an analog input module. It is only required with digital input modules when you need the instantaneous status of the inputs. In the figure below, the 8 input bits of PIB 22 are always moved to input image byte IB 22. From that point in the program, the input image bits reflect this latest update.

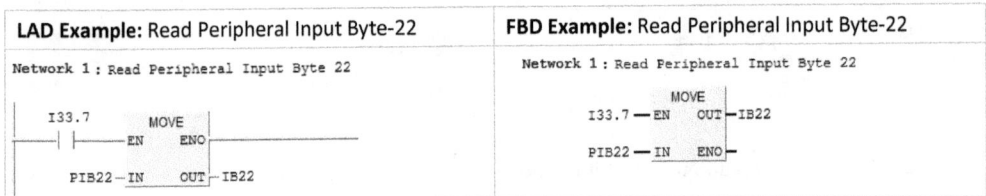

LAD Example: Read Peripheral Input Byte-22	FBD Example: Read Peripheral Input Byte-22
Network 1: Read Peripheral Input Byte 22 I33.7 MOVE —| |—— EN ENO PIB22 — IN OUT — IB22	Network 1: Read Peripheral Input Byte 22 MOVE I33.7 — EN OUT — IB22 PIB22 — IN ENO

Figure 5-8. MOVE operation to access Peripheral Input Memory for immediate read of digital inputs

To read an input module, you specify a peripheral input address as the source (IN) of the MOVE operation. Digital inputs may be addressed as peripheral input byte (PIB), word (PIW), and double word (PID). In the example above, for instance, peripheral input byte 22 (PIB 22) is read and then mapped onto its corresponding input image byte (IB 22). The bit addresses of IB 22 are now updated.

Each analog input channel is represented in a 16-bit word, and is therefore specified as a peripheral input word (PIW). To access a particular analog input, you simply specify the starting byte address of the analog input channel. In the following example, the analog input PIW256 is always read and is moved to the bit memory location MW44. From there it is available for further processing.

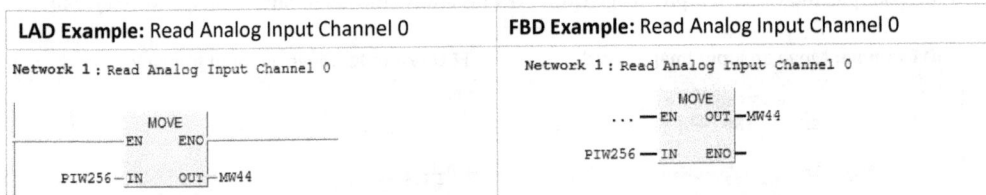

LAD Example: Read Analog Input Channel 0	FBD Example: Read Analog Input Channel 0
Network 1: Read Analog Input Channel 0 MOVE EN ENO PIW256 — IN OUT — MW44	Network 1: Read Analog Input Channel 0 MOVE ... — EN OUT — MW44 PIW256 — IN ENO

Figure 5-9. MOVE operation to access Peripheral Input Memory to read an analog input channel

Moving Peripheral Output Memory (PQ)

Peripheral output memory (PQ) is used with the Move operation to write data to an analog output module or to a digital output module directly via the I/O bus. See Chapter 1 *Absolute Addressing of S7 Memory Areas* for further detail on this topic. Peripheral output memory may only be specified as the destination **OUT** of the Move operation. When using peripheral output memory as a Move destination, you may specify PQB (byte), PQW (word), and PQD (double word). In the following example, the 16-bit hexadecimal constant of F0FF is always written to peripheral output word PQW34.

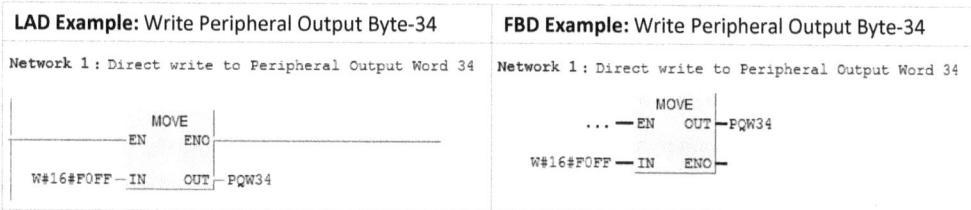

LAD Example: Write Peripheral Output Byte-34	FBD Example: Write Peripheral Output Byte-34
Network 1: Direct write to Peripheral Output Word 34	Network 1: Direct write to Peripheral Output Word 34

Figure 5-10. MOVE to access Peripheral Output Memory for immediate write to digital outputs

To write to an output module, you specify a peripheral output address as the destination (**OUT**) of the MOVE operation. Digital outputs may be addressed as peripheral output byte (PQB), word (PQW), and double word (PQD). In the example above, for instance, the hexadecimal value of F0FF is written directly to peripheral output byte 34 (PQB34)—thereby immediately updating the output devices normally controlled by the output image byte QB34.

Each analog output channel is represented in a 16-bit word, and is therefore specified as a peripheral output word (PQW). To access a particular analog output, you simply specify the starting byte address of the analog output channel. In the following example the value currently contained in the bit memory word MW44 is always transferred out to the analog output channel PQW260.

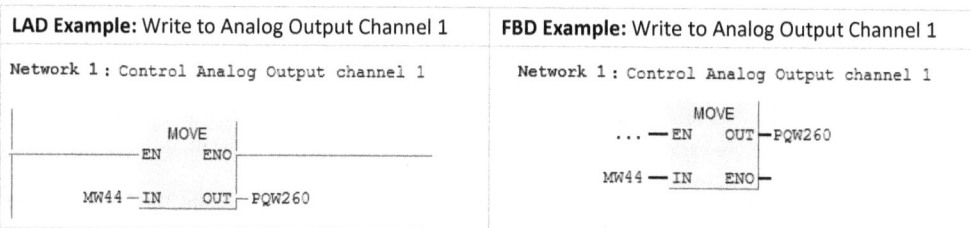

LAD Example: Write to Analog Output Channel 1	FBD Example: Write to Analog Output Channel 1
Network 1: Control Analog Output channel 1	Network 1: Control Analog Output channel 1

Figure 5-11. MOVE operation to access peripheral output memory to write to an analog output

Moving Local Memory (L)

Local memory may be specified as the source **IN** or the destination **OUT**. You may specify temporary variable names or absolute address locations as byte (LB), word (LW), or double word (LD) locations. Recall that local memory is temporary and only available while the block in which the local memory is allocated is executing. Moving a local value might be used to pass a temporary value to bit memory or to a data block location so that another block may have access to the value before it is overwritten. In the example below, the local variable #ERR_BYTE is always transferred to data location DB10.DBW2.

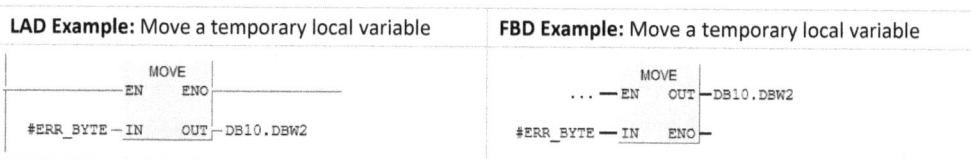

LAD Example: Move a temporary local variable	FBD Example: Move a temporary local variable

Figure 5-12. MOVE operation to copy Local Memory

Moving a Counter Value (Cn)

The MOVE operation can be used to transfer the current counter value to a new location. Moving a counter value to another location may be required, for example, prior to comparing the count to an intermediate counter value. A counter operand may be specified only as the source IN of the MOVE operation. In the following example, whenever the RLO = 1 at EN, the current counter value of counter C24 is transferred to bit memory word MW62 as a decimal integer.

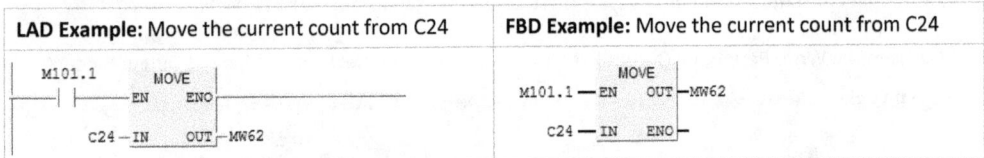

LAD Example: Move the current count from C24	FBD Example: Move the current count from C24
| ```
  M101.1        MOVE
———| |————EN    ENO——————————————————
         C24—IN    OUT⌐MW62
``` | ```
                MOVE
    M101.1 — EN    OUT—MW62
    C24 — IN    ENO⌐
``` |

Figure 5-13. MOVE operation to copy the current counter value.

Moving a Timer Value (Tn)

The MOVE operation can be used to transfer the remaining time of a timer to a new location. Moving the remaining time to another location, for example, may be required prior to comparing the value to an intermediate set point. A timer operand may be specified only as the source (IN) of the MOVE operation. In the following example, whenever the RLO = 1 at EN, the remaining time of timer T5 is transferred to bit memory word MW34 as a decimal integer.

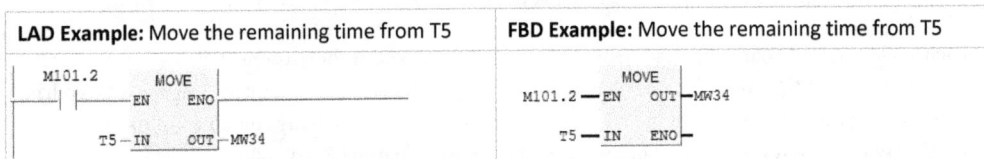

LAD Example: Move the remaining time from T5	FBD Example: Move the remaining time from T5
| ```
  M101.2        MOVE
———| |————EN    ENO——————
         T5—IN    OUT⌐MW34
``` | ```
                MOVE
    M101.2 — EN    OUT—MW34
    T5 — IN    ENO⌐
``` |

Figure 5-14. MOVE operation to copy the remaining time value

Moving Data Elements (DB)

The data elements of a data block may be specified as the source IN or the destination OUT of the **MOVE** operation. You may specify byte (DBB), word (DBW), or double word (DBD) locations.

When moving data elements in LAD/FBD, the data element may be addressed using the fully-addressed method or partially-addressed method. With the partially-addressed method, you must use the OPN DB instruction in a network prior to using the MOVE operation. With fully-addressed operands, as illustrated in the following example, you must specify the data block number as part of the byte, word, or double word address. The open data block operation is not required. Examples of this method include DB22.DBB10 (byte), DB22.DBW20 (word), or DB22.DBD12 (double word).

In the following example, when the RLO = 1 at EN, the contents of data word 20 of DB20 is moved to data word 20 of DB30. See Chapter 2 *Data Access with Partial Addressing* for an example of moving data elements with partial addressing.

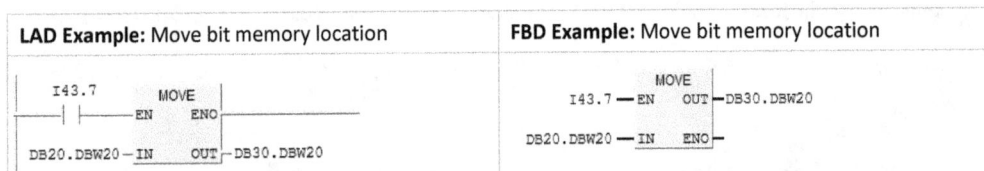

LAD Example: Move bit memory location	FBD Example: Move bit memory location
| ```
   I43.7         MOVE
———| |————EN    ENO——————
 DB20.DBW20—IN    OUT⌐DB30.DBW20
``` | ```
                MOVE
    I43.7 — EN    OUT—DB30.DBW20
  DB20.DBW20 — IN    ENO⌐
``` |

Figure 5-15. MOVE operation to copy Bit Memory

Moving a Pointer Value

The MOVE operation may be used to move both the area-internal and the area-crossing pointer. Since both of these pointer formats occupy 32-bits, you may specify a pointer constant of either format or a 32-bit memory location as the source of the MOVE operation. For the MOVE destination you must specify a 32-bit memory location or variable. The constant notation of the area-internal pointer is P#byte.bit, and for the area-crossing pointer is P#Abyte.bit—both formats are used with register-indirect addressing. The area-internal pointer may also be used with memory-indirect addressing.

In the first of the examples below, the pointer value of byte-12, bit-2 is moved to the pointer location MD24. The constant notation P#byte.bit is used to specify an area-internal pointer. In the second of the examples below, the area-crossing pointer of byte-16, bit-2 is moved to the pointer location MD24. The memory area 'Q' for output image table is encoded in MD24 as binary 010 in bits 26, 25, and 24 respectively. The constant notation P#Abyte.bit for the double word pointer is used to specify an area-crossing pointer.

LAD Example: Move area-internal pointer P#12.2	FBD Example: Move area-internal pointer P#12.2
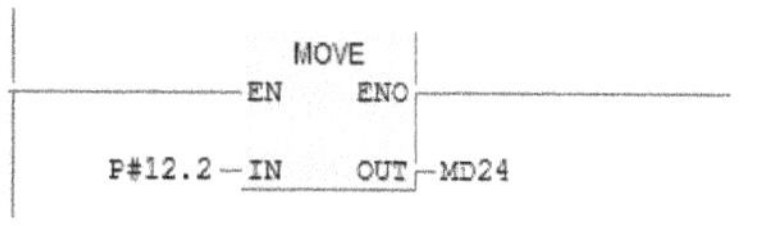 MOVE — EN ENO — P#12.2 — IN OUT — MD24	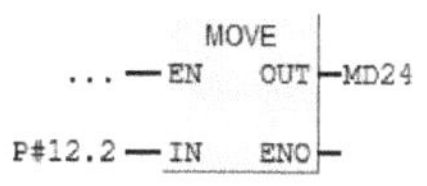 MOVE … — EN OUT — MD24 P#12.2 — IN ENO —
LAD Example: Move area-crossing pointer P#Q16.2	FBD Example: Move area-crossing pointer P#Q16.2
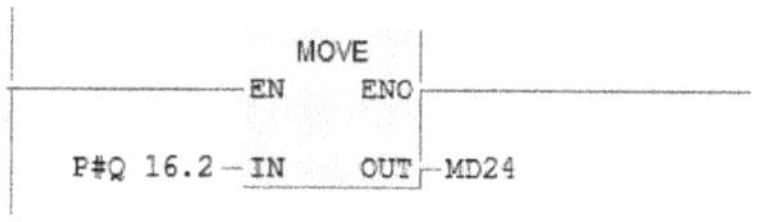 MOVE — EN ENO — P#Q 16.2 — IN OUT — MD24	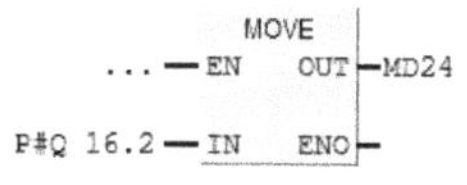 MOVE … — EN OUT — MD24 P#Q 16.2 — IN ENO —

Figure 5-16. MOVE operation to move with pointer

Chapter 6

LAD/FBD
Timer Operations

In this chapter you'll be introduced to LAD/FBD Timer Operations. In LAD and in FBD, each timer type is available as a box function and as a simple instruction. The box timer type and the simple timer type each have its own advantages.

Although timers have many program applications, they are often used to enable or disable devices or process events after a timed delay. Timers might also be used to determine whether a device or process fault has occurred based on whether the timer has exceeded the preset delay, or whether the timer has expired prematurely.

In addition to the STEP 7 timer instructions covered in this chapter, standard IEC timer functions are reviewed in Chapter 28 *Developing Functions and Function Blocks*. These callable Function Block operations, which include SFB 3 (TP—Pulse Generation), SFB 4 (TON—Timer On-Delay), and SFB 5 (TOF—Timer Off-Delay), are available in the STEP 7 library, and are integrated in some S7 CPUs.

Main Topics

- Box Timers Overview
- Box Timer Descriptions
- Programming Box Timers in LAD
- Programming Box Timers in FBD
- Simple Timers Overview
- Simple Timer Descriptions
- Programming Simple Timers in LAD
- Programming Simple Timers in FBD

BOX TIMER OPERATIONS OVERVIEW

STEP 7 timers, are based on the STEP 5 implementation of timers, and as such are referred to as S5 Timers. The timer operations set, listed in Table 6-1, includes five timer types—*on-delay, off-delay, stored on-delay, pulse,* and *extended pulse.* Each timer is implemented as a box function, as shown in Table 6-1; and as a simple timer, as shown later in Table 6-3. Box timers support the full functionality of the timer, allowing each timer type to be loaded with a preset value, started, reset, and monitored.

As mentioned in the introduction, you can learn about standard IEC timer functions that you also have access to in your program—these operations are reviewed in Chapter 28 *Developing Functions and Function Blocks.*

Table 6-1. LAD/FBD Box Timer Operations Summary

LAD	FBD	Brief Description
??? S_ODT —S Q— —TV BI— —R BCD—	??? S_ODT —S BI— —TV BCD— —R Q—	**ON-Delay.** If the start input **S** transitions from 0 -to- 1, the on-delay timer starts and will continue to time while the **S** input is '1'. While the Start input remains '1', the timer runs until the preset time delay expires. When the preset time expires, the output **Q** is activated and remains activated until the timer is reset or the RLO = 0 at the start(**S**) input or the RLO = 1 at the reset (**R**) input.
??? S_OFFDT —S Q— —TV BI— —R BCD—	??? S_OFFDT —S BI— —TV BCD— —R Q—	**OFF-Delay.** If the start input **S** transitions from 0 -to- 1, the off-delay output **Q** is immediately activated. If the Start input **S** transitions from 1-to- 0, the timer starts timing and runs until the time delay expires, unless reset (**R**). After the timed delay expires, the timer output **Q** is de-activated.
??? S_ODTS —S Q— —TV BI— —R BCD—	??? S_ODTS —S BI— —TV BCD— —R Q—	**ON-Delay Stored.** If the Start input **S** transitions from 0 -to- 1, the retentive on-delay timer starts and continues to time regardless of the Start signal **S**. If **S** input transitions to '0' and back to '1' before the timer expires, the timer restarts. The output **Q** is activated after the preset time delay expires, and remains activated, regardless of the start input, and until the RLO = 1 at the reset (**R**) input.
??? S_PULSE —S Q— —TV BI— —R BCD—	??? S_PULSE —S BI— —TV BCD— —R Q—	**Pulse.** If the Start input **S** transitions from 0 -to- 1, the pulse timer starts and continues to time for as long as the **S** input is '1', or until the time delay expires. While the timer is timing, output **Q** is activated for as long as the Start input **S** is activated. If the **S** input returns to '0' before the preset time delay expires, the timer is stopped and the output **Q** is de-activated.
??? S_PEXT —S Q— —TV BI— —R BCD—	??? S_PEXT —S BI— —TV BCD— —R Q—	**Extended Pulse.** If the Start input **S** transitions from 0 -to- 1, the extended pulse timer starts and continues to time until the programmed delay expires, even if the **S** input returns to '0'. The output **Q** is activated only while the timer is running. If the RLO = 1 at the reset input (**R**), the timer is reset.

BOX TIMER DESCRIPTIONS

Each of the five box timers operates slightly different and is identified by a unique mnemonic label at the top of the timer box. The mnemonic for the *on-delay timer* is **S_ODT**; for an *off-delay timer,* **S_OFFDT**; for a *retentive on-delay timer,* also called a *stored on-delay timer,* **S_ODTS**; for a *pulse timer,* **S_PULSE**; and for an *extended pulse timer,* **S_PEXT**. Operation of each of the timers is described in the following discussions. First let's take a look at the parameters that must be specified as you program any box timer.

Box Timer Parameters

As seen in the following table, a box timer has a parameter where you will specify the timer address—for example **T0**, and **T1**; a start (**S**) line, where you will insert control logic that starts the timer; **TV**, where you must specify a preset value; and a reset (**R**) logic line, where you may insert logic to reset the timer. On the output side is the timer status signal (**Q**), which can be used to check the running status of the timer; and **BI** and **BCD**, both of which allow you to specify a word location where the remaining time is kept. The remaining time is output in decimal at the location you specify at BI and in BCD at the location you specify at BCD. Not all timer parameters must be specified.

Table 6-2. LAD/FBD Timer box parameters—Note that entry of all timer parameters is not required

Parmeter	Data Type	Description	Required?	Box Timer	
				LAD	FBD
???		Timer Address < Tn >	YES	???	???
S		Timer Start impulse input line	YES	S_ODT	S_ODT
TV	S5TIME	Timer Preset Value	YES	—S Q—	—S BI—
R	BOOL	Timer Reset logic input line	NO	—TV BI—	—TV BCD—
Q	BOOL	Timer Status signal	NO	—R BCD—	—R Q—
BI	WORD	Remaining Time in Decimal	NO		
BCD	WORD	Remaining Time in BCD	NO		

You may specify the preset time TV as a constant, in S5TIME format, where the range is S5T#0h_0m_0s_10ms to S5T#2h_46m_30s_0ms, and **h = hour, m = minutes, s = seconds,** and **ms = milliseconds**. You only need to specify the required units. You may also reference an S5TIME variable, or a word-width memory location that contains the preset—for example MW 102. Operation of the status signal Q varies based on the timer type. You may connect Q to a coil/output referenced with any unused bit address. The status Q can then be checked using a NO or NC contact (LAD), or binary inputs in FBD. You may specify BI and BCD as WORD variables, or as word-width memory locations.

ON-DELAY Timer: S_ODT

If the start-input **S** transitions from 0-to-1, the *on-delay timer* starts timing; if the start-input remains at '1', the timer times until the preset time value **TV** expires—at which time the status signal **Q** is activated (1). After the timer is expired, the status signal **Q** remains '1' while the start-input **S** is '1', and is de-activated (0) if the start-input **S** returns to '0'. If the start-input **S** transitions from 1-to-0 while the timer is running and prior to reaching the programmed duration, the timer will stop.

The remaining time, output at **BI** or **BCD**, is maintained until the timer is reset or restarted. Whenever the RLO = 1 on the reset line **R**, the remaining time at **BI** (decimal) and at **BCD** (binary coded decimal), and the status signal **Q** are all reset. Figure 6-2 illustrates a timing diagram of the on-delay box timer.

In the example below, the ON-delay timer T24 starts timing with a preset value of 500 milliseconds when input I 5.2 **OR** input I 5.3 transitions from 0-to-1—the timer is reset whenever I 5.4 is '1'. The remaining time is output as a decimal value in MW60 and as a BCD value in MW62.

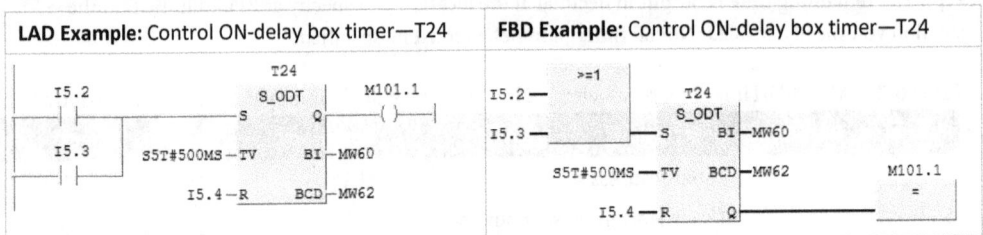

LAD Example: Control ON-delay box timer—T24	**FBD Example:** Control ON-delay box timer—T24

Figure 6-1. Illustration of LAD/FBD ON-delay timer box operation

The behavior of the on-delay timer is illustrated in the following timing diagram.

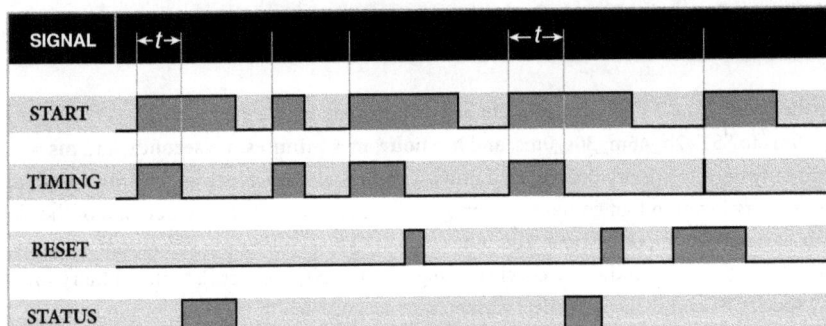

Figure 6-2. Timing diagram for the ON-Delay timer

OFF-DELAY Timer: S_OFFDT

If the start-input **S** transitions from 0-to-1, the output status **Q** of the *off-delay timer* is activated (1). If the start-input **S** transitions from 1-to-0, the timer starts and times while the start-input is '0' and until the preset time value **TV** expires. When the off-delay expires, the status **Q** is de-activated (0). The status signal **Q** is '1' while the Start-input is '1'or while the timer is timing. If the start-input **S** transitions from 0-to-1 while the timer is timing, the timer is reset and the status **Q** is activated (1).

The remaining time, output at **BI** or **BCD**, is maintained until the timer is reset or restarted. Only a new falling-edge transition of 1-to-0 at the Start-input **S** will restart the timer. Whenever the RLO = 1 on the reset line **R**, the remaining time at **BI** (decimal) and at **BCD** (binary coded decimal) is reset; the status signal **Q** is also reset. Figure 6-4 illustrates a timing diagram of the off-delay box timer.

In the example below, the OFF-delay timer T25 starts timing with a preset value of 55 seconds when the logic combination of I 6.2 AND I 6.3 transitions from logic 1-to-0, and is reset whenever I 6.4 is logic 1. The remaining time is output as a decimal value in MW64 and as a BCD value in MW66.

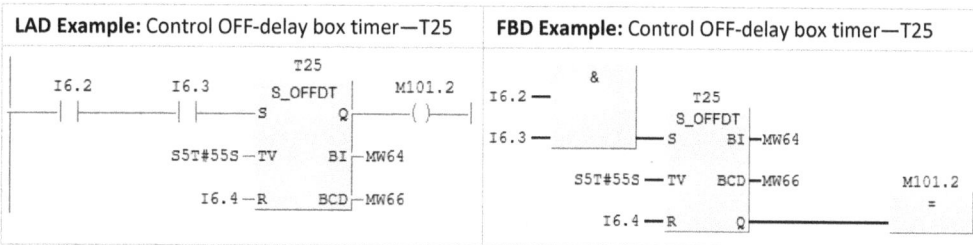

LAD Example: Control OFF-delay box timer—T25	**FBD Example:** Control OFF-delay box timer—T25

Figure 6-3. Illustration of LAD/FBD OFF-delay box timer operation

The behavior of the off-delay timer is illustrated in the following timing diagram.

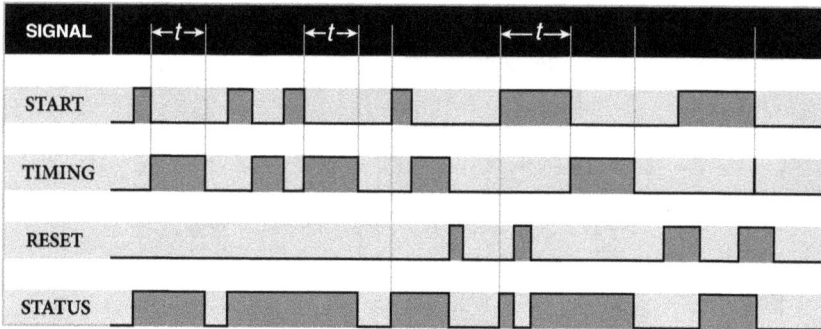

Figure 6-4. Timing diagram for the OFF-Delay timer

STORED ON-DELAY Timer: S_ODTS

If the start-input **S** transitions from 0-to-1, the *stored on-delay timer* starts and times until the preset time value **TV** expires, even if the start-input **S** transitions from '1' to '0' before the timer expires. The status signal **Q** outputs '1'when the time delay expires and remains at '1' until the timer is reset. If the start-input **S** transition from 1-to-0, while the timer is timing, and again from 0-to-1, the timer is restarted with the initial preset delay value and continues until the preset delay expires.

While timing and until reset, the remaining time is output at the location specified at **BI** and at **BCD**. Whenever the RLO = 1 on the reset line **R**, the remaining time at **BI** (decimal) and at **BCD** (binary coded decimal) is reset; the status signal **Q** is also reset. Figure 6-6 illustrates a timing diagram of the stored on-delay box timer.

In the example below, the ON-delay stored timer T26 starts timing with a preset value of 25 seconds when the logic combination of I 7.2 OR I 7.3 transitions from logic 0 to 1, and is reset whenever I 7.4 is logic 1. The remaining time is output as a decimal value in MW68 and as a BCD value in MW70.

LAD Example: Stored ON-delay box timer—T26	**FBD Example:** Stored ON-delay box timer—T26

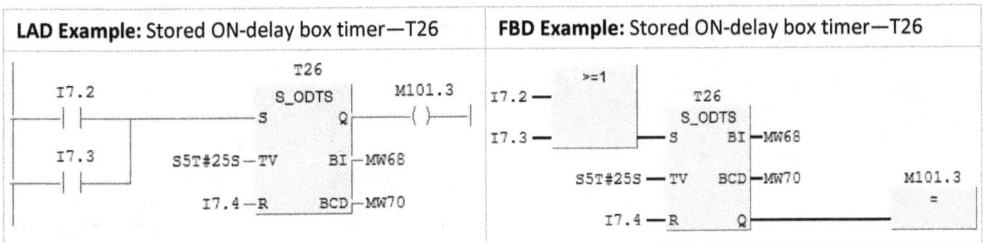

Figure 6-5. Illustration of a LAD/FBD Stored ON-delay timer box operation

The behavior of the stored on-delay timer is illustrated in the following timing diagram.

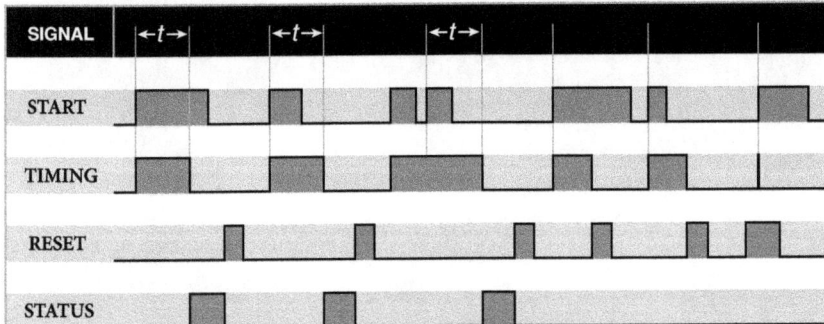

Figure 6-6. Timing diagram for the Stored ON-Delay timer

PULSE Timer: S_PULSE

If the start-input **S** transitions from 0-to-1, the *pulse timer* starts and continues for as long as the start-input is '1', or until the preset time value **TV** expires. After the start input transitions to '1', the status signal **Q** will output '1'for as long as the start-input **S** is '1' (pulse output) and the time delay is not expired. If the start-input **S** transitions from 1-to-0, while the timer is timing, the timer stops and the status signal **Q** returns to '0'.

The remaining time, output at **BI** or **BCD**, is maintained until the timer is reset or restarted. Whenever the RLO = 1 on the reset line **R**, the remaining time at **BI** (decimal) and at **BCD** (binary coded decimal) is reset; the status signal **Q** is also reset. Figure 6-8 illustrates a timing diagram of the pulse box timer.

In the example below, the PULSE timer T27 starts timing with a preset value of 5 seconds when I 8.2 transitions from 0-to-1, and is reset whenever the I 8.4 is '1'. The PULSE timer will run and M101.4 will be '1' as long as the RLO = 1 at I 8.2, or until the preset time expires. The remaining time is output as a decimal value in MW72 and as a BCD value in MW74.

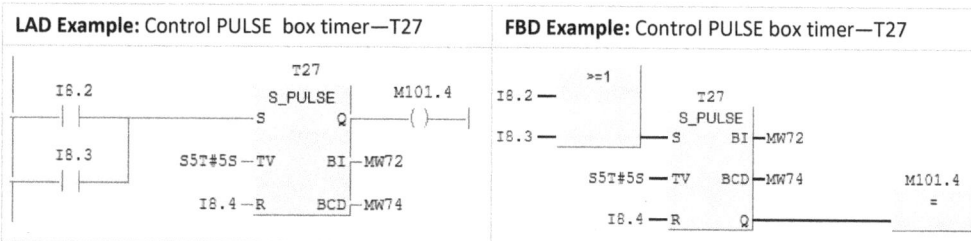

LAD Example: Control PULSE box timer—T27	FBD Example: Control PULSE box timer—T27

```
              T27
  I8.2       S_PULSE    M101.4
  --| |----S        Q---( )---|
  I8.3
  --| |---- S5T#5S--TV   BI --MW72

         I8.4--R    BCD --MW74
```

```
            >=1
  I8.2---           T27
                  S_PULSE
  I8.3---        S      BI--MW72
         S5T#5S --TV   BCD--MW74    M101.4
         I8.4 --R   Q                 =
```

Figure 6-7. Illustration of LAD/FBD Pulse box timer operation

The behavior of the pulse timer is illustrated in the following timing diagram.

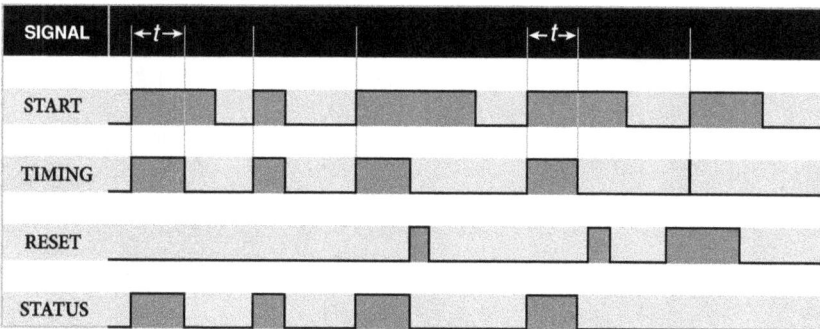

Figure 6-8. Timing diagram for the Pulse timer

EXTENDED PULSE Timer: S_PEXT

If the start-input **S** transitions from 0-to-1, the *extended pulse timer* starts and continues until the preset time value **TV** expires, regardless of any signal change at the start-input (extended pulse). After the start-input **S** transitions from 0-to-1, the status signal **Q** outputs '1' until the time delay expires. If the **S** input transitions again from 0-to-1, while the timer is still timing, the timer will restart and continue until the programmed delay expires.

Whenever the RLO = 1 on the reset line **R**, the remaining time at **BI** (decimal) and at **BCD** (binary coded decimal) is reset; the status signal **Q** is also reset. Figure 6-10 illustrates a timing diagram of the extended pulse box timer.

In the example below, the EXTENDED PULSE timer T28 starts timing with a preset of 10 seconds when I 9.2 OR I 9.3 transitions from 0-to-1, and is reset whenever I 9.4 is '1'. The EXTENDED PULSE timer will run, and M101.5 will be '1', until the preset time expires, even if the RLO at the **S** input transitions from '1' to '0'. The remaining time is output as a decimal value in MW76 and as a BCD value in MW78.

LAD Example: EXTENDED PULSE box timer—T28	FBD Example: EXTENDED PULSE box timer—T28

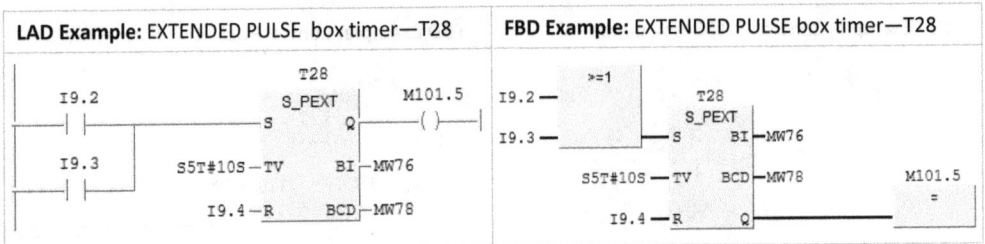

Figure 6-9. Illustration of LAD/FBD Extended Pulse box timer operation

The behavior of the extended pulse timer is illustrated in the following timing diagram.

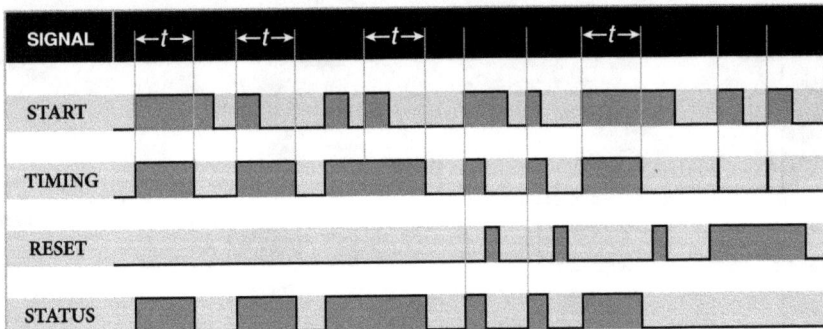

Figure 6-10. Timing diagram for the Extended Pulse timer

PROGRAMMING BOX TIMERS IN LAD

After inserting a new network, programming a timer is first a matter of selecting the required timer type from the Program Elements tab, based on the name at the inside top of the box, and then dropping it onto the LAD network. For example, **S_ODT** is an on-delay timer.

Selecting and Addressing a Box Timer

With the new network inserted and selected, you can double-click on the desired timer object in the Program Elements folder, or you can drag and drop the object onto the network. Next, enter the timer address—for example **T0, T1**, or **T2**, just above the timer element. You may also use a symbolic address—for example **Horn_Delay**, if it is already defined. Each timer you specify references a single word location in the S7 timer memory.

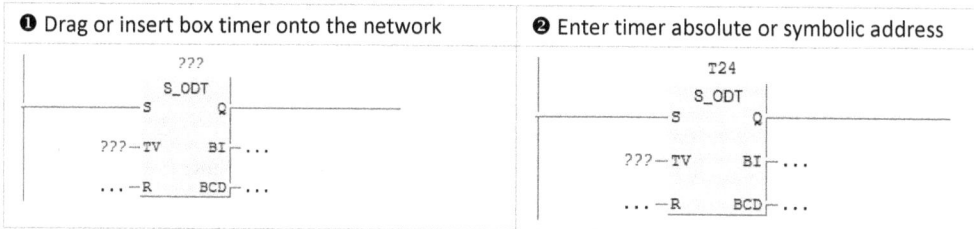

❶ Drag or insert box timer onto the network	❷ Enter timer absolute or symbolic address

Figure 6-11. Inserting and addressing a box timer in LAD

Setting, Starting, and Resetting a Box Timer

Each timer is started based on the timer type, and the start logic you program at the **S** input. Timers, except for the off-delay, start timing on a 0-to-1 transition of the RLO. The off-delay is only enabled on an RLO transition from 0-to-1, and starts timing on a 1-to-0 transition of the RLO. With the timer inserted, click on the **S** line and enter the logic—for example a parallel circuit ORing I5.2 with I5.3.

Click on the **TV** field and specify an S5TIME constant, where the range is from S5T#0h_0m_0s_10ms to S5T#2h_46m_30s_0ms (h = hour, m = minutes, s = seconds, and ms = milliseconds). Specify only the units of the preset that are required. **TV** may also reference a declared variable of type S5TIME, or a word-width memory location from which the value will be loaded—for example MW 102, or IW 10. The preset time specified below is 500 milliseconds.

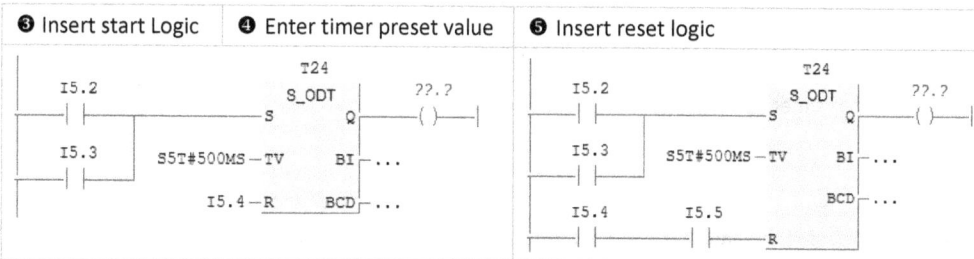

❸ Insert start Logic	❹ Enter timer preset value	❺ Insert reset logic

Figure 6-12. Inserting the start logic, preset value, and reset logic for a box timer in LAD

Although each timer is different in what causes it to start and to continue timing, the reset line **R** affects all timers in the same way. Whenever the RLO = 1 on the reset line **R,** both the remaining time values (**BI** and **BCD**), and the timer status **Q** are reset to zero. Click on the **R** line and enter the required driving logic. In the figure shown above, **T24** is reset whenever **I** 5.4 is at logic 1. If more than a single binary input will control the reset line, then simply enter the binary logic circuit using the desired series/parallel combination as shown in the second cell above.

Evaluating Box Timer Remaining Time

When writing code to check the timer value, remember that S5 timers time down, and that the remaining time is output in binary (the equivalent of decimal) at the location you specify at **BI**; and in binary coded decimal, at the location you specify at **BCD**. You do not have to specify either of these output locations, but either one or both may be entered. You would use the **BCD** output parameter, for instance **QW62** in the example below, to direct the remaining time to a BCD display device or to compare to a BCD setpoint. You might also use these output parameters in a compare operation to check the remaining time at intermediate checkpoints.

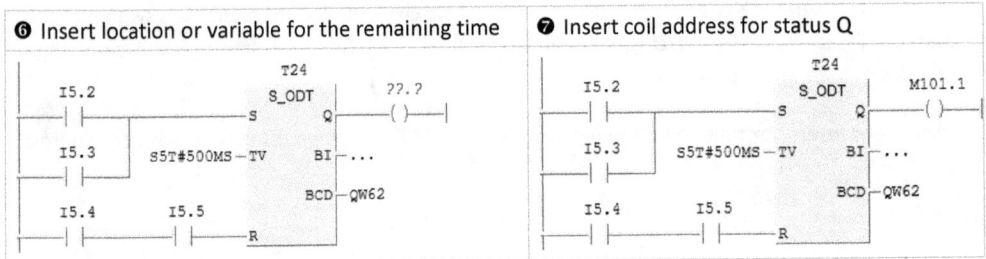

Figure 6-13. Inserting logic to monitor the remaining time and status of a box timer in FBD

You may also access the remaining time directly by referencing the timer operand (timer address) as the source location in a MOVE operation, as shown below. The remaining time is available as a decimal value, the same as would be taken from the **BI** output of the timer box function.

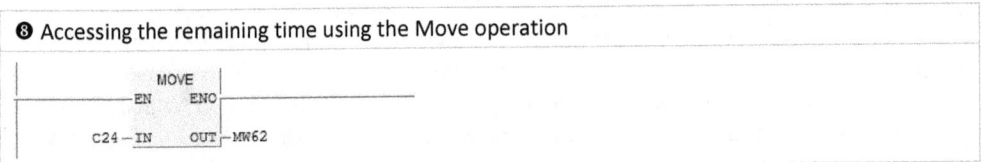

Figure 6-14. Alternate method for accessing the remaining time in LAD

Evaluating Box Timer Status

Finally, the status signal **Q**, in conjunction with the start signal **S**, may be checked to determine whether the timer is enabled or done timing. You can examine the timer status in your program using normally-open and normally-closed contacts that reference the address of **Q**. Remember that the exact behavior of the status signal **Q** varies according to the timer type.

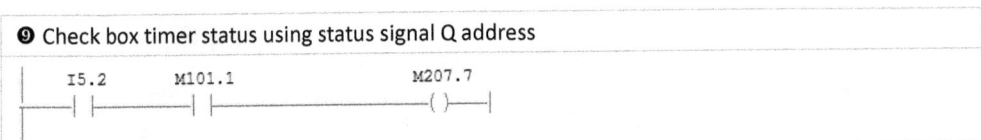

Figure 6-15. Using the status signal Q to monitor the status of a box timer in LAD

The timer address operand (address) can also be used in bit operations to check the timer's running status as shown below.

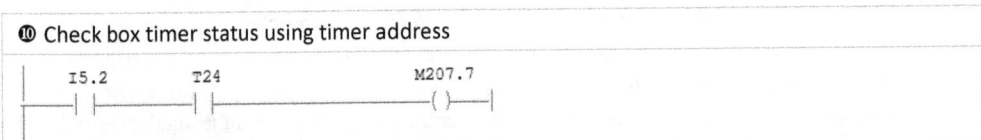

Figure 6-16. Using the timer address to monitor the status of a box timer in LAD

PROGRAMMING BOX TIMERS IN FBD

After inserting a new network, programming a timer is first a matter of selecting the required timer type from the Program Elements tab, based on the name at the inside top of the box, and then dropping it into the FBD network frame. For example, **S_ODT** is an on-delay timer.

Selecting and Addressing a Box Timer

With the new network inserted and selected, you can double-click on the desired timer object in the Program Elements folder, or you can drag and drop the object into the network frame. Next, enter the timer address—for example **T0, T1,** or **T2,** just above the timer element. You may also use a symbolic address—for example **Horn_Delay,** if it is already defined. Each timer you specify references a single word location in the S7 timer memory.

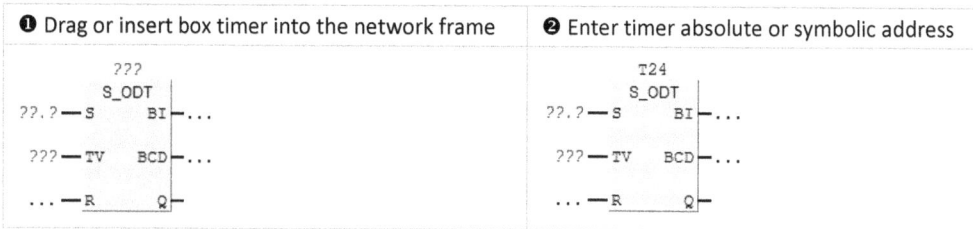

❶ Drag or insert box timer into the network frame	❷ Enter timer absolute or symbolic address
??? S_ODT ??.? — S BI ⊢... ??? — TV BCD ⊢... ... — R Q ⊢	T24 S_ODT ??.? — S BI ⊢... ??? — TV BCD ⊢... ... — R Q ⊢

Figure 6-17. Inserting and addressing a box timer in FBD

Setting, Starting, and Resetting a Box Timer

Each timer is started based on the timer type, and by the start logic you program at the **S** input. Timers, except for the off-delay, start timing on a 0-to-1 transition of the RLO. The off-delay is only enabled on an RLO transition from 0-to-1, and starts timing on a 1-to-0 transition of the RLO. With the timer inserted, click on the **S** line and enter the driving logic—for example, click on the toolbar and insert an OR gate or AND gate, add additional binary inputs if required, and assign the addresses.

Click on the TV field and specify a constant using the S5TIME format, where the range is from S5T#0h_0m_0s_10ms to S5T#2h_46m_30s_0ms (h = hour, m = minutes, s = seconds, and ms = milliseconds). Specify only the units of the preset that are required. **TV** may also reference a declared variable of type S5TIME, or a word-width memory location from which the value will be loaded—for example MW 102, or IW 10. The preset time specified below is 500 milliseconds.

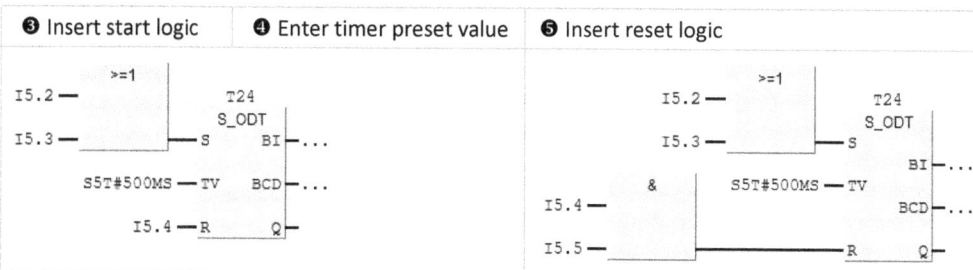

❸ Insert start logic ❹ Enter timer preset value	❺ Insert reset logic

I5.2 — ⎤ >=1
 I5.3 — ⎦ — S   T24
         S_ODT
 S5T#500MS — TV   BI ⊢...
        I5.4 — R   BCD ⊢...
                Q ⊢
	I5.2 — ⎤ >=1
	 I5.3 — ⎦ — S T24
	             S_ODT
	 I5.4 — ⎤ & S5T#500MS — TV BI ⊢...
	 I5.5 — ⎦ BCD ⊢...
	                  R Q ⊢

Figure 6-18. Inserting the start logic, preset value, and reset logic for a box timer in FBD

Although each timer is different in what causes it to start and to continue timing, the reset line **R** affects all timers in the same way. Whenever the RLO = 1 on the reset line **R,** both the remaining time values (**BI** and **BCD**), and the timer status **Q** are reset to zero. Click on the **R** line and enter the required driving logic. In the first cell shown above, **T24** is reset whenever **I** 5.4 is at logic 1. If additional combination logic is required, simply click on the toolbar OR gate or AND gate, add additional binary inputs if required, and assign the addresses as shown in the second cell above.

Evaluating Box Timer Remaining Time

When writing code to check the timer value, remember that S5 timers time down, and that the remaining time is output in binary (the equivalent of decimal) at the location you specify at **BI**; and in binary coded decimal, at the location you specify at **BCD**. You do not have to specify either of these output locations, but either one or both may be entered. You would use the **BCD** output parameter, for instance **QW62** in the example below, to direct the remaining time to a BCD display device or to compare to a BCD setpoint. You might also use these output parameters in a compare operation to check the time at intermediate checkpoints.

❻ Insert location or variable for remaining time	❼ Insert output address for status Q

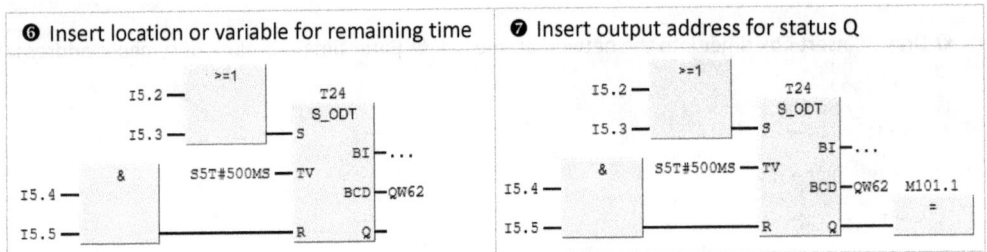

Figure 6-19. Inserting logic to monitor the remaining time and status of a box timer in FBD

You may also access the remaining time directly by referencing the timer operand (timer address) as the source location in a Move operation, as shown below. The remaining time is available as a decimal value, the same as would be taken from the **BI** output of the box timer function.

❽ Accessing the remaining time using the Move operation

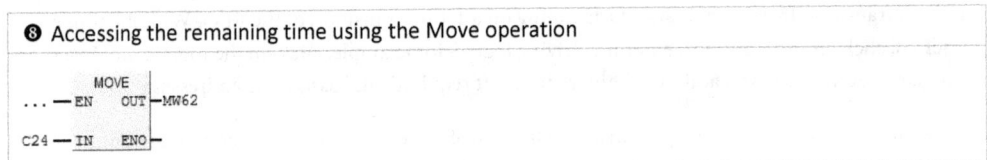

Figure 6-20. Alternate method for accessing the remaining time in FBD

Evaluating Box Timer Status

Finally, the status signal **Q**, in conjunction with the start signal **S**, may be checked to determine whether the timer is enabled or done timing. Click on the Q line, and then on the toolbar click the Assignment element, and then enter the output address. You can examine the timer status in your program using binary inputs or negated binary inputs that reference the address of **Q**. Remember that the exact behavior of the status signal **Q** varies according to the timer type.

❾ Check timer status using status signal Q address	❿ Check timer status using timer address

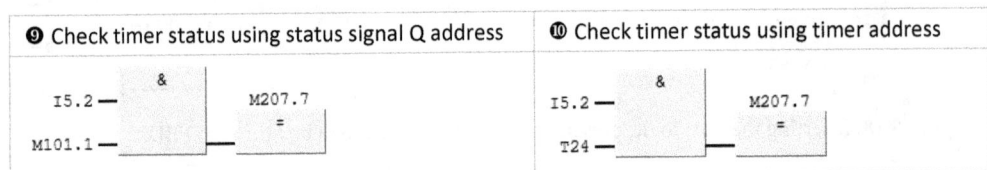

Figure 6-21. Using status signal Q or the timer address to examine the status of a box timer in FBD

As shown in the second cell above, the timer operand (address) can also be used in bit operations to check the timer's running status.

SIMPLE TIMER OPERATIONS OVERVIEW

STEP 7 timer instructions also include the five S5 timer types implemented as simple timer instructions. In LAD these timers are referred to as timer coils; in FBD as timer outputs. A simple timer provides the same time delay action as its box timer counterpart; however simple timers only implement the start timer and load preset functions of the box timer. You will need separate instructions to reset a simple timer, to monitor the remaining time, and to check the timer status to see if the timer is done timing.

Table 6-3. LAD/FBD Simple Timer Operations Summary

LAD	FBD	Brief Description
??? —(SD)— ???	??? SD ??.? — ??? — TV	**ON Delay.** Start the specified ON-Delay timer with the specified preset value. The timer address is Tn—for example, T0 -to- T255.
??? —(SF)— ???	??? SF ??.? — ??? — TV	**OFF Delay.** Start the specified OFF-Delay timer with the specified preset value. The timer address is Tn—for example, T0 -to- T255.
??? —(SS)— ???	??? SS ??.? — ??? — TV	**Retentive ON-Delay.** Start the specified Retentive ON-Delay timer with the specified preset. The timer address is Tn—for example, T0 -to- T255.
??? —(SP)— ???	??? SP ??.? — ??? — TV	**Pulse.** Start the specified Pulse timer with the specified preset value. The timer address is Tn—for example, T0 -to- T255.
??? —(SE)— ???	??? SE ??.? — ??? — TV	**Extended Pulse.** Start the specified Extended Pulse timer with the specified preset value. The timer address is Tn—for example, T0 -to- T255.

SIMPLE TIMER DESCRIPTIONS

Each of the five simple timers operates slightly different and is identified by a unique mnemonic label inside the coil/output element. The mnemonic for the simple *on-delay timer* is **SD**; for an *off-delay timer*, **SF**; for a *retentive on-delay timer*, also called a *stored on-delay timer*, **SS**; for a *pulse timer*, **SP**; and for an *extended pulse timer*, **SE**.

The simple timer address **Tn** is specified above the timer element. For example, a CPU may have timers T0 to T255. The timer preset is entered below the element, and, like with the box timer, is specified as a constant using the S5TIME format. In this format the range is S5T#0h_0m_0s_10ms to S5T#2h_46m_30s_0ms, and h = hour, m = minutes, s = seconds, and ms = milliseconds. You only need to specify the required units. You may also reference a variable of type S5TIME, or a word-width memory location (e.g., MW 102) that contains the preset. Each of the simple timer operations is described in the following discussions.

ON-DELAY Timer: SD

If the driving logic transitions from '0' to '1', the *on-delay timer coil* starts and times until the preset value specified below the timer element expires. The on-delay timer only continues while the driving logic remains at '1'. If the driving logic remains satisfied (RLO = 1), a check of the timer status produces a logic 1 after the time delay expires (on-delay).

If the driving logic transitions from '1' to '0' while the timer is timing, the timer will stop; a check of the on-delay timer status produces '0'.

In the following example when the logic combination of M 22.3 AND NOT Q 3.7 transitions from '0' to '1', then the ON-Delay timer T33 is started with a preset value of 1 minute and 30 seconds.

LAD Example: Simple ON-delay timer—T33	FBD Example: Simple ON-delay timer—T33

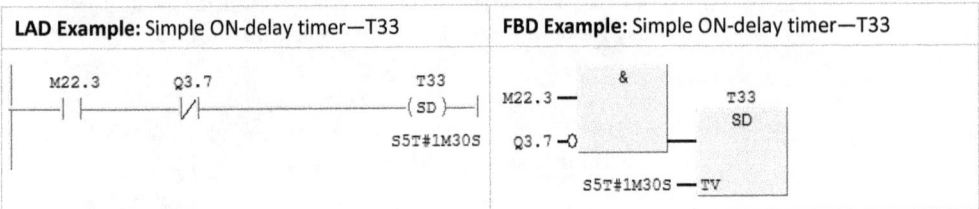

Figure 6-22. Illustration of LAD/FBD simple ON-delay timer operation

OFF DELAY Timer: SF

If the driving logic transitions from '1' to '0', the *off-delay timer coil* starts and times until the preset time value, specified below the timer element, expires. The timer only continues while the driving logic remains at '0'. A check of the status of an off-delay timer will produce '1' while the driving logic is at '1', or while the timer is running.

If the driving logic transitions from '0' to '1' while the timer is timing, the timer will stop; a check of the off-delay timer status produces '1'. Only a new falling-edge transition of the driving logic, from '1' to '0', will restart the timer.

In the following example when the logic combination of I 5.6 AND NOT Q 3.6 transitions from '1' to '0', then the OFF-Delay timer T34 is started with a preset value of 25 minutes.

LAD Example: Simple OFF-delay timer—T34	FBD Example: Simple OFF-delay timer—T34

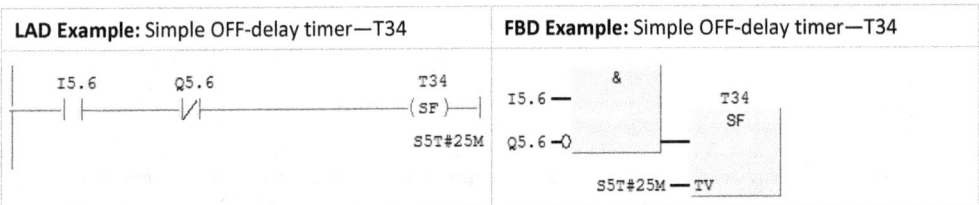

Figure 6-23. Illustration of LAD/FBD simple OFF-delay timer operation

STORED ON-Delay Timer: SS

If the driving logic transitions from '0' to '1', the *stored on-delay timer coil* starts and times until the preset value specified below the timer element expires—even if the driving logic transitions from '1' to '0' before the timer expires (retentive). A check of the retentive on-delay timer status will yield '1' after the time delay expires and until the timer is reset.

If the driving logic transitions from '1' to '0' while the timer is timing, and again from '0' to '1', the timer will restart and continue until the programmed delay expires.

In the following example when I 6.2 transitions from '0' to '1', then the Stored ON-Delay timer T35 is started with a preset value of 2 minutes and 30 seconds.

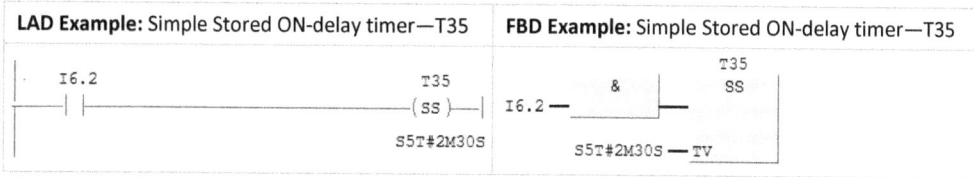

LAD Example: Simple Stored ON-delay timer—T35	FBD Example: Simple Stored ON-delay timer—T35

Figure 6-24. Illustration of LAD/FBD simple Stored ON-delay (Retentive) timer operation

PULSE Timer: SP

If the driving logic transitions from '0' to '1', the *pulse timer coil* starts and times until the preset value specified below the timer element expires. The timer only continues while the driving logic is at '1' (RLO = 1). For as long as the driving logic signal is '1', or until the time delay expires, a status check of the pulse timer signal will produce '1' (pulse output).

If the driving logic does transition from '1' to "0 before the timer expires, the timer stops and maintains its last value. A check of the timer status then produces '0'.

In the following example when I 6.4 transitions from '0' to '1', then the PULSE timer T36 is started with a preset value of 5 seconds.

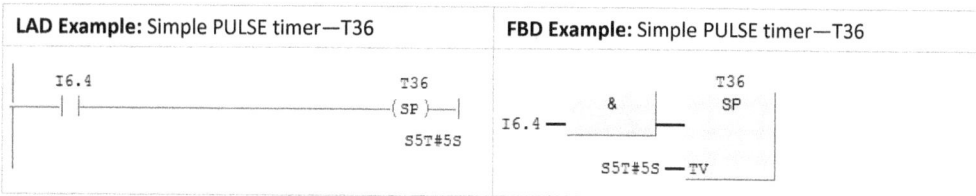

LAD Example: Simple PULSE timer—T36	FBD Example: Simple PULSE timer—T36

Figure 6-25. Illustration of LAD/FBD simple Pulse timer operation

EXTENDED PULSE Timer: SE

If the driving logic transitions from '0' to "1, the *extended pulse timer coil* starts and times until the preset value specified below the timer element expires—regardless of any change in the driving logic signal. From the time the driving logic transitions from '0' to '1', a status check of the extended pulse timer will produce '1' until the time delay expires (extended pulse).

If the driving logic does transition from '0' to '1' again, while the timer is still timing, the timer will restart, and continue until the programmed delay expires.

In the following example when the logic combination of I 6.5 OR I 6.6 transitions from '0' to '1', then the EXTENDED PULSE timer T37 is started with a preset value of 5 seconds.

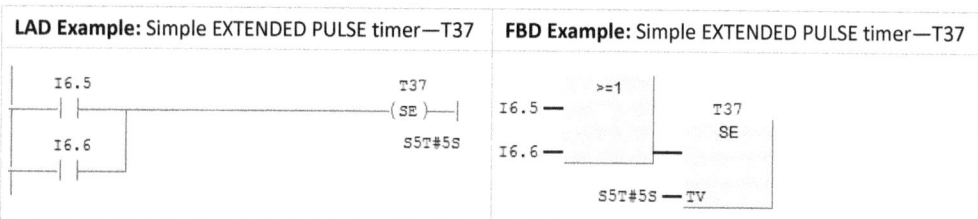

LAD Example: Simple EXTENDED PULSE timer—T37	FBD Example: Simple EXTENDED PULSE timer—T37

Figure 6-26. Illustration of LAD/FBD simple Extended Pulse timer operation

PROGRAMMING SIMPLE TIMERS IN LAD

After inserting a new network, programming a coil timer is first a matter of selecting the required timer from the Program Elements tab. The timer, identified by the mnemonic name inside a coil, can be dragged onto the new LAD network. For example, **SD** is an on-delay timer coil.

By using a simple timer, you are able to isolate the operations of a box timer. Loading a preset value, starting the timer, monitoring the remaining time value or status signal, and resetting the timer are all handled with separate operations. These operations are entered as separate networks and preferably in the sequence shown below.

- Specify timer preset value before starting the timer
- Reset the timer (R)
- Check timer status (binary check)
- Check remaining time value (in decimal), using MOVE and Compare
- Check remaining time (in BCD), using MOVE, integer-BCD conversion, and Compare

Selecting and Addressing a Simple Timer

With the new network inserted, you can double-click on the desired timer object in the Program Elements folder, or drag and drop the object onto the network. The timer address must be entered just above the timer coil, using the absolute address—for example **T0, T1, T2**. You may also use a symbolic address—for example **Sys_Delay**, if it is already defined. Each timer you specify references a single word in the S7 timer memory.

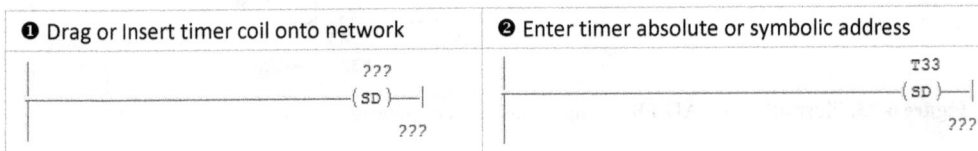

❶ Drag or Insert timer coil onto network	❷ Enter timer absolute or symbolic address
??? ———(SD)—| ???	T33 ———(SD)—| ???

Figure 6-27. Inserting and addressing a simple timer in LAD

Setting and Starting a Simple Timer

The timer coil is started based on the type of timer and based on the logic you insert to start the timer. All timer types, except for the off-delay timer, start timing when the RLO of the timer start logic transitions from 0-to-1. The off-delay starts timing on a 1-to-0 transition of the RLO. Click on the simple timer network and enter the required timer start logic.

The timer preset value, entered just below the timer coil, is specified as a constant using the S5TIME format, where the range is S5T#0h_0m_0s_10ms to S5T#2h_46m_30s_0ms. The preset units of h = hour, m = minutes, s = seconds, and ms = milliseconds are only specified if required. For example, the preset constant S5T#1m30s is 1 minute and 30 seconds. For the timer preset, you may also reference a declared variable of type S5TIME, or a word-width memory location.

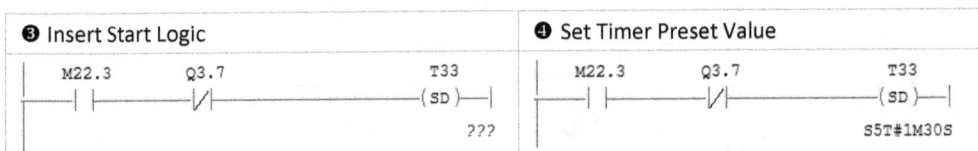

❸ Insert Start Logic	❹ Set Timer Preset Value
M22.3 Q3.7 T33 ——| |——|/|————————(SD)—| ???	M22.3 Q3.7 T33 ——| |——|/|————————(SD)—| S5T#1M30S

Figure 6-28. Inserting the start logic and preset value of a simple timer

Resetting a Simple Timer

Any coil timer you program is reset in a separate network, using a simple reset operation that references the timer address. While the timer is running, or even if it is not running, a transition from 0-to-1 of the reset logic resets both the remaining time value and the timer status to 0.

In the figure shown below, when the logic combination of I 6.6 AND T33 transitions from '0' to '1' M133.7 goes high (1) for one scan and thereby resets the timer T33.

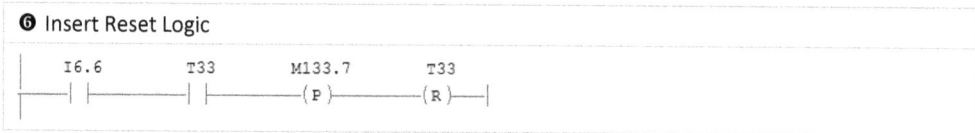

---

❻ Insert Reset Logic

```
    I6.6        T33        M133.7       T33
  ──┤ ├─────────┤ ├─────────( P )────────( R )──┤
```

Figure 6-29. Inserting the reset logic of a simple timer in LAD

Evaluating Simple Timer Remaining Time

When working with S5 timers, remember that these timers time down and that you are monitoring the remaining time. When writing code to check the timer value for simple timers, you will access the time directly by referencing the timer operand (timer address) as the source location in a MOVE operation, as shown below. In this operation, you will access the remaining time as an integer (decimal) value.

If the remaining time will be directed to a BCD display device or compared to a BCD setpoint, then you will need to perform a conversion Integer-to-BCD (INT_BCD). This conversion is shown in the second cell in the following illustration. The remaining time value is output in QW54.

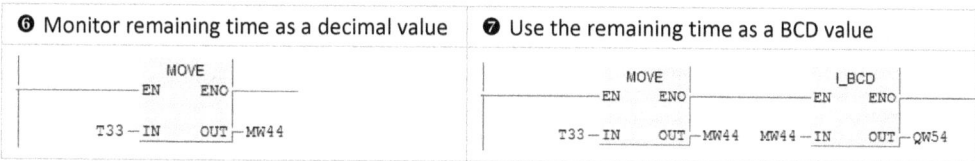

---

❻ Monitor remaining time as a decimal value	❼ Use the remaining time as a BCD value
``` MOVE ─── EN ENO ─── T33 ─ IN OUT ─ MW44 ```	``` MOVE I_BCD ── EN ENO ── ── EN ENO ── T33 ─ IN OUT ─ MW44 MW44 ─ IN OUT ─ QW54 ```

Figure 6-30. Accessing the remaining time of a simple timer in LAD

Evaluating Simple Timer Status

Finally, the status signal **Q,** in conjunction with the start signal **S,** can be checked to determine whether the timer is enabled or done timing. You can examine the timer status in your program using normally-open and normally-closed contacts that reference the address of **Q.** Remember that the exact behavior of **Q** varies according to the timer type.

---

❼ Check timer running/enable status

```
    I6.6        T33        M100.1      M100.2
  ──┤ ├─────────┤ ├─────────( P )────────( S )──┤
```

Figure 6-31. Checking the status of a simple timer in LAD

PROGRAMMING SIMPLE TIMERS IN FBD

After inserting a new network, programming a coil timer is first a matter of selecting the required timer from the Program Elements tab, according to function. The timer, identified by a mnemonic name inside an output, can be dragged onto the FBD network. For example, **SD** is an on-delay timer.

By using a simple timer, you are able to isolate the operations of a box timer. Loading a preset value, starting the timer, monitoring the remaining time or status signal, and resetting the timer are all handled with separate operations. These operations are entered as separate networks and preferably in the sequence shown below.

- Specify timer preset value before starting the timer

- Reset the timer (R)

- Check timer status (binary check)

- Check remaining time value (in decimal), using MOVE and Compare

- Check remaining time (in BCD), using MOVE, integer-BCD conversion, and Compare

Selecting and Addressing a Simple Timer

With the new network inserted and selected, you can double-click on the desired timer object in the Program Elements folder, or drag and drop the object onto the network. The timer address must be entered just above the timer coil, using the absolute address—for example **T0, T1, T2**. You may also use a symbolic address—for example **Sys_Delay**, if it is already defined.

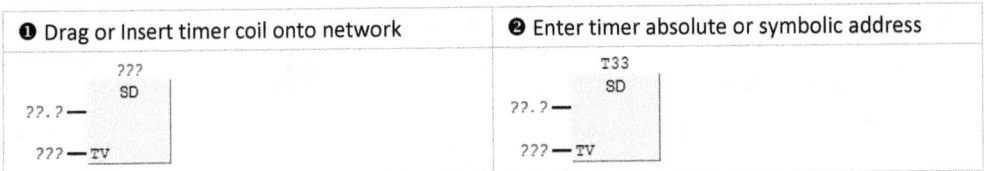

❶ Drag or Insert timer coil onto network	❷ Enter timer absolute or symbolic address

       ???
       SD
??.? —
 	
??? — TV
	
	       T33
	       SD
	??.? —
	
	??? — TV

Figure 6-32. Inserting and addressing a simple timer in FBD

Setting and Starting a Simple Timer

The timer coil is started based on the type of timer and based on the logic you insert to start the timer. All timer types, except for the off-delay timer, start timing when the RLO of the timer start logic transitions from 0-to-1. The off-delay starts timing when the RLO transitions from 1-to-0.

The timer preset value, entered just below the timer element, is specified as a constant using the S5TIME format, where the range is S5T#0h_0m_0s_10ms to S5T#2h_46m_30s_0ms. The preset units of h = hour, m = minutes, s = seconds, and ms = milliseconds are only specified if required. For example, the preset constant S5T#1m30s is 1 minute and 30 seconds. For the timer preset, you may also reference a declared variable of type S5TIME, or a word-width memory location.

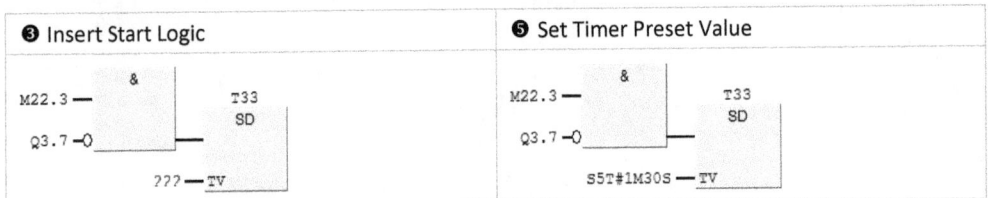

❸ Insert Start Logic	❺ Set Timer Preset Value

           &
M22.3 —   T33
               SD
Q3.7 —○
 	
       ??? — TV
	
	           &
	M22.3 — T33
	               SD
	Q3.7 —○
	
	   S5T#1M30S — TV

Figure 6-33. Inserting the start logic and preset value of a simple timer in FBD

Resetting a Simple Timer

Any simple timer you program is reset in a separate network using a simple reset operation that references the timer address. If the reset logic has an RLO = 1, while the timer is running or even if it is not running, both the remaining time value and the timer status are reset to 0.

In the figure shown below, when the logic combination of I 6.6 AND T33 transitions from '0' to '1' M133.7 goes high (1) for one scan and thereby resets the timer T33.

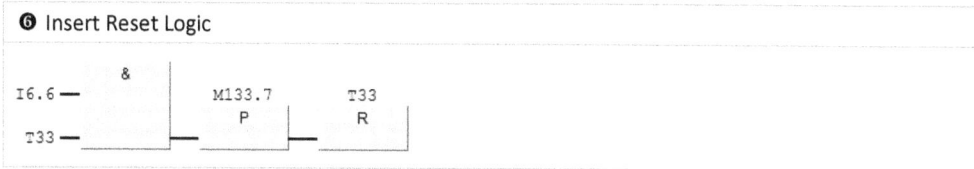

Figure 6-34. Inserting the reset logic of a simple timer in FBD

Evaluating Simple Timer Remaining Time

When working with S5 timers, remember that these timers time down and that you will be monitoring the remaining time. When writing code to check the timer value for simple timers, you will access the time directly by referencing the timer operand (timer address) as the source location in a MOVE operation, as shown below. In this operation, you will access the remaining time as a decimal value.

If the remaining time will be directed to a BCD display device or compared to a BCD setpoint, then you will need to perform an Integer-to-BCD (INT_BCD) conversion. This conversion is shown in the second cell of the following illustration. The remaining time in BCD is output in QW54.

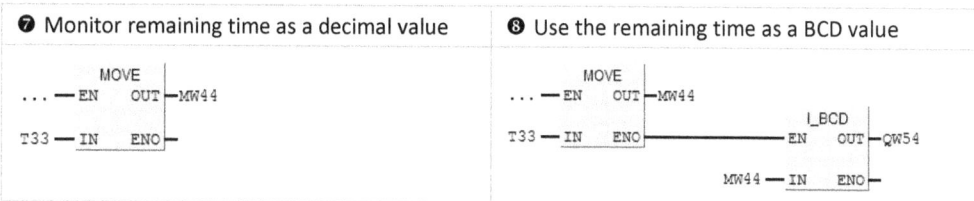

Figure 6-35. Accessing the remaining time of a simple timer in FBD

Evaluating Simple Timer Status

Finally, the status of a simple timer, accessed using the timer operand, can be checked to determine whether the timer is enabled or done timing. You can examine the timer status in your program using either a positive or negated binary input that references the address of the simple timer. Remember that the exact behavior of the timer status varies according to the timer type.

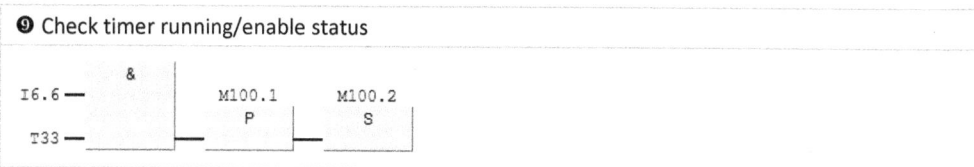

Figure 6-36. Checking the status of a simple timer in FBD

Chapter 7

LAD/FBD
Counter Operations

In this chapter you'll be introduced to LAD/FBD Counter Operations. In LAD and in FBD, each counter type is available as a box function and as a simple instruction. The box counter type and the simple counter type each have its own advantages.

Although counters have many program applications, they are generally used to enable or disable a control circuit or process event after a predetermined count is reached; or in production accounting and other event tracking applications.

In addition to the standard STEP 7 counter instructions, covered in this chapter, standard IEC counter functions are reviewed in Chapter 28 *Developing Functions and Function Blocks*. These callable Function Block operations, which include SFB 0 (CTU—Up Counter), SFB 1 (CTD—Down Counter), and SFB 2 (CTD—Up/Down Counter), are available in the STEP 7 library, and are integrated in some S7 CPUs.

Main Topics

- Counter Operations Overview

- Box Counter Descriptions

- Programming Box Counters in LAD

- Programming Box Counters in FBD

- Simple Counter Descriptions

- Programming Simple Counters in LAD

- Programming Simple Counters in FBD

COUNTER OPERATIONS OVERVIEW

STEP 7 LAD counter instructions include the *Up counter*, *Down counter*, and the *Up/Down counter*. UP and Down counters are implemented as box functions and as simple instructions. Box counters support the full functionality of the S7 counter, allowing the count to be incremented or decremented, a preset value to be loaded, and for the counter to be monitored, and reset.

The simple UP and Down counter operations only support counting up or down. Separate operations are used to load a preset, and to reset a simple counter. The MOVE and Compare operation are used in conjunction to check the current count value of a simple counter.

As mentioned in the introduction, you can learn about standard IEC counter functions that you also have access to in your program—these operations are reviewed in Chapter 28 *Developing Functions and Function Blocks.*

Table 7-1. LAD/FBD Counter Operations Summary (Box and Simple Operations)

LAD	FBD	Brief Description
??? —(SC)— ???	??? SC ??.?— ???—PV	**Set Counter Coil/Output.** Whenever the driving logic transitions from logic 0 to logic 1, load the specified preset value **PV** to the addressed counter. **PV** = 000—to—999 (C#000—C#999). Example counter addresses: C0 to C256.
??? —(CU)—	??? CU ??.?—	**UP Counter Coil/Output.** Whenever the driving logic transitions from logic 0 -to- logic 1, increment the specified counter by one count. Example counter addresses: C0 to C256.
??? —(CD)—	??? CD ??.?—	**DOWN Counter Coil/Output.** Whenever the driving logic transitions from logic 0 -to- logic 1, decrement the specified counter by one count. Example counter addresses: C0 to C256.
??? S_CU —CU Q— —S CV— —PV CV_BCD— —R	??? S_CU —CU —S CV— —PVCV_BCD— —R Q—	**UP Counter Box.** Whenever the logic input line **CU** transitions from logic 0 -to- logic 1, increment the specified counter by one count. Logic 1 on the reset input line **R** resets the counter to zero. Example counter addresses: C0 to C256.
??? S_CD —CD Q— —S CV— —PV CV_BCD— —R	??? S_CD —CD —S CV— —PVCV_BCD— —R Q—	**Down Counter Box.** Whenever the logic input line **CD** transitions from logic 0 -to- logic 1, decrement the specified counter by one count. Logic 1 on the reset input line **R** resets the counter to zero. Example counter addresses: C0 to C256.
??? S_CUD —CU Q— —CD CV— —S CV_BCD— —PV —R	??? S_CUD —CU —CD —S CV— —PVCV_BCD— —R Q—	**UP-Down Counter.** For each logic 0 -to- logic 1 transition on the **CU** input line, increment the specified counter by +1; for each logic 0 -to- logic 1 transition on the **CD** input line, decrement the count by +1. Logic 1 on the reset input line **R** resets the counter to zero. Example counter addresses: C0 to C256.

Note: Simple counters are listed in the first three rows; the remaining operations are box counters.

BOX COUNTER DESCRIPTIONS

Each of the three box counters is identified by a unique mnemonic label at the top of the counter box. The mnemonic for the *up-counter* is **S_CU**; for a *down-counter*, **S_CD**; and for an *up/down counter*, **S_CUD.** Each of the counter box operations is described in the following discussions. First we'll take a look at the parameters of Table 7-2—these are the parameters that you must specify as you program any box counter.

Box Counter Parameters

As seen in the table below, a box counter has a parameter above the counter where you specify the counter address—for example C1. There is the count line CU or CD, where you will enter the control logic for counting (up/down); a set line S, which you may use to insert logic that causes the preset value, specified at PV, to be loaded to the counter word as the current or new start value; and a reset line R, which you may use to insert logic to reset the counter.

Table 7-2. LAD/FBD Box counter parameters. Note that entry of all parameters is not required.

Parameter	Data Type	Description	Required?	Box Counter	
				LAD	**FBD**
-		Counter Address < Cn >	YES		
CU/CD		Count Signal Logic Input Line	YES		
S	BOOL	Load Counter Preset Signal Line	NO		
PV	WORD	Counter Preset Value	YES		
R	BOOL	Reset Logic Line	NO		
Q	BOOL	Counter Status Signal	NO		
CV	WORD	Current Count in Decimal	NO		
CV_BCD	WORD	Current Count in BCD	NO		

Note: Logic must be inserted on the **S** input line if a value is entered for **PV**.

On the output side, the counter status signal Q is '1' whenever the count is greater than zero and is otherwise '0'. You can check the status Q using contacts in LAD, or gate logic in FBD. The outputs CV and CV_BCD both allow you to specify a word location where the current count is output. The current count is output in decimal at the location you specify at CV and in BCD at the location you specify at CV_BCD. If you specify CV and CV_BCD, you may use declared WORD variables, or word-width memory locations—for example MW 102, QW 34, or DB 10.DBW 4.

The counter preset value PV, though not required with the UP counter, is specified for a DOWN counter as the initial value from which the counter counts down. You may specify PV as a BCD constant, as a declared WORD variable, or as a word-width memory location that contains the value. The valid range of PV is C#001 to C#999. The counter status signal Q may be left un-programmed, or may be connected to an output that you reference with any unused bit address. The status Q can then be checked using a NO or NC contact in LAD or in FBD using a binary input or negated binary input.

UP Counter Box: S_CU

For each transition from '0' to '1' at the CU input, the *up counter* increments the specified counter by +1, except for when the count has reached the maximum count of 999. A '0' to '1' transition at the set (S) input causes the count to be preset with the value you specify at PV. The preset value may be specified as a BCD (C#000 to C#999) or Hex (W#16#0000 to W#16#0999) constant, or a word location containing a valid count value.

The current count is output as a decimal value in the word location specified at CV, and in BCD in the location specified at CV_BCD. Whenever the RLO = 1 on the reset line R, the counter value is cleared to zero and the status signal Q is reset.

In the following example, the UP Counter increments by +1 when the logic combination of I 5.1 AND NOT M 3.7 transitions from '0' to '1'. Whenever the logic combination of I 5.3 OR M 101.4 transitions from '0' to '1', the counter preset value is set to 100. The counter is reset whenever the logic combination of I 5.4 OR M 101.7 is True (1).

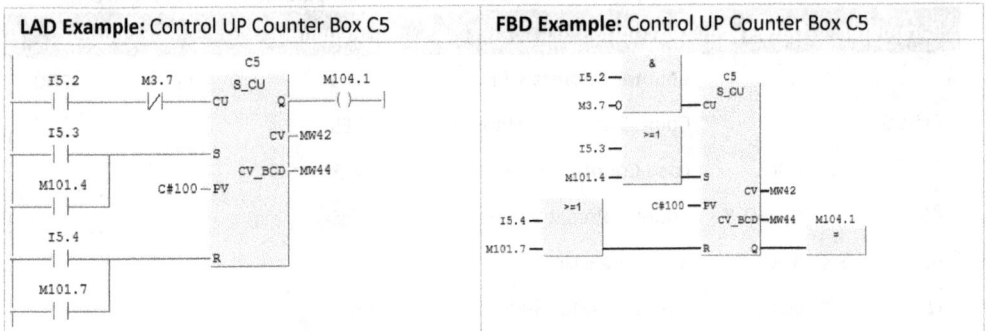

Figure 7-1. Illustration of LAD/FBD UP Counter box operation

DOWN Counter Box: S_CD

For each transition from '0' to '1' at the CD input, the *down counter* decrements the specified counter by +1, except for when the count has reached the minimum count of 000. A '0' to '1' transition at the set (S) input causes the count to be preset with the value you specify at PV. The preset may be specified as a BCD (C#000 to C#999) or Hex (W#16#0000 to W#16#0999) constant, or a word location containing a valid count value.

The current count is output as a decimal value in the word location specified at CV, and in BCD at the location specified at CV_BCD. Whenever the RLO = 1 on the reset line R, the counter value is cleared to zero and the status signal Q is reset.

In the following example, the Down Counter decrements by +1 when the logic combination of M 30.3 AND NOT M 3.6 transitions from '0' to '1'. Whenever the logic combination of I 5.3 OR M 101.4 transitions from '0' to '1', the counter preset value is set to 100. The counter is reset whenever the logic combination of I 5.4 OR M 101.7 is True (1).

LAD Example: Control DOWN Counter Box C6	FBD Example: Control DOWN Counter Box C6

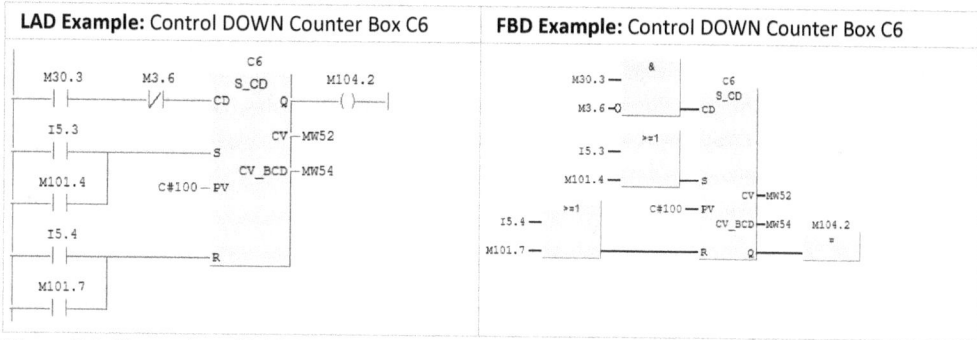

Figure 7-2. Illustration of LAD/FBD DOWN Counter box operation

UP-Down Counter Box: S_CUD

The *up-down counter* increments the specified counter by +1 on each '0' to '1' transition, at the **CU** input, except for when the count has reached the maximum count of 999. The counter decrements by +1 on each transition from '0' to '1' at the **CD** input, except for when at the minimum count of 000. A transition from '0' to '1' at the set (**S**) input causes the count to be preset with the value specified at **PV**. The preset may be specified as a BCD(C#000 to C#999) or Hex constant (W#16#0000 to W#16#0999), or as a word location containing a valid count value.

The current count is output in decimal in the word location specified at **CV**, and in BCD at the location specified at **CV_BCD**. Whenever the RLO = 1 on the reset line **R**, the counter value is cleared to zero and the status signal **Q** is reset.

In the following example, the Up-Down Counter increments by +1 when the logic combination of I 5.1 OR M 7.1 transitions from logic 0 to logic 1; and decrements by +1 when the logic combination of I 5.2 OR M 10.7 transitions from logic 0 to logic 1. Whenever the logic combination of I 5.3 AND M 101.4 transitions from logic 0 to logic 1, the counter preset value is set to 550. The counter is reset whenever the logic combination of I 5.4 AND NOT M 10.0 is True (1).

LAD Example: Control UP/DN Counter Box C7	FBD Example: Control UP/DN Counter Box C7

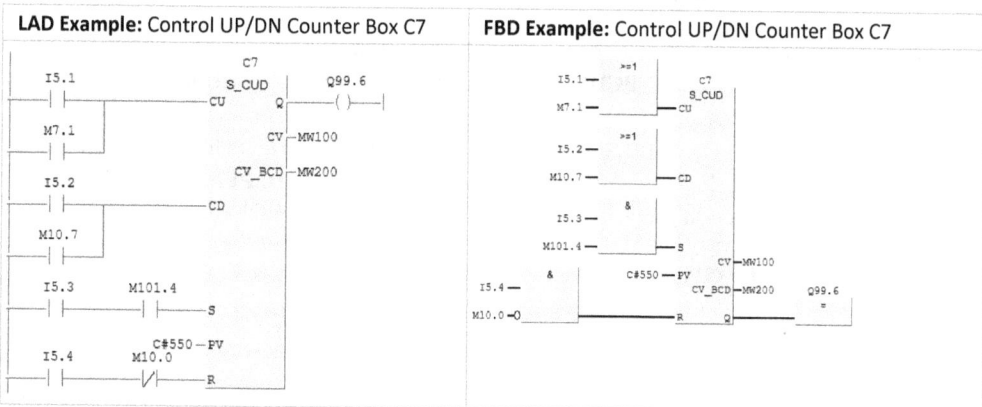

Figure 7-3. Illustration of LAD/FBD UP-Down Counter box operation

PROGRAMMING BOX COUNTERS IN LAD

After inserting a new network, programming a box counter is first a matter of selecting the required counter type from the Program Elements tab, based on the name at the inside top of the box. For example, **S_CU** is an UP counter. Then, as required, you may insert the driving logic for counter inputs (**CU** or **CD**, **S**, and **R**), and enter the appropriate addresses for the status signal (**Q**), and for outputting the current count in decimal (**CV**) or in BCD (**CV_BCD**).

Selecting and Addressing the Box Counter

With the new network inserted and selected, you may drag and drop the desired counter object from the Program Elements folder to the network. The counter address—for example **C0**, **C1**, or **C2** is entered just above the counter. You may also use a symbolic address—for example **AmTotal**, if it is already defined. Each counter you specify references a single word location in the S7 counter memory.

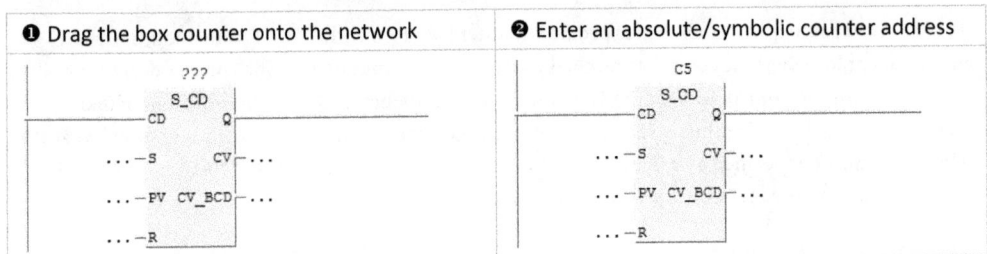

Figure 7-4. Inserting and addressing a box counter in LAD

Controlling a Box Counter

A counter, depending on the type, will either increment (**CU**) or decrement (**CD**) the counter value by +1 on each '0' to '1' transition of the count line. For the illustration below, click on the **CD** input line, then use the **F2** and **F3** keys to insert a normally-open an normally-closed contact in series; enter the appropriate addresses for the contact elements. If you are programming an UP/Down counter, you would need to insert the control logic for both **CU** and **CD** lines.

The reset line **R** affects all counters in the same way. Whenever the RLO = 1 on the reset line **R**, both the current count values (**CV** and **CV_BCD**) and the counter status **Q** are reset to zero. Click on the **R** line and enter the required reset logic. In the figure shown below, **C5** is reset whenever the RLO= 1 for the logic combination of I 5.4 OR M 101.7. To insert this logic, click on the **R** input line and use toolbar buttons open branch and close branch to insert a parallel circuit; then use the **F2** and **F3** keys to insert the normally-open contacts in parallel; enter the appropriate addresses for the contacts.

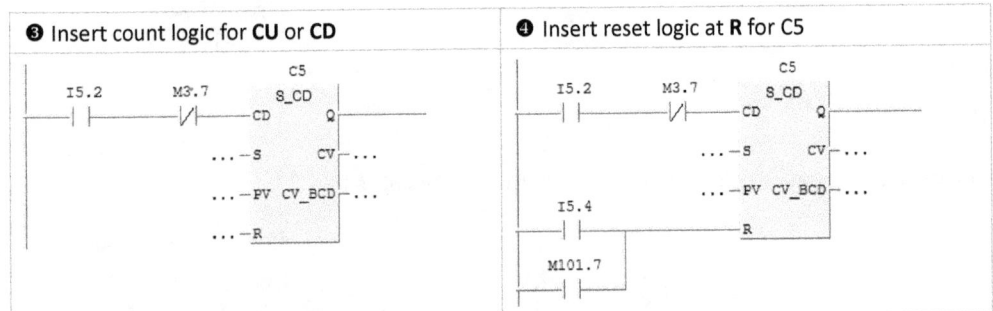

Figure 7-5. Inserting the count logic, and reset logic for a box counter in LAD

Setting Box Counter Preset Value

The counter preset value **PV**, though not required, is specified if you wish to have a value transferred to the counter as an initial start value or as an override value. In a down counter, for example, the initial value would provide the value from which the counter counts down. You may specify **PV** as a BCD constant—for example C#100; as a declared WORD variable; or as a word-width memory location from which the value is taken—for example MW 100. Click on the **S** input line to insert binary logic that controls when the value at **PV** is loaded to the counter. In the illustration below, a value of 100 is loaded to the counter value when I 5.3 transitions from '0' to '1'.

❺ Enter counter preset value—**PV**	❻ Insert load counter preset logic as—**S**

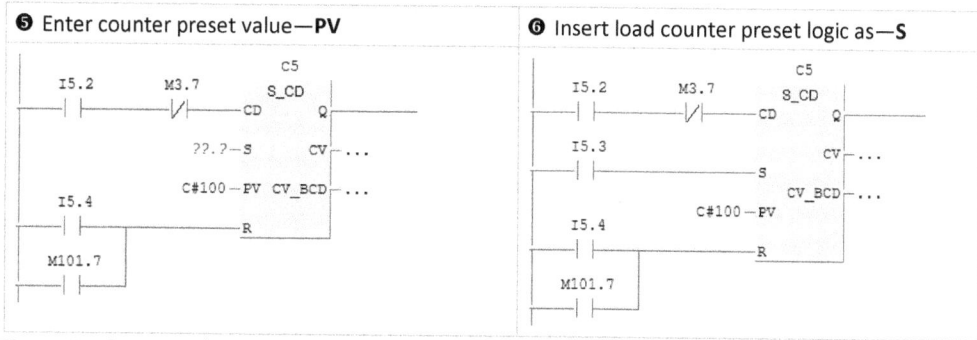

Figure 7-6. Inserting the preset value and the set preset value logic for a box counter in LAD

Evaluating Box Counter Status and Count

When writing code to check the counter value, recall that the current count is output in binary (the equivalent of decimal) at the memory location or variable you specify at **CV**; and in binary coded decimal, at the location or variable you specify at **CV_BCD**. You do not have to specify either of these locations, but either one or both may be entered. You would use **CV_BCD**—for instance QW54 in the example below, to direct the count to a BCD display device or to compare to a BCD setpoint. The value output at MW 54 might be used in a calculation of data type INT.

The counter status signal **Q** is logic 1 whenever the count is greater than zero, and is logic 0 whenever the count is zero. You may connect **Q** to a coil that you reference with any unused bit address—for example M 104.1. You may then check the status **Q** using contact elements that reference the coil address assigned to **Q**. The counter address may also be referenced to check the counter status.

❼ Specify **CV** and **CV_BCD** for the current count	❽ Insert coil and assign address for status—**Q**
❾ Check status using status signal **Q** address	❿ Check status using the counter address—C5

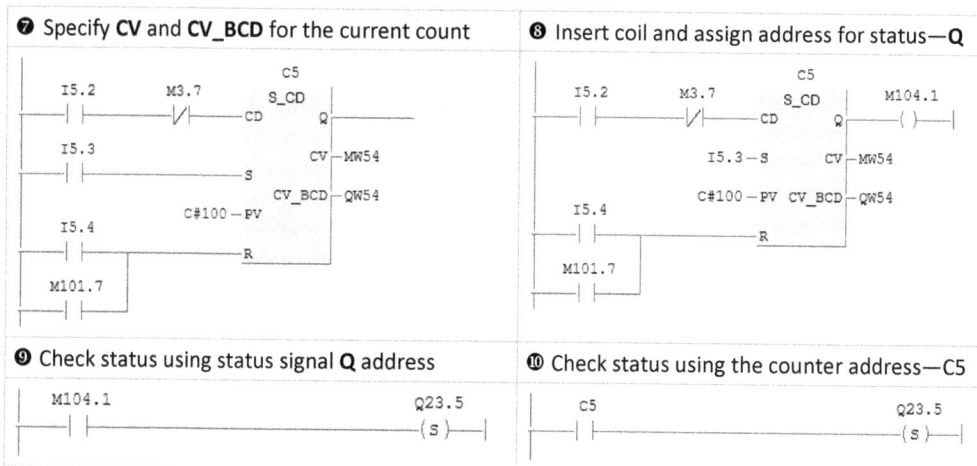

Figure 7-7. Inserting logic to monitor the count value and the status of a box counter in LAD

PROGRAMMING BOX COUNTERS IN FBD

After inserting a new network, programming a box counter is first a matter of selecting the required counter type from the Program Elements tab, based on the name at the inside top of the box. For example, **S_CU** is an UP counter. Then, as required, you may insert the driving logic for counter inputs (**CU** or **CD**, **S**, and **R**), and enter the appropriate addresses for the status signal (**Q**), and for outputting the current count in decimal (**CV**) or in BCD (**CV_BCD**).

Selecting and Addressing a Box Counter

With the new network inserted and selected, you may drag and drop the desired counter object from the Program Elements folder and to the network. The counter address—for example **C0**, **C1**, or **C2** is entered just above the counter element. You may also use a symbolic address—for example **AmTotal**, if it is already defined. Each counter specified, references a single word location in S7 counter memory.

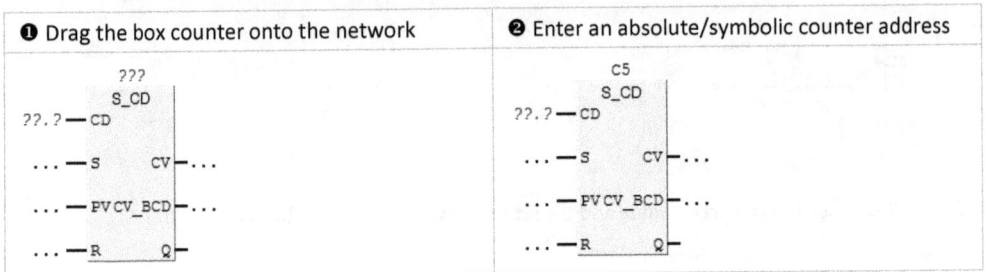

Figure 7-8. Inserting and addressing a box counter in FBD

Controlling a Box Counter

A counter, depending on the type, will either increment (**CU**) or decrement (**CD**) the counter value by +1 on each '0' to'1' transition of the count line. For the illustration below, click on the **CD** input line, then click on the toolbar AND gate; click on the input line of the second input, then on the toolbar click the button for negate binary input; finally enter the addresses at the binary inputs. If you were programming an UP/Down counter, you would need to insert control logic for the **CU** and **CD** lines.

The reset line **R** affects all counters in the same way. Whenever the RLO = 1 on the reset line **R,** both the current count values (**CV** and **CV_BCD**) and the counter status **Q** are reset to zero. Click on the **R** line and enter the required reset logic. In the figure shown below, **C5** is reset whenever the RLO= 1 for the logic combination of I 5.4 OR is M 101.7. To insert this logic, click on the **R** input line, then click on the toolbar OR gate button, and then enter the addresses for the binary inputs.

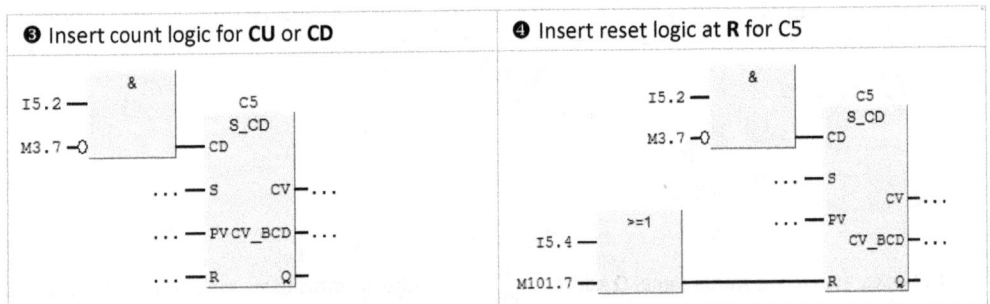

Figure 7-9. Inserting the count logic, and reset logic for a box counter in FBD

Setting Box Counter Preset Value

The counter preset value **PV**, though not required, is specified if you wish to have a value transferred to the counter as an initial start value or as an override value. In a down counter, for example, the initial value would provide the value from which the counter counts down. You may specify **PV** as a BCD constant—for example C#100; as a declared WORD variable; or as a word-width memory location from which the value is taken—for example MW 100. Click on the **S** input line to insert binary logic that controls when the value at **PV** is loaded to the counter. In the illustration below, a value of 100 is loaded to the counter value when I 5.3 transitions from '0' to '1'.

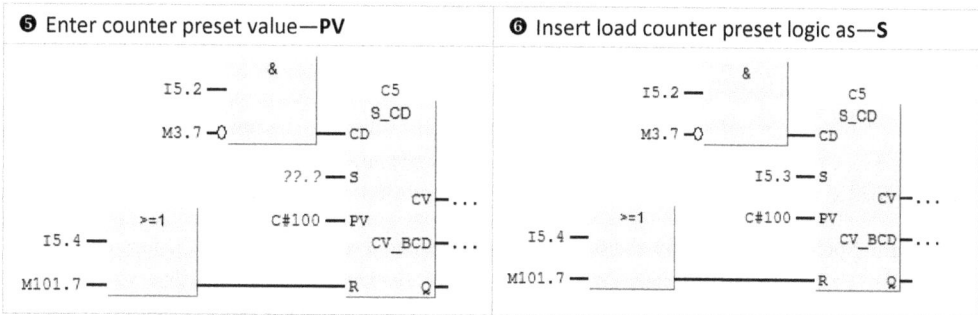

Figure 7-10. Inserting the preset value and the set preset value logic for a box counter in FBD

Evaluating Box Counter Status and Count

When writing code to check the counter value, recall that the current count is output in binary (the equivalent of decimal) at the memory location or variable you specify at **CV**; and in binary coded decimal, at the location or variable you specify at **CV_BCD**. You do not have to specify either of these locations, but either one or both may be entered. You would use **CV_BCD**—for instance **QW54** in the example below, to direct the count to a BCD display device or to compare to a BCD setpoint. The value output at **MW 54** might be used in a calculation of data type INT.

The counter status signal **Q** is '1' whenever the count is greater than zero, and is '0' whenever the count is zero. You may connect **Q** to an output that you reference with any unused bit address—for example M 104.1. You may then check the status **Q** using positive and negated binary inputs, referenced with the appropriate address. The counter address may also be referenced to check the counter status.

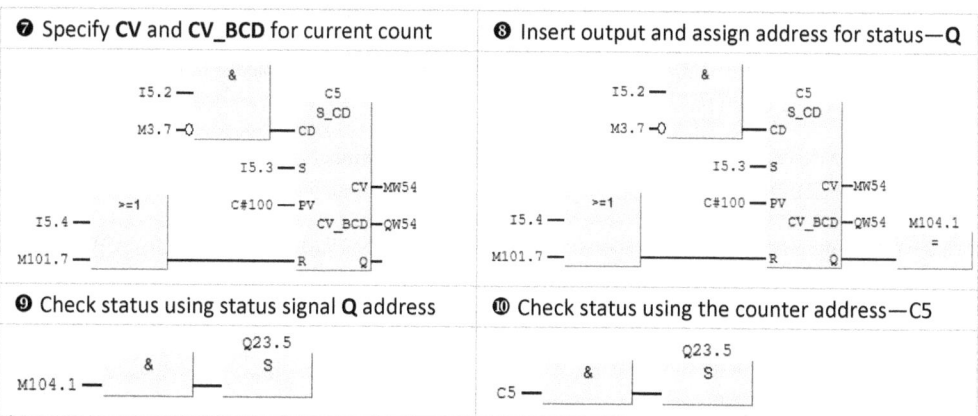

Figure 7-11. Inserting logic to monitor the count value and the status of a box counter in FBD

SIMPLE COUNTER DESCRIPTIONS

As shown back in Table 7-1, the UP and DOWN count functions are also implemented as a single LAD/FBD operation. In this book, these operations are referred to as simple counters. Simple counter operations are implemented using coil elements in LAD, or output elements in FBD. With simple counters, the operations of a box counter are isolated. Counting up or down, loading an initial value, monitoring the counter value or status signal, and resetting the counter are all separate operations.

Each of the simple counters is identified by a unique mnemonic label inside the LAD coil or FBD output instruction. The mnemonic for the simple *up-counter* is **CU**; and for the *down-counter* is **CD**. Each of the simple counter operations is described in the following discussions.

Set Counter Coil/Output: SC

The simple *set counter coil/output* operation is used to load a value into the counter address you specify above the set counter element. Only when the preceding logic makes a 0-to-1 transition is the specified counter loaded with the preset value specified below the set counter element.

The counter preset, entered below the simple counter, is entered as a constant in BCD format (C#000-C#999) or as a hexadecimal constant (W#16#0000 –to– W#16#0999). You may specify a word-with memory location containing the preset, or a variable of type WORD. The preset values may then be modified by writing new values to the specified word location or variable.

In the figure below, when the logic combination of I5.3 **OR** M101.4 transitions from '0' to '1', then the counter C5 is loaded with a preset value of 100.

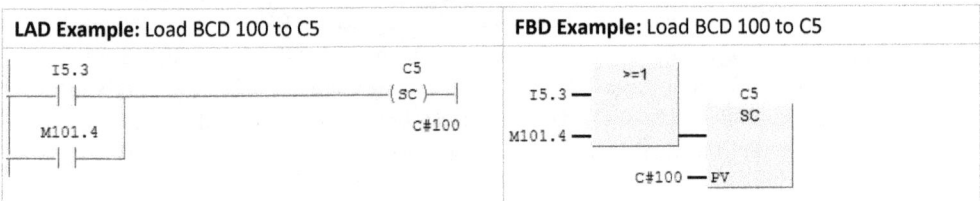

LAD Example: Load BCD 100 to C5	FBD Example: Load BCD 100 to C5

Figure 7-12. Illustration of LAD/FBD simple Set Counter operation

UP Counter Coil/Output: CU

The simple *up counter coil/output* is used when a simple UP count function is required—especially if it is desirable to separate the control logic of the UP count from the other operations of the UP count box function. Whenever the driving logic makes a '0' to '1' transition, the specified counter increments by +1, except when the count has reached the maximum count of 999.

The counter status can be checked by referencing the counter number (**Cn**). The status of the counter **C5,** for example, is '1' whenever the counter value is other than zero.

In the following, whenever the logic combination of I5.2 **AND NOT** M3.7 transitions from '0' to '1', the simple counter C5 is incremented by +1.

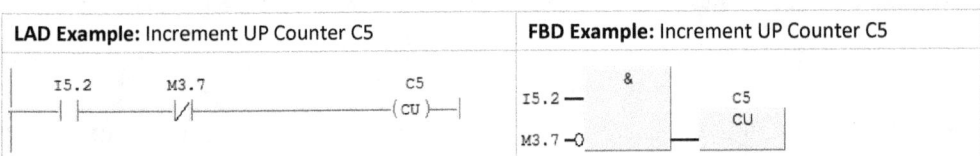

LAD Example: Increment UP Counter C5	FBD Example: Increment UP Counter C5

Figure 7-13. Illustration of LAD/FBD simple UP Counter operation

Down Counter Coil/Output: CD

The simple *down counter coil/output* is used when a simple Down count function is required—especially if it is desirable to separate the control logic of the Down count from the other operations of the Down counter box function. Whenever the driving logic makes a '0' to '1' transition, the specified counter decrements by +1, except when the count has reached the minimum count of 000.

The counter status can be checked by referencing the counter number (**Cn**). The status of the counter **C6**, for example, is '1' whenever the counter value is other than zero.

In the figure below, whenever the logic combination of I5.2 **AND NOT** M3.7 transitions from '0' to '1', the simple counter C6 is decremented by +1.

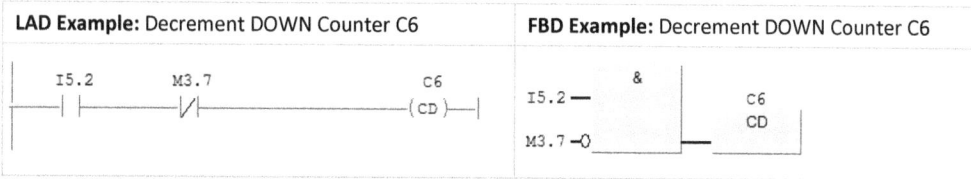

LAD **Example:** Decrement DOWN Counter C6	FBD **Example:** Decrement DOWN Counter C6
I5.2 M3.7 C6 ─┤├──┤/├────(CD)─┤	I5.2 —& M3.7 ─O C6 CD

Figure 7-14. Illustration of LAD/FBD simple Down Counter operation

PROGRAMMING SIMPLE COUNTERS IN LAD

After inserting a new network, programming a coil counter is first a matter of selecting the desired counter from the Program Elements tab. The simple counter, identified by the mnemonic name inside the coil, can be dragged onto the new LAD network. For example, **S_CU** is an UP Counter coil.

By using a simple counter, you are able to isolate the operations of a box counter. Counting up or down, loading an initial value, monitoring the current counter value or status signal, and resetting the counter are all handled with separate operations. These operations are entered as separate networks and preferably in the following sequence.

- Set Counter by specifying a counter preset value
- Count Up (CU) or Count Down (CD)
- Reset counter (R)
- Check counter status (binary check)
- Check counter value (in decimal) using MOVE and Compare
- Check counter value (in BCD) using MOVE, integer-BCD conversion, and Compare

Setting Simple Counter Preset Value

A counter preset value, although not required, is loaded into the counter word using the set counter operation **SC**. First, insert a new network and drag the **SC** coil from the Program Elements folder onto the network. Next, enter the counter address above the coil and the preset value just below the coil. The preset value might be used as an initial start value or as an override value. You may specify the start value as a BCD or Hexadecimal constant—for example C#100 or W#16#0100; as a declared variable, or as a word-width location like MW 100.

Finally, click on the **SC** network to insert binary logic to control when the preset is loaded to the counter. In the illustration below, a preset value of 100 is loaded to the counter whenever the combination of I5.3 **OR** M101.4 transitions from '0' to '1'.

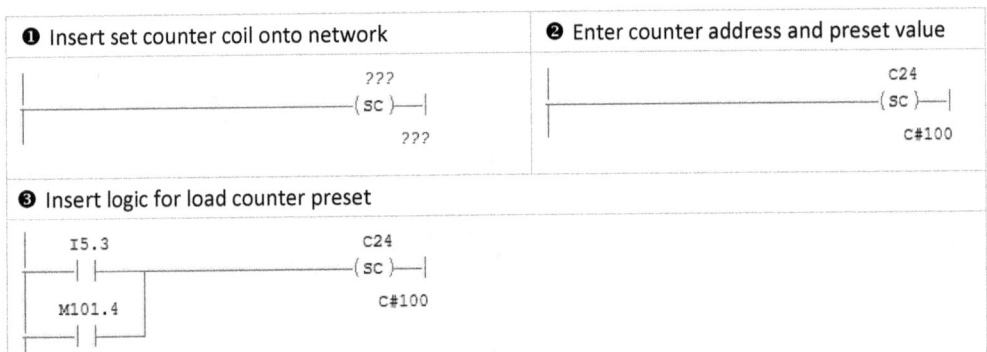

❶ Insert set counter coil onto network	❷ Enter counter address and preset value
??? —(SC)—| ???	C24 —(SC)—| C#100

❸ Insert logic for load counter preset
I5.3   C24 —| |———————(SC)—| C#100 M101.4 —| |—

Figure 7-15. Inserting a simple counter, the preset value, and logic for load preset value in LAD

Selecting and Addressing a Simple Counter

With the new network inserted, you can double-click on the desired counter object in the Program Elements folder, or drag and drop the object onto the network. The counter address is entered just above the counter coil, using the absolute address—for example **C0, C1, C2**. You may also use a symbolic address—for example **AM_TOTAL**, if it is already defined. Each counter you specify references a single word location in the S7 counter memory.

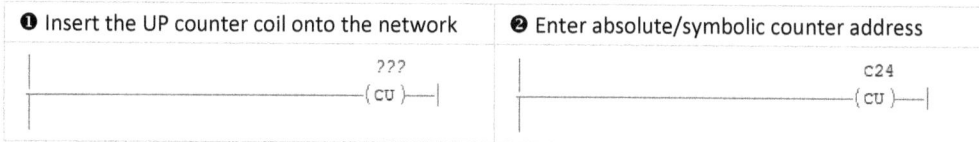

❶ Insert the UP counter coil onto the network	❷ Enter absolute/symbolic counter address
??? ——(CU)—|	C24 ——(CU)—|

Figure 7-16. Inserting and addressing a simple counter in LAD

Controlling a Simple Counter

A simple counter is triggered by the control logic that precedes the counter coil. The up counter (**CU**) will increment (count up by 1), and the down counter (**CD**) will decrement (count down by 1), for each 0-to-1 transition of the RLO of the count logic. Click on the **CU** or **CD** network and insert the control logic that triggers the up count or down count.

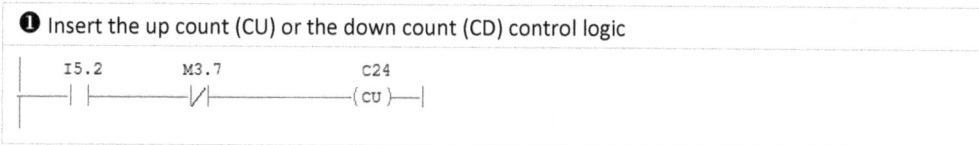

❶ Insert the up count (CU) or the down count (CD) control logic

```
   I5.2        M3.7             C24
 ——| |————————|/|———————————( CU )—|
```

Figure 7-17. Inserting the count up or count down logic of a simple counter in LAD

Resetting a Simple Counter

Any simple counter you program is reset in a separate network, using a standard reset coil operation. To reset a specific counter, first insert the reset network. Just above the reset coil, enter the counter address for the counter you wish to reset; then, click on the network and enter the required reset logic. In the figure below, **C24** is reset whenever the RLO= 1 for the logic combination of I5.4 OR M101.4.

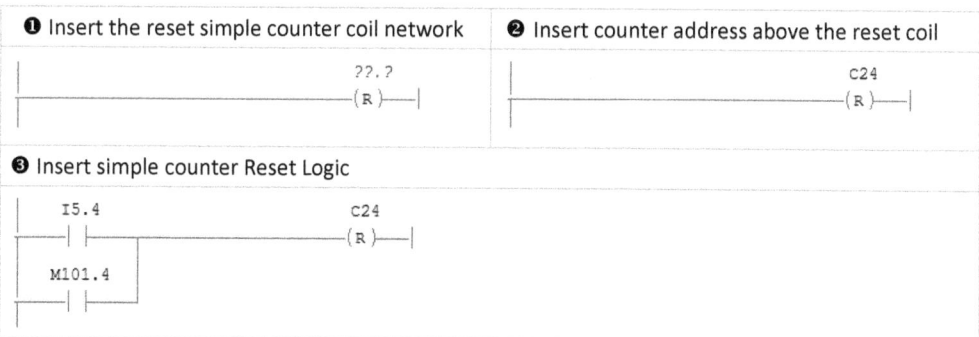

❶ Insert the reset simple counter coil network	❷ Insert counter address above the reset coil
??.? ——(R)—|	C24 ——(R)—|

❸ Insert simple counter Reset Logic

```
   I5.4              C24
 ——| |————————————( R )—|
   M101.4
 ——| |—
```

Figure 7-18. Inserting the counter reset logic of a simple counter in LAD

Checking the Simple Counter Status

Finally, the status of a simple counter may be checked in your program using normally-open and normally-closed contacts that reference the address of the simple counter. The counter status will be '1' when the current count is greater than zero, and '0' when the current count is equal zero.

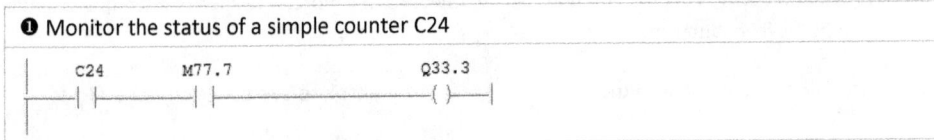

```
❶ Monitor the status of a simple counter C24

      C24        M77.7                    Q33.3
   ──┤ ├────────┤ ├────────────────────( )──┤
```

Figure 7-19. Inserting logic to check the status of a simple counter in LAD

Evaluating Simple Counter Count

When writing code to check the current count of a simple counter, you will access the count by referencing the counter address as the source location in a MOVE operation, as shown in the first cell below. In this operation the current count is accessed as an integer (decimal) value.

To direct the current count to a BCD display device, you will first need to perform an integer-to-BCD conversion on the counter value. This conversion is shown in the second cell of the following illustration. The current count of C24 is output toMW34 as a decimal value, using the Move operation; the current count is then converted to BCD in the I_BCD operation; and then the converted (BCD) value is directed to output image word QW34.

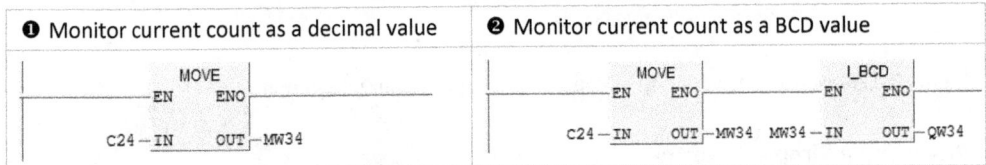

```
❶ Monitor current count as a decimal value    | ❷ Monitor current count as a BCD value

          MOVE                                 |        MOVE                  I_BCD
        EN    ENO                              |      EN    ENO             EN    ENO
                                               |
  C24 ─ IN   OUT ─ MW34                        | C24 ─ IN  OUT ─ MW34  MW34 ─ IN  OUT ─ QW34
```

Figure 7-20. Inserting logic to check the current counter value of a simple counter in LAD

PROGRAMMING SIMPLE COUNTERS IN FBD

After inserting a new network, programming a simple counter is first a matter of selecting the desired counter from the Program Elements tab. The simple counter, identified by the mnemonic name inside the output box, can be dragged onto the new FBD network. For example, **S_CU** is an UP Counter.

By using a simple counter, you are able to isolate the operations of a box counter. Counting up or down, loading an initial value, monitoring the current counter value or status signal, and resetting the counter are all handled with separate operations. These operations are entered as separate networks and preferably in the following sequence.

- Set Counter by specifying a counter preset value
- Count Up (CU) or Count Down (CD)
- Reset counter (R)
- Check counter status (binary check)
- Check counter value (in decimal) using MOVE and Compare
- Check counter value (in BCD) using MOVE, integer-BCD conversion, and Compare

Setting Simple Counter Preset Value

A counter preset value, although not required, it is loaded into the counter word using the set counter operation **SC**. First, insert a new network and drag the **SC** output from the Program Elements folder onto the network. Next, enter the counter address above the output and the preset value at the **PV** input. The preset value might be used as an initial start value or as an override value. You may specify the start value as a BCD or Hexadecimal constant—for example C#150 or W#16#0150; as a declared variable, or as a word-width location like MW 100.

Finally, click on the **SC** network to insert binary logic to control when the preset is loaded to the counter. In the illustration below, a preset value of 150 is loaded to the counter whenever the combination of I5.3 **OR** M101.4 transitions from '0' to '1'.

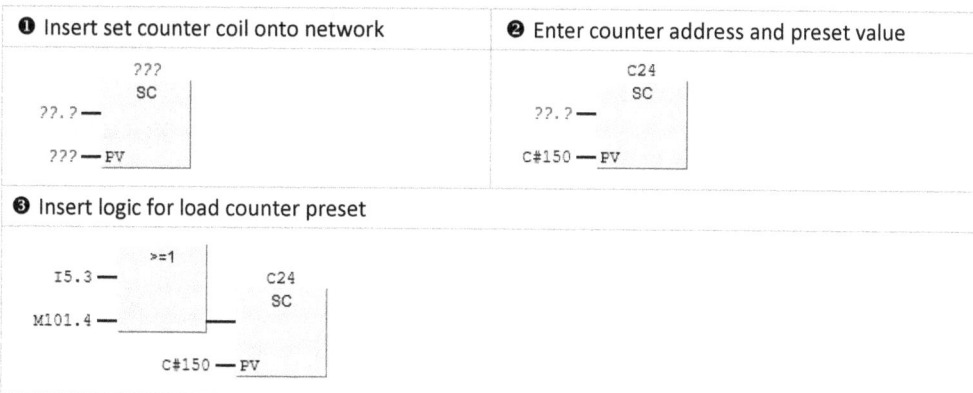

❶ Insert set counter coil onto network	❷ Enter counter address and preset value

``
          ???
          SC
 ??.? —
 	
 ??? — PV
``
	``
	          C24
	          SC
	 ??.? —
	
	 C#150 — PV
	``

❸ Insert logic for load counter preset	

``
          >=1
 I5.3 —| C24
 	
	SC
	 M101.4 —|_____|
	            C#150 — PV
	``	

Figure 7-21. Inserting a simple counter, the preset value, and load preset value logic in FBD

Selecting and Addressing a Simple Counter

With the new network inserted, you can double-click on the desired counter object in the Program Elements folder, or drag and drop the object onto the network. The counter address is entered just above the counter output, using the absolute address—for example **C0, C1, C2**. You may also use a symbolic address—for example **AM_TOTAL**, if it is already defined. Each counter you specify references a single word location in the S7 counter memory.

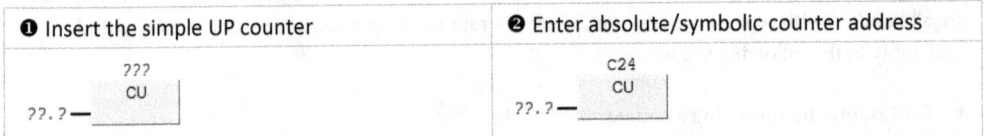

❶ Insert the simple UP counter	❷ Enter absolute/symbolic counter address
??? CU ??.? —	C24 CU ??.? —

Figure 7-22. Inserting and addressing a simple counter in FBD

Controlling a Simple Counter

A simple counter is triggered by the control logic that precedes the counter output. The up counter (**CU**) will increment (count up by 1) and the down counter (**CD**) will decrement (count down by 1) for each 0-to-1 transition of the RLO of the count logic. Click on the **CU** or **CD** network and insert the control logic that triggers the up count or down count. Click on the OR gate or AND gate to insert additional gate logic.

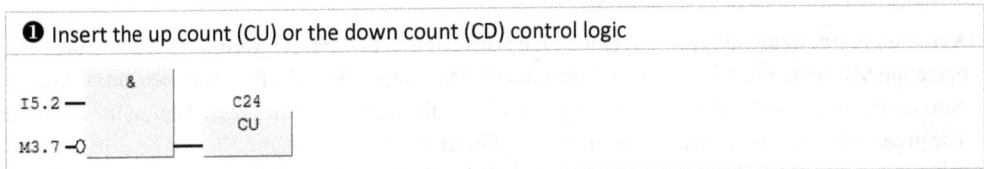

❶ Insert the up count (CU) or the down count (CD) control logic
I5.2 — & M3.7 —O   C24 CU

Figure 7-23. Inserting the count up or count down logic of a simple counter in FBD

Resetting a Simple Counter

Any simple counter you program is reset in a separate network using a standard reset output operation. To reset a particular counter, first insert the reset network. Just above the reset output, enter the counter address for the counter you wish to reset. Then, click on the binary input line and enter the required reset logic. With the binary input line selected, click on the OR gate or AND gate to insert additional gate logic.

In the figure below, **C24** is reset when the RLO= 1 for the logic combination of **I** 5.4 OR is **M** 101.4.

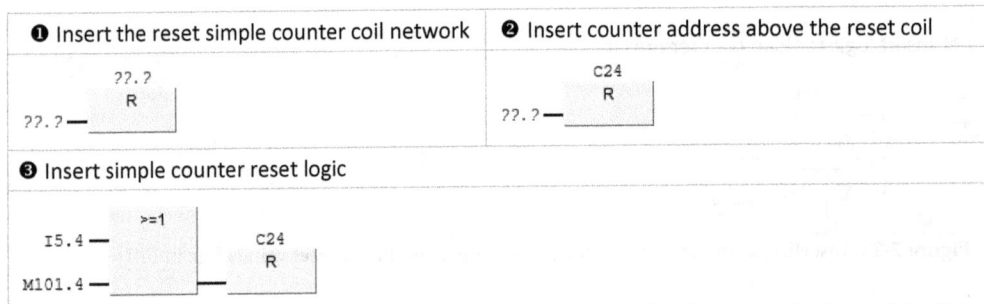

❶ Insert the reset simple counter coil network	❷ Insert counter address above the reset coil
??.? R ??.? —	C24 R ??.? —
❸ Insert simple counter reset logic	
I5.4 — >=1 M101.4 — C24 R	

Figure 7-24. Inserting the counter reset logic of a simple counter in LAD

Checking Simple Counter Status

Finally, the status of a simple counter may be checked in your program using a positive or negated binary input that references the address of the simple counter. The counter status will be '1' when the current count is greater than zero, and '0' when the count is equal zero.

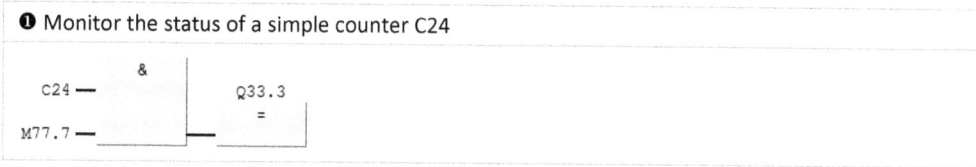

❶ Monitor the status of a simple counter C24

```
            &
C24 —               Q33.3
                      =
M77.7 —
```

Figure 7-25. Inserting logic to check the status of a simple counter in FBD

Evaluating Simple Counter Count

When writing code to check the current count of a simple counter, you will access the count by referencing the counter address as the source location in a MOVE operation, as shown in the first cell below. In this operation, you will access the current count as an integer (decimal) value.

To direct the current count to a BCD display device, you will first need to perform an integer-to-BCD conversion on the counter value. This conversion is shown in the second cell of the following illustration. The current count of C24 is output to MW34 as a decimal value using the Move operation; the current count is then converted to BCD in the I_BCD operation; and then the converted (BCD) value is directed to output image word QW 34.

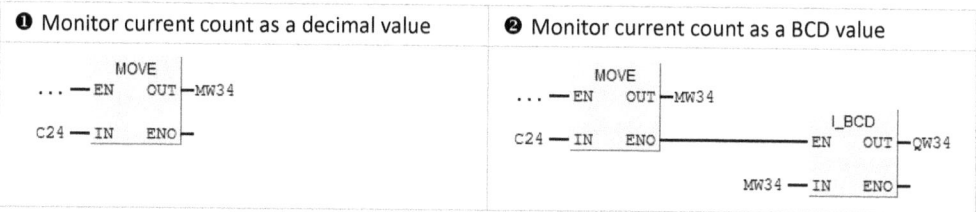

❶ Monitor current count as a decimal value	❷ Monitor current count as a BCD value
| ```
        MOVE
... —EN    OUT —MW34
C24 — IN   ENO
``` | ```
        MOVE
... —EN    OUT —MW34
                          I_BCD
C24 — IN   ENO           EN    OUT —QW34
                   MW34 — IN    ENO
``` |

Figure 7-26. Inserting logic to check the current counter value of a simple counter in FBD

Chapter 8

LAD/FBD
Conversion Operations

This chapter introduces LAD/FBD Conversion Operations. Conversions are necessary when it comes to changing from one data type or format to another. Often a conversion is required before a value contained in a memory location or variable may be used in a compare or an arithmetic operation. For example, the conversion of an integer variable would be required prior to outputting it to a BCD device. The various S7 data types and formats are covered in Chapter 1 *Memory and Addressing*.

Main Topics

- Conversion Operations Overview

- Conversion Descriptions

- Programming Conversions in LAD

- Programming Conversions in FBD

CONVERSION OPERATIONS OVERVIEW

STEP 7 *conversion operations* are listed and briefly described in Table 8-1. Conversion operations support conversion of variables of data type INT (integer) and DINT (double integer), BCD numbers, and REAL numbers. There are also four operations for converting a REAL number to the DINT format by rounding. These four conversion operations differ in how the number is rounded with respect to the fractional part of the floating-point number.

Using conversion instructions, you can change the format or data type of the contents of a memory location or declared block variable. For instance, a value input as BCD must be converted before use in an integer arithmetic operation. Conversely, if the integer result must be output to a BCD display, then the value would first require conversion to BCD. STEP 7 arithmetic operations support integer, double integer and real numbers—a variable that does not conform to one of these types would require the appropriate conversion.

Table 8-1. LAD/FBD Conversion Operations Summary

LAD	FBD	Brief Description
BCD_I — EN ENO — — IN OUT —	BCD_I — EN OUT — — IN ENO —	**BCD to Integer.** When **EN** is at logic 1, convert the 3-digit BCD value supplied at **IN** to a 16-bit INT value. The converted result is placed in the location specified at **OUT**.
I_BCD — IN OUT —	I_BCD — EN OUT — — IN ENO —	**Integer to BCD.** When **EN** is at logic 1, convert the 16-bit INT value supplied at **IN** to a 3-digit BCD value from 000-999. The converted result is placed in the location specified at **OUT**.
I_DI — IN OUT —	I_DI — EN OUT — — IN ENO —	**Integer to Double Integer.** When **EN** is at logic 1, convert the 16-bit INT value supplied at **IN** to a 32-bit DINT value. The converted result is placed in the location specified at **OUT**.
BCD_DI — IN OUT —	BCD_DI — EN OUT — — IN ENO —	**BCD to Double Integer.** When **EN** is at logic 1, convert the 7-digit BCD value supplied at **IN** to a 32-bit DINT value. The converted result is placed in the location specified at **OUT**.
DI_BCD — IN OUT —	DI_BCD — EN OUT — — IN ENO —	**Double Integer to BCD.** When **EN** is at logic 1, convert the 32-bit DINT value supplied at **IN** to a 7-digit BCD value. The converted result is placed in the location specified at **OUT**.
DI_R — IN OUT —	DI_R — EN OUT — — IN ENO —	**Double Integer to REAL.** When **EN** is at logic 1, convert the 32-bit DINT value supplied at **IN** to a REAL value (floating-point). The converted result is placed in the location specified at **OUT**.

Note: When EN is '1' ENO will also be '1' except on error; then while EN is '1', ENO is '0'.

Table 8-1. LAD/FBD Conversion Operations Summary (continued)

LAD	FBD	Brief Description
INV_I —IN OUT—	INV_I —EN OUT— —IN ENO—	**Invert Integer.** When **EN** is logic 1, invert or complement the integer value supplied at **IN**. The converted result is placed in the location specified at **OUT**.
INV_DI —IN OUT—	INV_DI —EN OUT— —IN ENO—	**Invert Double Integer.** When **EN** is logic 1, invert or complement the double integer value at **IN**. The converted result is placed in the location specified at **OUT**.
NEG_I —IN OUT—	NEG_I —EN OUT— —IN ENO—	**Negate Integer.** When **EN** is logic 1, negate the integer value at **IN** (change sign of value; positive-to-negative, negative-to-positive). The converted result is placed in the location specified at **OUT**.
NEG_DI —IN OUT—	NEG_DI —EN OUT— —IN ENO—	**Negate Double Integer.** When **EN** is logic 1, negate the double integer value at **IN** (positive-to-negative, negative-to-positive). The converted result is placed in the location specified at **OUT**.
NEG_R —IN OUT—	NEG_R —EN OUT— —IN ENO—	**Negate REAL.** When **EN** is logic 1, negate the REAL value supplied at **IN** (positive-to-negative, negative-to-positive). The converted result is placed in the location specified at **OUT**.
ROUND —IN OUT—	ROUND —EN OUT— —IN ENO—	**Round.** When **EN** is logic 1, convert the REAL value at **IN** by Rounding to the nearest Double Integer value. The converted result is placed in the location specified at **OUT**.
TRUNC —IN OUT—	TRUNC —EN OUT— —IN ENO—	**Truncate.** When **EN** is logic 1, convert the REAL value at **IN** to a DINT value by Truncating the fractional part of the REAL value. The truncated result is placed in the location specified at **OUT**.
CEIL —IN OUT—	CEIL —EN OUT— —IN ENO—	**Ceiling.** When **EN** is logic 1, convert the REAL value at **IN** to a DINT value by Rounding to the lowest DINT value greater-than-or-equal to **IN**. The rounded result is placed in the location specified at **OUT**.
FLOOR —IN OUT—	FLOOR —EN OUT— —IN ENO—	**Floor.** When **EN** is logic 1, convert the REAL value at **IN** to a DINT value by Rounding to the highest DINT value less-than-or-equal to **IN**. The rounded result is placed in the location specified at **OUT**.

Note: When EN is '1', ENO will also be '1' except on error; then while EN is '1'; ENO is '0'.

CONVERSION DESCRIPTIONS

Each conversion operation is identified by a unique mnemonic. For example, the mnemonic for the instruction that converts a 7-decade BCD value to a double integer value is **BCD_DI**. Each conversion operation is described in the following discussion. First we'll take a look at the parameters of Table 8-2. These are the parameters that must be specified when you program any conversion operation.

Conversion Instruction Parameters

As shown in the table below, the conversion box has an input parameter **IN**, where you specify the value to be converted, and an output parameter **OUT** where you must specify where the converted result is stored. The parameter **IN** may be a declared variable of the appropriate data type, or a memory location of correct width. The location you specify for **OUT** can also be a declared variable of the correct data type, or a memory location of the correct width.

The enable input line **EN** is where you may insert logic that determines when the conversion is executed. On the output side, you may connect the **ENO** signal to drive a LAD coil/FBD output or the EN input line of another operation. If an error occurs while the RLO = 1 at the EN input, **ENO** is set to '0'; otherwise, the status of **ENO** follows the status of EN. Although each conversion operation has EN/ENO signals, all do not necessarily signal an error.

Table 8-2. LAD/FBD Conversion Instruction Parameters

Parameter	Data Type	Description	Required?	Conversion Box	
EN		Enable Input Logic Line	YES	**LAD**	**FBD**
IN	WORD	Value to be Converted	YES	BCD_I —EN ENO—	BCD_I —EN OUT—
ENO		Enable Output Logic Line	NO	—IN OUT—	—IN ENO—
OUT	INT	Converted Result	YES		

BCD to Integer: BCD_I

When the RLO = 1 at the **EN** input, the *BCD to Integer* operation converts the 3-digit BCD value at **IN** to a 16-bit integer value (decimal) in the range of 0 to ± 999. The converted result is placed in the word location you specify at **OUT**.

At the input **IN,** you may reference a word-width memory location, or specify a variable of type WORD. At **OUT** you may specify a word-width memory location—for example MW70, or a variable of type INT. An invalid BCD value at **IN**—for example 10-15 (A-F_{16}), will result in a parameter assignment error; an error is reported and synchronous error OB 121 is called. The CPU is interrupted and transitions to STOP, if OB 121 is not loaded in the CPU. Status bits are not set on this error.

In the example below, the WORD variable #BCD3_In is interpreted as a 3-digit BCD value that is converted on each execution, to a 3-digit decimal number of the same value. The result is placed in bit memory word MW70.

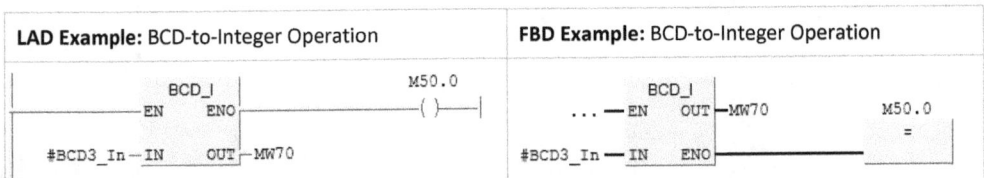

LAD Example: BCD-to-Integer Operation	FBD Example: BCD-to-Integer Operation
BCD_I —EN ENO——————( )—— #BCD3_In—IN OUT—MW70 M50.0	BCD_I ... —EN OUT—MW70 M50.0 #BCD3_In — IN ENO——————— =

Figure 8-1. Illustration of LAD/FBD instruction BCD-to-Integer

BCD to Double Integer: BCD_DI

When the RLO = 1 at the **EN** input, the *BCD to Double Integer* instruction converts the 7-digit BCD value at **IN** to a 32-bit integer value (decimal) in the range of 0 to ± 9 999 999. The converted result is placed in the DINT variable or double word location you specify at **OUT**.

At the input **IN,** you may reference a double word location, or specify a variable of type DWORD—for example BCD_In. At **OUT** you may specify a double word location—for example MD130, or specify a variable of type DINT. An invalid BCD value at **IN**—for example 10-15 (A-F$_{16}$), results in a parameter assignment error; an error is reported and synchronous error OB 121 is called. The CPU is interrupted and will transition to STOP, if OB 121 has not been programmed. Status bits are not set on this error.

In the following example the WORD variable #BCD7_In is interpreted as a 7-digit BCD value that is converted to a 7-digit decimal number of the same value on each execution. The result is placed in bit memory double word MD130.

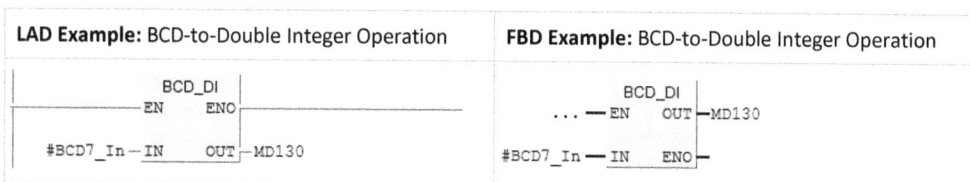

LAD Example: BCD-to-Double Integer Operation	FBD Example: BCD-to-Double Integer Operation
| ```
         BCD_DI
      EN      ENO

 #BCD7_In — IN   OUT — MD130
``` | ```
          BCD_DI
 ... — EN    OUT — MD130

 #BCD7_In — IN   ENO
``` |

Figure 8-2. Illustration of LAD/FBD instruction BCD-to-Double Integer

Integer to BCD: I_BCD

When the RLO = 1 at the **EN** input, the *Integer to BCD* instruction converts the 16-bit integer value (decimal) at **IN** to a 3-digit BCD value, and places the result in the word location you specify at **OUT**.

At the input **IN,** you may reference a word-width location—for example MW72, or specify a variable of type INT. At **OUT** you may specify a word-width location, or a variable of type WORD—for example BCD3_Out. If the integer input value is greater than 999, then the OV and OS status bits are set and the conversion operation is aborted. You may use OS and OV status bits to check for errors.

In the following example, the INT (16-bit integer) value in the bit memory word MW72 is converted to a 3-digit BCD value on each execution. The result is placed in the WORD variable #BCD3_Out. The data format for a 3-digit BCD number is shown in Figure 2-27.

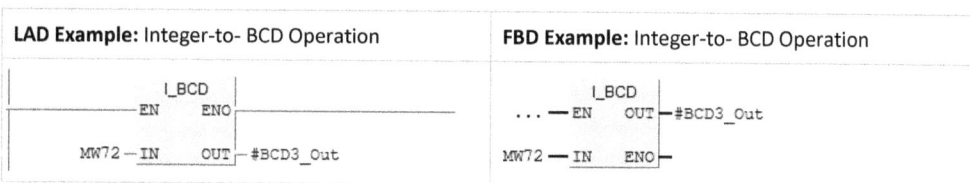

LAD Example: Integer-to- BCD Operation	FBD Example: Integer-to- BCD Operation
| ```
         I_BCD
      EN      ENO

  MW72 — IN    OUT — #BCD3_Out
``` | ```
          I_BCD
 ... — EN    OUT — #BCD3_Out

 MW72 — IN    ENO
``` |

Figure 8-3. Illustration of LAD/FBD instruction Integer-to-BCD

Double Integer to BCD: DI_BCD

When the RLO = 1 at the **EN** input, the *Double Integer to BCD* instruction converts the 32-bit integer value (decimal) at **IN** to a 7-digit BCD value in the range of 0 to ±9 999 999. The converted result is placed in the double word location you specify at **OUT**.

At the input **IN** you may reference a double word location—for example MD74, or specify a variable of type DINT. At **OUT** you may specify a double word location, or specify a variable of type DWORD—for example BCD7_Out. If the value at **IN** is greater than 9,999,999, then the OV and OS status bits are set and the conversion operation is aborted. You may use OS and OV status bits to check for errors.

In the following example the DINT (32-bit integer) value in bit memory double word MW74 is converted to a 7-digit BCD value on each execution. The result is placed in the DWORD variable #BCD7_Out. The data format for a 7-digit BCD number is shown in Figure 2-30.

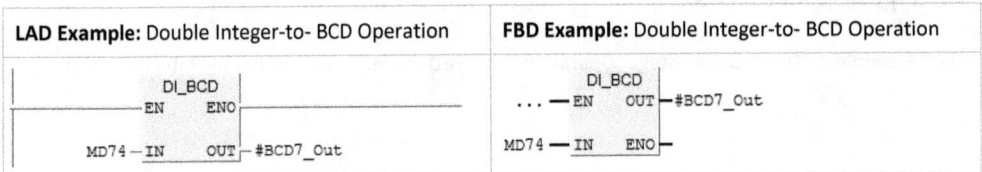

LAD Example: Double Integer-to- BCD Operation	FBD Example: Double Integer-to- BCD Operation
DI_BCD / EN ENO / MD74 — IN OUT — #BCD7_Out	DI_BCD / ... — EN OUT — #BCD7_Out / MD74 — IN ENO

Figure 8-4. Illustration of LAD/FBD instruction Double Integer-to-BCD

Integer to Double Integer: I_DI

When the RLO = 1 at the **EN** input, the *Integer to Double Integer* instruction converts the 16-bit integer value at **IN** to a 32-bit integer format and puts the result in the double word location you specify at **OUT**. The 16-bit value at **IN** is moved right-justified to the location you specify at **OUT**. The original sign bit (bit-15) value is placed in bits 16-31.

At the input **IN** you may reference a word-width location—for example MW78, or specify a variable of type INT. At **OUT** you may specify a double word location—for example MD134, or a variable of type DINT. No errors are signaled with this conversion operation.

In the following example, the INT (16-bit integer) value in memory word MW78 is converted, on each program cycle, to a DINT (32-bit integer) value. The result is placed in memory double word MD134.

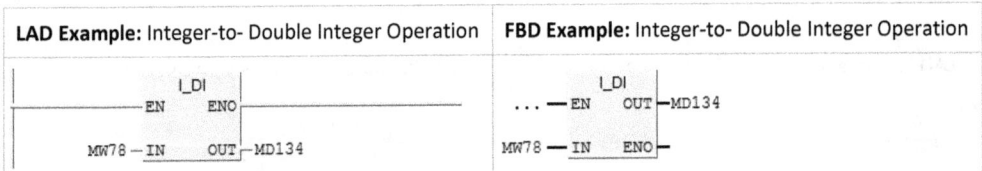

LAD Example: Integer-to- Double Integer Operation	FBD Example: Integer-to- Double Integer Operation
I_DI / EN ENO / MW78 — IN OUT — MD134	I_DI / ... — EN OUT — MD134 / MW78 — IN ENO

Figure 8-5. Illustration of LAD/FBD instruction Integer-to-Double Integer

Double Integer to REAL: DI_R

When the RLO = 1 at the **EN** input, the *Double Integer to Real* instruction converts the 32-bit integer value at **IN** to a REAL value in floating-point number, and places the result in the double word location you specify at **OUT**. Because of its greater accuracy, the double integer result may be rounded to the next highest whole number when converted to REAL (See ROUND instruction).

At the input at **IN** you may reference a double word location—for example MD134, or specify a variable of type DINT. At **OUT** you may specify a double word location—for example MD234, or specify a variable of type REAL. No errors are signaled with this conversion operation.

In the following example, the DINT (32-bit integer) value in memory double word MD134 is converted, on each execution, to a 32-bit floating-point value. The result is placed in memory double word MD234.

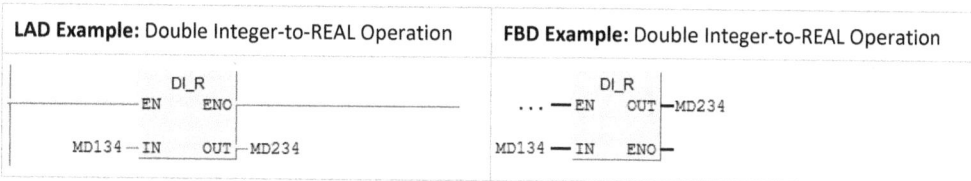

LAD Example: Double Integer-to-REAL Operation	**FBD Example:** Double Integer-to-REAL Operation
DI_R —EN ENO— MD134 — IN OUT ⊢ MD234	DI_R ... — EN OUT ⊢ MD234 MD134 — IN ENO ⊢

Figure 8-6. Illustration of LAD/FBD instruction Double Integer-to-Real

Round to Nearest Integer: ROUND

When the RLO = 1 at the **EN** input, the *ROUND* instruction converts the floating-point value at **IN** to a double integer value and places the result in the double word location specified at **OUT**. The result is the nearest whole number that is greater than or equal to the original whole number. If the input value lies halfway between the lower and upper integer values, then the result returns the even integer value. For example, if **IN** = 2.5 then **OUT** = 2; if **IN** = 2.59 then **OUT** = 3.

At the input at **IN,** you must specify a variable, or reference a double word location that contains a floating-point value. At **OUT** you may specify a double word location—for example MD210, or a variable of type DINT. If the REAL input value at **IN** is invalid or is outside of the permissible range of a double integer value, then status bits OV and OS are set and the conversion operation is aborted.

In the following example, the floating point value in bit memory double word MD110 is rounded, on each execution, to the nearest whole DINT (32-bit integer) value. The result is placed in bit memory double word MD210.

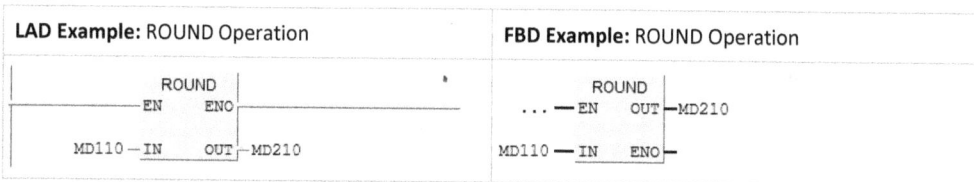

LAD Example: ROUND Operation	**FBD Example:** ROUND Operation
ROUND —EN ENO— MD110 — IN OUT ⊢ MD210	ROUND ... — EN OUT ⊢ MD210 MD110 — IN ENO ⊢

Figure 8-7. Illustration of LAD/FBD instruction Round to Double Integer

Truncate: TRUNC

When the RLO = 1 at the **EN** input, the *TRUNC* instruction converts the floating-point value at **IN** to a double integer and places the result in the double word location specified at **OUT**. The result will reflect the original floating-point number specified at **IN**, after truncating or dropping the fractional part of the floating-point number. For example, if **IN** = 2.95 then **OUT** = 2.

At the input **IN,** you must specify a variable of type REAL, or a double word location that contains floating-point value. At **OUT** you may specify a double word location—for example MD214, or a variable of type DINT. If the REAL input value at **IN** is invalid, or is outside of the permissible range of a double integer value, then status bits OV and OS are set and the conversion operation is aborted.

In the following example, the floating point value in bit memory double word MD110 is truncated, on each execution. The DINT (32-bit integer) value result is placed in bit memory double word MD214.

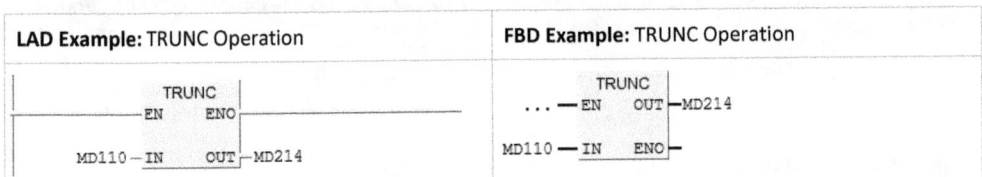

LAD Example: TRUNC Operation	FBD Example: TRUNC Operation
TRUNC EN ENO MD110 — IN OUT — MD214	TRUNC ... — EN OUT — MD214 MD110 — IN ENO —

Figure 8-8. Illustration of LAD/FBD instruction Truncate

Ceiling: CEIL

When the RLO = 1 at the **EN** input, the *CEIL* instruction converts the floating-point value at **IN** to a double integer and places the result in the double word location specified at **OUT**. The result will reflect the original floating-point number specified at **IN**, rounded to the next higher integer value that is greater than or equal to the original number. For example, if **IN** = 2.3 then **OUT** = 3. In short, this conversion rounds the value at IN to the next higher whole number.

At the input **IN,** you must specify a variable of type REAL, or a double word location that contains floating-point value. At **OUT** you may specify a double word location—for example MD218, or a variable of type DINT. If the REAL input value at **IN** is invalid, or is outside of the permissible range of a double integer value, then status bits OV and OS are set and the conversion operation is aborted.

In the following example, the floating-point value in bit memory double word MD110 is rounded, on each execution, to the next higher whole DINT (32-bit integer) value. The result is placed in memory double word MD218.

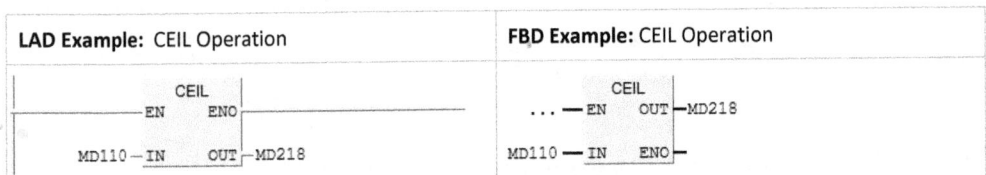

LAD Example: CEIL Operation	FBD Example: CEIL Operation
CEIL EN ENO MD110 — IN OUT — MD218	CEIL ... — EN OUT — MD218 MD110 — IN ENO —

Figure 8-9. Illustration of LAD/FBD instruction Ceiling

Floor: FLOOR

When the RLO = 1 at the **EN** input, the *FLOOR* instruction converts the floating-point value at **IN** to a double integer and places the result in the double word location specified at **OUT**. The result will reflect the original floating-point number specified at **IN**, rounded to the next lower integer value that is less than or equal to the original number. For example, if **IN** = 2.95 then **OUT** = 2. In short, this conversion rounds the value at **IN** to the next lower whole number.

At the input **IN,** you must specify a variable of type REAL, or a double word location that contains floating-point value. At **OUT** you may specify a double word location—for example MD222, or a variable of type DINT. If the REAL input value at **IN** is invalid, or is outside of the permissible range of a double integer value, then status bits OV and OS are set and the conversion operation is aborted.

In the following example, the floating point value in double word MD110 is rounded to the next lower whole DINT (32-bit integer) value, on each execution. The result is placed in double word MD222.

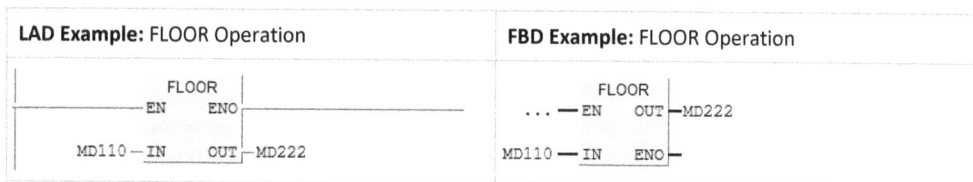

LAD Example: FLOOR Operation	FBD Example: FLOOR Operation
| ```
              FLOOR
        ─EN       ENO─
  MD110 ─IN       OUT─MD222
``` | ```
              FLOOR
      ... ─EN    OUT─MD222
  MD110 ─ IN     ENO─
``` |

Figure 8-10. Illustration of LAD/FBD instruction Floor.

Invert Integer: INV_I

When the RLO = 1 at the **EN** input, the *Invert Integer* instruction inverts the 16-bit integer value at **IN** using a one's complement operation, which simply inverts each bit according to the NOT truth table. Upon execution, each '1' of the initial value at **IN** is changed to '0', and each '0' is changed to '1', as shown below. The result is placed in the word-width location you specify at **OUT**. No errors are signaled with this conversion operation.

NOT Truth Table		Example: Invert Integer at IN	
IN	**OUT**	1010 0101 1111 1100	IN (INT) = IW22
0	1		
1	0	0101 1010 0000 0011	OUT = MW22

Figure 8-11. Illustration of Ones Complement Integer operation

At the input **IN,** you may reference a word-width location—for example IW22, or a variable of type INT. At **OUT** you may specify a word-width location—for example MW22, or a variable of type INT. No errors are signaled with this conversion. In the example below, when M100.1 is '1' AND Q24.7 is '0', the integer value in input word IW22 is inverted; the result is placed in bit memory word MW22.

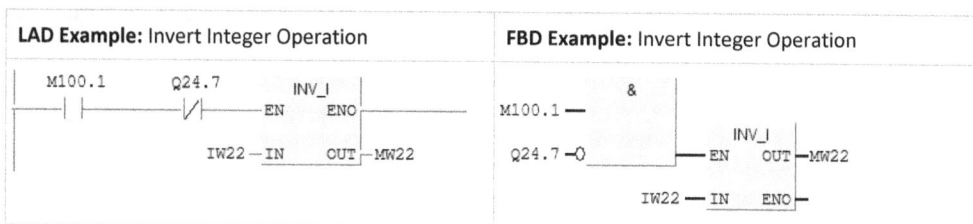

LAD Example: Invert Integer Operation	FBD Example: Invert Integer Operation
| ```
  M100.1    Q24.7      INV_I
  ─┤ ├───────┤/├────EN    ENO─
               IW22 ─IN   OUT─MW22
``` | ```
               &
  M100.1 ─
                           INV_I
  Q24.7 ─○          ─EN    OUT─MW22
              IW22 ─ IN    ENO─
``` |

Figure 8-12. Illustration of LAD/FBD instruction Double Integer-to-Real

Invert Double Integer: INV_DI

When the RLO = 1 at the **EN** input, the *Invert Double Integer* instruction inverts the 32-bit double integer value at **IN** using the one's complement operation, which simply inverts each bit according to the NOT truth table. Upon execution, each '1' of the original value at **IN** is changed to '0', and each '0' is changed to '1'. The result is placed in the double word location you specify at **OUT**.

NOT Truth Table		Example: Invert Double Integer at IN	
IN	OUT	1010 0101 1111 1100 1010 0101 1111 1100	IN (DINT) = ID42
0	1		
1	0	0101 1010 0000 0011 0101 1010 0000 0011	OUT = ID142

Figure 8-13. Illustration of one's complement to invert integer

At input **IN,** you may reference a double word location containing a 32-bit integer, or a variable of type DINT. At **OUT** you may specify a double word location or a variable of type DINT. In the example below, the DINT value of input double word ID42 is inverted when M100.2 is '1' AND Q14.7 is '0'; the result is placed in bit memory double word MD142. Errors are not reported in this operation.

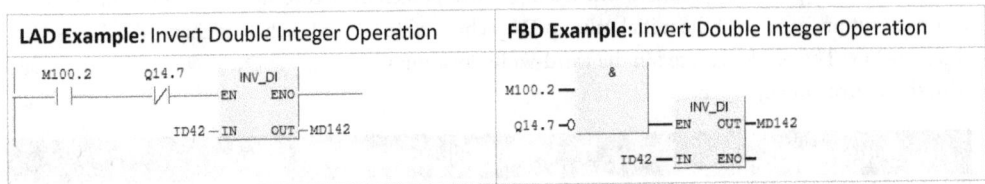

LAD Example: Invert Double Integer Operation	FBD Example: Invert Double Integer Operation

Figure 8-14. Illustration of LAD/FBD instruction Invert Double Integer

Negate Integer: NEG_I

When the RLO = 1 at **EN**, the *Negate Integer* instruction changes the sign of the 16-bit integer at **IN** by performing a two's complement. This operation changes the sign of a positive value to negative and a negative value to positive. The result is placed in the location specified at **OUT**. As shown below, the 2's complement is computed by first performing a 1's complement on the original value and then adding +1 to the complemented result. At the input at **IN** you may reference a word location containing a 16-bit integer, or a variable of type INT. At **OUT** you may specify a word location or a variable of type INT.

Example: Negate using Two's Complement	
0000 0000 0000 0101	IN = + 5
1111 1111 1111 1010	Complement (1s to 0s and 0s to 1s)
0000 0000 0000 0001	Add +1 to complemented value
1111 1111 1111 1011	OUT =—5

Figure 8-15. Illustration of two's complement to negate integer

In the following example, when M77.6 closes, the integer value in word MW230 is negated. The result is placed in memory word MW232. Upon executing of the NEG_I operation, status bits CC0, CC1, OV, and OS are set; and may be checked to determine if the result is zero, positive, negative, or extreme negative. See Table 13-3.

LAD Example: Negate Double Integer Operation	FBD Example: Negate Double Integer Operation
| ```
M77.6      NEG_I
 ┤ ├──────EN   ENO──
    MW230─IN   OUT─MW232
``` | ```
              NEG_I
M77.6 ─ EN    OUT─MW232
MW230 ─ IN    ENO─
``` |

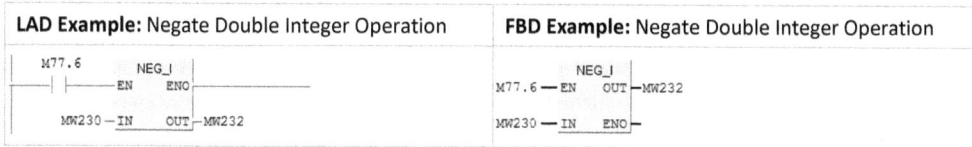

Figure 8-16. Illustration of LAD/FBD instruction Negate Integer

Negate Double Integer: NEG_DI

When the RLO = 1 at the **EN** input, the *Negate Double Integer* instruction changes the sign of the 32-bit double integer value at **IN** by performing a two's complement operation. This operation simply changes the sign of the value. Positive values are changed to negative, negative values are changed to positive. The operation is equivalent to multiplying by -1. The result is placed in the double word location you specify at **OUT**.

At the input at **IN,** you may reference a double word location containing a 32-bit double integer value (-2,147,483,648 to 2,147,483,647), or a variable of type DINT. At **OUT** you may specify a double word location or a variable of type DINT. Upon executing the NEG_DI, CPU status bits CC0, CC1, OV, and OS are set; and may be checked to determine if the result is zero, positive, negative, or extreme negative (-2,147,483,648). See Table 13-3.

In the following example, when M77.7 closes 1, the DINT (32-bit integer) value in memory double word MD234 is negated. The result is placed in memory double word MD238.

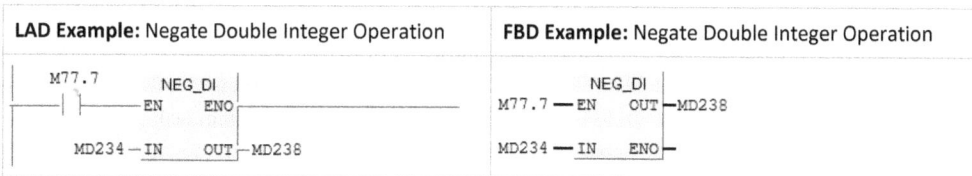

LAD Example: Negate Double Integer Operation	FBD Example: Negate Double Integer Operation
| ```
M77.7      NEG_DI
 ┤ ├──────EN   ENO──
    MD234─IN   OUT─MD238
``` | ```
              NEG_DI
M77.7 ─ EN    OUT─MD238
MD234 ─ IN    ENO─
``` |

Figure 8-17. Illustration of LAD/FBD instruction Negate Double Integer

Negate REAL: NEG_R

When the RLO = 1 at the **EN** input, the *Negate Real* instruction negates the floating-point value at **IN**, by changing the sign of the number. Positive values are changed to negative and negative values are changed to positive. The operation is equivalent to multiplying by -1. The Negate Real instruction changes the sign of the mantissa of the original value. The negated result is placed in the double word location you specify at **OUT**.

At the input **IN,** you may specify a double word location that contains a positive or negative floating-point value (±1.75495e-38 to ±3.402823+e38), or a variable of type REAL. At **OUT,** you may specify a double word location, or a variable of type REAL. No errors are signaled with the NEG_R operation.

In the following example, when M77.7 closes, the REAL (32-bit floating-point) value in memory double word MD234 is negated. The result is placed in the memory double word MD242.

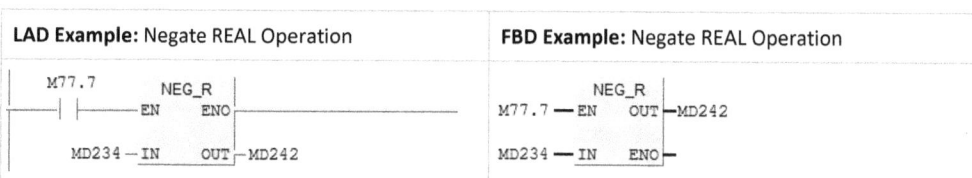

LAD Example: Negate REAL Operation	FBD Example: Negate REAL Operation
| ```
M77.7      NEG_R
 ┤ ├──────EN   ENO──
    MD234─IN   OUT─MD242
``` | ```
              NEG_R
M77.7 ─ EN    OUT─MD242
MD234 ─ IN    ENO─
``` |

Figure 8-18. Illustration of LAD/FBD instruction Negate Real

PROGRAMMING CONVERSIONS IN LAD

After inserting a new network, programming a conversion operation is a matter of first selecting the desired instruction from the Program Elements tab. The operation can be dragged and placed onto the LAD network. For example, **BCD_DI** is used to convert Double Integer to BCD.

Assigning Conversion Parameters

With the new network inserted and selected, you can double-click on the desired conversion object, or drag and drop the object onto the network. You must specify the parameter **IN** as the value to be converted, and the parameter **OUT** as the location where the result is to be stored. You may also use symbolic addresses—for example BCD7_IN, if they are already defined.

When specifying the conversion parameters, remember that, based on the operation, the data type of the variables at **IN** and at **OUT** may differ both in type and bit-width. For example, in the integer to double integer conversion, the input is 16-bit (INT); the output is 32-bit (DINT).

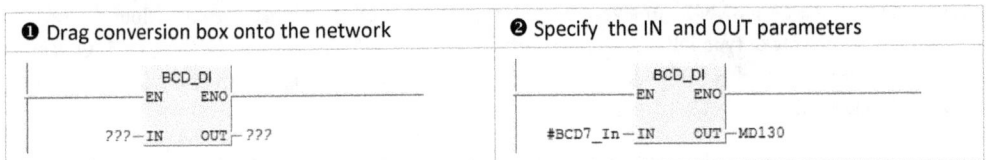

Figure 8-19. Select and insert conversion operation and specify the parameters IN and OUT in LAD

The next step is to insert the driving logic that will enable the conversion. On the **EN** line, you may insert any series or parallel logic combination using normally-open and normally-closed contacts, as well as other operations, as shown below. If the RLO = 1 at **EN**, the conversion is executed; if the RLO = 0 at **EN**, the conversion is not executed. You may also leave the **EN** input programmed without any driving conditions, as shown in the second cell above. If logic conditions are not programmed, the conversion will always execute.

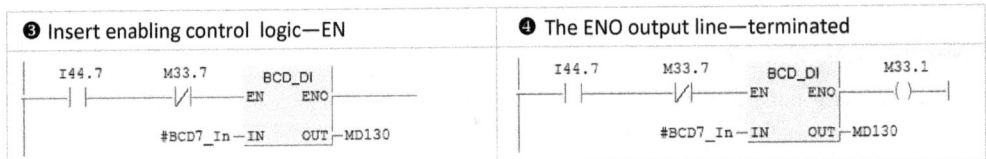

Figure 8-20. Insert enable (EN) logic and terminate the ENO output line in LAD

The **ENO** output line may be left un-terminated as in the first cell above, or terminated with a coil output instruction, as shown in the second cell. Series and parallel logic may also be inserted after the enable output **ENO** of the conversion operation, but only if the operation is terminated.

Conversion boxes may be inserted in series with other conversions, or with other operations, by connecting the **ENO** line of one conversion box to the **EN** line of the following box operation. The following box is evaluated only when the preceding box processes without error.

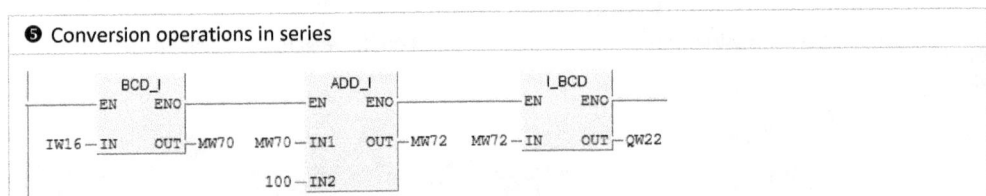

Figure 8-21. Using EN/ENO mechanism to program conversion operations in series in LAD

Conversion boxes may also be inserted in parallel if placed in a branch that begins directly connected to the left rail of the network. When placing conversion boxes in parallel, you may connect directly to the left rail without preceding logic conditions, or you may insert contact logic in line with **EN** and following **ENO**. When conversion boxes are in parallel, the network must be terminated with a coil. The coil may serve as part of the logic used to provide error evaluation ,or simply as a dummy output.

❻ Conversion operations in parallel

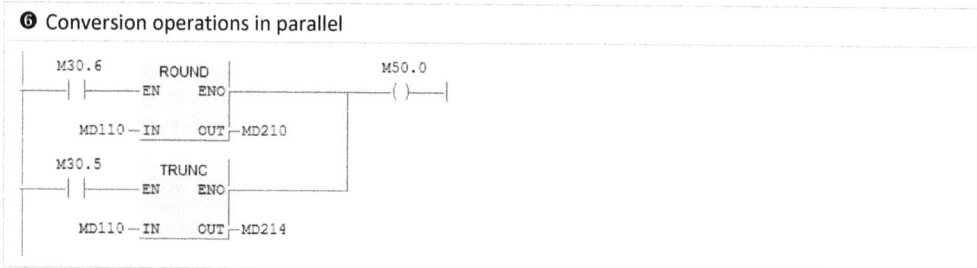

Figure 8-22. Inserting conversion operations in parallel in LAD

If a network arrangement involves two or more parallel branches, each of which has two or more conversion boxes in series, boxes are evaluated from left-to-right starting with the first branch, then the second branch, and so on.

❼ Conversion operations—parallel/series branches

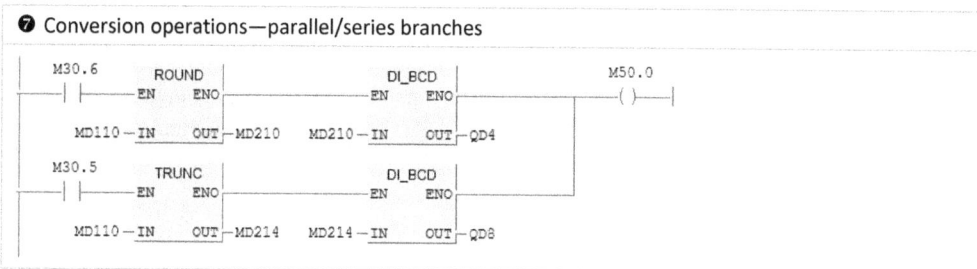

Figure 8-23. Inserting multiple branches of series conversion operations in LAD

PROGRAMMING CONVERSIONS IN FBD

After inserting a new network, programming a conversion operation is a matter of first selecting the desired instruction from the Program Elements tab. The operation can be dragged and placed onto the LAD network. For example, **BCD_DI** is used to convert Double Integer to BCD.

Assigning Conversion Parameters

With the network inserted and selected, double-click on the desired conversion object, or drag and drop the object onto the network. Specify the input parameter **IN** as the value to be converted, and the parameter **OUT** as the place where the result is stored. Use symbolic addresses—for example BCD7_IN, if they are already defined. When specifying the conversion parameters, remember that the data type of the variables at **IN** and at **OUT** may differ both in type and bit-width. For instance, in the integer to double integer conversion the input is 16-bit (INT); the output is 32-bit (DINT).

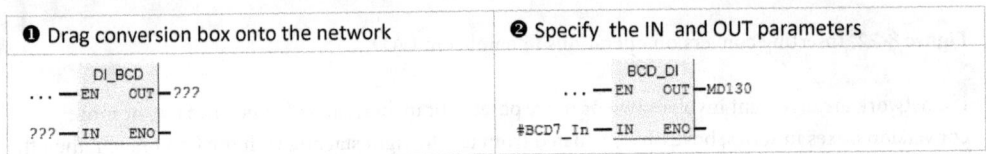

❶ Drag conversion box onto the network	❷ Specify the IN and OUT parameters
DI_BCD ... — EN OUT — ??? ??? — IN ENO —	BCD_DI ... — EN OUT — MD130 #BCD7_In — IN ENO —

Figure 8-24. Select and insert conversion operation and specify the parameters IN and OUT in FBD

The next step is to insert the enabling logic that drives the conversion. On the **EN** line, you may insert any gate logic combination, using positive or negated binary inputs, as well as other operations as shown below. If the RLO = 1 at **EN**, the conversion is executed; if the RLO = 0 at **EN**, the conversion is not executed. You may also leave the **EN** input programmed without any driving conditions, as shown in the second cell above. If logic conditions are not programmed, the conversion will always execute.

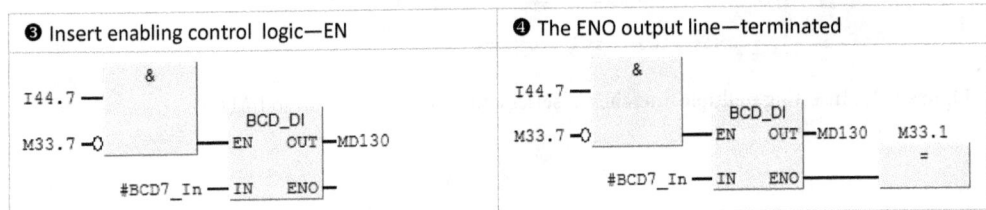

❸ Insert enabling control logic—EN	❹ The ENO output line—terminated
I44.7 — & M33.7 —O ——— BCD_DI EN OUT — MD130 #BCD7_In — IN ENO —	I44.7 — & M33.7 —O ——— BCD_DI EN OUT — MD130 M33.1 #BCD7_In — IN ENO ———— =

Figure 8-25. Insert enable (EN) logic and terminate the ENO output line in FBD

The **ENO** output line may be left un-terminated, as in the first cell above, or terminated with a binary output instruction as shown in the second cell. You may insert AND/OR gate logic after the enable output **ENO** of the conversion operation, but only if the operation is terminated.

Conversion boxes may be inserted in series with other conversions, or with other operations, by connecting the ENO line of one conversion box to the **EN** line of the following box operation. The following box is evaluated only when the preceding box processes without error.

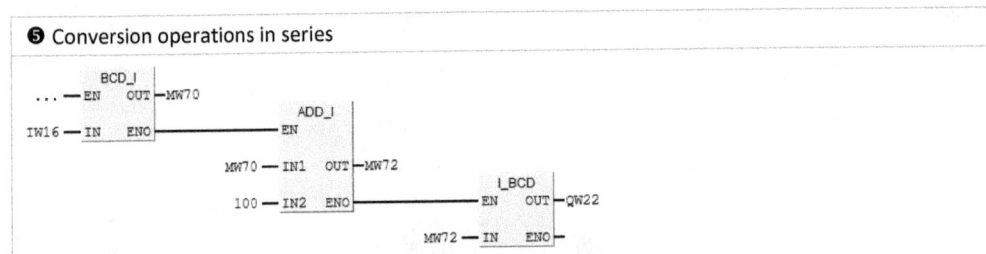

❺ Conversion operations in series
BCD_I ... — EN   OUT — MW70 IW16 — IN   ENO ————— ADD_I EN MW70 — IN1   OUT — MW72 100 — IN2   ENO ————— I_BCD EN   OUT — QW22 MW72 — IN   ENO —

Figure 8-26. Using EN/ENO mechanism to program conversion operations in series in FBD

160

Conversion boxes may also be inserted in an OR gate as shown below. In this arrangement, you may insert the box without preceding logic conditions, or you may insert logic in line with **EN** and following **ENO**. The logic string must be terminated with an assign output.

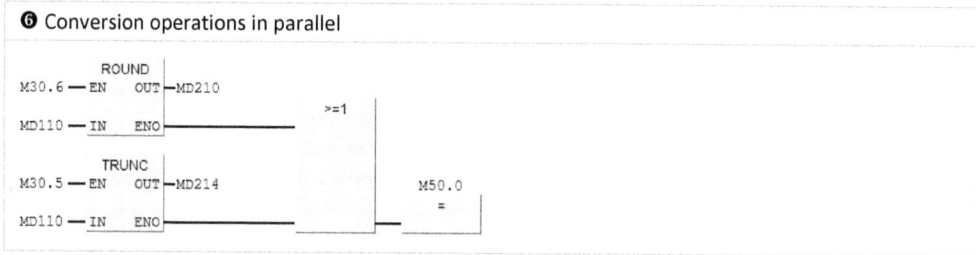

❻ Conversion operations in parallel

```
            ROUND
M30.6 ─ EN    OUT ─MD210
                              >=1
MD110 ─ IN    ENO

            TRUNC
M30.5 ─ EN    OUT ─MD214
                                    M50.0
MD110 ─ IN    ENO                     =
```

Figure 8-27. Inserting conversion operations in parallel in FBD

If a network has two or more parallel branches, each of which has two or more box operations in series, boxes are evaluated from left-to-right, starting with the first branch, then the second branch, and so on.

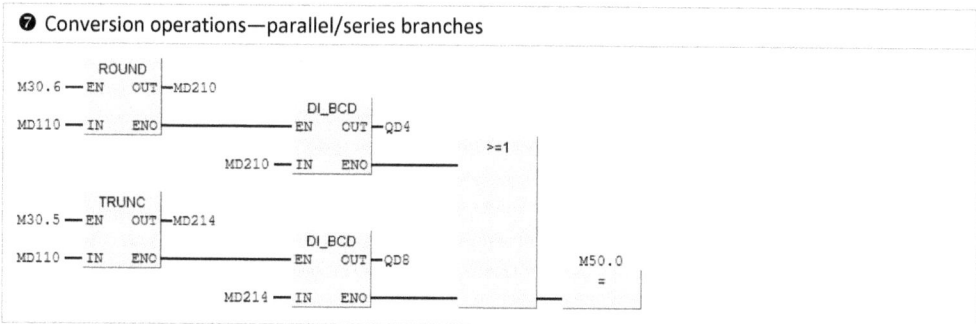

❼ Conversion operations—parallel/series branches

```
            ROUND
M30.6 ─ EN    OUT ─MD210
                              DI_BCD
MD110 ─ IN    ENO           ─ EN    OUT ─QD4
                                                >=1
                    MD210 ─ IN    ENO

            TRUNC
M30.5 ─ EN    OUT ─MD214
                              DI_BCD
MD110 ─ IN    ENO           ─ EN    OUT ─QD8
                                                      M50.0
                    MD214 ─ IN    ENO                   =
```

Figure 8-28. ORing multiple branches of series conversion operations in FBD

Chapter 9

LAD/FBD
Compare Operations

This chapter introduces LAD/FBD Compare Operations. The various compare operations compare two word, or double word locations, or variables. The outcome of the compare test is indicated by setting the result of logic operations to '1' if the result is True, and to '0' if the result is False.

In your control program, the compare result may be the basis of many other decisions. For example, the compare might determine if atimer duration, counter value, arithmetic calculation, or other process value has reached a setpoint or is within range. Depending on the outcome of the compare, the program may activate or de-activate a device, or enable or disable a process event. The compare result might also determine whether a block is called, or if an iterative process should be terminated.

Main Topics

- Compare Operations Overview
- Compare Operations Descriptions
- Programming Comparisons in LAD
- Programming Comparisons in FBD

COMPARE OPERATIONS OVERVIEW

STEP 7 *compare operations* are listed in Table 9-1. These operations support six types of compare tests, each of which may be performed with 16-bit integers (data type INT), 32-bit integers (data type DINT), and floating-point numbers (data type REAL). As shown in the table below, these tests include *equal, not equal, less than, greater than, less than or equal,* and *greater than or equal*.

Table 9-1. LAD/FBD Compare Operations Summary

Compare Test	INT -32,768 to +32,767	DINT -2,147,483,648 to +2,147,483,647	REAL ± 1.75495e-38 to ± 3.402823+e38
Equal	CMP = = I	CMP = = D	CMP = = R
Not Equal	CMP < > I	CMP <> D	CMP < > R
Less Than	CMP < I	CMP < D	CMP < R
Greater Than	CMP > I	CMP > D	CMP > R
Less Than or Equal	CMP < = I	CMP < = D	CMP < = R
Greater Than or Equal	CMP > = I	CMP > = D	CMP > = R

Typical uses of compare operations include range checking, setpoint control, and calculation limit checks. *Compare operations* might be used, for instance, to check the remaining time of a timer or to determine whether a counter has reached a certain intermediate or final value. In operations that involve a loop or an iterative calculation—like the averaging of analog readings, then the compare may be used to check if the process is complete, or if the loop should be terminated. The compare might also be used to perform machine/process diagnostics—for example if a machine has 32 known states at a given step of its cycle, then the known values might be compared to the 32-bits of the actual states.

COMPARE OPERATIONS DESCRIPTIONS

Each of the six compare tests supports comparison of integer (**I**), double integer (**D**), and Real (**R**) values. Each operation is identified by a mnemonic label of the box instruction. For example, to compare two integer values for non-equality you would use **CMP < > I**; you would use **CMP < > D** to compare two double integer values; and for two Real values, **CMP < > R**. In the following discussions, the six basic compare operations are described for integer (INT) values. The operations described would be the same for comparing DINT or REAL values—the mnemonic instructions would change respectively to **D** or **R**.

Compare Instruction Parameters

The parameters that must be specified as you work with compare operations are listed and briefly described in Table 9-2. The compare does not have an enable EN line and is not affected by logic that precedes the operation. Instead, the compare functions like a normally-open contact. It either allows or interrupts power flow based on the outcome of the compare. Logic that you insert on the unlabeled input line, or unlabeled output line, is logically combined with the compare result of logic '0' or '1'.

Table 9-2. LAD/FBD Compare Box Parameters

Parameter	Data Type	Description	Required?	Compare Box
		Unlabeled Input Line		LAD/FBD
IN1	INT/DINT/REAL	Value 1 for Compare	YES	CMP ==D
IN2	INT/DINT/REAL	Value 2 for Compare	YES	??? — IN1
		Unlabeled Output Line (Signal result = 1 > Compare = True) (Signal result = 0 > Compare = False)	YES	??? — IN2

At parameters IN1 and IN2, you must specify the values to be compared. On the output side, the unlabeled line should drive a binary output to signal the True/False outcome. Logic '1' represents a true outcome; logic '0' represents a false outcome. You may use the unlabeled output line to drive a specific output address or variable of type BOOL.

When specifying IN1 and IN2, you must specify two variables of the same type—for example you may specify two variables of type INT (16-bit), DINT (32-bit) or REAL (32-bit), based on the operation. You may also specify two memory locations that contain the values to be compared. The specified locations must be of the same width—for example MW22 and IW22 for an INT compare; MD 46 and MD60 for a DINT compare. A variable or memory location may also be compared to a constant; (typically a constant is compared to a variable)—the constant needs to be in the permissible range of the data type to which it is being compared.

Compare Equal: CMP ==I

When processed the *Compare Equal* instruction evaluates whether the value at **IN1** is *equal to* the value at **IN2**. If the values at **IN1** and **IN2** are equal, then the RLO = 1 on the box output line, and power flow is passed out of the compare box.

At the inputs **IN1** and **IN2,** you may specify a constant in the range of -32768 to +32767, variables of data type INT, or you may reference a memory location of the correct width.

In the following example, the test is whether '1000' is equal to the value contained in bit memory word MW62. If the contact I 45.1 is closed AND the compare result is TRUE, then M101.1 is activated.

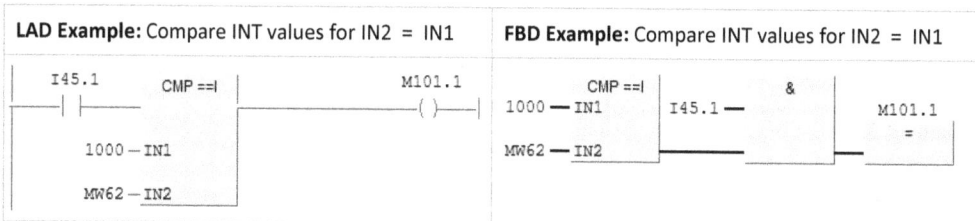

LAD Example: Compare INT values for IN2 = IN1

FBD Example: Compare INT values for IN2 = IN1

Figure 9-1. LAD/FBD illustration of Compare Equal instruction, using integers

Compare Not Equal: CMP <>I

When processed, the *Compare Not Equal* instruction evaluates whether the value at **IN1** is *not equal to* the value at **IN2**. If the values at **IN1** and **IN2** differ, then the RLO = 1 on the box output line, and power flow is passed out of the compare box.

At the inputs **IN1** and **IN2,** you may specify a constant in the range of -32768 to +32767, variables of data type INT, or you may reference a memory location of the correct width.

In the following example, the test is whether data element DB12.DBW10 is not equal to data element DB12.DBW12. If the input contact I 45.2 is closed AND the compare is TRUE, then M101.2 is activated.

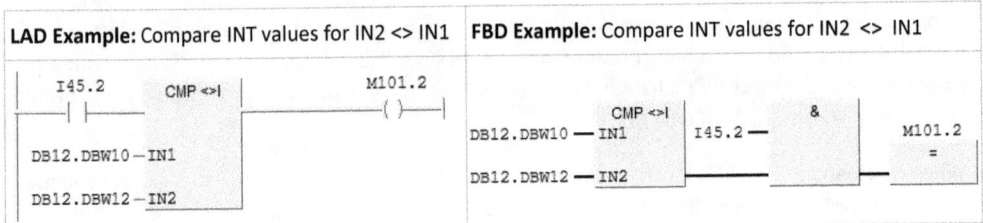

LAD Example: Compare INT values for IN2 <> IN1	FBD Example: Compare INT values for IN2 <> IN1
I45.2 CMP <>I M101.2 —| |——————————————( )—— DB12.DBW10 — IN1 DB12.DBW12 — IN2	CMP <>I & M101.2 DB12.DBW10 — IN1 I45.2 — = DB12.DBW12 — IN2

Figure 9-2. LAD/FBD illustration of Compare Not Equal instruction, using integers

Compare Greater Than: CMP >I

When processed, the *Compare Greater Than* instruction evaluates whether the value at **IN1** is *greater than* the value at **IN2**. If the value at **IN1** is greater than the value at **IN2**, then the RLO = 1 on the box output line, and power flow is passed out of the compare box.

At the inputs **IN1** and **IN2,** you may specify a constant in the range of -32768 to +32767, variables of data type INT, or you may reference a memory location of the correct width.

In the following example, the test is whether the value of #Speed_1 is greater than 1500. If the compare result is TRUE, then M101.3 is activated.

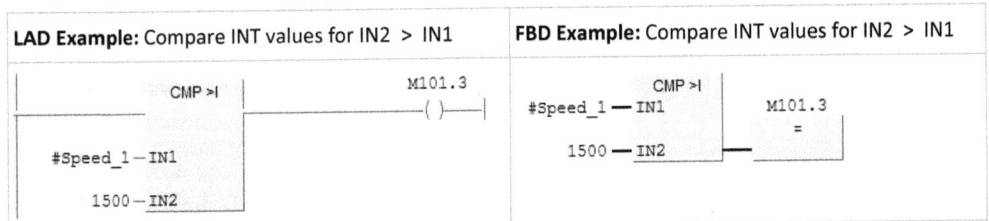

LAD Example: Compare INT values for IN2 > IN1	FBD Example: Compare INT values for IN2 > IN1
CMP >I M101.3 ——————————————( )—— #Speed_1 — IN1 1500 — IN2	CMP >I M101.3 #Speed_1 — IN1 = 1500 — IN2

Figure 9-3. LAD/FBD illustration of Compare Greater Than instruction, using integers

Compare Less Than: CMP <I

When processed, the *Compare Less Than* instruction evaluates whether the value at **IN1** is *less than* the value at **IN2**. When evaluated, if the value at **IN1** is less than the value at **IN2**, then the RLO = 1 on the box output line, and power flow is passed out of the compare box.

At the inputs **IN1** and **IN2,** you may specify a constant in the range of -32768 to +32767, variables of data type INT, or you may reference a memory location of the correct width.

In the following example, the test is whether the value of #Speed_1 is less than the value of #Speed_2. If the compare result is TRUE, then M101.4 is activated.

LAD Example: Compare INT values for IN2 < IN1	FBD Example: Compare INT values for IN2 < IN1

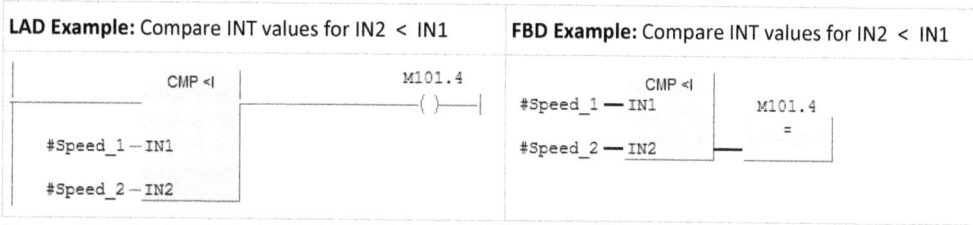

Figure 9-4. LAD/FBD illustration of Compare Less Than instruction, using integers

Compare Greater Than or Equal: CMP >=I

When processed, the *Compare Greater Than or Equal* instruction evaluates if the value at **IN1** is *greater than or equal to* the value at **IN2**. If the value at **IN1** is greater than or equal to the value at **IN2**, then the RLO = 1 on the box output line, and power flow is passed out of the compare box.

At the inputs **IN1** and **IN2,** you may specify a constant in the range of -32768 to +32767, variables of data type INT, or you may reference a memory location of the correct width.

In the following example, the test is whether the value of #Speed_4 is greater than or equal to 3500. If the contact I 45.3 is closed AND the compare result is TRUE, then M101.5 is activated.

LAD Example: Compare INT values for IN2 >= IN1	FBD Example: Compare INT values for IN2 >= IN1

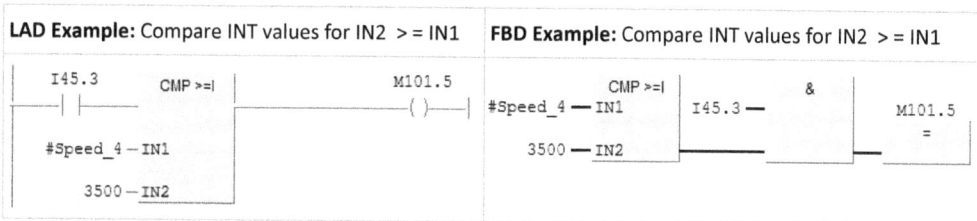

Figure 9-5. LAD/FBD illustration of Compare Greater Than or Equal instruction, using integers

Compare Less Than or Equal: CMP <=I

When processed, the *Compare Less Than or Equal* instruction evaluates if the value at **IN1** is *less than or equal to* the value at **IN2**. If the value at **IN1** is less than or equal to the value at **IN2**, then the RLO = 1 on the box output line, and power flow is passed out of the compare box.

At the inputs **IN1** and **IN2,** you may specify a constant in the range of -32768 to +32767, variables of data type INT, or you may reference a memory location of the correct width.

In the following example, the test is whether the value of #Speed_5 is less than or equal to '0'. If the contact I 45.4 is open AND the compare result is TRUE, then M101.6 is activated.

LAD Example: Compare INT values for IN2 <= IN1	FBD Example: Compare INT values for IN2 <= IN1

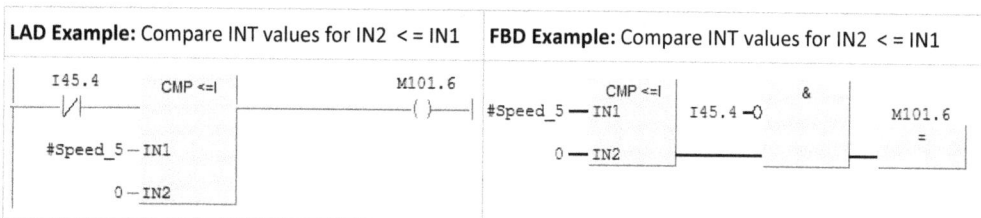

Figure 9-6. LAD/FBD illustration of Compare Less Than or Equal instruction, using integers

PROGRAMMING COMPARISONS IN LAD

After inserting a new network, programming a compare is a matter of first selecting the desired instruction from the Program Elements tab, based on the name. Each of the six compare types is available for comparing integer (I), double integer (D), and real (R) values. In Table 9-3, compare for equality is shown for integer (**CMP==I**), double integer (**CMP==D**), and REAL (**CMP==R**).

Table 9-3. Typical LAD Compare Instructions with possible parameter assignments

IN1/IN2	INT	DINT	REAL
	CMP ==I ??? — IN1 ??? — IN2	CMP ==D ??? — IN1 ??? — IN2	CMP ==R ??? — IN1 ??? — IN2
Constants (in permissible range)	-32,768 to +32,767	-2,147,483,648 to +2,147,483,647	± 1.75495e-38 to ± 3.402823+e38
S7 Memory Locations (Absolute or Symbolic)	IW 28 QW 42 MW 54 DB6.DBW12	ID 28 QD 42 MD 54 DB6.DBD12	ID 28 QD 42 MD 54 DB6.DBD12
Declared Variables (of correct data type)	#Value_1 #Value_2	#Value_1 #Value_2	#Value_1 #Value_2

Note: Symbolic addresses may be substituted for absolute addresses.

Assigning Compare Parameters

With the new network selected, you can drag and drop the desired object onto the network. Next, the two values to be compared, **IN1** and **IN2**, must be specified as absolute or symbolic addresses. The table above gives examples of values that can be compared. Remember the variable at **IN1** and at **IN2** must be of the same type. For instance, if using CMP = D, both values must be 32-bit DINT variables.

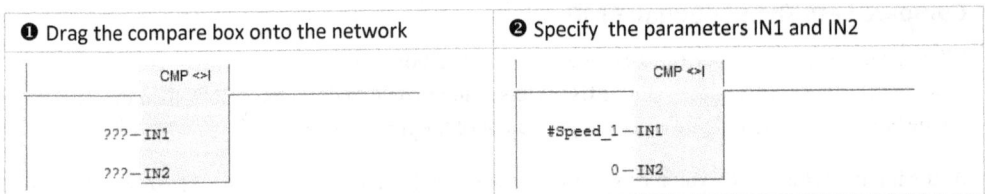

❶ Drag the compare box onto the network	❷ Specify the parameters IN1 and IN2
CMP <>I ??? — IN1 ??? — IN2	CMP <>I #Speed_1 — IN1 0 — IN2

Figure 9-7. Select and insert compare operation and specify the parameters IN1 and IN2 in LAD

The next step, as shown in the first cell below, is to insert a binary output variable to indicate the compare outcome. Here you may use any valid binary output instruction—for example set or reset coils. In cell ❸ below, if the integer variable Speed_1 is not equal to '0', M33.2 is activated.

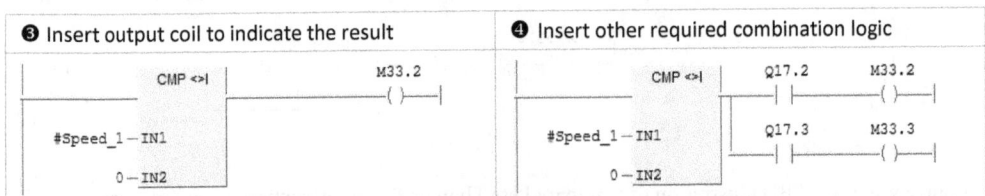

❸ Insert output coil to indicate the result	❹ Insert other required combination logic

CMP <>I   M33.2
                        —( )—
#Speed_1 — IN1
0 — IN2
	CMP <>I Q17.2 M33.2
	            —| |— —( )—
	#Speed_1 — IN1 Q17.3 M33.3
	            —| |— —( )—
	0 — IN2

Figure 9-8. Insert an output and combine other required logic combinations in LAD

As shown in cell ❹ of the previous figure, you may insert any combination of series/parallel logic either before or after the compare box. Remember that the compare behaves similar to a normally-open contact. If the compare is successful, power flows through the compare; if the test fails, power flow is interrupted. Any logic that you insert before or after the compare is logically combined with the compare result using the logical AND. Any logic placed in parallel with the compare is combined with the compare result using the logical OR.

As shown in ❺, multiple compare operations may also be placed in series, with preceding or following logic. If in series, all of the compare tests must be satisfied to complete logic continuity.

❺ Compare operations in series

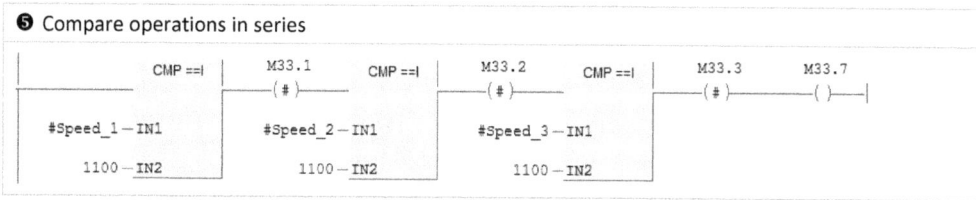

Figure 9-9. Inserting compare operations in series in LAD

Compare boxes may also be placed in parallel. If single compare operations are in parallel, a '1' result is produced for the branch when any compare in the branch is satisfied. In the first example ❻ below, M33.6 is energized if either of the compare tests is satisfied. In the second example ❼ below, M33.6 is energized if both compare tests are satisfied in either of the two branches.

❻ Compare operations in parallel

❼ Compare with—parallel/series branches

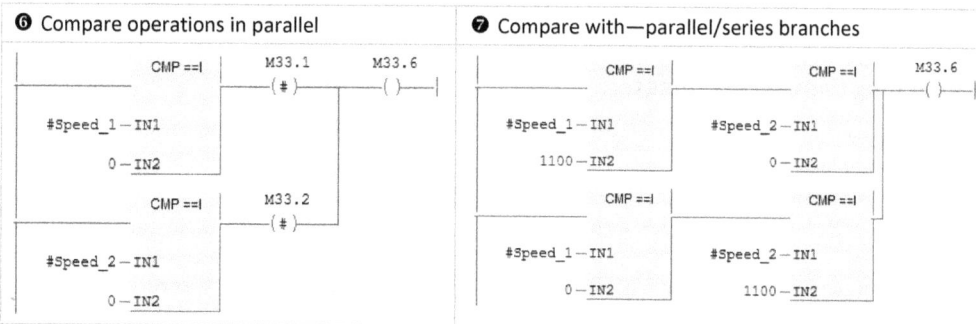

Figure 9-10. Inserting compare operations in parallel in LAD

PROGRAMMING COMPARISONS IN FBD

After inserting a new network, programming a compare operation is a matter of first selecting the desired operation from the Program Elements tab, based on the name. Each of the six compare types is available for comparing integer (I), double integer (D), and real (R) values. In the Table 9-4, compare for equality is shown for integer (**CMP = =I**), double integer (**CMP = =D**), and REAL (**CMP ==R**).

Table 9-4. Typical FBD Compare Instructions with possible parameter assignments

IN1/IN2	INT	DINT	REAL
	CMP ==I — IN1 — IN2	CMP ==D — IN1 — IN2	CMP ==R — IN1 — IN2
Constants (in permissible range)	-32,768 to +32,767	-2,147,483,648 to +2,147,483,647	± 1.75495e-38 to ± 3.402823+e38
S7 Memory Locations (Absolute or Symbolic)	IW 28 QW 42 MW 54 DB6.DBW12	ID 28 QD 42 MD 54 DB6.DBD12	ID 28 QD 42 MD 54 DB6.DBD12
Declared Variables (of correct data type)	#Value_1 #Value_2	#Value_1 #Value_2	#Value_1 #Value_2

Note: Symbolic addresses may be substituted for absolute addresses.

Assigning Compare Parameters

With the new network selected, you can drag and drop the desired object onto the network. Next, the two values to be compared, **IN1** and **IN2**, must be specified as absolute or symbolic addresses. The table above gives examples of values that can be compared. Remember the variables at **IN1** and at **IN2** must be of the same type. For instance, if using CMP = D, both values must be 32-bit DINT variables.

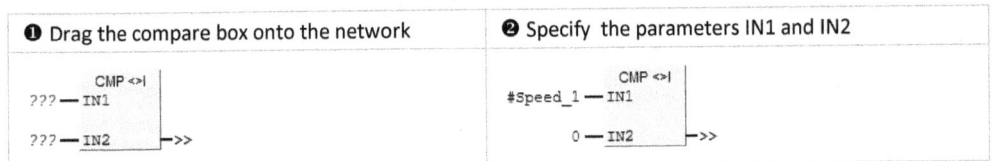

❶ Drag the compare box onto the network	❷ Specify the parameters IN1 and IN2
CMP <>I ??? — IN1 ??? — IN2 └—>>	CMP <>I #Speed_1 — IN1 0 — IN2 └—>>

Figure 9-11. Select and insert compare operation and specify the parameters IN1 and IN2 in FBD

The next step, as shown in the first cell below, is to insert a binary output variable to indicate the compare outcome. Here you may use any valid binary output instruction—for example set or reset outputs. In cell ❸ below, if the integer variable Speed_1 is not equal to '0', M33.2 is activated.

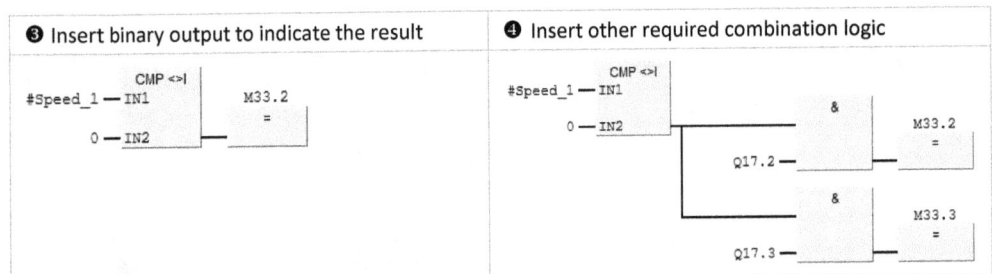

❸ Insert binary output to indicate the result	❹ Insert other required combination logic
CMP <>I #Speed_1 — IN1 0 — IN2	CMP <>I #Speed_1 — IN1 0 — IN2 Q17.2 — & ... M33.2 Q17.3 — & ... M33.3

Figure 9-12. Insert an output and combine other required logic combinations in FBD

As shown in cell ❹ of the previous figure, you may insert any combination of AND/OR logic either before or after the compare box. Remember that the compare behaves similar to a positive binary input. If the test is successful, power flows through the compare; if the test fails, power flow is interrupted. Any logic inserted before or after the compare is logically combined with the result using the logical AND. Any logic placed in parallel with the compare is combined with the result using the logical OR.

As shown below❺, multiple compare operations may also be placed in series, with preceding or following logic. If in series, all of the compare tests must be satisfied to complete logic continuity.

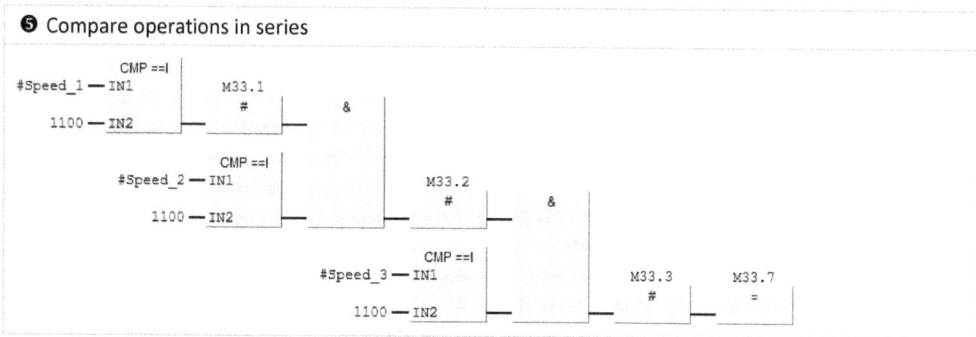

Figure 9-13. AND compare operations in FBD

Compare boxes may also be placed in parallel. If the operations are in parallel, a '1' result is produced for the branch when any compare in the branch is satisfied. In the first example ❻ below, M33.6 is activated if either of the compare tests is satisfied. In the second example ❼ below, M33.6 is activated if both compare tests are satisfied in either one of the three AND gates.

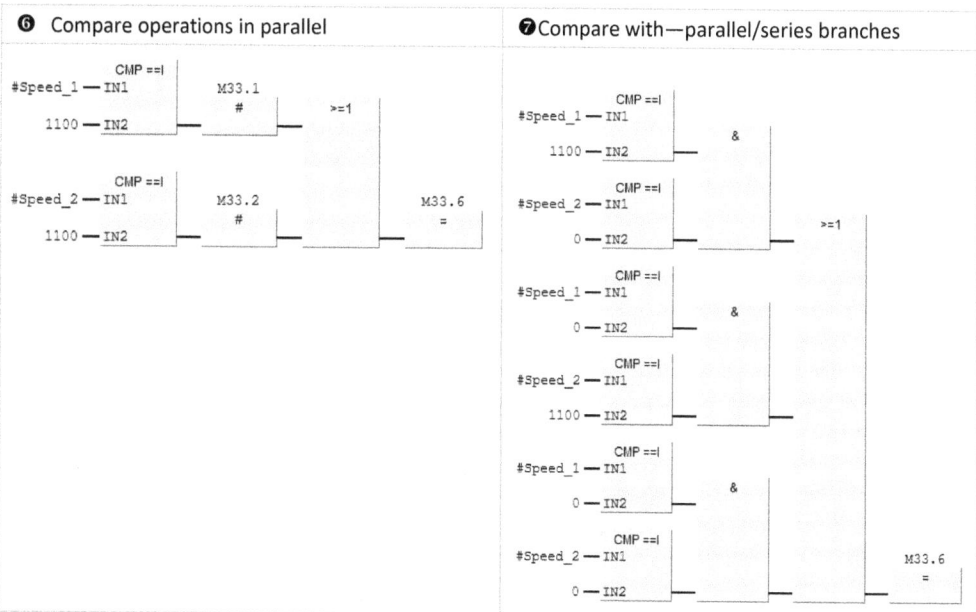

Figure 9-14. OR compare operations in FBD; OR compare operations that are ANDed in FBD

Chapter 10

LAD/FBD
Integer and Real Operations

This chapter introduces LAD/FBD Integer and REAL operations. These operations support simple to complex calculations using 16-bit integer (INT), 32-bit double integer (DINT), and 32-bit floating-point (REAL) values. In STEP 7, these operations represent basic arithmetic operations, as well as standard trigonometric and mathematical functions. In LAD and FBD these operations may be combined in series and in parallel to form basic calculations, control algorithms, or other numerical solutions.

An important part of Integer and Real operations is that information pertaining to each result, whether valid or in error, is available for evaluation in the S7 status bits CC0, CC1, OV, and OS. How arithmetic and standard functions influence these status bits is described in Chapter 13 *LAD/FBD Status Bit Operations*.

Main Topics

- Arithmetic Operations Overview
- Integer Arithmetic Descriptions
- Double Integer Arithmetic Descriptions
- REAL Arithmetic Descriptions
- Programming Arithmetic Operations in LAD
- Programming Arithmetic Operations in FBD
- Standard Math Functions Overview
- Standard Math Functions Descriptions

ARITHMETIC OPERATIONS OVERVIEW

LAD/FBD *integer and real arithmetic operations* are listed in Table 10-1. These operations support the four basic arithmetic operations of *addition*, *subtraction*, *multiplication*, and *division* on numbers of data type INT (integer), DINT (double-integer), and REAL (floating-point). Production accounting, control algorithms, diagnostic calculations, and other routines that involve arithmetic are typical applications of these operations.

Table 10-1. LAD/FBD Integer and REAL Operations Summary.

Operation	INT -32,768 to +32,767	DINT -2,147,483,648 to +2,147,483,647	REAL ± 1.75495e-38 to ± 3.402823+e38
Addition	ADD_I	ADD_DI	ADD_R
Subtraction	SUB_I	SUB_DI	SUB_R
Multiplication	MUL_I	MUL_DI	MUL_R
Division (Normal Quotient Result)	DIV_I	DIV_DI	DIV_R
Division (Remainder Result)		MOD_DI	

INTEGER ARITHMETIC DESCRIPTIONS

Each integer arithmetic operation is identified by a mnemonic at the top of the box instruction. For example the mnemonic of the instruction used to add two integer values is **ADD_I**. Integer operations are described in the following discussions. First let's take a look at the parameters you must specify as you program any integer arithmetic operation.

Integer Arithmetic Box Parameters

Entry of each integer arithmetic operation will require you to specify source values for IN1 and IN2, the two values used in the calculation; and a third location, OUT, to designate where the result is stored. At the enable input line EN, you may insert logic to determine when the operation is executed. When EN is logic 1, the operation is executed and the result placed in the location or variable specified at the parameter OUT. If there is an error, ENO produces logic 0.

Table 10-2. LAD/FBD Arithmetic Box Parameters with INT (integer) instruction shown

Parameter	Data Type	Description	Required?	INT Arithmetic Box	
				LAD	FBD
EN	-	Enable Input	-		
IN1	INT	Value 1 Input	YES		
IN2	INT	Value 2 Input	YES		
ENO	-	Enable Output	NO		
OUT	INT	Value Output	YES		

For IN1 and IN2, you may specify integer constants, variables of type INT, or memory locations of the correct width—for example MW 42, DB22.DW24. For the parameter OUT, you may specify a variable of type INT, or a memory location of the correct width—for example MW 44, or DB10.DW26.

Add Integer: ADD_I

When the **EN** input is logic 1, the *Add Integer* operation uses the integer value at IN1 and at **IN2** to process the addition as **IN1 + IN2**. The result is stored in the destination you specify at **OUT**. At **IN1** and **IN2**, you may specify an integer constant, a variable of type INT, or a word-width memory location that contains an integer value. At **OUT**, specify a word-width memory location or variable of type INT.

After each operation you can check the results using status bit operations. For example, status bits OV and OS are set to '1' if the range for INT values is exceeded. You may also check if the result is greater than zero, equal to zero, or less than zero. See Chapter 13 *Status Bit Operations*, and Table 13-3.

In the example below, when the RLO = 1 at **EN**, the ADD_I operation is performed as bit memory location MW12 **plus** the integer value of +100. The result is placed in bit memory word MW14.

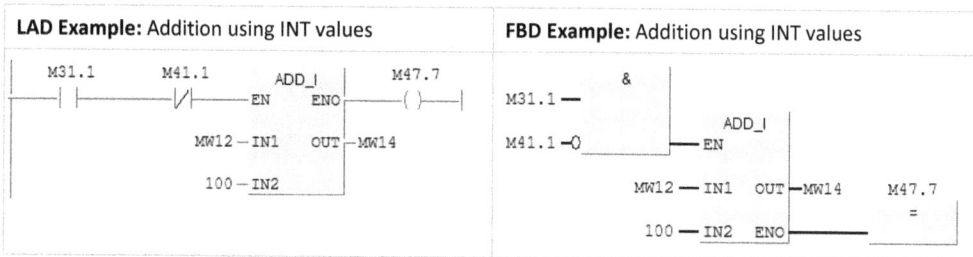

LAD Example: Addition using INT values	FBD Example: Addition using INT values

Figure 10-1. Illustration of LAD/FBD Add Integer instruction

Subtract Integer: SUB_I

When the **EN** input is logic 1, the *Subtract Integer* operation uses the integer value at **IN1** and at **IN2** to process the subtraction as **IN1—IN2**. The result is stored in the destination you specify at **OUT**. At **IN1** and **IN2**, you may specify an integer constant, a variable of type INT, or a word-width memory location that contains an integer value. At **OUT**, specify a word-width memory location or a variable of type INT.

After each operation you can check the results using status bit operations. For example, status bits OV and OS are set to '1' if the range for INT values is exceeded. You can also check the result for greater than zero, equal zero, or less than zero. See Chapter 13 *Status Bit Operations*, and Table 13-3.

In the example, whenever the RLO = 1 at **EN**, the SUB_I operation is performed as the integer variable #Value_1 **minus** the integer variable #Value_2. The result is placed in the integer variable #Total_1.

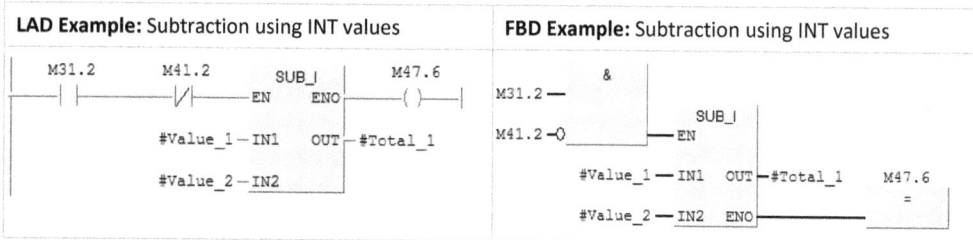

LAD Example: Subtraction using INT values	FBD Example: Subtraction using INT values

Figure 10-2. Illustration of LAD/FBD Subtract Integer instruction

Multiply Integer: MUL_I

When the **EN** input is logic 1, the *Multiply Integer* operation uses the integer value at **IN1** and at **IN2** to process the multiplication as **IN1** * **IN2**. The result is placed in the destination you specify at **OUT**. At **IN1** and **IN2**, you may specify an integer constant, a variable of type INT, or a word-width location that contains an integer value. At **OUT**, specify a word-width location or a variable of type INT.

After each operation you can check the results using status bit operations. For example, status bits OV and OS are set to '1' if the permissible range for INT values is exceeded. You can also check the result for greater than zero, equal zero, or less than zero. See Chapter 13 *Status Bit Operations*.

In the following example, whenever the RLO = 1 at **EN,** the MUL_I operation is performed as word DBW12 of DB11 **multiplied by** the integer value of +2. The result is placed in word DBW12 of DB21.

LAD Example: Multiplication using INT values	FBD Example: Multiplication using INT values

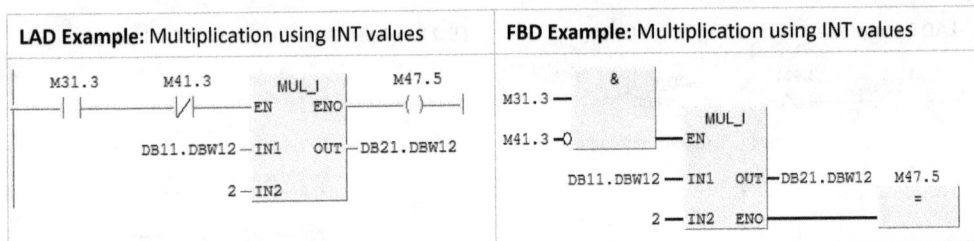

Figure 10-3. Illustration of LAD/FBD Multiply Integer instruction

Divide Integer: DIV_I

When the **EN** input is logic 1, the *Divide Integer* operation uses the integer value at IN1 and at **IN2** to process the division as **IN1** / **IN2**. The quotient result is stored in the destination you specify at **OUT**. Any fractional part of the result is truncated. At **IN1** and **IN2,** you may specify an integer constant, a variable of type INT, or a word-width memory location that contains an integer value. At **OUT**, specify a word-width location or a variable of type INT.

After each operation you can use status bit operations to check if the result is within the permissible range, or if there was an attempt to divide by zero. For example, status bits OV and OS are set to '1' if the permissible range for INT values is exceeded. You can also determine if the result is greater than zero, equal to zero, or less than zero. See Chapter 13 *Status Bit Operations*, and Table 13-3.

In the following example, whenever the RLO = 1 at **EN,** the DIV_I operation is performed as word DBW12 of DB21 **divided by** the integer value of +10. The result is placed in bit memory word MW14.

LAD Example: Division using INT values	FBD Example: Division using INT values

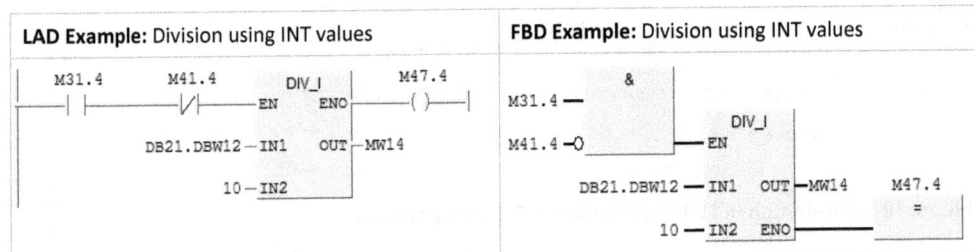

Figure 10-4. Illustration of LAD/FBD Divide Integer instruction

DOUBLE INTEGER ARITHMETIC DESCRIPTIONS

Each double integer arithmetic operation is identified by a mnemonic at the top of the box instruction. For example, the mnemonic of the instruction used to add two double integer values is **ADD_DI**. Double integer operations are described in the following discussions. First let's take a look at the parameters you must specify as you program double integer operations.

Double Integer Arithmetic Parameters

Entry of each double integer arithmetic operation will require you to specify source locations for **IN1** and **IN2,** the two values used in the calculation; and a third location, **OUT,** to designates where the result is to be stored. At the enable input line **EN,** you may insert logic to control when the operation is executed. When **EN** is logic 1, the operation is executed; the result is placed in the location you specify at the parameter **OUT**. If there is an error, **ENO** produces logic 0.

Table 10-3. LAD/FBD Arithmetic Box Parameters with DINT (double integer) instruction shown

Parameter	Data Type	Description	Required?	DINT Arithmetic Box	
				LAD	**FBD**
EN	-	Enable Input	-		
IN1	DINT	Value 1 Input	YES		
IN2	DINT	Value 2 Input	YES		
ENO	-	Enable Output	-		
OUT	DINT	Value Output	YES		

For IN1 and IN2, you may specify valid integer constants, declared variables, or memory locations of the correct width—for example MD44, or DB22.DD24. The source values must be of the data type DINT. For the parameter **OUT,** you may specify a declared variable of type DINT or a memory location of the correct width—for example MD46, or DB22.DD28.

Add Double Integer: ADD_DI

When the **EN** input is logic 1, the *Add Double Integer* operation uses the double integer value at **IN1** and at **IN2** to process the addition as **IN1** + **IN2**. The result is stored in the destination you specify at **OUT**. At **IN1** and **IN2,** you may specify a double integer constant, a variable of type DINT, or a double word memory location that contains a double integer value. At **OUT,** specify a double word memory location or variable of type DINT.

After each operation you can check the results using status bit operations. For example, status bits OV and OS are set to '1' if the 32-bit integer range is exceeded. You can also check the result for greater than zero, equal zero, or less than zero. See Chapter 13 *Status Bit Operations*, and Table 13-3.

In the following example, whenever the RLO = 1 at **EN,** the ADD_DI operation is performed as the value in MD72 **plus** the long integer value of +100. The result is placed in the double word MD74.

LAD Example: Addition using DINT values	FBD Example: Addition using DINT values

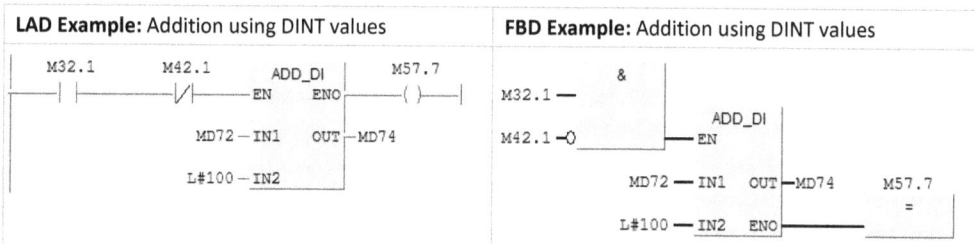

Figure 10-5. Illustration of LAD/FBD Add Double Integer instruction

Subtract Double Integer: SUB_DI

When the **EN** input is logic 1, the *Subtract Double Integer* operation uses the double integer value at **IN1** and at **IN2** to process the subtraction as **IN1—IN2**. The result is stored in the variable you specify at **OUT**. At **IN1** and **IN2**, you may specify a double integer constant, a variable of type DINT, or a double word location that contains a double integer value. At **OUT**, specify a double word location or a variable of type DINT.

After each operation you can check the result using status bit operations. For example, status bits OV and OS are set to '1' if the DINT number range is exceeded. You can also determine if the result is less than zero, equal to zero, or greater than zero. See Chapter 13 *Status Bit Operations*, and Table 13-3.

In the example, whenever the RLO = 1 at **EN**, the SUB_DI operation is performed as the double integer variable #Value_3 **minus** the double integer variable #Value_4. The result is in placed the double integer variable #Total_2.

LAD Example: Subtraction using DINT values	FBD Example: Subtraction using DINT values

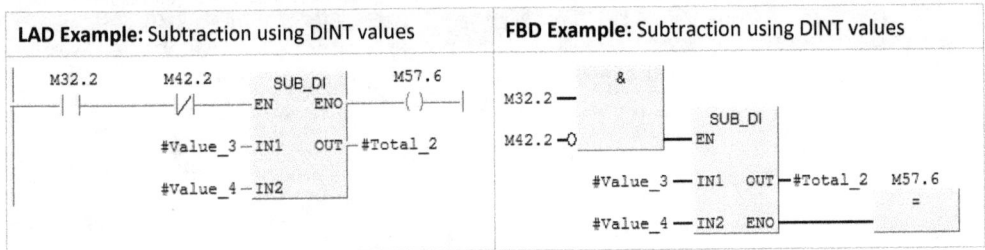

Figure 10-6. Illustration of LAD/FBD Subtract Double Integer instruction

Multiply Double Integer: MUL_DI

When the **EN** input is logic 1, the *Multiply Double Integer* operation uses the double integer value at **IN1** and at **IN2** to process the multiplication as **IN1 * IN2**. The result is placed in the destination you specify at **OUT**. At **IN1** and **IN2**, you may specify a double integer constant, a variable of type DINT, or a double word memory location that contains a double integer value. At **OUT**, specify a double word memory location or a variable of type DINT.

After each operation you can check the result using status bit operations. For example, status bits OV and OS are set to '1' if the permissible range for DINT numbers is exceeded. You can also check whether the result is greater than zero, equal to zero, or less than zero. See Chapter 13 *Status Bit Operations*, and Table 13-3.

In the example, whenever the RLO = 1 at **EN**, the MUL_DI operation is performed as the value in double word DBD22 of DB11 **multiplied by** the long integer value of +2. The result is placed in the double word DBD21 of DB22.

LAD Example: Multiplication using DINT values	FBD Example: Multiplication using DINT values

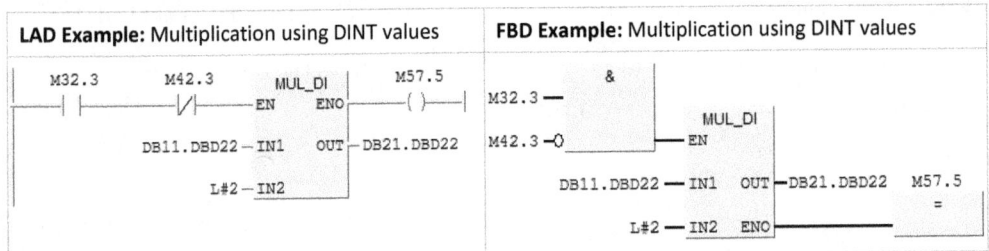

Figure 10-7. Illustration of LAD/FBD Multiply Double Integer instruction

Divide Double Integer: DIV_DI

When the **EN** input is logic 1, the *Divide Double Integer* operation uses the double integer value at **IN1** and at **IN2** to process the division as **IN1 / IN2**. Any fractional part of the result is truncated. The quotient result is stored in the destination you specify at **OUT**. At **IN1** and **IN2,** you may specify a valid double integer constant, a variable of type DINT, or a double word memory location that contains a double integer value. At **OUT,** specify a double word location or a variable of type DINT.

After each operation you can use status bit operations to check if the result is within the permissible range or if there was an attempt to divide by zero. For example, status bits OV and OS are set to '1' if the permissible range of DINT values is exceeded. You can also check whether the result is greater than zero, equal to zero, or less than zero. See Chapter 13 *Status Bit Operations*, and Table 13-3.

In the example, whenever the RLO = 1 at **EN,** the DIV_DI operation is performed as the value in double word DBD26 **divided by** the long integer value of +10. The quotient result is placed in MD74.

LAD Example: Division using DINT values	FBD Example: Division using DINT values

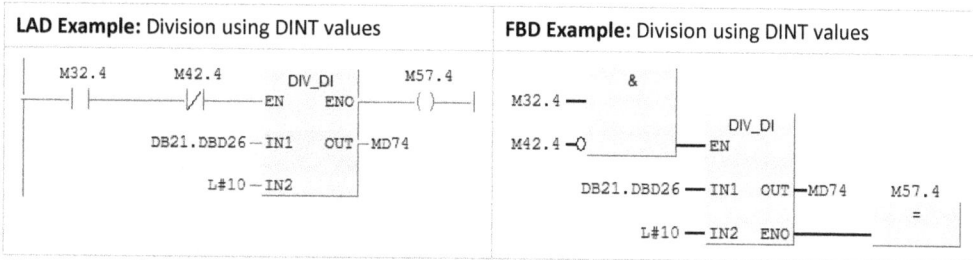

Figure 10-8. Illustration of LAD/FBD Divide Double Integer instruction

Modulus Double Integer: MOD_DI

When the **EN** input is logic 1, the *Modulus Double Integer* operation uses the double integer value at **IN1** and at **IN2** to process the division as **IN1 / IN2**. The quotient remainder (fractional part of the result) is stored in the destination you specify at **OUT**. At **IN1** and **IN2,** you may specify a double integer constant, a variable of type DINT, or a double word memory location that contains a double integer value. At **OUT,** specify a double word location or a variable of type DINT.

After each operation you can use status bit operations to check whether the result is within the permissible range or if there was an attempt to divide by zero. For example, status bits OV and OS are set to '1' if the permissible range of DINT values is exceeded. You can also check if the result is greater than zero, equal to zero, or less than zero. See Chapter 13 *Status Bit Operations*, and Table 13-3.

In the example, whenever the RLO = 1 at **EN,** the MOD_DI operation is performed as the double word DBD26 **divided by** the long integer value of +10. The remainder result is placed in MD78.

LAD Example: Modulus using DINT values	FBD Example: Modulus using DINT values

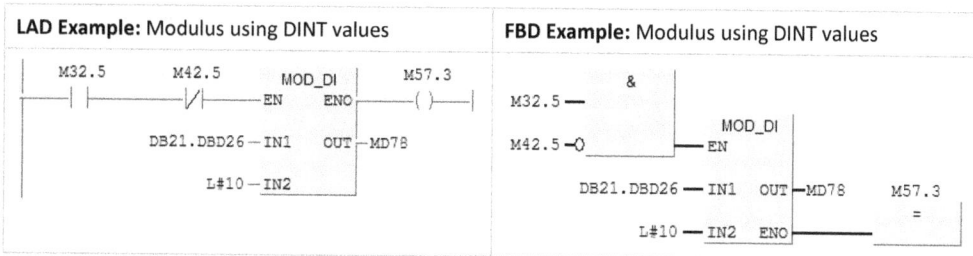

Figure 10-9. Illustration of LAD/FBD Modulus Double Integer instruction

REAL ARITHMETIC DESCRIPTIONS

Each REAL arithmetic operation is identified by a mnemonic at the top of the instruction. For example, the mnemonic of the instruction used to add two REAL values is **ADD_R**. REAL arithmetic operations are described in the following discussions, after a review of the parameters you will need to specify as you use REAL arithmetic operations in your program.

REAL Arithmetic Parameters

For each REAL arithmetic operation, you must specify the calculation source values for the parameters **IN1** and **IN2**; and the result parameter **OUT**. Valid entries for IN1 and IN2 include REAL constants in the permissible range ($\pm$ 1.75495e-38 to $\pm$ 3.402823e+38), declared variables of data type REAL, or double word locations—for example DB22.DD12 or MD 42. You may specify the parameter **OUT** as a variable of type REAL, or a double word location—for example DB22.DD16.

The logic inserted at the enable input line **EN** controls when the arithmetic operation is executed. When **EN** is logic 1, the operation is executed; the result is placed in the location specified at **OUT**. If there is an error, **ENO** produces logic 0.

Table 10-4. REAL Arithmetic Instruction Parameters

Parameter	Data Type	Description	Required?	REAL Arithmetic Box	
				LAD	**FBD**
EN		Enable Input			
IN1	REAL	Value 1 Input	YES		
IN2	REAL	Value 2 Input	YES		
ENO		Enable Output			
OUT	REAL	Value Output	YES		

Add REAL: ADD_R

When the **EN** input is logic 1, the *Add REAL* operation uses the floating-point value at **IN1** and at **IN2** to process the addition as **IN1 + IN2**. The sum is stored in the destination you specify at **OUT**. At **IN1** and **IN2**, you may specify a REAL constant, a variable of type REAL, or a double word location that contains the floating-point value. At **OUT**, specify a double word location or a variable of type REAL.

After each operation you can use status bit operations to check for errors such as the use of an invalid REAL number or if the result has exceeded the normal range. For example, status bits OV and OS are set to '1' if the REAL number range is exceeded. You can also determine if the result is greater than zero, equal to zero, or less than zero. See Chapter 13 *Status Bit Operations*.

In the following example, whenever the RLO = 1 at **EN,** the ADD_R operation is performed as the floating-point value in MD12 **plus** the value of +100.0. The result is in double word location MD18.

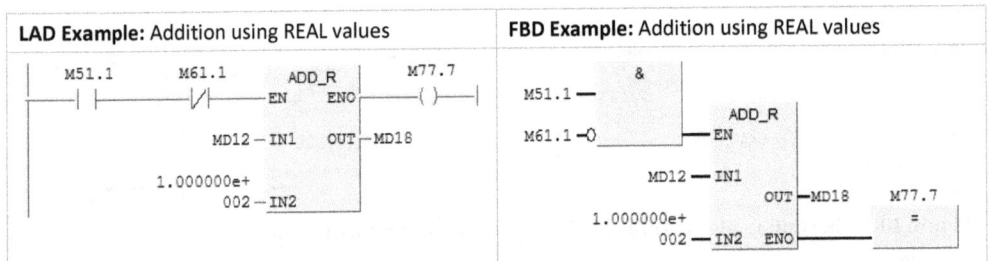

LAD Example: Addition using REAL values	FBD Example: Addition using REAL values

Figure 10-10. Illustration of LAD/FBD Add Real instruction

Subtract REAL: SUB_R

When the **EN** input is logic 1, the *Subtract Real* operation uses the floating-point value at **IN1** and at **IN2** to process the subtraction as **IN1—IN2**. The result is stored in the destination you specify at **OUT**. At **IN1** and **IN2**, you may specify a REAL constant, a variable of type REAL, or a double word memory location that contains the floating-point value. At **OUT,** specify a double word memory location or a variable of type REAL.

After each operation you can use status bit operations to check for errors such as the use of an invalid REAL number or if the result has exceeded the normal range. For example, status bits OV and OS are set to '1' if the REAL number range is exceeded. You can also determine if the result is greater than zero, equal to zero, or less than zero. See Chapter 13 *Status Bit Operations*, and Table 13-3.

In the following example, whenever the RLO = 1 at **EN**, the SUB_R operation is performed as the variable #REAL_V1 **minus** the variable #REAL_V2. The result is in the variable #R_TOTAL1.

LAD Example: Subtraction using REAL values	FBD Example: Subtraction using REAL values

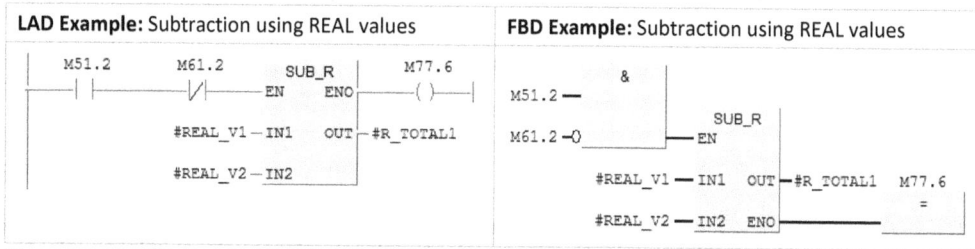

Figure 10-11. Illustration of LAD/FBD Subtract Real instruction

Multiply REAL: MUL_R

When the **EN** input is logic 1, the *Multiply Real* operation uses the floating-point value at **IN1** and at **IN2** to process the multiplication as **IN1 * IN2**. The product result is stored in the destination you specify at **OUT**. At **IN1** and **IN2,** you may specify a REAL constant, a variable of type REAL, or a double word memory location that contains the floating-point value. At **OUT,** specify a double word memory location or a variable of type REAL.

After each operation you can use status bit operations to check for errors such as the use of an invalid REAL number or if the result has exceeded the normal range. For example, status bits OV and OS are set to '1' if the REAL number range is exceeded. You can also determine if the result is greater than zero, equal to zero, or less than zero. See Chapter 13 *Status Bit Operations*, and Table 13-3.

In the following example, whenever the RLO = 1 at **EN,** the MUL_R operation is performed as the floating-point value in double word DBD12 of DB31 is **multiplied by** the floating-point value of +2.0. The result is in double word DBD12 of DB41.

LAD Example: Multiplication using REAL values	FBD Example: Multiplication using REAL values

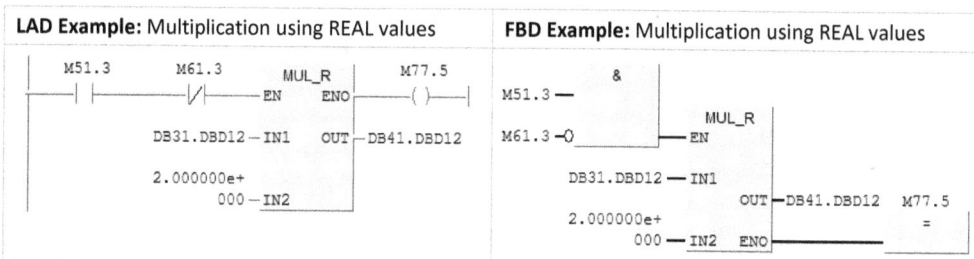

Figure 10-12. Illustration of LAD/FBD Multiply Real instruction

Divide REAL: DIV_R

When the **EN** input is logic 1, the *Divide Real* operation uses the floating-point value at **IN1** and at **IN2** to process the division as **IN1 / IN2**. The quotient result is stored in the destination you specify at **OUT**. At **IN1** and **IN2**, you may specify a REAL constant, a variable of type REAL, or a double word memory location. At **OUT**, specify a double word memory location or a variable of type REAL.

After each operation you can use status bit operations to check for errors such as the use of an invalid REAL number or if there was an attempt to divide by zero. For example, status bits OV and OS are set to '1' if the REAL number range is exceeded. You can also determine if the result is greater than zero, equal to zero, or less than zero. See Chapter 13 *Status Bit Operations*, and Table 13-3.

In the example below, whenever the RLO = 1 at **EN,** the DIV_R operation is performed as the value in double word DBD12 of DB41is **divided by** the REAL value of +10.0. The remainder result is in MD84.

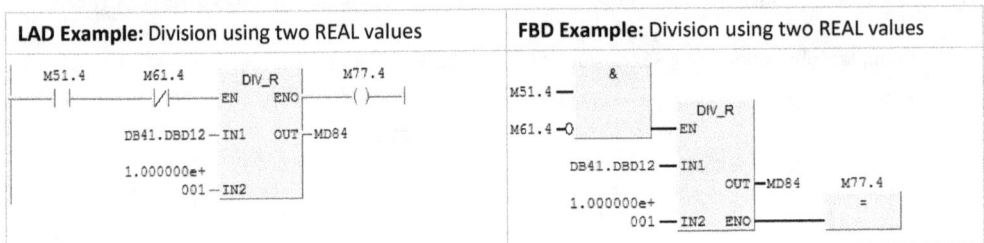

Figure 10-13. Illustration of LAD/FBD Divide Real instruction

PROGRAMMING ARITHMETIC OPERATIONS IN LAD

Programming arithmetic in LAD is first a matter of selecting the correct operation, based on the name. For example, **ADD_I** is Add Integers; **ADD_DI** is Add Double Integers, and **ADD_R** is Add REAL.

Generally, you will use integer instructions when whole number results in the range of -32,768 to + 32,767 are acceptable or required. If the results are expected to fall outside of this 16-bit range, then the 32-bit double integer range of -2,147,483,648 to +2,147,483,647 should be used. REAL arithmetic operations are appropriate when your process requires the accuracy of floating-point numbers.

Table 10-5. LAD Arithmetic operations with permissible number ranges and variables

IN1/IN2/OUT	INT	DINT	REAL
	ADD_I	ADD_DI	ADD_R
Constants (in permissible range)	- 32768 to +32767	-2,147,483,648 to +2,147,483,647	± 1.75495e-38 to ± 3.402823+e38
S7 Memory Locations (Absolute or Symbolic)	IW 28 QW 42 MW 54 DB6.DBW12	ID 28 QD 42 MD 54 DB6.DBD12	ID 28 QD 42 MD 54 DB6.DBD12
Declared Variables (of correct data type)	#Value_1 #Value_2 #Total	#Value_1 #Value_2 #Total	#Value_1 #Value_2 #Total

Assigning Arithmetic Parameters

With a new or existing network, you can drag and drop an arithmetic object onto the network. You must specify the **IN1** and **IN2** parameters as the two values to be operated on; and the output parameter **OUT** as the location to store the result. You may also use symbolic addresses.

When specifying arithmetic parameters, remember that the data type of the input and output variables is based on the operation. For instance, with double integer arithmetic you must specify variables of type DINT; memory operands must be double word locations.

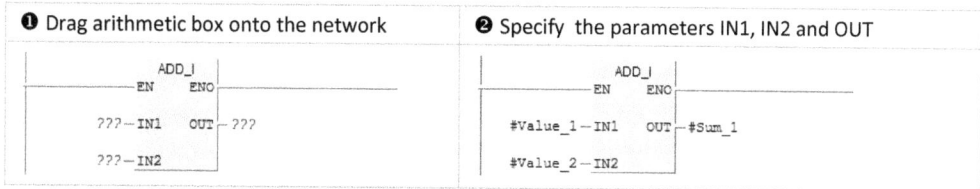

❶ Drag arithmetic box onto the network	❷ Specify the parameters IN1, IN2 and OUT
ADD_I EN ENO ??? — IN1 OUT — ??? ??? — IN2	ADD_I EN ENO #Value_1 — IN1 OUT — #Sum_1 #Value_2 — IN2

Figure 10-14. Select and insert arithmetic operation and enter parameters IN1, IN2, and OUT in LAD

The next step, if required, is to insert the enabling logic that drives the operation. On the **EN** line, you may insert any series or parallel logic combination as shown below. You may also leave the **EN** line without conditional logic, as in the second cell above—in this case, the operation will always execute.

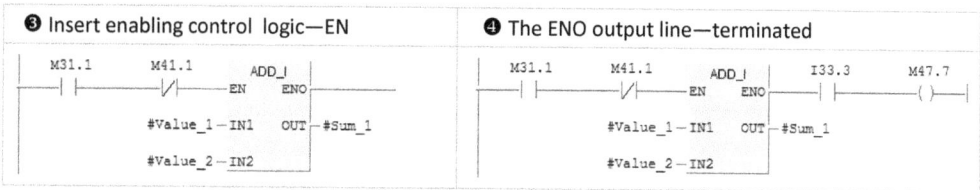

❸ Insert enabling control logic—EN	❹ The ENO output line—terminated
M31.1 M41.1 ADD_I —┤ ├——┤/├——EN ENO #Value_1 — IN1 OUT — #Sum_1 #Value_2 — IN2	M31.1 M41.1 ADD_I I33.3 M47.7 —┤ ├——┤/├——EN ENO——┤ ├——( )— #Value_1 — IN1 OUT — #Sum_1 #Value_2 — IN2

Figure 10-15. Insert enable (EN) logic and terminate the ENO output line in LAD

The **ENO** output line may be left unterminated, as in the first cell above, or terminated with a binary output instruction as shown in the second cell. Control logic may also be placed after the enable output **ENO**, but in such a case the network must be terminated.

For longer calculations, you may place arithmetic boxes in series as required. The **ENO** line of one box is connected to the **EN** line of the following operation. Each subsequent operation is evaluated only if the preceding box processes without error.

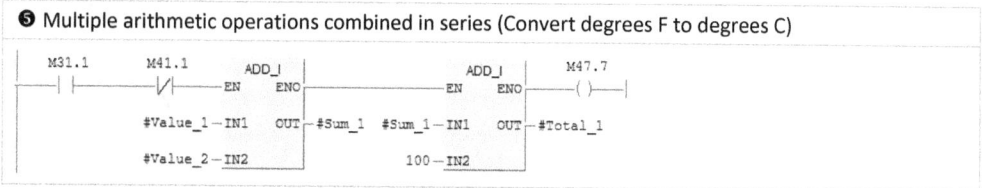

❺ Multiple arithmetic operations combined in series (Convert degrees F to degrees C)
M31.1   M41.1   ADD_I   ADD_I   M47.7 —┤ ├——┤/├——EN   ENO————EN   ENO——( )— #Value_1 — IN1   OUT — #Sum_1   #Sum_1 — IN1   OUT — #Total_1 #Value_2 — IN2   100 — IN2

Figure 10-16. Using EN/ENO mechanism to program arithmetic operations in series in LAD

Arithmetic boxes may also be inserted in parallel if placed in a branch that connects directly to the left rail of the network. When arithmetic boxes are placed in parallel, the network must be terminated with an output coil—for example, assign output. This output coil may serve as part of the logic used to provide error evaluation, or simply as a dummy output.

❻ Arithmetic operations combined in parallel

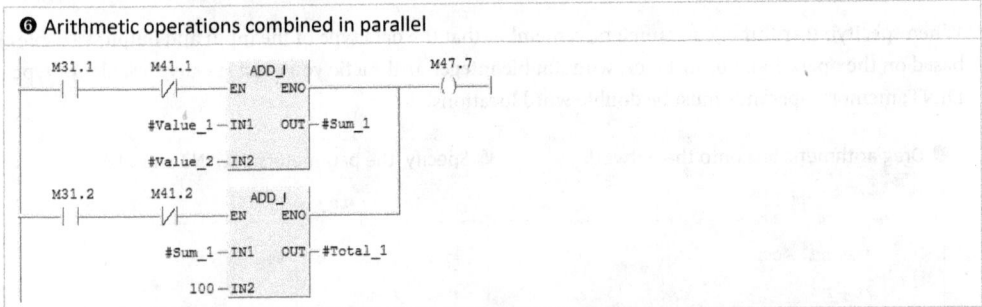

Figure 10-17. Inserting arithmetic operations in parallel in LAD

If a network has two or more parallel branches, each of which has two or more arithmetic boxes in series, then each branch is evaluated from left-to-right starting with the first branch, then the second branch, and so on.

❼ Arithmetic operations—parallel branches with series operations

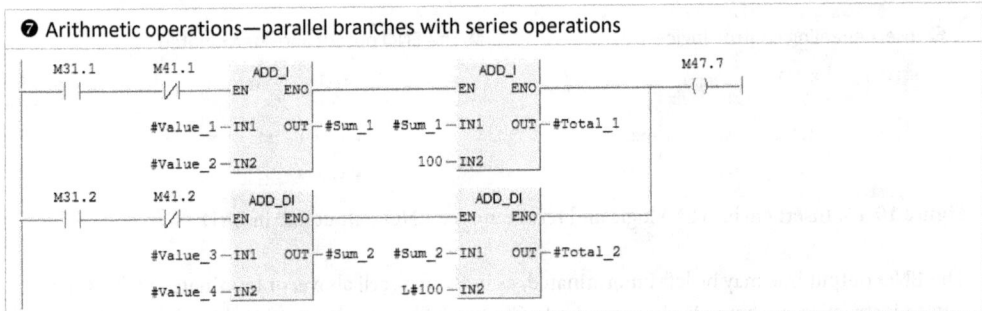

Figure 10-18. Inserting multiple branches of series arithmetic operations in LAD

PROGRAMMING ARITHMETIC OPERATIONS IN FBD

Programming arithmetic in FBD is first a matter of selecting the correct operation, based on the name. For example, **ADD_I** is Add Integers; **ADD_DI** is Add Double Integers, and **ADD_R** is Add REAL.

Generally, you will use integer instructions when whole number results in the range of -32,768 to + 32,767 are acceptable or required. If the results are expected to fall outside of this 16-bit range, then the 32-bit double integer range of -2,147,483,648 to +2,147,483,647 should be used. REAL arithmetic operations are appropriate when whole number results are not acceptable or when your process requires the accuracy of floating-point numbers.

Table 10-6. FBD Arithmetic operations with permissible number ranges and variables

IN1/IN2/OUT	INT	DINT	REAL
	ADD_I — EN — IN1 OUT — IN2 ENO	ADD_DI — EN — IN1 OUT — IN2 ENO	ADD_R — EN — IN1 OUT — IN2 ENO
Constants (in permissible range)	- 32768 to +32767	-2,147,483,648 to +2,147,483,647	± 1.75495e-38 to ± 3.402823+e38
S7 Memory Locations (Absolute or Symbolic)	IW 28 QW 42 MW 54 DB6.DBW12	ID 28 QD 42 MD 54 DB6.DBD12	ID 28 QD 42 MD 54 DB6.DBD12
Declared Variables (of correct data type)	#Value_1 #Value_2 #Total	#Value_1 #Value_2 #Total	#Value_1 #Value_2 #Total

Assigning Arithmetic Parameters

With a new or existing network, you can drag and drop an arithmetic object onto the network. You must specify the **IN1** and **IN2** parameters as the two values to be operated on; and the output parameter **OUT** as the location to store the result. You may also use symbolic addresses.

When specifying arithmetic parameters, remember that the data type of the input and output variables is based on the operation. For instance, with double integer arithmetic you must specify variables of type DINT; memory operands must be double word locations.

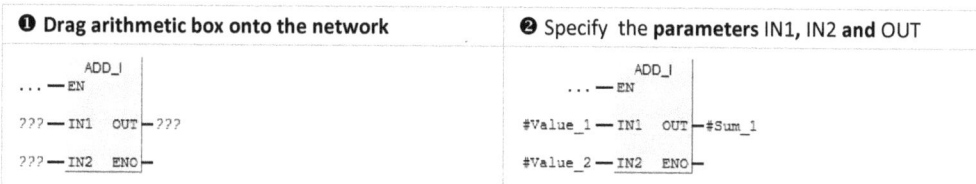

❶ Drag arithmetic box onto the network	❷ Specify the **parameters** IN1, IN2 **and** OUT
ADD_I ... — EN ??? — IN1 OUT — ??? ??? — IN2 ENO	ADD_I ... — EN #Value_1 — IN1 OUT — #Sum_1 #Value_2 — IN2 ENO

Figure 10-19. Select and insert arithmetic operation and enter parameters IN1, IN2, and OUT

The next step, if required, is to insert the driving logic that enables the operation. On the **EN** line, you may insert any series or parallel logic combination as shown below. You may also leave the **EN** line without driving conditions, as in the second cell above—in this case, the operation will always execute.

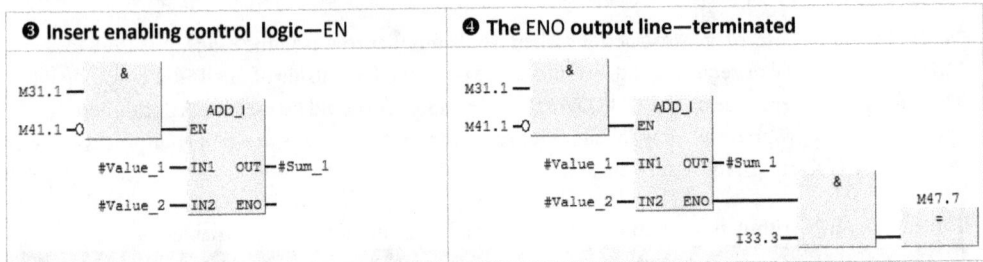

Figure 10-20. Insert enable (EN) logic and terminate the ENO output line

The **ENO** output line may be left unterminated, as in the first cell above, or terminated with a binary output instruction as shown in the second cell. Control logic may also be placed after the enable output **ENO**, but in such a case the network must be terminated.

For longer calculations, you may place arithmetic boxes in series as required. The **ENO** line of one box is connected to the **EN** line of the following operation. Each subsequent operation is evaluated only if the preceding box processes without error.

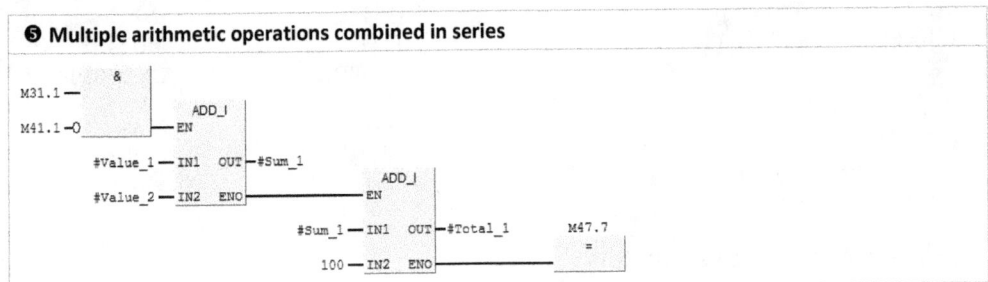

Figure 10-21. Using EN/ENO mechanism to program arithmetic operations in series

Arithmetic boxes may also be inserted in parallel. When arithmetic boxes are placed in parallel, the network must be terminated with an output—for example, assign output. This output may serve as part of the logic used to provide error evaluation, or simply as a dummy output.

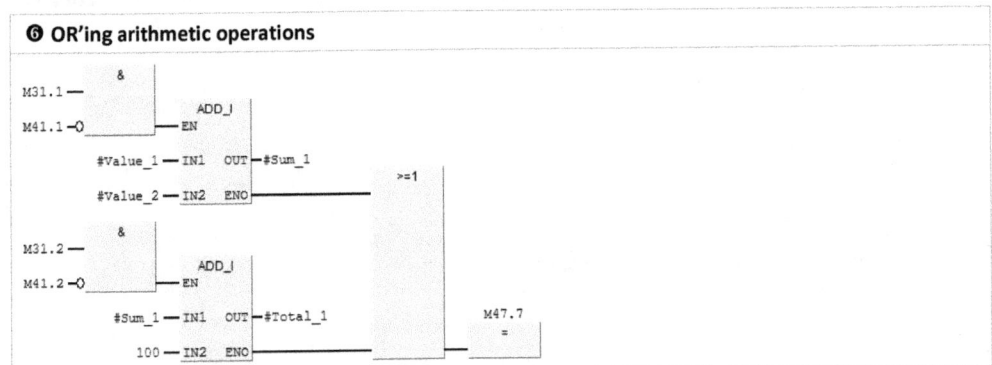

Figure 10-22. Inserting arithmetic operations in parallel

If a network has two or more parallel branches, each of which has two or more arithmetic boxes in series, then the branches are evaluated from left-to-right starting with the first branch.

❼ **OR'ing of multiple arithmetic operations combined using AND**

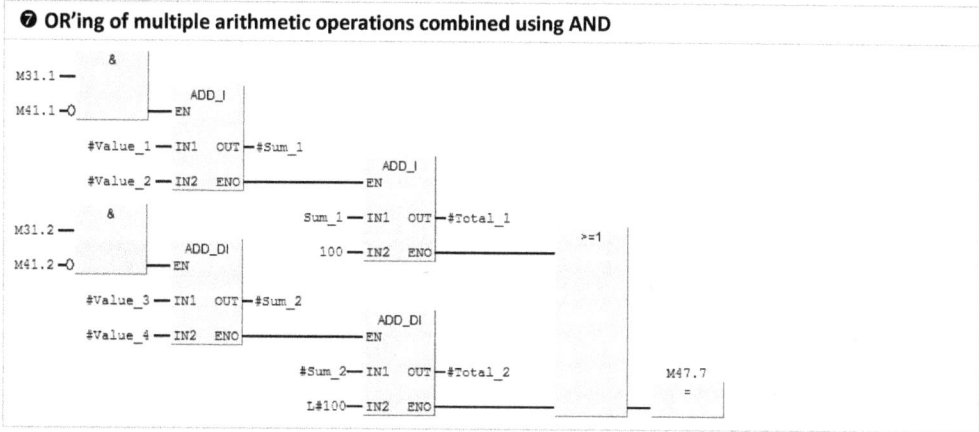

Figure 10-23. ORing multiple branches of series arithmetic operations

STANDARD MATH FUNCTIONS OVERVIEW

In addition to the four basic arithmetic operations, LAD and FBD instruction sets include the *standard mathematical functions* listed in the table below. These operations include trigonometric and arc functions for finding angle and radian measurements; functions to determine absolute value, square root, square of a number, natural log of a number, and a function to determine the exponent of a number with a radix of base e.

In STEP 7, these operations are located in the Floating-Point folder of the Program Elements tab.

Table 10-7. LAD/FBD Standard Mathematical Functions (REAL number operations) Summary

LAD	FBD	Brief Description
SIN — EN ENO — IN OUT	SIN — EN OUT — IN ENO	**Sine.** When RLO = 1 at **EN**, find the *sine* of the REAL value (radian angle) at **IN**. Put the result in the variable specified at **OUT**.
COS — EN ENO — IN OUT	COS — EN OUT — IN ENO	**Cosine.** When RLO = 1 at **EN**, find the *cosine* of the REAL value (radian angle) at **IN**. Put result in the variable specified at **OUT**.
TAN — EN ENO — IN OUT	TAN — EN OUT — IN ENO	**Tangent.** When RLO = 1 at **EN**, find the *tangent* of the REAL value (radian angle) at **IN**. Put the result in the variable specified at **OUT**.
ASIN — EN ENO — IN OUT	ASIN — EN OUT — IN ENO	**Arc Sine.** When RLO = 1 at **EN**, find the *arc sine* of the value at **IN** (-1 to +1). Put angle result (radians) in variable specified at **OUT**.
ACOS — EN ENO — IN OUT	ACOS — EN OUT — IN ENO	**Arc Cosine.** When RLO = 1 at **EN**, find *arc cosine* of the value at **IN** (-1 to +1). Put angle result (radians) in the variable specified at **OUT**.
ATAN — EN ENO — IN OUT	ATAN — EN OUT — IN ENO	**Arc Tangent.** When RLO = 1 at **EN**, find the *arc tangent* of the value at **IN**. Put angle result (radians) in the variable specified at **OUT**.
ABS — EN ENO — IN OUT	ABS — EN OUT — IN ENO	**Absolute Value.** When RLO = 1 at **EN**, find the *absolute value* of the REAL value at **IN**. Put the result in the variable specified at **OUT**.
SQRT — EN ENO — IN OUT	SQRT — EN OUT — IN ENO	**Square Root.** When the RLO = 1 at **EN**, find the *square root* of the REAL value at **IN**. Put the result in the variable specified at **OUT**.
SQR — EN ENO — IN OUT	SQR — EN OUT — IN ENO	**Square.** When RLO = 1 at **EN**, find the *square* of the REAL value at **IN**. Put the result in the variable specified at **OUT**.
LN — EN ENO — IN OUT	LN — EN OUT — IN ENO	**Natural LOG.** When RLO = 1 at **EN**, find the *natural log* of the REAL value at **IN**. Put the result in the variable specified at **OUT**.
EXP — EN ENO — IN OUT	EXP — EN OUT — IN ENO	**Exponent.** When RLO = 1 at **EN**, find the *exponent* of the REAL value at **IN**. Put the result in the variable specified at **OUT**.

Note: When EN is '1', ENO will also be '1', except on error; then while EN is '1', ENO is '0'.

STANDARD MATH FUNCTIONS DESCRIPTIONS

Standard math functions in STEP 7 include trigonometric and other *standard functions*. Each math function is identified by a unique mnemonic label. For example, **ABS** is the mnemonic of the absolute value instruction. Standard math functions are described in the following discussions. First let's take a look at the parameters that must be specified when using these operations.

Standard Math Functions Parameters

You may insert logic in the enable input line **EN** to control when the math function is executed. When **EN** is at logic 1, the operation is executed and the result is placed in the location specified at **OUT**. If the operation executes without error, **ENO** is set to '1'. If an error occurs, **ENO** is set to '0'.

For each standard math function, you must specify the source value at parameter **IN** and the result variable at the parameter **OUT**. Valid entries for both **IN** and **OUT** include variables of data type REAL, or double word memory locations like DB22.DD12 and MD44. As with basic arithmetic operations, the outcome of standard math functions affect status bits CC0, CC1, OV, and OS. You can learn more of status bit operations, in Chapter 13 *LAD/FBD Status Bit Operations*.

Table 10-8. Standard Mathematical Function Instruction Parameters

Parameter	Data Type	Description	Required?	REAL Math Functions Box	
				LAD	**FBD**
EN		Enable Input Line	YES		
IN	REAL	Value Input for Operation	YES		
ENO		Enable Output Line	NO		
OUT	REAL	Numeric Value Resultant	YES		

While math functions each operate on real numbers, the nature of some of the operations impose restrictions on the permissible range of input values, as shown in the table below.

Table 10-9. Valid Range at IN and OUT in Math Functions

Function	IN Value	OUT Result
Sine	Full range	-1 to + 1
Cosine	Full range	-1 to + 1
Tangent	Full range	Full range
Arc Sine	-1 to + 1	$-\Pi/2$ to $+\Pi/2$
Arc Cosine	-1 to + 1	0 to $+\Pi$
Arc Tangent	Full range	$-\Pi/2$ to $+\Pi/2$

Note: Pi = 3.141592. For the trigonometric functions sine, cosine, and tangent, the greatest accuracy is achieved when the source value at **IN** is greater than -2Π and less than 2Π.

Sine: SIN

When the input **EN** is logic 1, the *Sine* instruction reads the value specified at **IN** (an angle in radians), and returns the sine in the REAL variable specified at **OUT**. At the input **IN**, you may specify a variable of type REAL, a valid floating-point constant, or reference a double word memory location. At **OUT**, specify a double word memory location or a REAL variable. The result in **OUT** is always greater than or equal to -1, and less than or equal to +1.

If at execution time the input value at **IN** is an invalid REAL number, an invalid REAL number is returned by the function and status bits CC0, CC1, OV, and OS are set to '1'.

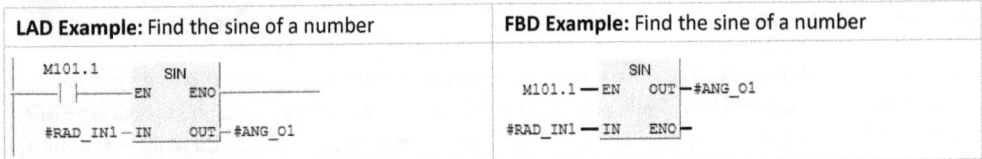

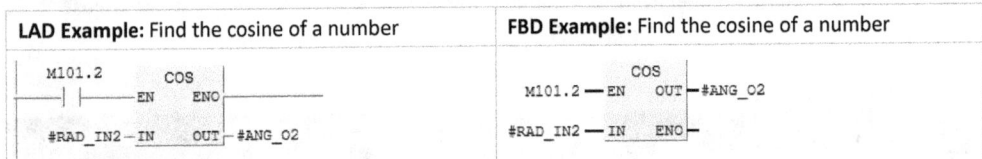

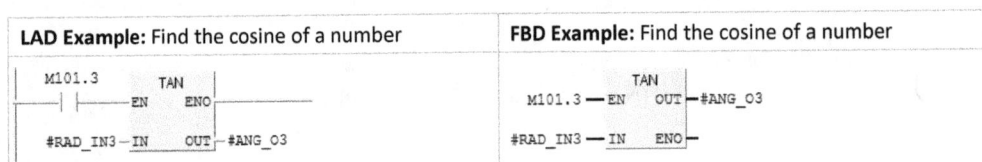

LAD Example: Find the sine of a number	FBD Example: Find the sine of a number
M101.1 SIN EN ENO #RAD_IN1 — IN OUT — #ANG_O1	SIN M101.1 — EN OUT — #ANG_O1 #RAD_IN1 — IN ENO

Figure 10-24. Illustration of LAD/FBD Sine instruction

Cosine: COS

When the input **EN** is logic 1, the *Cosine* instruction reads the value specified at **IN** (an angle in radians), and returns the cosine in the REAL variable specified at **OUT**. At the input **IN**, you may specify a variable of type REAL, enter a valid floating-point constant, or reference a double word memory location. At **OUT**, specify a double word memory location or a REAL variable. The result in **OUT** is always greater than or equal to -1, and less than or equal to +1.

If at execution time the input value at **IN** is an invalid REAL number, an invalid REAL number is returned by the function and status bits CC0, CC1, OV, and OS are set to '1'.

LAD Example: Find the cosine of a number	FBD Example: Find the cosine of a number
M101.2 COS EN ENO #RAD_IN2 — IN OUT — #ANG_O2	COS M101.2 — EN OUT — #ANG_O2 #RAD_IN2 — IN ENO

Figure 10-25. Illustration of LAD/FBD Cosine instruction

Tangent: TAN

When the input **EN** is logic 1, the *Tangent* operation reads the value specified at **IN** (an angle in radians), and returns the tangent in the REAL variable specified at **OUT**. At the input **IN**, you may specify a REAL variable, a valid floating-point constant, or a double word memory location. At **OUT**, specify a double word memory location or a REAL variable. The result in **OUT** is a real number.

If at execution time the input value at **IN** is an invalid REAL number, an invalid REAL number is returned by the function, and status bits CC0, CC1, OV, and OS are set to '1'.

LAD Example: Find the cosine of a number	FBD Example: Find the cosine of a number
M101.3 TAN EN ENO #RAD_IN3 — IN OUT — #ANG_O3	TAN M101.3 — EN OUT — #ANG_O3 #RAD_IN3 — IN ENO

Figure 10-26. Illustration of LAD/FBD Tangent instruction

Arc Sine: ASIN

When the **EN** input is logic 1, the *Arc Sine* instruction reads the value specified at **IN**, and returns its Arc sine (an angle in radians) in the REAL variable specified at **OUT**. At the input **IN**, you may specify a variable of type REAL, a floating-point constant in the range of -1 <= **IN** <= +1, or reference a double word location. At **OUT**, specify a double word memory location or a REAL variable. The result in **OUT** is always greater than or equal to -Π/2, and less than or equal to + Π/2 (pi = 3.141593 e +00).

If the input value at **IN** is outside of the permissible range for this arc function, an invalid REAL number is returned and status bits CC0, CC1, OV, and OS are set to '1'.

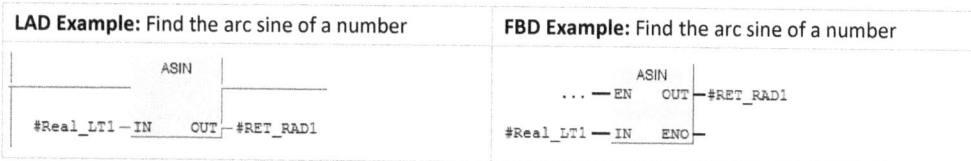

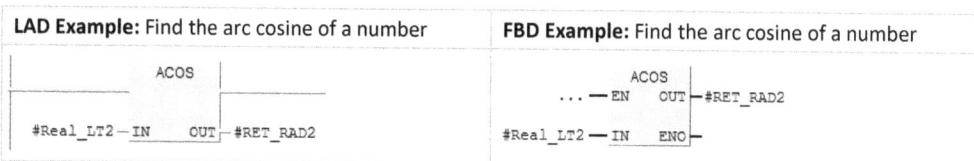

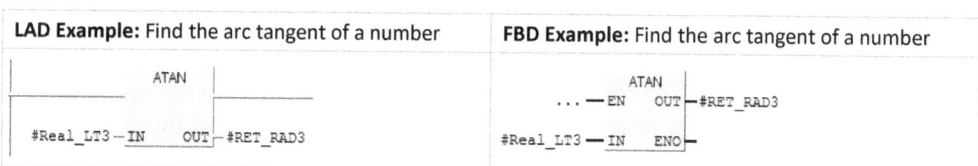

LAD Example: Find the arc sine of a number	**FBD Example:** Find the arc sine of a number
ASIN #Real_LT1 — IN OUT — #RET_RAD1	ASIN ... — EN OUT — #RET_RAD1 #Real_LT1 — IN ENO

Figure 10-27. Illustration of LAD/FBD Arc Sine instruction

Arc Cosine: ACOS

When the **EN** input is logic 1, the *Arc Cosine* operation reads the value at **IN** (in radians), and returns its Arc cosine (an angle in radians) in the REAL variable specified at **OUT**. At the input **IN**, you may specify a variable of type REAL, a valid floating-point constant in the range of -1 <= **IN** <= +1, or a double word location. At **OUT**, specify a double word memory location or a REAL variable. The result in **OUT** is always greater than or equal to zero, and less than or equal to + Π (pi = 3.141593 e +00).

If the input value at **IN** is outside of the permissible range for this arc function, an invalid REAL number is returned and status bits CC0, CC1, OV, and OS are set to '1'.

LAD Example: Find the arc cosine of a number	**FBD Example:** Find the arc cosine of a number
ACOS #Real_LT2 — IN OUT — #RET_RAD2	ACOS ... — EN OUT — #RET_RAD2 #Real_LT2 — IN ENO

Figure 10-28. Illustration of LAD/FBD Arc Cosine instruction

Arc Tangent: ATAN

When the **EN** input is logic 1, the *Arc Tangent* instruction reads the value specified at **IN**, and returns its Arc tangent (an angle in radians) in the variable specified at **OUT**. At the input **IN**, you may specify a variable of type REAL, a valid floating-point constant, or reference a double word memory location. At **OUT**, specify a double word memory location or a REAL variable. The result in **OUT** is always greater than or equal to -Π/2, and less than or equal to + Π/2.

If the input value at **IN** is outside of the permissible range for this arc function, an invalid REAL number is returned and status bits CC0, CC1, OV, and OS are set to '1'.

LAD Example: Find the arc tangent of a number	**FBD Example:** Find the arc tangent of a number
ATAN #Real_LT3 — IN OUT — #RET_RAD3	ATAN ... — EN OUT — #RET_RAD3 #Real_LT3 — IN ENO

Figure 10-29. Illustration of LAD/FBD Arc Tangent instruction

Absolute Value: ABS

When the **EN** input is logic 1, the *Absolute Value* instruction determines the absolute value of the floating-point value at **IN**, and stores the result in the variable specified at **OUT**. The result is always a positive value.

At the input **IN**, you may specify a variable of type REAL, a floating-point constant, or reference a double word memory location. At **OUT**, you may specify a double word memory location or a variable of type REAL. The ABS operation does not influence the status bits.

Absolute Value		LAD: Find absolute value of a number	FBD: Find absolute value of a number
IN	**OUT**		
-5.2	5.2		
-7.35	7.35		
50.0	50.0		

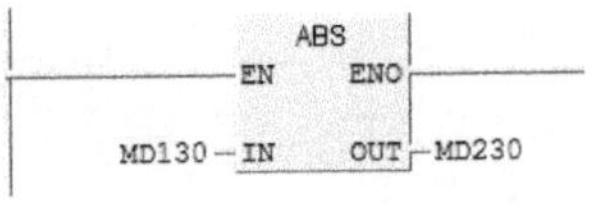

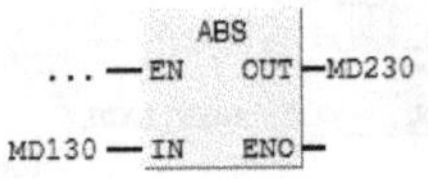

Figure 10-30. Illustration of LAD/FBD Absolute Value instruction

Square Root: SQRT

When the **EN** input is logic 1, the *Square Root* instruction returns the square root of the floating-point value at **IN**, and stores the result in the variable specified at **OUT**.

At the input at **IN**, you may enter a valid floating-point constant, or reference a double word location that contains a floating point value in the permissible range. At **OUT**, you may specify a double word location or a variable of type REAL.

The square root always returns a positive result when the input at **IN** is greater than zero. If the input at **IN** is minus zero, then minus zero is returned. An attempt to extract the square root of a negative value other than zero returns an invalid REAL number, and status bits CC0, CC1, OV, and OS are all set to '1'.

LAD Example: Find the square root of a number	FBD Example: Find the square root of a number

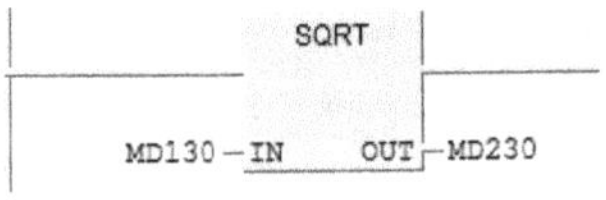

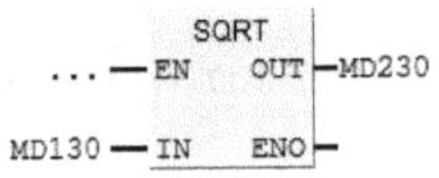

Figure 10-31. Illustration of LAD/FBD Square Root instruction

Square: SQR

When the **EN** input is logic 1, the *Square* instruction returns the square of the floating-point value at **IN** and stores the result in the REAL variable specified at **OUT**. At the input **IN**, you may specify a variable of type REAL, enter a valid floating-point constant in the permissible range, or reference a double word memory location. At **OUT**, you may specify a double word memory location or a variable of type REAL.

If the input value at **IN** is an invalid REAL number, an invalid REAL number is returned by the function and status bits CC0, CC1, OV, and OS are set to '1'.

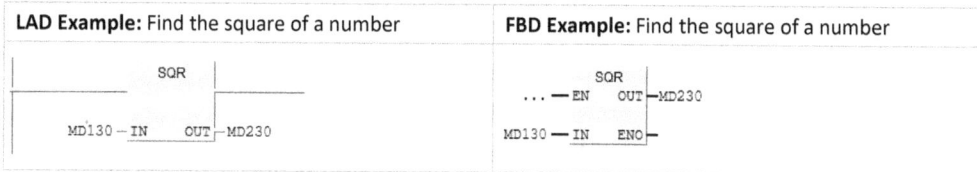

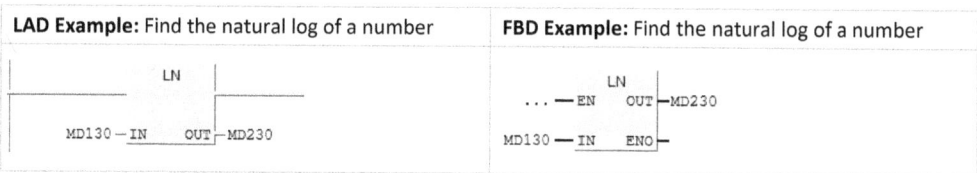

LAD Example: Find the square of a number	**FBD Example:** Find the square of a number
SQR MD130 — IN OUT — MD230	SQR ... — EN OUT — MD230 MD130 — IN ENO

Figure 10-32. Illustration of LAD/FBD Square instruction

Natural Log: LN

When the **EN** input is logic 1, the *Natural Log* instruction returns the natural logarithm (base e, base e = 2.718282e+00) of the floating-point value at **IN,** and stores the result in the REAL variable specified at **OUT.**

At the input **IN,** you may specify a variable of type REAL, a valid floating-point constant greater than zero, or you may reference a double word memory location. At **OUT,** you may specify a double word memory location or a variable of type REAL.

If the input value at **IN** is less than or equal to zero, an invalid REAL number is returned by the function, and status bits CC0, CC1, OV, and OS are set to '1'.

LAD Example: Find the natural log of a number	**FBD Example:** Find the natural log of a number
LN MD130 — IN OUT — MD230	LN ... — EN OUT — MD230 MD130 — IN ENO

Figure 10-33. Illustration of LAD/FBD Natural Log instruction

Exponent: EXP

When the **EN** input is logic 1, the *Exponential Value* instruction returns the exponential value. The value is computed from base e (= 2.718282e+0) and the floating-point value at **IN,** using the expression (e^{IN} (= **OUT**). The result is placed in the REAL variable specified at **OUT.**

At the **IN** input, you may specify a variable of type REAL, or reference a double word memory location. At **OUT,** you may specify a variable of type REAL or a double word memory location.

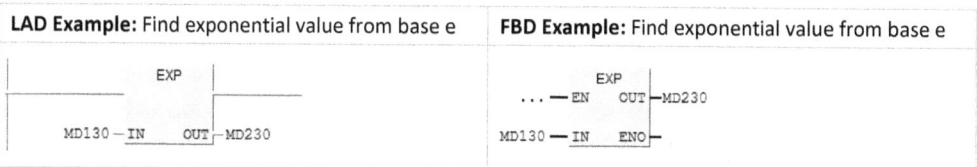

LAD Example: Find exponential value from base e	**FBD Example:** Find exponential value from base e
EXP MD130 — IN OUT — MD230	EXP ... — EN OUT — MD230 MD130 — IN ENO

Figure 10-34. Illustration of LAD/FBD Exponential Value instruction

Chapter 11

LAD/FBD
Shift-Rotate Operations

In this chapter you'll be introduced to LAD/FBD Shift-Rotate operations. In your S7 program development, shift-rotate operations are used to manipulate the contents of a memory location or variable by moving the data by some number of bit positions to the left or to the right. The shift and rotate operations are often used in tracking applications. Shift operations, specifically, may also be used to perform quick multiplications and divisions by two, or by any multiple of two.

Main Topics

- Shift-Rotate Operations

- Shift-Rotate Operations Descriptions

- Programming Shift-Rotate Operations in LAD

- Programming Shift-Rotate Operations in FBD`

SHIFT-ROTATE OPERATIONS

LAD and FBD *shift-rotate instructions* are listed in Table 11-1. With these operations, you can move the bit contents of a specified word or double word variable to the left or right by a specified number of positions. A variety of techniques may be accomplished with these instructions. Shift-rotate operations are often used in conveying applications where information associated with discrete parts is tracked for some distance, or for a fixed number of zones, prior to performing a secondary control operation. A value of '1', for example, may indicate 'passed'—a value of '0' may indicate 'failed'.

Table 11-1. LAD/FBD Shift-Rotate Operations Summary

LAD	FBD	Brief Description
SHL_W — EN ENO — IN OUT — N	SHL_W — EN — IN OUT — N ENO	**Shift Left Word.** If **EN** is logic 1, shift the 16-bits of the Word value specified at **IN**, by **N**-bits to the left. The shifted result is placed in **OUT**.
SHR_W — EN ENO — IN OUT — N	SHR_W — EN — IN OUT — N ENO	**Shift Right Word.** If **EN** is logic 1, shift the 16-bits of the Word value specified at **IN**, by **N**-bits, to the right. The shifted result is placed in **OUT**.
SHL_DW — EN ENO — IN OUT — N	SHL_DW — EN — IN OUT — N ENO	**Shift Left Double Word.** If **EN** is logic 1, shift the 32-bits of the Double-Word value specified at **IN**, by **N**-bits, to the left. The shifted result is placed in **OUT**.
SHR_DW — EN ENO — IN OUT — N	SHR_DW — EN — IN OUT — N ENO	**Shift Right Double Word.** If **EN** is logic 1, shift 32-bits of the Double-Word value at **IN**, by **N**-bits, to the right. The shifted result is placed in **OUT**.
ROL_DW — EN ENO — IN OUT — N	ROL_DW — EN — IN OUT — N ENO	**Rotate Left Double Word.** If **EN** is logic 1, rotate the 32-bits of the Double-Word value specified at **IN**, by **N**-bits, to the left. The result is placed in **OUT**.
ROR_DW — EN ENO — IN OUT — N	ROR_DW — EN — IN OUT — N ENO	**Rotate Right Double Word.** If **EN** is logic 1, rotate the 32-bits of the Double-Word value specified at **IN**, by **N**-bits, to the right. The rotated result is placed in **OUT**.
SHR_I — EN ENO — IN OUT — N	SHR_I — EN — IN OUT — N ENO	**Shift Right Integer.** If **EN** is logic 1, shift the 16-bits of the Integer value specified at **IN**, by **N**-bits, to the right, while maintaining the sign bit. The shifted result is placed in **OUT**.
SHR_DI — EN ENO — IN OUT — N	SHR_DI — EN — IN OUT — N ENO	**Shift Right Double Integer.** If **EN** is logic 1, shift the 32-bits of the Double Integer value specified at **IN**, by **N**-bits, to the right, while maintaining the sign bit. The shifted result is placed in **OUT**.

SHIFT-ROTATE OPERATIONS DESCRIPTIONS

In a *shift operation*, bits vacated on the shift end are filled with zeros; bits shifted out on the opposite end are lost. In a *rotate operation*, bits shifted out on the far end, are rotated into the vacated positions.

Whereas shift and rotate instructions generally operate on a variable without concern for its numeric content, the *shift integer* and *shift double integer* operations shift the bits of a 16-bit or 32-bit integer, thereby changing the integer value, without altering the original sign. Shift and rotate operations are described in the following discussions—but first let's take a look at the parameters that must be specified when using any of these operations.

Shift-Rotate Box Parameters

The parameters for LAD shift-rotate operations are shown below. The enable input line EN allows you to control when the operation is executed. When EN is at logic 1, the variable, specified at the parameter IN, is shifted or rotated by the number of bit-positions specified at N. The result is placed in the variable specified at OUT. The value at IN is unchanged.

Table 11-2. Shift/Rotate Parameters

Parameter	Data Type	Description	Required?	Shift/Rotate Box	
EN		Enable Input line		**LAD**	**FBD**
IN	WORD	Variable to Shift/Rotate	YES		
N			YES		
ENO		Enable Output Signal			
OUT	WORD	Shift/Rotate Result	YES		

In shift and rotate operations, the variables you specify at the parameters IN and OUT must be of the same memory width or data type. In a shift or rotate operation on 32-bits, for example, you must specify a variable of type DWORD at IN and OUT, or a double word memory location such as ID24 or MD42. At the parameter IN, you may also specify a DWORD constant.

Shift Left Word: SHL_W

When the **EN** input is at logic 1, the *Shift Left Word* shifts the 16-bits of the variable you specify at **IN**, bit-by-bit by **N**-positions to the left. The shifted result is placed in the variable specified at **OUT**, the positions vacated on the right are filled with zeros, and the last bit shifted is assigned to status bit CC1.

At **IN,** you may specify a WORD constant—for example W#16#3, a variable of type WORD, or a word-width location. At **OUT,** specify a word-width location—for example MW28, or a variable of type WORD. The number of shift positions is defined by the WORD variable, or constant (0-16₁₆) you specify at the input **N**. If **N** = 0, the operation is not executed; if **N** > 0 then status bits CC0 and OV are set to '0'; and if **N** > 16, then zero is placed in **OUT,** and status bits CC0 and OV are set to '0'.

In the following example, the contents of MW20, at **IN**, is shifted left by 3 positions. The 3-bits vacated on the right are filled with '0'. A shift to the left by 3 is equivalent to multiplying by 2^3 or +8. Let's say that MW20 contains a value of +6; the result would be equivalent to the multiplication of +6 by +8, which is equal to +48.

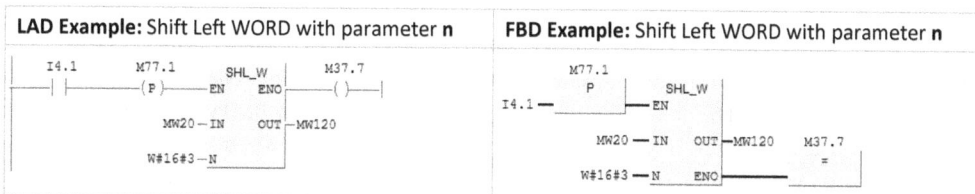

LAD Example: Shift Left WORD with parameter **n**	**FBD Example:** Shift Left WORD with parameter **n**

The example shown above is illustrated in the following figure:

Example: Shift Left Word	
0000 0000 0000 0110	IN = MW20 = 6
	N = 3 positions
0000 0000 0011 0000	OUT = MW120 = 48

Figure 11-1. Illustration of Shift Left Word instruction

Shift Left Double Word: SHL_DW

When the **EN** input is at logic 1, the *Shift Left Double Word* shifts the 32-bits of the variable specified at **IN**, by N-positions, to the left. The shifted result is placed in the variable specified at **OUT**, the vacated positions on the right are filled with zeros, and the last bit shifted is assigned to the status bit CC1.

At **IN**, you may specify a DWORD constant—for example DW#16#38FB, a variable of type DWORD, or a double word memory location. At **OUT**, specify a double word location—for example MD28, or a variable of type DWORD. The number of shift positions is defined by the WORD variable, or constant (0-32₁₆), specified at the input **N**. If **N** = 0, the operation is not executed; if **N** > 0 then status bits CC0 and OV are set to '0'; and if **N** > 32, zero is placed in **OUT**, and status bits CC0 and OV are set to '0'.

In the following example, the value at **IN** is shifted left by 8 positions—the 8-bits vacated on the right are filled with '0'.

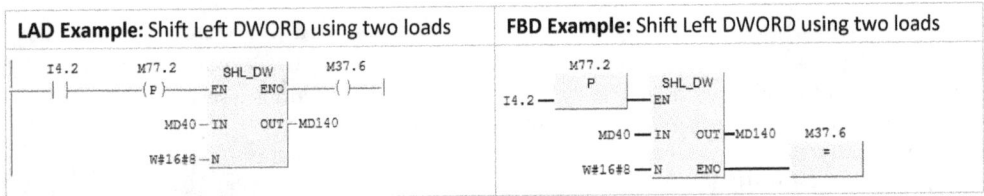

LAD Example: Shift Left DWORD using two loads	FBD Example: Shift Left DWORD using two loads

The previous example is illustrated in the following figure:

Example: Shift Left Double Word	
0000 0000 0000 0000 0011 1000 1111 1011	IN = MD40
	N = 8 positions
0000 0000 0011 1000 1111 1011 0000 0000	OUT MD140

Figure 11-2. Illustration of Shift Left Double Word instruction

Shift Right Word: SHR_W

When the **EN** input is at logic 1, the *Shift Right Word* shifts the 16-bits of the variable you specify at **IN**, bit-by-bit by N-positions to the right. The shifted result is placed in the variable specified at **OUT**; the positions vacated on the left filled with zeros; and the last bit shifted is assigned to status bit CC1.

At **IN**, you may specify a WORD constant—for example W#16#F080, a variable of type WORD, or a word-width location. At **OUT**, you may specify a word-width location—for example MW42, or a variable of type WORD. The number of shift positions is defined by the WORD variable, or constant (0-16₁₆), specified at the input **N**. If **N** = 0, the operation is not executed; if **N** > 0, then status bits CC0 and OV are set to '0'; and if **N** > 16, zero is placed in **OUT**, and status bits CC0 and OV are set to '0'.

In the following example, the value at **IN** is shifted right by 2 positions—the 2-bits vacated on the left are filled with '0'.

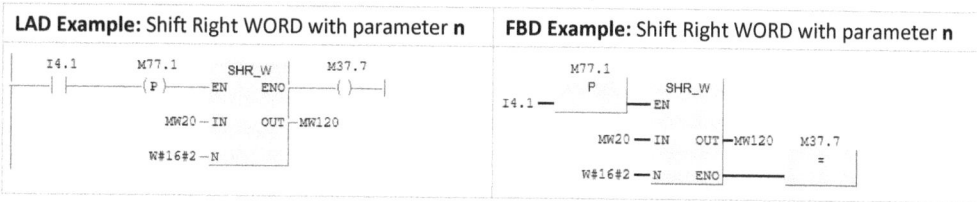

LAD Example: Shift Right WORD with parameter **n**	FBD Example: Shift Right WORD with parameter **n**

The previous example is illustrated in the following figure:

Example: Shift Right Word	
1111 0000 1000 0000	**IN = MW20**
	N = 2 positions
0011 1100 0010 0000	**OUT = MD120**

Figure 11-3. Illustration of Shift Right Word instruction

Shift Right Double Word: SHR_DW

When the **EN** input is at logic 1, the *Shift Right Double Word* shifts the 32-bits of the variable you specify at **IN**, bit-by-bit by **N**-positions to the right. The shifted result is placed in the variable specified at **OUT**, the positions vacated on the left are filled with zeros, and the last bit shifted is assigned to status bit CC1.

At **IN,** you may specify a DWORD constant—for example DW#16#80000000, a variable of type DWORD, or a double word memory location. At **OUT,** you may specify a double word location—for example MD42, or a variable of type DWORD. The number of shift positions is defined by the WORD variable, or constant $(0-32_{16})$, you specify at the input **N**. If **N** = 0, the operation is not executed; if **N** > 0, then status bits CC0 and OV are set to '0'; and if **N** > 32, then zero is placed in **OUT,** and status bits CC0 and OV are set to '0'.

In the following example, the value at **IN** is shifted right by 16 positions—the 16-bits vacated on the left are filled with '0'.

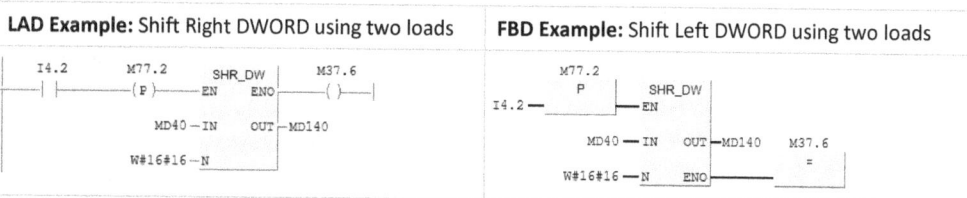

LAD Example: Shift Right DWORD using two loads	FBD Example: Shift Left DWORD using two loads

The previous example is illustrated below:

Example: Shift Right Double Word	
1000 0000 0000 0000 0000 0000 0000 0000	**IN = MD40 =**
	N = 16 positions
0000 0000 0000 0000 1000 0000 0000 0000	**OUT = MD140**

Figure 11-4. Illustration of Shift Right Double Word instruction

Shift Right Integer: SHR_I

When the **EN** input is at logic 1, the *Shift Right Integer* shifts the 16-bits of the variable you specify at **IN**, bit-by-bit by **N**-positions to the right. The shifted result is placed in the variable specified at **OUT**. Each position vacated on the left is filled with the sign bit (bit-15) value—'0' if positive, '1' if negative.

At **IN,** you may specify an INT constant or variable, or a word-width memory location. At **OUT**, specify a word-width location—for example MW66, or a variable of type INT. The number of shift positions is defined by the WORD variable, or constant (0-15₁₆), you specify at the input **N**. If **N** = 0, the operation is not executed; if **N** > 0, then status bits CC0 and OV are set to '0'; and if **N** ≥ 16, the vacated positions of the shift result are filled with '0' if the integer was positive, and '1' if the integer was negative.

In the following example, the integer value of +64 at **IN** is shifted right by 2 positions. A shift to the right by 2 is equivalent to dividing by 2^3 or +4. The result is equivalent to the division of +64 by +4, which is +16. Notice that each position vacated on the left is filled with the sign bit of '0'.

LAD Example: Shift Right Integer with parameter N	FBD Example Shift Right Integer with parameter N
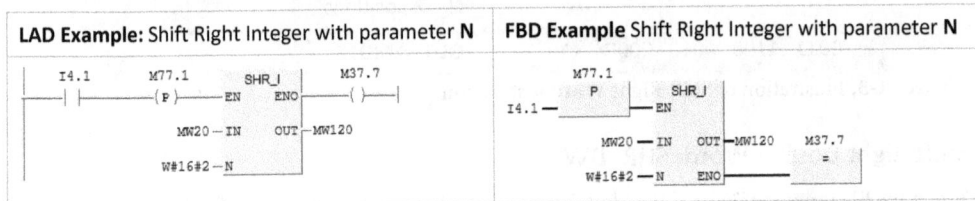	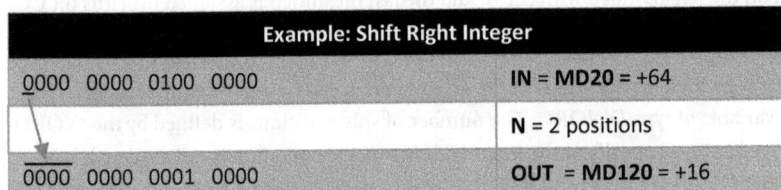

The previous example is illustrated below:

Example: Shift Right Integer	
0000 0000 0100 0000	IN = MD20 = +64
	N = 2 positions
0000 0000 0001 0000	OUT = MD120 = +16

Figure 11-5. Illustration of Shift Right Integer instruction

Shift Right Double Integer: SHR_DI

When the **EN** input is at logic 1, the *Shift Right Double Integer* shifts the 32-bits of the variable you specify at **IN**, bit-by-bit by **N**-positions to the right. The shifted result is placed in the variable specified at **OUT**. Each position vacated on the left is filled with the value of the sign bit (bit-31)—'0' if positive, '1' if negative.

At **IN,** you may specify a DINT constant or a variable, or a double word location. At **OUT,** specify a word-width location—for example MW68, or a variable of type DINT. The number of shift positions is defined by the WORD constant (0-31), or variable, you specify at the input **N**. If **N** = 0, the operation is not executed; if **N** > 0, then status bits CC 0 and OV are set to '0'; and if **N** ≥ 32, the vacated positions of the shift result are filled with '0' if the value was positive, and '1' if the value was negative.

In the following example, the double integer value at **IN** is shifted right by 4 positions—the 4-bits vacated on the left are filled with the sign bit of '1'.

LAD Example: Shift Right Double Integer

```
    I4.2      M77.2     SHR_DI        M37.6
 ───┤ ├───────( P )───┤EN    ENO├──────( )───┤
                      MD40─IN   OUT├─MD140
                    W#16#4─N
```

FBD Example: Shift Right Double Integer

```
              M77.2
               P     SHR_DI
      I4.2 ─         ─EN
                   MD40 ─ IN  OUT├─MD140   M37.6
                                            =
                 W#16#4 ─ N  ENO├
```

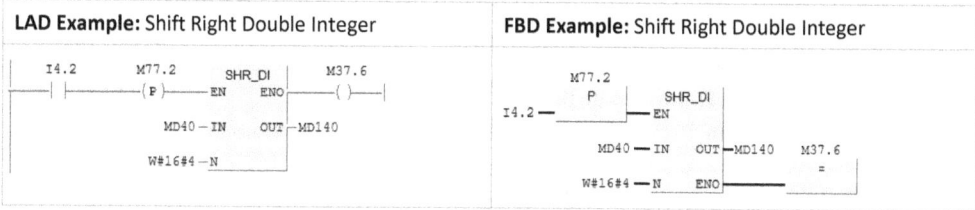

The previous example is illustrated below:

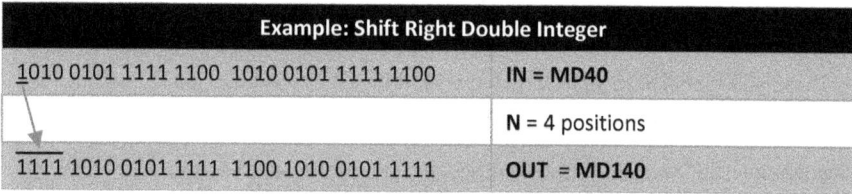

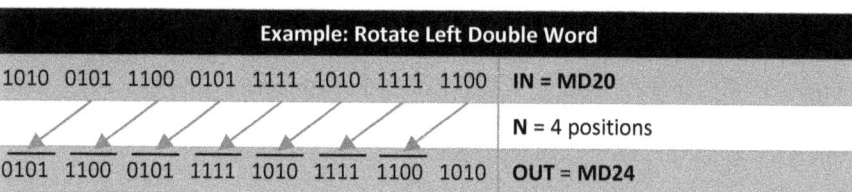

Example: Shift Right Double Integer	
1010 0101 1111 1100 1010 0101 1111 1100	**IN = MD40**
	N = 4 positions
1111 1010 0101 1111 1100 1010 0101 1111	**OUT = MD140**

Figure 11-6. Illustration of Shift Right Double Integer instruction

Rotate Left Double Word: ROL_DW

When the **EN** input is logic 1, the *Rotate Left Double Word* shifts the 32-bits of the variable you specify at **IN**, bit-by-bit by **N**-positions to the left. The rotated result is placed in the variable specified at **OUT**, and the N-positions vacated on the right are filled with the N-bit values rotated out on the left.

At **IN,** you may specify a DWORD constant—for example DW#16#A5C5FAFC, a variable of type DWORD, or a double word memory location. At **OUT,** specify a double word location—for example MD44, or a variable of type DWORD. The number of rotate positions is defined by the WORD variable, or constant (0-32), that you specify at input **N**. If **N** = 0, the operation is not executed; if **N** is > 32, the last bit shifted is moved to the CC1 status bit, and accumulator 1 is shifted based on modulo 32—for example the shift variable is shifted by 1 position, if n = 33; shifted by 2 positions, if n = 34; shifted by 3 positions, if n = 35; and so on. The result zero is placed in **OUT**.

In the following example, the value at **IN** is rotated left by 4 positions—the 4-bits vacated on the right are filled with 4-bits rotated out on the left.

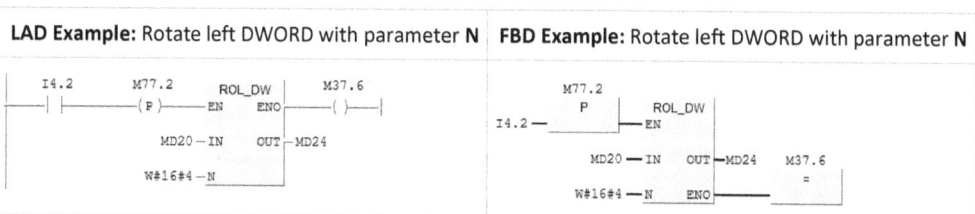

LAD Example: Rotate left DWORD with parameter **N**

```
    I4.2      M77.2     ROL_DW        M37.6
 ───┤ ├───────( P )───┤EN    ENO├──────( )───┤
                      MD20─IN   OUT├─MD24
                    W#16#4─N
```

FBD Example: Rotate left DWORD with parameter **N**

```
              M77.2
               P     ROL_DW
      I4.2 ─         ─EN
                   MD20 ─ IN  OUT├─MD24   M37.6
                                           =
                 W#16#4 ─ N  ENO├
```

The previous example is illustrated below:

Example: Rotate Left Double Word	
1010 0101 1100 0101 1111 1010 1111 1100	**IN = MD20**
	N = 4 positions
0101 1100 0101 1111 1010 1111 1100 1010	**OUT = MD24**

Figure 11-7. Illustration of Rotate Left Double Word instruction

Rotate Right Double Word: ROR_DW

When the **EN** input is at logic 1, the *Rotate Right Double Word* rotates the 32-bits of the variable you specify at **IN**, bit-by-bit by **N**-positions to the right. The rotated result is placed in the variable specified at **OUT**, and the N-positions vacated on the left are filled with the N-bit values rotated out on the right.

At **IN**, you may specify a DWORD constant—for example DW#16# A5C5FAFC, a variable of type DWORD, or a double word memory location. At **OUT**, specify a double word location—for example MD28, or a variable of type DWORD. The number of rotate positions is defined by the WORD variable, or constant (0-32), that you specify at input **N**. If **N** = 0, the operation is not executed; if **N** is > 32, the last bit shifted is moved to the CC1 status bit, and accumulator 1 is shifted based on modulo 32—for example the shift variable is shifted by 1 position, if n = 33; shifted by 2 positions, if n = 34; shifted by 3 positions, if n = 35; and so on. The result zero is placed in **OUT**.

In the following example, the value at **IN** is rotated right by 4 positions—the 4-bits vacated on the left are filled with the 4-bits rotated out on the right.

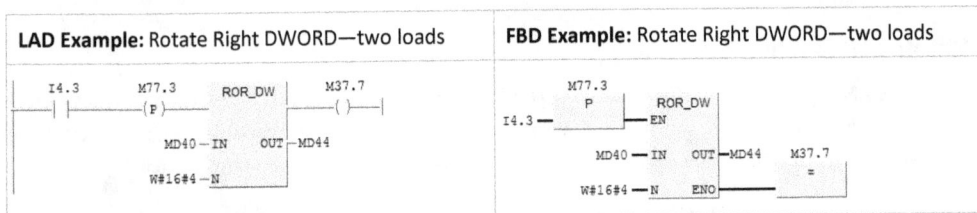

LAD Example: Rotate Right DWORD—two loads	FBD Example: Rotate Right DWORD—two loads

The previous example is illustrated below:

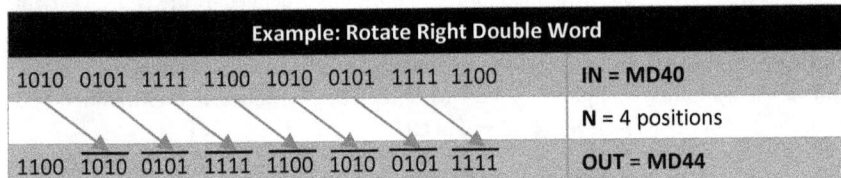

Figure 11-8. Illustration of Rotate Right Double Word instruction

PROGRAMMING SHIFT-ROTATE OPERATIONS IN LAD

Shift-rotate operations move the contents of a source variable by a specified number of positions to the left or right. The result is placed in a destination variable. Programming these operations in LAD is a matter of selecting an operation, based on the name. For example, SHL_W is Shift Left Word; **ROR_DW** is Rotate Right Double Word; and **SHR_W** is Shift Right Word.

A typical application of the shift operation is to multiply by a factor of two, for each position shifted to the left, or to divide by a factor of two, for each position shifted to the right. Each shift represents a factor of 2^n, where n is represents the number of positions shifted—for example, $2^1 = 2$, $2^2 = 4$, $2^3 = 8$, $2^4 = 16$, $2^5 = 32$, and so on. So, a variable is multiplied by 2 by shifting to the left by one. A shift by three positions to the left multiplies by 8. Conversely, shifting by 1 position to the right, results in division by 2; a shift of 3 places to the right results in division by 8. The shift operation might be used in this manner to index a value in a memory location that is used to indirectly point to an indirect address.

Assigning Shift-Rotate Parameters

With the new network inserted and selected, you can drag and drop the desired object onto the network. You must specify the input parameter **IN** as the value to be shifted or rotated, the output parameter

OUT as the location where the result is stored, and the value **N** as the number of positions to shift or rotate. You may also use symbolic addresses, if they are already defined.

When specifying the input and output variables of the shift or rotate, remember that the data type of the variables at **IN** and at **OUT** must be of the same type based on the operation—for example WORD, DWORD, INT, or DINT. You will also have to ensure the correct bit-width if you are specifying memory operands. The shift variable **N** is always a WORD variable or constant.

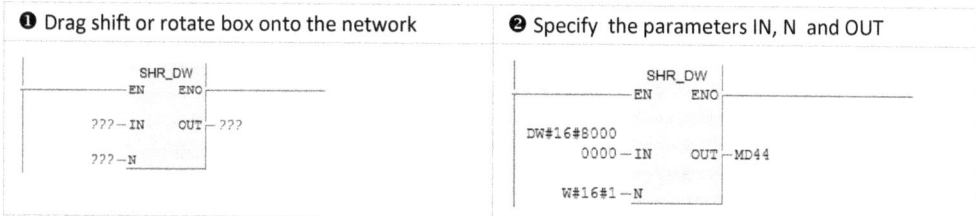

❶ Drag shift or rotate box onto the network	❷ Specify the parameters IN, N and OUT

Figure 11-9. Select and insert shift-rotate operation and specify parameters IN, N, and OUT in LAD.

The next step is to insert logic that triggers the shift or rotate. On the **EN** line, you may insert control logic as required, like that shown in the first cell below. If logic 1 is present at **EN,** the shift is executed; logic 0 at **EN** means the shift-rotate is not executed. In the figure below, a 0-to-1 transition of I 4.1 causes M77.1 to produce a one-shot pulse to trigger the shift.

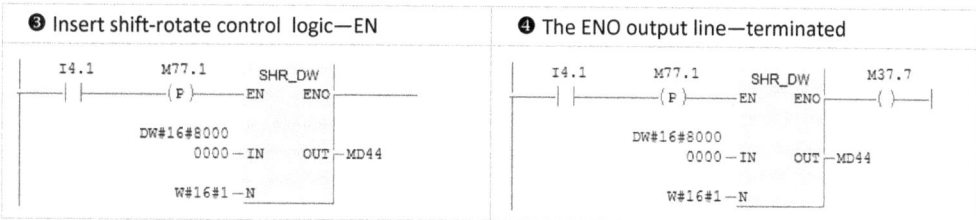

❸ Insert shift-rotate control logic—EN	❹ The ENO output line—terminated

Figure 11-10. Insert enable (EN) logic and terminate the ENO output line in LAD.

The ENO output line may be left unterminated as in the first cell above, or terminated with any valid binary output instruction as shown in the second cell. Any logic combination may be inserted after the enable output **ENO** of the operation when the operation is terminated.

Shift-Rotate operations may be inserted in series by connecting the ENO line of one operation to the **EN** line of the next operation. The following box is evaluated only when the preceding box processes without error.

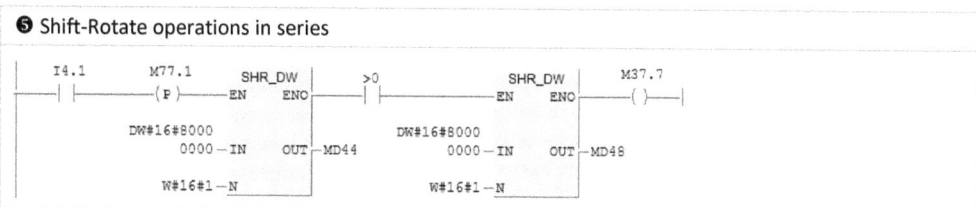

❺ Shift-Rotate operations in series

Figure 11-11. Using EN/ENO mechanism to program shift-rotate operations in series in LAD

As illustrated in the following network, shift-rotate boxes may also be inserted in parallel if placed in a branch that begins directly connected to the left rail of the network. When shift-rotate operations are placed in parallel, the network must be terminated with an output. This output may serve as part of the logic used to evaluate the results, or simply as a dummy output.

If a network arrangement involves two or more parallel branches, each of which has two or more shift-rotate operations in series, boxes are evaluated from left-to-right starting with the first branch, then the second, and so on.

❻ Shift-Rotate operations in parallel or parallel/series branches

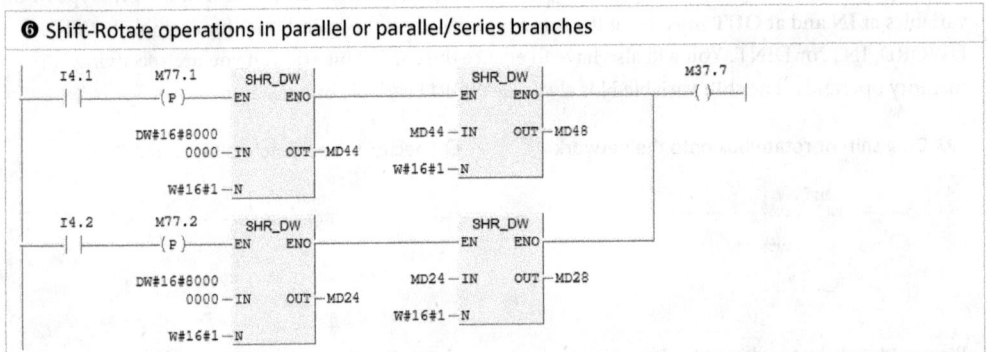

Figure 11-12. Inserting shift-rotate operations in parallel or parallel/series in LAD

PROGRAMMING SHIFT-ROTATE OPERATIONS IN FBD

Shift-rotate operations move the contents of a source variable by a specified number of positions to the left or right. The result is placed in a destination variable. Programming these operations in FBD is a matter of selecting an operation, based on the name. For example, SHL_**W** is Shift Left Word; **ROR_DW** is Rotate Right Double Word; and **SHR_W** is Shift Right Word.

A typical application of the shift operation is to multiply by a factor of two for each position shifted to the left, or to divide by a factor of two for each position shifted to the right. Each shift represents a factor of 2^n, where n is the number of positions shifted. In other words $2^1 = 2$, $2^2 = 4$, $2^3 = 8$, $2^4 = 16$, $2^5 = 32$, and so on. So, a variable is multiplied by 2 simply by shifting to the left by one position. A shift by three positions to the left multiplies by 8. Conversely, shifting by 1 position to the right results in division by 2; 2 places to the right is division by 4. The shift operation is often used in this manner, to index a memory pointer or some other value.

Assigning Shift-Rotate Parameters

With the new network inserted and selected, you can drag and drop the desired object onto the network. You must specify the input parameter **IN** as the value to be shifted or rotated, the output parameter **OUT** as the location where the result is stored, and the value **N** as the number of positions to shift or rotate. You may also use symbolic addresses, if they are already defined.

When specifying the input and output variables of the shift or rotate, remember that the data type of the variables at **IN** and at **OUT** must be of the same type based on the operation—for example WORD, DWORD, INT, or DINT. You will also have to ensure the correct bit-width if you are specifying memory operands. The shift variable **N** is always a WORD variable or constant.

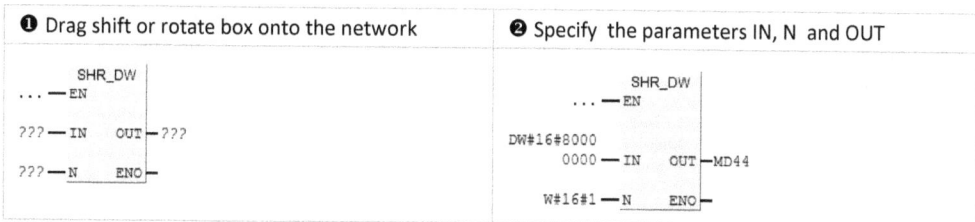

Figure 11-13. Select and insert shift-rotate operation and specify parameters IN, N, and OUT in FBD

The next step is to insert logic that triggers the shift or rotate. On the **EN** line, you may insert control logic as required, like that shown in the first cell below. If logic 1 is present at **EN,** the shift is executed; logic 0 at **EN** means the shift-rotate is not executed. In the figure below, a 0-to-1 transition of I 4.1 causes M77.1 to produce a one-shot pulse to trigger the shift.

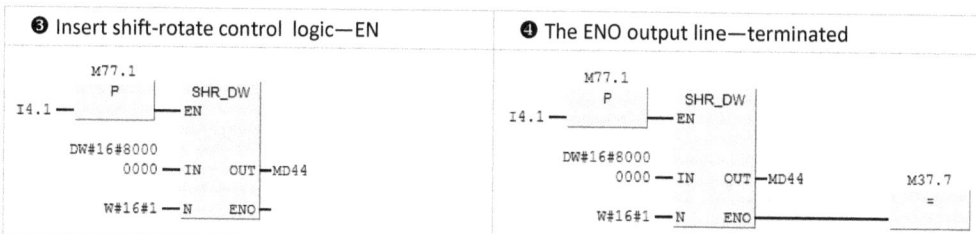

Figure 11-14. Insert enable (EN) logic and terminate the ENO output line in FBD

The ENO output line may be left unterminated as in the first cell above, or terminated with any valid binary output instruction as shown in the second cell. Any logic combination may be inserted after the enable output **ENO** of the operation when the operation is terminated.

Shift-Rotate operations may be inserted in series by connecting the ENO line of one operation to the **EN** line of the following operation. The following box is evaluated only when the preceding box processes without error.

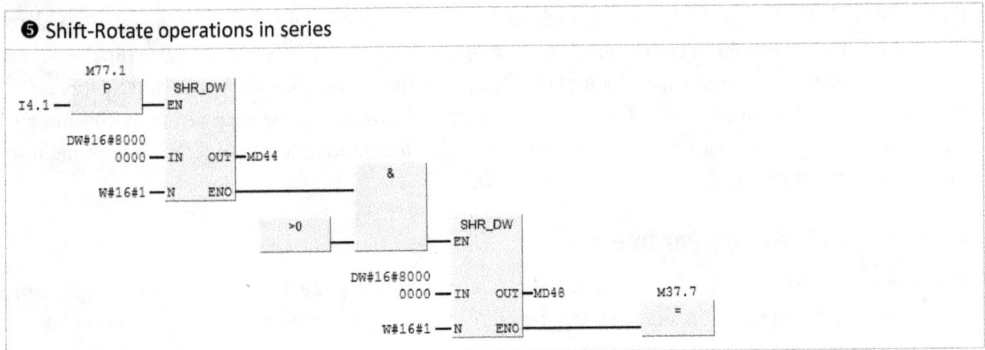

Figure 11-15. Using EN/ENO mechanism to program shift-rotate operations in series in FBD

Shift-Rotate boxes may also be inserted in parallel if placed in a branch that begins directly connected to the left rail of the network. When shift-rotate operations are placed in parallel, the network must be terminated with an output. This output may serve as part of the logic used to evaluate the results, or simply as a dummy output.

If a network arrangement involves two or more parallel branches, each of which has two or more shift-rotate operations in series, boxes are evaluated from left-to-right starting with the first branch, then the second, and so on.

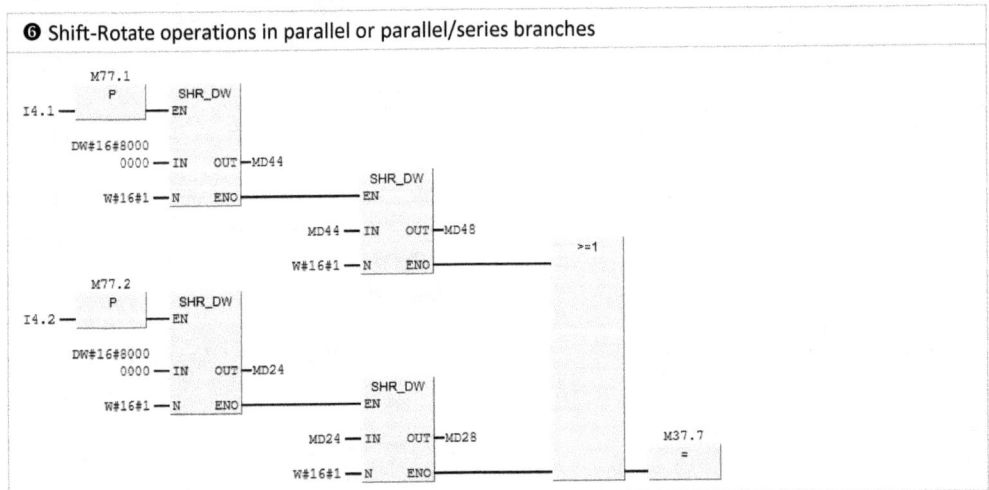

Figure 11-16. Inserting shift-rotate operations in parallel or parallel/series in FBD

Chapter 12

LAD/FBD
Word Logic Operations

This chapter introduces LAD/FBD Word Logic instructions. In S7 program development, these operations are used to perform bit-wise Boolean logic combinations of S7 memory locations or variables. Using Word Logic operations, you can perform standard Boolean operations of AND, OR and Exclusive OR on two 16-bit (word) or two 32-bit (double word) locations or variables.

Word Logic operations generally serve as a mechanism for applying various masking techniques that let you manipulate the values of specific bit positions in word or double word bit patterns.

Main Topics

- Word Logic Operations Overview

- Word Logic Descriptions

- Programming Word Logic in LAD

- Programming Word Logic in FBD

- Word Logic Applications

WORD LOGIC OPERATIONS OVERVIEW

LAD and FBD *Word logic instructions* allow you to perform standard Boolean operations of AND, OR, and Exclusive OR in a bit-wise fashion on two word (16-bit) locations, or on two double word (32-bit) locations. These instructions are useful in performing diagnostic checks on machine or process states, or when you need to ensure that certain bit positions are set to '1' or to '0', or simply inverted.

Take a few moments and review the brief descriptions of the LAD/FBD word logic operations listed in Table 12-1. Some typical word logic applications are described later in this chapter.

Table 12-1. LAD/FBD Word Logic Operations Summary

LAD	FBD	Brief Description
WAND_W —EN ENO— —IN1 OUT— —IN2	WAND_W —EN —IN1 OUT— —IN2 ENO—	**AND Word.** When the **EN** signal is at logic 1, perform a bit-by-bit logical AND on the two 16-bit values supplied at (**IN1/IN2**). The result is output to the variable specified at **OUT**.
WOR_W —EN ENO— —IN1 OUT— —IN2	WOR_W —EN —IN1 OUT— —IN2 ENO—	**OR Word.** When the **EN** signal is at logic 1, perform a bit-by-bit logical OR on the two 16-bit values supplied at (**IN1/IN2**). The result is output to the variable specified at **OUT**.
WXOR_W —EN ENO— —IN1 OUT— —IN2	WXOR_W —EN —IN1 OUT— —IN2 ENO—	**Exclusive OR Word.** When the **EN** signal is at logic 1, perform a bit-by-bit logical XOR on the two 16-bit values supplied at (**IN1/IN2**). The result is output to the variable specified at **OUT**.
WAND_DW —EN ENO— —IN1 OUT— —IN2	WAND_DW —EN —IN1 OUT— —IN2 ENO—	**AND Double-Word.** When the **EN** signal is at logic 1, perform a bit-by-bit logical AND on the two 32-bit values supplied at (**IN1/IN2**). The result is output to the variable specified at **OUT**.
WOR_DW —EN ENO— —IN1 OUT— —IN2	WOR_DW —EN —IN1 OUT— —IN2 ENO—	**OR Double-Word.** When the **EN** signal is at logic 1, perform a bit-by-bit logical OR on the two 32-bit values supplied at (**IN1/IN2**). The result is output to the variable specified at **OUT**.
WXOR_DW —EN ENO— —IN1 OUT— —IN2	WXOR_DW —EN —IN1 OUT— —IN2 ENO—	**Exclusive OR Double-Word.** When the **EN** signal is at logic 1, perform a bit-by-bit logical XOR on the two 32-bit values supplied at (**IN1/IN2**). The result is output to the variable specified at **OUT**.

Note: When EN is '1', ENO will also be '1' except on error; then while EN is '1', ENO is '0'.

WORD LOGIC DESCRIPTIONS

Each word logic operation is identified by a mnemonic label, where **W** indicates word and **DW** indicates double word. For example, **WAND_W** applies the Boolean AND on two word locations; **WAND_DW** applies the AND to two double words. The operation is performed on the two source locations, based on the associated truth table, and places the result in a third location.

Word logic operations are described in the following discussions—but first let's take a look at the parameters that must be specified when using any of these operations.

Word Logic Box Parameters

Entry of each word logic operation requires you to specify source locations for IN1 and IN2, the values used in the logic operation; and a third location, OUT that designates where the result is to be stored. You may insert logic at the enable input line EN to control when the operation is executed. When EN is at logic 1, the bit-by-bit logic operation is performed on IN1 and IN2; the result is placed in the corresponding bit positions in the location specified at OUT. The enable output ENO line signals execution of the operation. IN1 and IN2 are left unchanged.

Table 12-2. Word Logic Instruction Parameters

Parameter	Data Type	Description	Required?	Word Logic Operation	
				LAD	**FBD**
EN		Enable Input line		WOR_W —EN ENO—	WOR_W —EN
IN1	WORD	Bit-Pattern 1 for logic operation	YES	—IN1 OUT—	—IN1 OUT—
IN2	WORD	Bit-Pattern 2 for logic operation	YES	—IN2	—IN2 ENO—
ENO		Signal instruction execution			
OUT	WORD	Logic operation resultant	YES		

Note: The data type is DWORD for 32-bit operations AD, OD, and XOD.

For 16-bit operations, WORD constants, variables of data type WORD, or word-width locations such as IW 8, MW 28, or DB22.DW12. For 32-bit operations may be specified, DWORD constants or variables, or double word locations such as ID 14, MD 44, and DB22.DD12 may be specified. The variable specified at **OUT** must be the same data type or memory width as specified at **IN1** and **IN2**.

AND Word: WAND_W

When the RLO = 1 at the **EN** input, the *AND Word* operation applies a bit-wise logic AND function, using the WORD variables at **IN1** and **IN2**, according to the AND truth table. The result is placed in the corresponding bit positions of the WORD variable specified at **OUT**. According to the AND truth table, the output is '1' only if both inputs are '1', and is '0' if either of the inputs is '0'.

In the following example, the AND Word serves as a 'mask-out' (set bits to '0') operation. To mask-out a bit from a source word (**IN1** in this case) requires a '0' in the corresponding bit position of the variable you choose as your mask word (**IN2** in this case). Notice that only the lower eight bits of **IN1** are moved to **OUT**. The upper eight bits are masked out. To leave specific bits of the source word unaltered, when using the bitwise AND, you place a value of '1' in the corresponding bit position of the mask word.

The masking characteristics of the AND function, as described above, are based on the fact that '0' AND any value is '0', and '1' AND any value is that value. Stated another way, '0' **AND** 'X' = 0; and '1' **AND** 'X' = X. You can review the AND Truth Table of Figure 12-1 to verify these rules.

AND Truth Table			Example: AND Word IW22 and MW22	
IN1	**IN2**	**OUT**	1010 0101 1011 1100	WORD—IW22 (IN1)
0	0	0	**AND**	
0	1	0	0000 0000 1111 1111	WORD—MW22 (IN2)
1	0	0	=	
1	1	1	0000 0000 1011 1100	WORD—OUT = RESULT

Figure 12-1. Illustration of AND Truth Table and AND WORD operation

The LAD/FBD code for this example is shown below:

In the following example, when the RLO = 1 at **EN,** the 16-bits of the input word IW22 are combined bit-by-bit with the 16-bits of bit memory word MW22 using the logical AND operation.

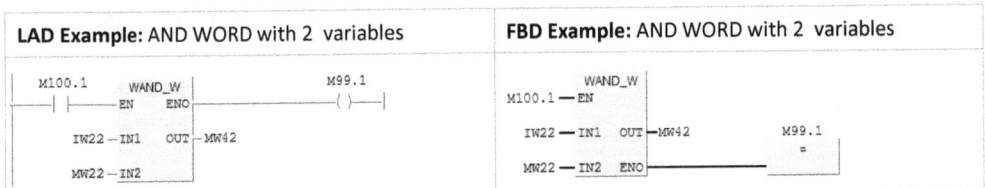

LAD Example: AND WORD with 2 variables	FBD Example: AND WORD with 2 variables
``` M100.1 WAND_W M99.1 ─┤├──── EN ENO ────( )──┤ IW22 ─ IN1 OUT ─ MW42 MW22 ─ IN2 ```	``` WAND_W M100.1 ─ EN IW22 ─ IN1 OUT ─ MW42 M99.1 MW22 ─ IN2 ENO ─────────── = ```

Figure 12-2. Illustration of LAD/FBD AND word operation—using two 16-bit word locations

In addition to specifying WORD variables or word-width locations, as shown in the example above, you may also specify a constant value at either **IN1** or at **IN2**. In the example shown below, the mask value of MW22, from our example in Figure 12-1, is entered as a constant at **IN2**.

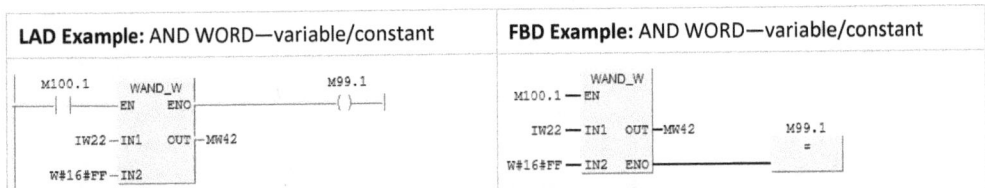

LAD Example: AND WORD—variable/constant	FBD Example: AND WORD—variable/constant
``` M100.1 WAND_W M99.1 ─┤├──── EN ENO ────( )──┤ IW22 ─ IN1 OUT ─ MW42 W#16#FF ─ IN2 ```	``` WAND_W M100.1 ─ EN IW22 ─ IN1 OUT ─ MW42 M99.1 W#16#FF ─ IN2 ENO ──────── = ```

Figure 12-3. Illustration of LAD/FBD AND word operation—using a 16-bit constant

OR Word: WOR_W

When the RLO = 1 at the **EN** input, the *OR Word* operation applies a bit-wise logic OR function using the WORD variables at **IN1** and at **IN2**, according to the OR truth table. The result is placed in the corresponding bit positions of the WORD variable specified at **OUT**. According to the OR truth table, the output is '1' if any one of the inputs is '1', and is '0' if all inputs are '0'.

In the following example, the OR Word serves as a 'mask-in' operation. To mask-in a bit from a source word (**IN1** in this case), even if the bit is '0', requires a '1' in the corresponding bit position of the variable you choose as your mask word (**IN2** in this case). The upper eight bits of **IN1** are masked in. To leave specific bits of the source variable unaltered when using the bitwise OR, a value of '0' is placed in the corresponding bit position of the mask word.

The masking characteristics of the OR function, as described above, are based on the fact that '1' OR any value is '1', and '0' OR any value is that value. Stated another way, '1' **OR** 'X' = 1; and '0' **OR** 'X' = X. You can use the OR Truth Table of Figure 12-4 to verify these rules.

OR Truth Table			Example: OR Word IW22 and MW22	
IN1	**IN2**	**OUT**		
0	0	0	1010 0101 1101 1001	WORD—IW22 (IN1)
0	1	1	OR	
1	0	1	1111 1111 0000 0000	WORD—MW22 (IN2)
1	1	1	=	
			1111 1111 1101 1001	WORD—OUT = RESULT

Figure 12-4. Illustration of OR Truth Table and OR WORD operation

The LAD/FBD code for this example is shown below:

In the following example, when the RLO = 1 at **EN,** the 16-bits of the input word IW22 are combined bit-by-bit with the 16-bits of bit memory word MW22 using the logical OR operation.

LAD Example: OR WORD with 2 variables | **FBD Example:** OR WORD with 2 variables

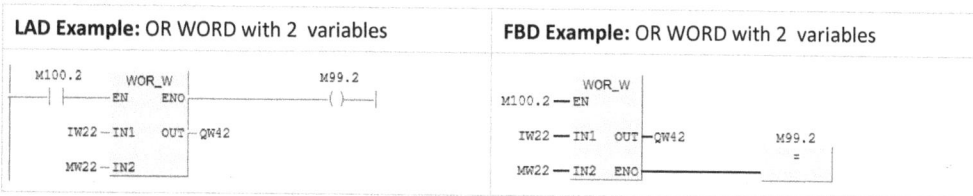

Figure 12-5. Illustration of LAD/FBD OR word operation—using two 16-bit word locations

In addition to specifying WORD variables or word-width locations, as shown in the example above, you may also specify a constant value at either **IN1** or at **IN2**. In the example shown below, the mask value of MW22, from our example in Figure 12-4, is entered as a constant at **IN2**.

LAD Example: OR WORD—variable/constant | **FBD Example:** OR WORD—variable/constant

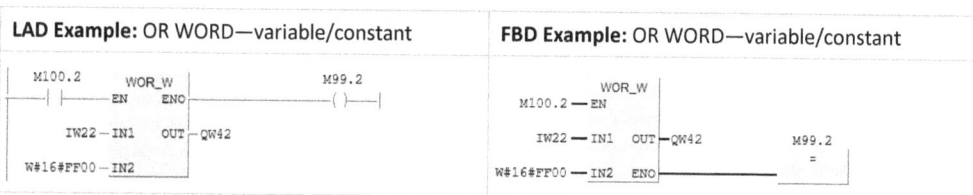

Figure 12-6. Illustration of LAD/FBD OR word operation—using a 16-bit constant

Exclusive OR Word: WXOR_W

When the RLO = 1 at the **EN** input, the *XOR Word* operation applies a bit-wise logic XOR function using the WORD variables at **IN1** and **IN2**, according to the XOR truth table. The bit-wise result is placed in the corresponding bit positions of the WORD variable specified at **OUT**. According to the XOR truth table, the output is always '0' if both inputs are the same; and is '1' if either input is '1', but not both—that is to say that the result is '1' whenever the inputs are different.

In some masking situations, the need is not to cause a bit value to be set to '0' or '1', but instead to invert the current state. The XOR function returns '1' if and only if there is an odd number of 1's in the corresponding bit positions of the source (**IN1** in this case) and in the mask (**IN2** in this case). If two corresponding bits are '1', the result is always '0'; if only one of the values is '1', the result is always '1'. Therefore inversion of a bit position is done by XORing that position with '1'. If the original bit was '1', it returns '0'—XOR '1' = 0; if the original bit was '0' it returns '1'—XOR '0' = 1.

This masking result, as described above, is based on the fact that '0' XOR any value is that value, and '1' XOR any value is that value inverted. Stated another way, '0' **XOR** 'A' = A; and '1' **XOR** 'A' = Ā. You can verify these rules using the XOR Truth Table of Figure 12-7.

XOR Truth Table			Example: XOR Word IW22 and IW24	
IN1	**IN2**	**OUT**	1110 0101 1011 1100	WORD—IW22 (IN1)
0	0	0	XOR	
0	1	1	1111 1111 1111 0000	WORD—IW24 (IN2)
1	0	1	=	
1	1	0	0001 1010 0100 1100	WORD—OUT = RESULT

Figure 12-7. Illustration of XOR Truth Table and XOR WORD operation

The LAD/FBD code for this example is shown below:

In the following example, when the RLO = 1 at **EN,** the 16-bits of the input word IW22 are combined bit-by-bit with the 16-bits of input word IW24 using the logical XOR operation.

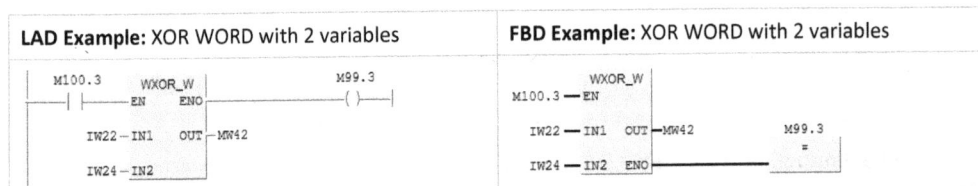

LAD Example: XOR WORD with 2 variables	**FBD Example:** XOR WORD with 2 variables

Figure 12-8. Illustration of LAD/FBD XOR word operation—using two 16-bit word locations

In addition to specifying WORD variables or word-width locations, as shown in the example above, you may also specify a constant value at either **IN1** or at **IN2**. In the example shown below, the mask value of IW24, from our example in Figure 12-7, is entered as a constant at **IN2**.

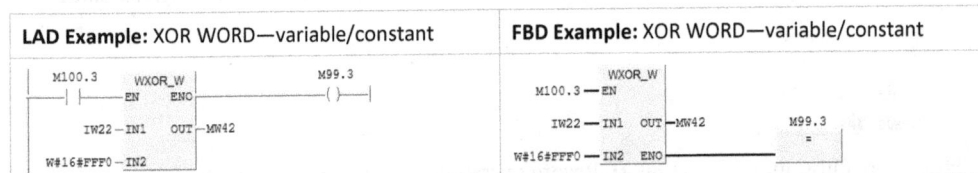

LAD Example: XOR WORD—variable/constant	**FBD Example:** XOR WORD—variable/constant

Figure 12-9. Illustration of LAD/FBD XOR word operation—using a 16-bit constant

212

AND Double Word: WAND_DW

When the RLO = 1 at the **EN** input, the *AND Double Word* operation applies a bit-wise logic AND function using the DWORD variables at **IN1** and **IN2**, according to the AND truth table. The result is placed in the corresponding bit positions of the DWORD variable specified at **OUT**. According to the AND truth table, the output is '1' only if both inputs is '1', and is '0' if either of the inputs is '0'.

In the following example, the result contains a '1' only in those positions where a '1' occurs in the corresponding positions of both **IN1** and **IN2**.

AND Truth Table			Example: AND Double Word ID22 and ID42	
IN1	**IN2**	**OUT**	1010 0101 1111 1100 1010 0111 0001 1110	DWORD—ID22 (IN1)
0	0	0	AND	
0	1	0	1111 0000 0000 0011 1111 0110 0101 0011	DWORD—ID42 (IN2)
1	0	0	=	
1	1	1	1010 0000 0000 0000 1010 0110 0001 0010	DWORD - OUT = RESULT

Figure 12-10. Illustration of AND Truth Table and AND DWORD operation

The LAD/FBD code for this example is shown below:

In the following example, when the RLO = 1 at **EN,** the 32-bits of the input double word ID22 are combined bit-by-bit with the 32-bits of input double word ID42 using the logical AND operation.

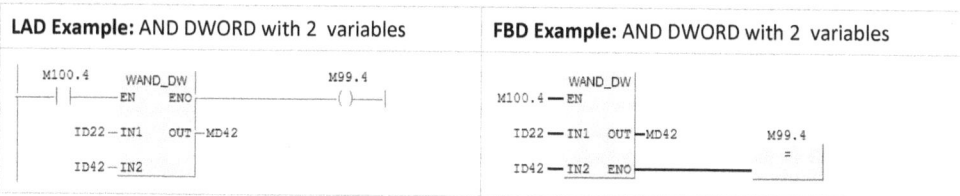

LAD Example: AND DWORD with 2 variables	FBD Example: AND DWORD with 2 variables

Figure 12-11. Illustration of LAD/FBD AND double word operation—using two 32-bit word locations

In addition to specifying DWORD variables or double word locations, as shown in the example above, you may also specify a constant value at either **IN1** or at **IN2**. In the example shown below, the mask value of ID42, from our example in Figure 12-10, is entered as a constant at **IN2**.

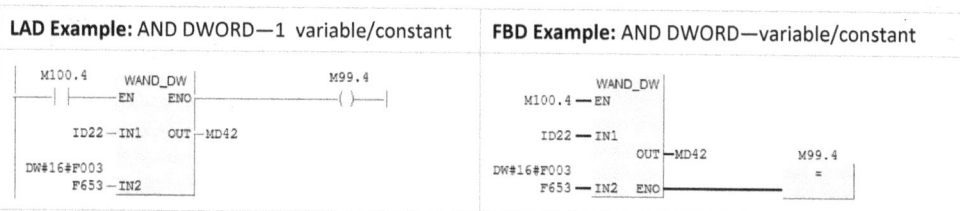

LAD Example: AND DWORD—1 variable/constant	FBD Example: AND DWORD—variable/constant

Figure 12-12. Illustration of LAD/FBD AND double word operation—using a 32-bit constant

OR Double Word: WOR_DW

When the RLO = 1 at the **EN** input, the *OR Double Word* operation applies a bit-wise logic OR function using the DWORD variables at **IN1** and at **IN2**, according to the OR truth table. The result is placed in the corresponding bit positions of the DWORD variable specified at **OUT**. According to the OR truth table, the output is '1' if any one of the inputs is '1', and is '0' if all inputs are '0'.

In the following example, the result contains a '1' in those positions where a '1' occurs in the corresponding position in either **IN1** or in **IN2**, or in both.

OR Truth Table			Example: OR Double Word ID22 and ID42	
IN1	IN2	OUT	1010 0101 1111 1100 1010 0101 0000 0011	DWORD—ID22 (IN1)
0	0	0	OR	
0	1	1	1111 1000 0010 0011 1111 0000 0000 0100	DWORD—ID42 (IN2)
1	0	1	=	
1	1	1	1111 1101 1111 1111 1111 0101 0000 0111	DWORD—OUT = RESULT

Figure 12-13. Illustration of OR Truth Table and OR DWORD operation

The LAD/FBD code for this example is shown below:

In the following example, when the RLO = 1 at **EN,** the 32-bits of the input double word ID22 are combined bit-by-bit with the 32-bits of input double word ID42 using the logical OR operation.

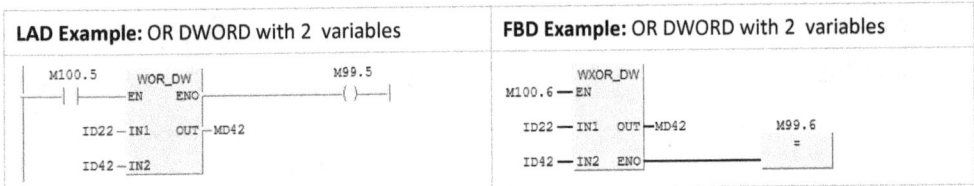

LAD Example: OR DWORD with 2 variables	FBD Example: OR DWORD with 2 variables

Figure 12-14. Illustration of LAD/FBD OR double word operation—using two 32-bit word locations

In addition to specifying DWORD variables or double word locations, as shown in the example above, you may also specify a constant value at either **IN1** or at **IN2**. In the example shown below the mask value of ID42, from our example in Figure 12-13, is entered as a constant at **IN2**.

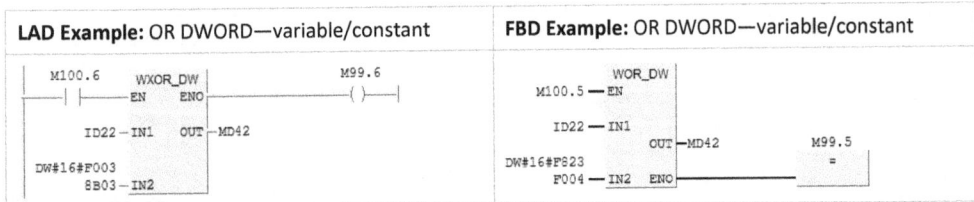

LAD Example: OR DWORD—variable/constant	FBD Example: OR DWORD—variable/constant

Figure 12-15. Illustration of LAD/FBD OR double word operation—using a 32-bit constant

XOR Double Word: WXOR_DW

When the RLO = 1 at the **EN** input, the *XOR Double Word* operation applies a bit-wise logic XOR function using the DWORD variables at **IN1** and **IN2**, according to the XOR truth table. The bit-wise result is placed in the corresponding bit positions of the DWORD variable specified at **OUT**. According to the XOR truth table, the output is always '0' if both inputs are the same; and is '1' if either input is '1', but not both—that is to say that the result is '1' whenever the inputs are different.

In the following example, the result contains a '1' only in those positions where either **IN1** or **IN2** is '1', but not both.

IN1	IN2	OUT	Example: XOR Double Word ID22 and ID42	
0	0	0	1010 0101 1111 1100 1010 0101 1111 1100	DWORD—ID22 (IN1)
0	1	1	**XOR**	
1	0	1	1111 0000 0000 0011 1000 1011 0000 0011	DWORD—ID42 (IN2)
1	1	0	=	
			0101 0101 1111 1111 0000 1010 1111 1111	DWORD - OUT = RESULT

Figure 12-16. Illustration of XOR Truth Table and XOR DWORD operation

The LAD/FBD code for this example is shown below:

In the following example, when the RLO = 1 at **EN,** the 32-bits of the input double word ID22 are combined bit-by-bit with the 32-bits of input double word ID42 using the logical XOR operation.

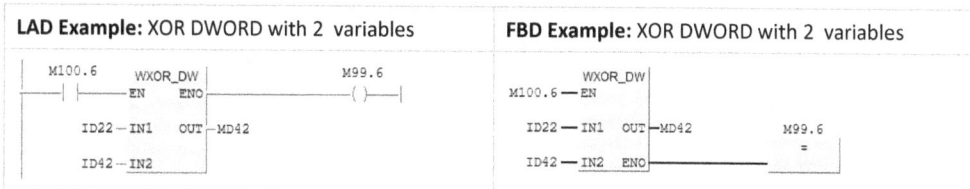

Figure 12-17. Illustration of LAD/FBD XOR double word operation—using two 32-bit word locations

In addition to specifying DWORD variables or double word locations, as shown in the example above, you may also specify a constant value at either **IN1** or at **IN2**. In the example shown below, the mask value of ID42, from our example in Figure 12-16, is entered as a constant at **IN2**.

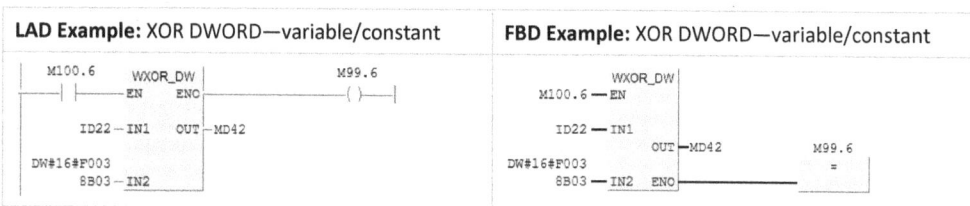

Figure 12-18. Illustration of LAD/FBD XOR double word operation—using a 32-bit constant

PROGRAMMING WORD LOGIC IN LAD

Programming word logic operations is a matter of selecting the operation, from the catalog, based on the name. For example, **WOR_W** is OR Word; **WOR_DW** is OR Double Word.

Assigning Word Logic Parameters

With a new or existing network, you can drag and drop an arithmetic object onto the network. You must specify the **IN1** and **IN2** parameters as the two values to be operated on; and the output parameter **OUT** as the location to store the result. You may also use symbolic addresses.

Remember that the data type of **IN1** and **IN2**, as well as the output **OUT**, depend on whether a word or double word operation is being performed. All three must be of the same type. For **IN1** and **IN2**, your entry may either be a hexadecimal constant (typical of a mask word), a word or double word location, or a WORD or DWORD variable. **OUT** will be specified according to the operation as either a WORD or DWORD variable. Symbolic addresses may also be used.

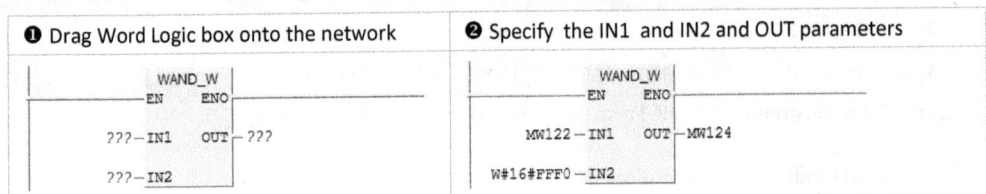

Figure 12-19. Select and insert word logic operation and specify parameters IN1, IN2, and OUT

The next step, if required, is to insert the driving logic that enables the operation. On the **EN** line, you may insert any series or parallel logic combination, as shown below. You may also leave the **EN** line without driving conditions, as in the second cell above—in this case, the operation is always processed.

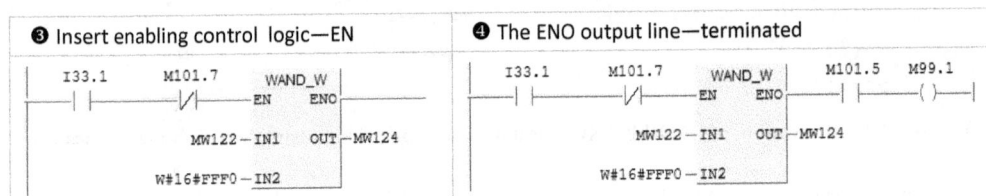

Figure 12-20. Insert enable (EN) logic and terminate the ENO output line

The **ENO** output line may be left unterminated, as in the first cell above, or terminated with a binary output instruction as shown in the second cell. Control logic may also be placed after the enable output **ENO**, but in such a case the network must be terminated.

To perform subsequent word logic operations, you may place word logic boxes in series as required. The **ENO** line of one box is connected to the **EN** line of the following operation. Each subsequent operation is evaluated only if the preceding box processes without error.

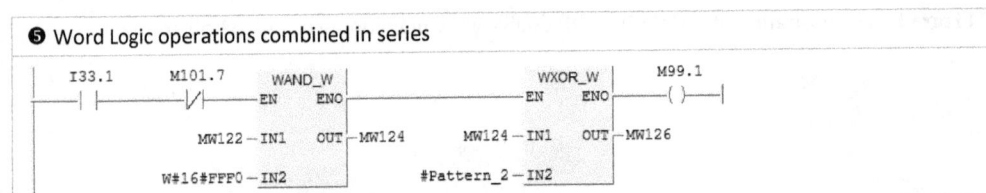

Figure 12-21. Using EN/ENO mechanism to program word logic operations in series

Word logic boxes may also be inserted in parallel if placed in a branch that connects directly to the left rail of the network. When word logic boxes are placed in parallel, the network must be terminated with an output coil. This output coil may serve as part of the logic used to evaluate the outcome of the operation, or simply as a dummy output.

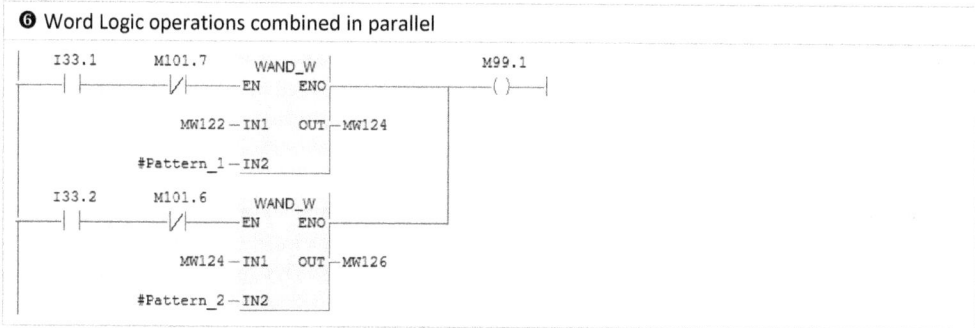

Figure 12-22. Inserting word logic operations in parallel

If a network has two or more parallel branches, each of which has two or more word logic boxes in series, then the branches are evaluated from left-to-right starting with the first branch.

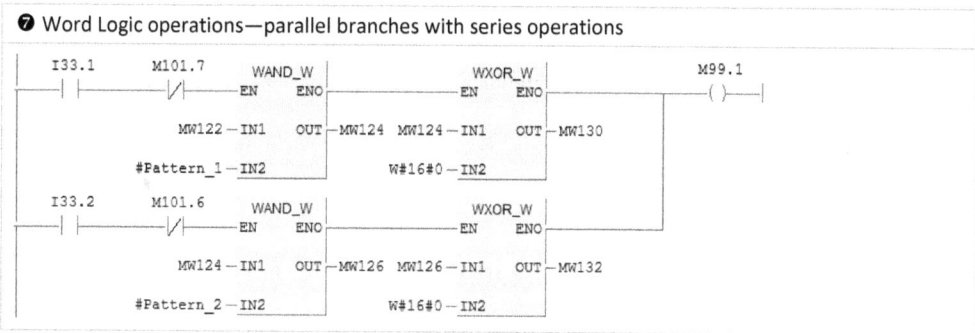

Figure 12-23. Inserting multiple branches of series word logic operations

PROGRAMMING WORD LOGIC IN FBD

Programming word logic operations is a matter of selecting the operation, from the catalog, based on the name. For example, **WOR_W** is OR Word; **WOR_DW** is OR Double Word.

Assigning Word Logic Parameters

With a new or existing network, you can drag and drop an arithmetic object onto the network. You must specify the **IN1** and **IN2** parameters as the two values to be operated on; and the output parameter **OUT** as the location to store the result. You may also use symbolic addresses.

Remember that the data type of **IN1** and **IN2**, as well as the output **OUT**, depend on whether a word or double word operation is being performed. All three must be of the same type. For **IN1** and **IN2**, your entry may either be a hexadecimal constant (typical of a mask word), a word or double word location, or a WORD or DWORD variable. **OUT** will be specified according to the operation as either a WORD or DWORD variable. Symbolic addresses may also be used.

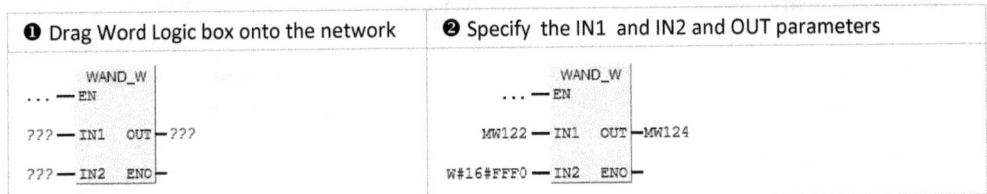

❶ Drag Word Logic box onto the network	❷ Specify the IN1 and IN2 and OUT parameters
| ```
        WAND_W
... — EN

??? — IN1   OUT — ???

??? — IN2   ENO
``` | ```
        WAND_W
... — EN

MW122 — IN1   OUT — MW124

W#16#FFF0 — IN2   ENO
``` |

Figure 12-24. Select and insert word logic operation and specify parameters IN1, IN2, and OUT

The next step, if required, is to insert the driving logic that enables the operation. On the **EN** line, you may insert any series or parallel logic combination as shown below. You may also leave the **EN** line without driving conditions, as in the second cell above—in this case, the operation will always execute.

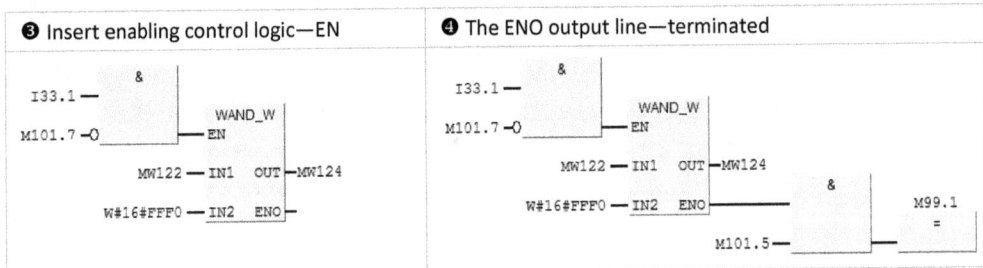

❸ Insert enabling control logic—EN	❹ The ENO output line—terminated
| ```
          &
I33.1 —
               WAND_W
M101.7 –O     — EN

        MW122 — IN1   OUT — MW124

        W#16#FFF0 — IN2   ENO
``` | ```
          &
I33.1 —
               WAND_W
M101.7 –O     — EN

        MW122 — IN1   OUT — MW124
                                    &
        W#16#FFF0 — IN2   ENO              M99.1
                                            =
                          M101.5 —
``` |

Figure 12-25. Insert enable (**EN**) logic and terminate the **ENO** output line

The **ENO** output line may be left unterminated, as in the first cell above, or terminated with a binary output instruction as shown in the second cell. Control logic may also be placed after the enable output **ENO**, but in such a case the network must be terminated.

To perform subsequent word logic operations, you may place word logic boxes in series as required. The **ENO** line of one box is connected to the **EN** line of the following operation. Each subsequent operation is evaluated only if the preceding box processes without error.

❺ Word Logic operations combined in series

```
                &
    I33.1 ──
                        WAND_W
    M101.7 ─O           EN

        MW122 ── IN1   OUT ─MW124
                                        WXOR_W
     W#16#FFF0 ── IN2   ENO              EN

                            MW124 ── IN1   OUT ─MW126    M99.1
                                                          ═
                        #Pattern_2 ── IN2   ENO
```

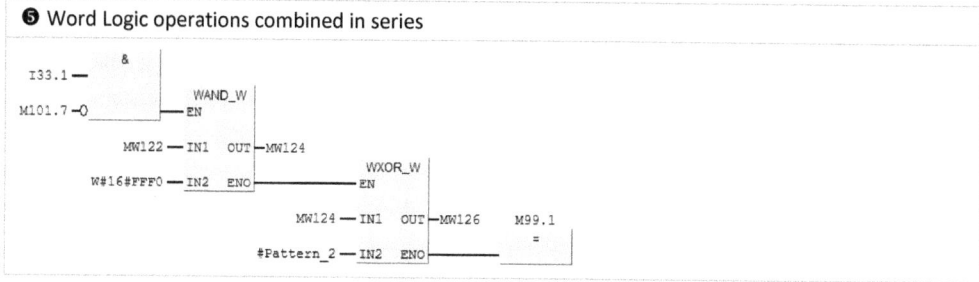

Figure 12-26. Using EN/ENO mechanism to program word logic operations in series

Word logic boxes may also be inserted in parallel if placed in a branch that connects directly to the left rail of the network. When word logic boxes are placed in parallel, the network must be terminated with an output coil. This output coil may serve as part of the logic used to evaluate the outcome of the operation, or simply as a dummy output.

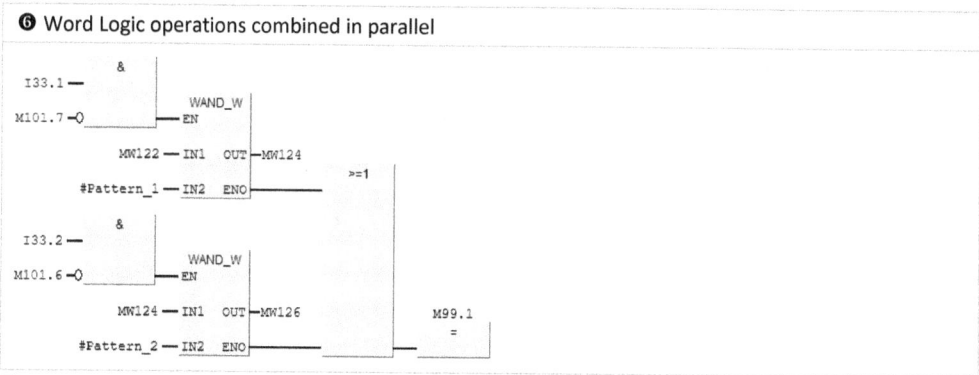

❻ Word Logic operations combined in parallel

```
                &
    I33.1 ──
                        WAND_W
    M101.7 ─O           EN

        MW122 ── IN1   OUT ─MW124
                                        >=1
     #Pattern_1 ── IN2   ENO

                &
    I33.2 ──
                        WAND_W
    M101.6 ─O           EN

        MW124 ── IN1   OUT ─MW126           M99.1
                                             ═
     #Pattern_2 ── IN2   ENO
```

Figure 12-27. Inserting word logic operations in parallel

If a network has two or more parallel branches, each of which has two or more word logic boxes in series, then the branches are evaluated from left-to-right starting with the first branch.

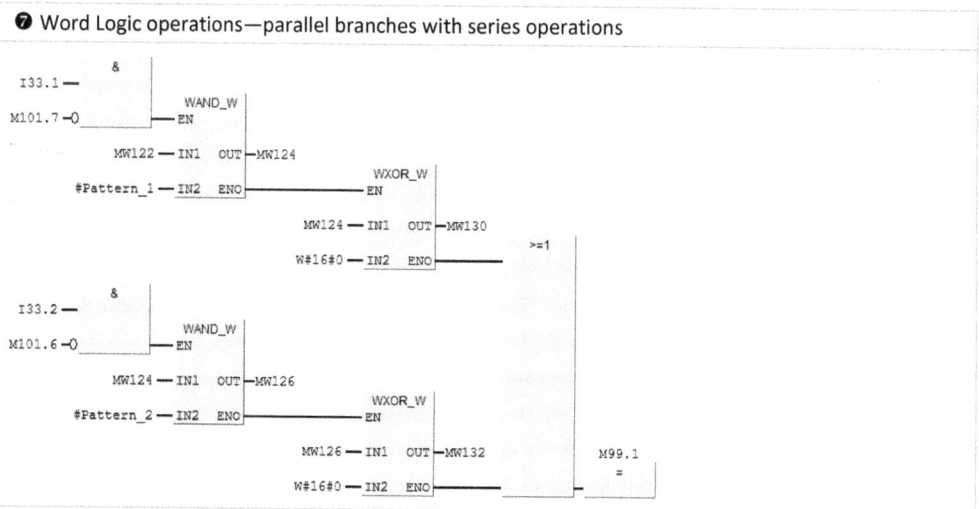

❼ Word Logic operations—parallel branches with series operations

```
                &
    I33.1 ──
                        WAND_W
    M101.7 ─O           EN

        MW122 ── IN1   OUT ─MW124
                                        WXOR_W
     #Pattern_1 ── IN2   ENO              EN

                            MW124 ── IN1   OUT ─MW130
                                                        >=1
                         W#16#0 ── IN2   ENO

                &
    I33.2 ──
                        WAND_W
    M101.6 ─O           EN

        MW124 ── IN1   OUT ─MW126
                                        WXOR_W
     #Pattern_2 ── IN2   ENO              EN

                            MW126 ── IN1   OUT ─MW132    M99.1
                                                          ═
                         W#16#0 ── IN2   ENO
```

Figure 12-28. ORing multiple branches of series word logic operations

WORD LOGIC APPLICATIONS

With a technique called *masking,* word logic operations let you manipulate specific bits within a word or double word variable, while leaving other bits unaffected. Masking in general is setting certain bit values to achieve a desired outcome. This ability is likely the most common application of Word Logic operations. Using the AND operation, you can force selected bit positions to '0'; with OR you can force selected bit positions to '1'; and with XOR, you can selectively invert specific bit positions.

An example of masking may be illustrated in a situation where you need to guarantee that the upper 8-bits of a 16-bit variable 'X' are all zero. To achieve this result you would AND the variable 'X' with a mask value in which the upper 8-bits are all '0'. The mask value would contain all '1s' in the lower 8-bits to cause these bits of the original variable to be unaffected. This bitwise AND, as shown below, forces the high-order 8-bits to '0' and passes the low-order 8-bits of 'X' unchanged. Masking principles for AND, OR, and XOR are illustrated below and described in the following discussions.

AND Function with Mask	OR Function with Mask	XOR Function with Mask
1111 1010 0101 1100 Var. 'X'	1111 1010 0101 1100 Var. 'X'	1111 1010 0101 1100 Var. 'X'
AND	OR	XOR
0000 0000 1111 1111 Mask	0000 0000 1111 1111 Mask	0000 0000 1111 1111 Mask
0000 0000 0101 1100 Result	0000 0000 1111 1111 Result	1111 1010 1010 0011 Result

Figure 12-29. Illustration of 'masking' with AND, OR, and XOR

Masking with AND

An important characteristic of the **AND** operation is that, with masking, you can selectively force a '0' result in multiple bit positions. Masking, as mentioned previously, involves setting one of the operand values of the operation in a way that causes a known result. For example, with the AND, if one of the operands is '0', the result is always '0' regardless of the other operand value. Conversely, if one operand is '1', the result is always the second operand value. Stated another way, '0' **AND** 'X' = 0, and '1' **AND** 'X' = X. You can apply these rules to set the bit values of a mask variable to affect a desired outcome.

Querying the status of a bit—Masking with AND Example

A typical application of the bitwise AND operation is to determine the status of a specific bit position of a given source word. The technique is referred to as "querying the status of a bit."

It is possible to use a bitmask to check the state of specific bits regardless of the other bits. The technique involves a first step of selectively forcing all bits off except for the bits in question (the queried bits); the other bits are turned off using the bitwise mask with AND, based on the previously described rule—'0' AND 'X' = 0. After performing the bitwise AND to the source and mask variables, the result is compared to '0'. A '0' result indicates that the bit in question was 'OFF'; if the result of the compare is any other value, then the bit in question was 'ON'. This technique is quite convenient since it is not necessary to determine the actual value of the result—just that the result is not equal to zero.

Example: Querying the status of one or more bit positions—with AND	
1010 0101 1111 1100 1010 0101 1111 1100	SOURCE VARIABLE
1000 0000 1000 0000 1000 0000 1000 0000	MASK VARIABLE
1000 0000 1000 0000 1000 0000 1000 0000	QUERIED RESULT

Figure 12-30. Illustration of masking with AND to query the status of a bit position(s)

Masking with OR

An important characteristic of the **OR** operation is that, with masking, you can selectively force a '1' result in multiple bit positions. Masking, as mentioned earlier, involves setting one of the operand values of the operation in a way that causes a known result. For example, with the OR, if one of the operands is set to '1', the result is always '1' regardless of the second operand value. Conversely, if one operand is set to '0', the result is always the second operand value. Stated another way, '1' **OR** 'X' = 1, and '0 **OR** 'X' = X. You can apply these rules to set the mask variable bit values to affect the outcome.

Setting the status of a bit to '1'—Masking with OR Example

A typical application of the OR Word operation is to simply ensure that one or more specific bit positions are set to '1'. Ensuring that a bit position is '1' is based on the rule '1' **OR** 'X' = 1.

When you define the mask value, '1' needs to be inserted for each bit position you wish to ensure is '1'. In the example below, the mask variable is set to ensure that the low-order 16-bits of the source variable are set to '1'. Notice that by placing '0' in the mask for bit positions, the high-order 16-bits are completely unaffected—this is true because of the rule with the bitwise OR, '0' **OR** 'X' = X.

Example: Inverting the status of one or more bit positions—with OR	
1010 0101 1111 1100 1010 0101 1011 1100	SOURCE VARIABLE
0000 0000 0000 0000 1111 1111 1111 1111	MASK VARIABLE
0101 0101 1111 1111 0000 1010 1111 1111	SET BIT POSITIONS RESULT

Figure 12-31. Illustration of masking with OR to set the status of a bit position(s) to '1'

Masking with XOR

An important characteristic of the **XOR** operation is that, with masking, you can selectively cause one or more bit positions to be inverted. Masking, as mentioned previously, involves setting the bit value of one of the operands (mask variable) in a way that produces a known result. For example, with the XOR, if one of the operands is '1', the XOR result is always the inverse of the other operand. If the other operand is '1', for example, the XOR result would be '0'; if the other operand is '0', the result would be '1'. In both cases the result is the second operand inverted. Stated as a rule, '1' **XOR** 'A' = Ā. Conversely, if one of the operands of an XOR operation is fixed at '0' (mask operand), then the XOR result is always equal to the second operand. Stated another way, '0' **XOR** 'A' = A.

Inverting the status of a bit—Masking with XOR Example

A typical application of the Exclusive OR Word operation is to invert the status of one or more bit positions—if the current value is '0', it is toggled to '1'; if the current value is '1', it is toggled to '0'.

Bit inversion using XOR is possible based on the rule '1' **XOR** 'A'= Ā. Ā is the inverted status of the original value 'A'. When you define the mask value, '1' needs to be inserted for each bit position you wish to invert. In the example below, every bit position needs to be inverted—therefore every bit position of the mask has a value of '1'.

Example: Inverting the status of one or more bit positions—with XOR	
1010 0101 1111 1100 1010 0101 0011 0001	SOURCE VARIABLE
1111 1111 1111 1111 1111 1111 1111 1111	MASK VARIABLE
0101 1010 0000 0011 0101 1010 1100 1110	INVERTED RESULT

Figure 12-32. Illustration of masking with XOR to invert the status of a bit position

Checking a variable against Itself—XOR Example

A typical application of the bitwise XOR is to verify that a specific bit pattern is what it is expected to be in a known situation. For example, imagine you need to verify whether the actual states of 32 inputs match the 32 expected states at a specific step of a machine or process. If the process is operating correctly as it should, the actual states of the 32 inputs should always match the 32 bit values of the expected states.

If all is as it should be, applying the XOR using the actual states (operand variable-1), and the expected states (operand variable-2), is in essence combining the source variable with itself (mask variable). An examination of the XOR Truth table (Figure 12-17), shows another XOR characteristic—if both operands are the same—both are '1' or both are '0', then the XOR result is always '0'. Stated another way, 'A' **XOR** 'A'= '0'.

In the following example, the actual states of 32 inputs are combined with the 32 expected states using the XOR function. In each bit position, a match of the actual and expected values is confirmed where the result of the XOR is '0'. Wherever a bit position has a result of '1', then the actual and expected bit values do not match.

Example: Using XOR to compare actual states (Source) with expected states (Mask)	
1010 0101 1111 1100 1010 0101 1011 0011	OPERAND VARIABLE—1 (Actual states)
1010 0101 1111 1100 1010 0101 0011 0001	OPERAND VARIABLE—2 (Expected states)
0000 0000 0000 0000 0000 0000 1000 0000	RESULT—differing state(s) include bit-7

Figure 12-33. Illustration of using with XOR to compare actual states to expected states

Chapter 13

LAD/FBD
Status Bit Operations

In this chapter, you'll be introduced to operations that provide access to the S7 CPU status word. The CPU status word contains control and indicator bits that are used and affected by the CPU as it processes your control program. Access to these status bits let you evaluate relevant information and take corresponding actions in your S7 program. Generally these operations are used to evaluate the numerical results, or respond to errors that may occur, in digital operations including arithmetic, standard functions, comparison, conversion, shift-rotate, and word-logic operations.

Main Topics

- Status Bit Operations Overview

- Status Bit Operations Descriptions

- Programming Status Bit Operations in LAD

- Programming Status Bit Operations in FBD

- Evaluating Digital Operations Results

STATUS BIT EXAMINE OPERATIONS OVERVIEW

S7 CPU's generate status bits as a result of executing each STEP 7 operation. Status bits are used by the CPU as it processes binary logic operations like AND/OR logic, and are set as it processes digital operations like compares, arithmetic, conversions, shift-rotate, and word logic. Relevant status bit information, derived from the so-called *CPU status word,* is accessible in your LAD/FBD user program using the *status bit examine operations* listed in Table 13-1. Take a moment to review these operations.

In LAD, each status bit examine operation is represented as a normally-open contact that checks for a status of '1'; and as a normally-closed contact that checks for a status of '0'. In FBD, each operation is represented as a positive binary input that checks the status bit for '1'—the positive condition; you may negate the positive operation in order to check for '0'—the negated condition.

Table 13-1. LAD/FBD Status Bit Examine Operations Summary. (See note following table)

LAD		FBD	Brief Description
NO	NC	Positive	
OV ⊣ ⊢	OV ⊣/⊢	ov	**Overflow.** Check for overflow in last digital operation. Check for '1', using NO contact; Check for '0', using the NC contact.
OS ⊣ ⊢	OS ⊣/⊢	os	**Overflow Stored.** Check for overflow in last series of math operations. Examine for '1', using NO contact; Check for '0', using the NC contact.
BR ⊣ ⊢	BR ⊣/⊢	BR	**Binary Result.** Check the binary result (BR) bit. Check for '1' using NO contact; Check for '0', using the NC contact.
UO ⊣ ⊢	UO ⊣/⊢	uo	**Unordered.** Check for invalid REAL number in last operation. Check for '1', using NO contact; Check for '0', using the NC contact.
>0 ⊣ ⊢	>0 ⊣/⊢	>0	**Greater Than Zero.** Check result of last operation for greater than zero. Examine for '1', using NO contact; Check for '0', using the NC contact.
<0 ⊣ ⊢	<0 ⊣/⊢	<0	**Less Than Zero.** Check last operation for less than zero. Check for '1', using NO contact; Check for '0', using the NC contact.
>=0 ⊣ ⊢	>=0 ⊣/⊢	>=0	**Greater Than or Equal Zero.** Check last operation for greater-than-or-equal zero. Check for '1', using NO contact; Examine for '0', using the NC contact.
<=0 ⊣ ⊢	<=0 ⊣/⊢	<=0	**Less Than or Equal Zero.** Check last operation for less than or equal zero. Check for '1', using NO contact; Examine for '0', using the NC contact.
==0 ⊣ ⊢	==0 ⊣/⊢	==0	**Equal Zero.** Check last operation for equal zero. Check for '1', using NO contact; Check for '0', using the NC contact.
<>0 ⊣ ⊢	<>0 ⊣/⊢	<>0	**Not Equal Zero.** Check last operation for not equal zero. Check for '1', using NO contact; Check for '0', using the NC contact.

Note: With the exception of OV, OS, and BR, all status bit operations in this table are derived from condition code status bits CC0 and CC1. S7 status bits are described in the following discussion.

S7 CPU Status Word Description

The CPU status word is divided into two groups—*binary flags that* include /FC, RLO, STA, OR, and BR, and are used and affected in binary operations; and *digital flags*, which include OS, OV, CC0, and CC1, and are used and affected in digital operations.

Bit	08	07	06	05	04	03	02	01	00
Name	BR	CC1	CC0	OV	OS	OR	STA	RLO	/FC

Figure 13-1. S7 CPU Status word

/FC—First Check

The status bit *first check (/FC)* indicates the first step in a binary logic string—it is a key mechanism that marks the beginning of a new logic string. The new logic string begins with the status bit /FC = 0 and the processing of a binary check instruction (first check)—for example a normally-open contact in LAD or a binary input to an AND gate in FBD. In other words, the first check in LAD corresponds to the first contact in a network. With the first check /FC is set to '1'. A binary operations string ends with a coil instruction—for example -( )- or -(R)-, or with a conditional jump or block termination, all of which set /FC = 0. With /FC = 0, a new binary logic string may be started.

RLO—Result of Logic Operation

In STEP 7 the completed binary logic result, and intermediate results after each evaluation, is called the *result of logic operations (RLO)*. In binary logic operations, the CPU uses the RLO to reflect the logic state at any node in a logic sequence. The RLO, for example, is updated after each binary check, and after each logic combination, to reflect whether the operation was fulfilled (at each node). The binary check yields '0' (RLO=0) if the operation is not fulfilled; and yields '1' (RLO=1) if the operation is fulfilled. So, if a LAD/FBD circuit has complete continuity, then the RLO = 1; otherwise, the RLO = 0. The RLO in turn is used to affect the binary logic outputs such as coil -( )-, Set -(S)-, and Reset -(R)-.

STA—Status

The *status bit (STA)* reflects the actual '0' or '1' status of a referenced address, as examined using any of the binary examination instructions. For example, a digital input may be checked and found to have a status of '1'. Once examined, for example during the evaluation of a logic network, the status bit STA is set to reflect the state of the referenced address. If the status is '0', STA is set to '0'; if the status is '1', then STA is set to '1'. Where the referenced address is a memory operation—for example set -(S)-, reset -(R)-, or the standard coil -( )-, then the '0' or '1' value assigned to STA is equivalent to the RLO.

OR—OR Status

In a logic combination involving two or more series circuits in parallel (AND-before-OR), the OR *status bit (OR)* buffers the result of a completed series circuit branch and indicates to a subsequent parallel circuit that a '1' result has already been determined. In such a case, the status bit OR is set to '1'. The status bit OR is reset to '0' by all other binary operations.

Binary Result—BR

The *binary result (BR) status bit* is used by STEP 7 in managing the graphic implementation of the EN/ENO mechanism used in LAD and FBD box operations. The function of the BR status bit in box operations may be seen by viewing a LAD/FBD box operation displayed in STL. If the functions of a box were developed in STL, there would be no need to implement the BR functionality—unless you wished to view the box in LAD/FBD, or simply if you wanted to save and later check the RLO status immediately following a digital operation—for example following a compare or word logic operation.

OV—Overflow

The *overflow status bit (OV)* indicates that the last operation has used or resulted in an invalid REAL number, or that the permissible range of the operation has been exceeded. Such an error might involve arithmetic operations, standard functions, conversion operations, or compares involving REAL numbers. When an overflow occurs, the OV bit is set, and is reset after the next error-free operation.

OS—Overflow Stored

The *overflow stored status bit (OS)* is set whenever the OV status bit is set—it actually stores the initial occurrence of the overflow. Whereas the OV bit is reset after the next error-free arithmetic operation, the OS status bit, on the other hand, remains set until a new block call. This characteristic of the OS status bit means, that in a series of calculations, you do have to evaluate for an error immediately after each operation (using the OV status bit)—instead you may evaluate the entire calculation for a numeric overflow, or use of an invalid REAL number by evaluating the OS status bit after the last operation. The OS status is evaluated in LAD using the NO and NC contacts for OS; in FBD evaluation is made using the positive OS contact or the negated OS contact.

CC0 and CC1—Condition Code Bits

The *condition code status bits (CC0/CC1))* provide information as to the numeric result, relative to zero, after arithmetic, standard function, comparison, conversion, word logic, and shift operations. In LAD/FBD, status bits CC0/CC1 are not accessed directly, but by using the equivalent examine operations listed in Table 13-2. For example, using the equivalent contact operations, you can check whether a numeric result is less than zero (negative), equal to zero, or greater than zero (positive). In each of these cases where the numeric result is within the permissible range, status bit OV is '0'.

Status bits CC0 and CC1 also combine with status bits OV and OS to report specific errors when the result is outside of the permissible number range, or other error conditions exist in a digital operation. These error results are listed in Table 13-3. In each of these cases where the numeric result is outside of the permissible range, status bits OV and OS are both set to '1'.

Table 13-2. Condition Code Status Bits CC0/CC1 Indications and Equivalent LAD/FBD Operations

Status Bits CC0 and CC1 Indication	LAD	FBD
	Equivalent Operation	
Result Equal Zero: (CC0 = 0; CC1 = 0)	==0 –| |–	==0
Greater Than Zero: (CC0 = 0 AND CC1 = 1)	>0 –| |–	>0
Result Less Than Zero: (CC0 = 1 AND CC1 = 0)	<0 –| |–	<0
Result Greater Than or Equal Zero: (CC0 = 0)	>=0 –| |–	>=0
Result Less Than or Equal Zero: (CC1 = 0)	<=0 –| |–	<=0
Result Not Equal Zero: (CC0 = 0 AND CC1 = 1) OR (CC0 = 1 AND CC1 = 0)	<>0 –| |–	<>0
Unordered (invalid REAL): (CC0 = 1 AND CC1 = 1)	UO –| |–	UO

Note: The UO status indicates that at least one source value in a previous REAL operation was invalid.

STATUS BIT EXAMINE OPERATIONS DESCRIPTIONS

CPU status bits indicate specific results after digital operations including compare, conversion, word logic, shift-rotate, arithmetic, and standard math functions. This information is available for use in your program using the status bit examine operations, which are briefly described in Table 13-1. These instructions, for example, let you determine if the result of an arithmetic operation or math function is zero, not zero, positive, or negative; or in shift operations, whether the last bit shifted out is '1' or '0'.

You may use status bit instructions alone to evaluate the result of a specific digital operation, or in conjunction with other binary operations as the basis of control logic—for example to determine if a user-written function that handles errors should be called.

Overflow: OV

If an invalid REAL number was used, or the permissible number range was exceeded in a previous operation, the *overflow status bit (OV)* is set to '1' and is not reset until the next error-free operation.

In LAD, examine for '1' using the **OV** normally-open contact; examine for '0' using the **OV** normally-closed contact. In FBD, examine for '1' using the **OV** binary input; examine for '0' using the **OV** negated binary input.

In the example, the integer value in MW12 is always multiplied by +100. In the subsequent network, the OV status bit is evaluated to check for an overflow. The OV status can always be checked directly after an operation that may result in overflow. If OV = 1, then the coil #Flag1 is energized.

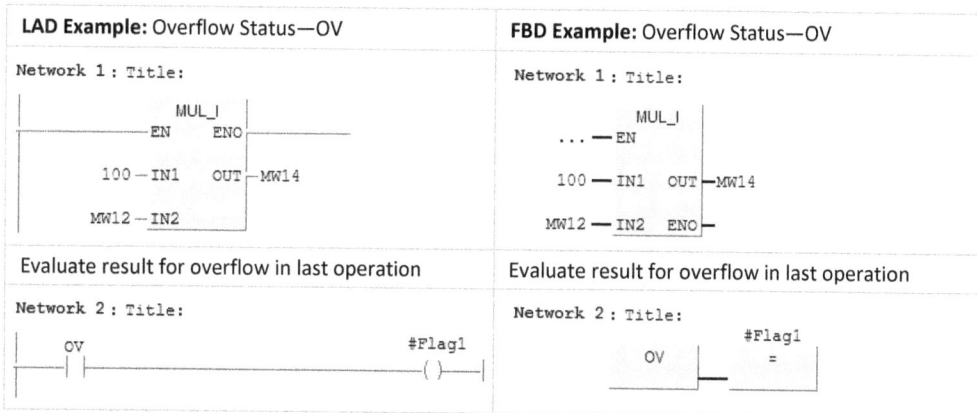

Figure 13-2. Illustration of checking for overflow, using the OV contact, immediately after operation

Overflow Stored: OS

Whenever the overflow status bit OV is set, the *overflow stored status bit (OS)* is also set. Whereas OV is reset with the next error-free operation, the OS bit is not reset until a new block call or the current block terminates. Leaving the OS bit set allows it to be used to check for an overflow after a series of operations instead of immediately after each operation. If the result of any one of the operations produces an overflow result, then the OS bit is set to '1'.

In LAD, examine for '1' using the **OS** normally-open contact; examine for '0' using the **OS** normally-closed contact. In FBD, examine for '1' using the **OS** binary input; examine for '0' using the **OS** negated binary input.

In the example below, the result in the addition operation is always subsequently multiplied in the following multiplication operation. The following network evaluates for an overflow in the completed calculation. The OS status =1 if either the addition or multiplication operation results in an overflow.

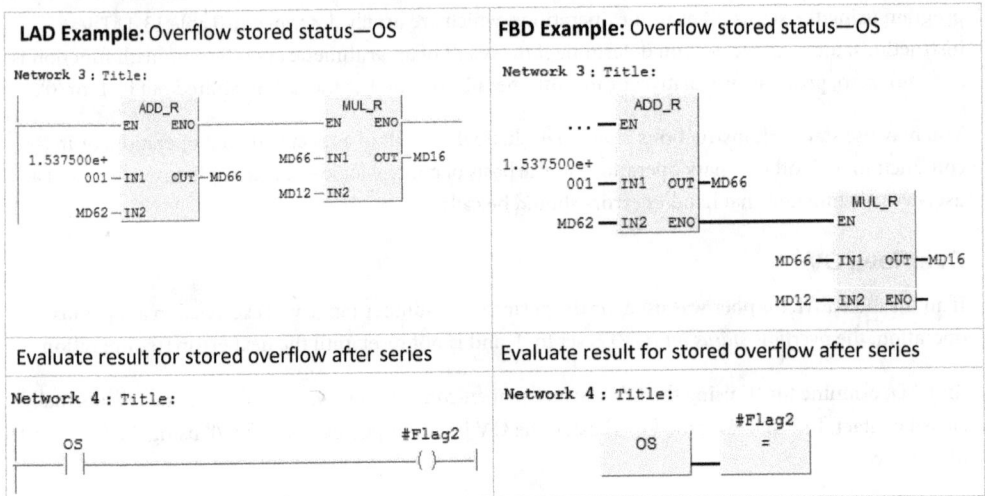

LAD Example: Overflow stored status—OS	**FBD Example:** Overflow stored status—OS
Network 3 : Title: ADD_R — EN ENO 1.537500e+001 — IN1 OUT — MD66 MD62 — IN2 MUL_R — EN ENO MD66 — IN1 OUT — MD16 MD12 — IN2	Network 3 : Title: ADD_R ... — EN 1.537500e+001 — IN1 OUT — MD66 MD62 — IN2 ENO MUL_R — EN MD66 — IN1 OUT — MD16 MD12 — IN2 ENO
Evaluate result for stored overflow after series	Evaluate result for stored overflow after series
Network 4 : Title: OS ————————( )—— #Flag2	Network 4 : Title: OS = #Flag2

Figure 13-3. Illustration of checking for overflow, using the OS contact, after a series of calculations

Unordered: UO

The *unordered status bit (UO) contacts* are used to determine if an invalid REAL number was used in the previous operation. For example, if either of the source values used in a REAL arithmetic or REAL compare operation is invalid, the UO bit is set to '1'; otherwise the bit is set to '0'. The UO bit also combines with status bits OV and OS to indicate other error results—see Table 13-3.

In LAD, examine for '1' using the **UO** normally-open contact; examine for '0' using the **UO** normally-closed contact. In FBD, examine for '1' using the **UO** binary input; examine for '0' using the **UO** negated binary input.

In the following example, the REAL compare is always processed. If the floating-point value in MD16 is greater than 3.2E+6, then M102.1 will energize; otherwise M102.1 will de-energize. The subsequent network checks for an invalid REAL value, and energizes #Flag3 if an invalid REAL value is reported.

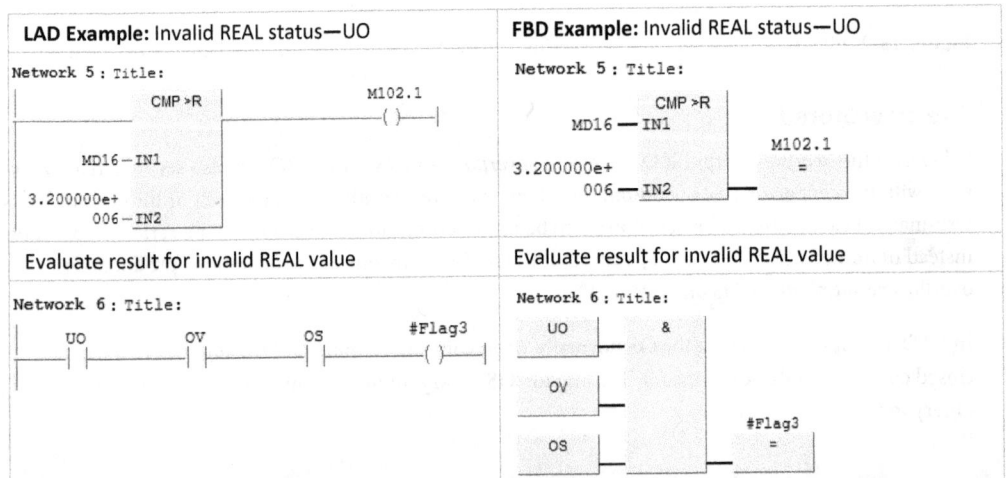

LAD Example: Invalid REAL status—UO	**FBD Example:** Invalid REAL status—UO
Network 5 : Title: CMP >R ————————( )—— M102.1 MD16 — IN1 3.200000e+006 — IN2	Network 5 : Title: CMP >R MD16 — IN1 3.200000e+006 — IN2 M102.1 =
Evaluate result for invalid REAL value	Evaluate result for invalid REAL value
Network 6 : Title: UO OV OS ————( )—— #Flag3	Network 6 : Title: UO & OV OS #Flag3 =

Figure 13-4. Illustration of the unordered (UO) status bit instruction in check for invalid REAL

Greater Than Zero: > 0

The *Greater Than Zero contacts* are used to determine if the result, after the last digital operation, is greater than zero. The > **0** contact operation (CC0/CC1 equivalent) also combines with status bits OV and OS to indicate specific error results—see Table 13-3.

In LAD, examine for '1' using the > **0** normally-open contact; examine for '0' using the > **0** normally-closed contact. In FBD, examine for '1' using the > **0** binary input; examine for '0' using the > **0** negated binary input.

In the example, the DINT addition is processed if I33.3 AND M31.5 is True (1). +Flag4 in Network 8 is activated if the result in MD66 is within the normal positive range.

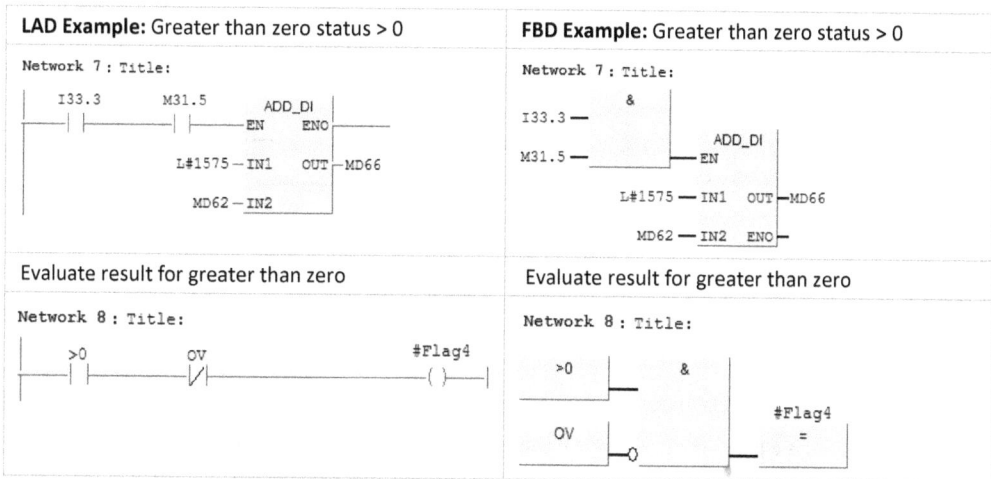

LAD Example: Greater than zero status > 0	**FBD Example:** Greater than zero status > 0

Network 7 : Title:
 	
I33.3   M31.5   ADD_DI
─┤ ├──────┤ ├──── EN   ENO ──
 	
L#1575 — IN1   OUT — MD66
 	
MD62 — IN2
	Network 7 : Title:
	
	       &
	I33.3 —
	           ADD_DI
	M31.5 — — EN
	
	L#1575 — IN1 OUT — MD66
	
	MD62 — IN2 ENO —

Evaluate result for greater than zero	Evaluate result for greater than zero

Network 8 : Title:
 	
 >0   OV   #Flag4
─┤ ├──────┤/├──────────( )──
	Network 8 : Title:
	
	 >0 &
	
	             #Flag4
	 OV =
	    O

Figure 13-5. Illustration of checking for greater than zero result after arithmetic operation

Less Than Zero: < 0

The *Less Than Zero contacts* are used to determine if the accumulator 1 result, after the last digital operation, is less than zero. The < **0** contact operation (CC0/CC1 equivalent) also combines with status bits OV and OS to indicate specific error results—see Table 13-3.

In LAD, examine for '1' using the < **0** normally-open contact; examine for '0' using the < **0** normally-closed contact. In FBD, examine for '1' using the < **0** binary input; examine for '0' using the < **0** negated binary input.

In the following example, the BOOL variable #CaseF1 is set if the result of a previous operations is within the normal negative range.

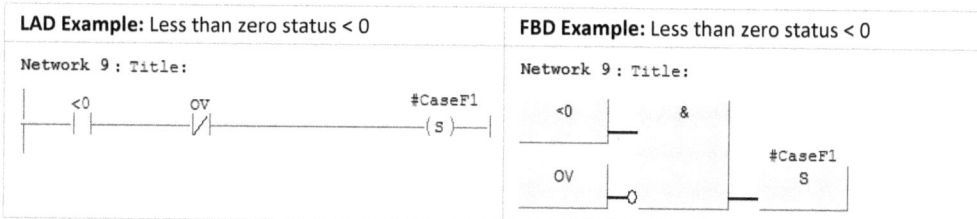

LAD Example: Less than zero status < 0	**FBD Example:** Less than zero status < 0

Network 9 : Title:
 	
 <0   OV   #CaseF1
─┤ ├──────┤/├──────────(S)──
	Network 9 : Title:
	
	 <0 &
	
	             #CaseF1
	 OV S
	    O

Figure 13-6. Illustration of checking for less than zero result

Less Than or Equal Zero: <= 0

The Less Than or Equal Zero examine operations are used to determine if the accumulator 1 result, after the last digital operation, is less than or equal zero. The < = 0 contact operation (CC0/CC1 equivalent) also combines with status bits OV and OS to indicate other error results—see Table 13-3.

In LAD, examine for '1' using the < = 0 normally-open contact; examine for '0' using the < = 0 normally-closed contact. In FBD, examine for '1' using the < = 0 binary input; examine for '0' using the < = 0 negated binary input.

In the following example, the binary flag #CaseF3 is set to '1' if the bit memory address M70.2 is '1' **AND** the less than or equal zero contact is '1' and the OV contact is '0'.

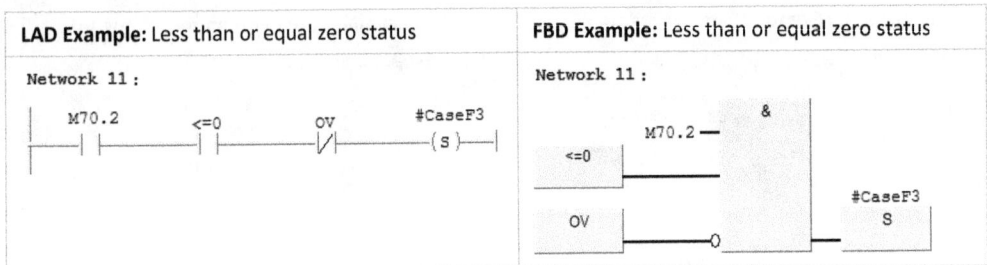

LAD Example: Less than or equal zero status	**FBD Example:** Less than or equal zero status
Network 11 :	Network 11 :

Figure 13-7. Illustration of checking for less than or equal to zero result

Equal Zero: == 0

The *Equal Zero* status examine operations are used to determine if the accumulator 1 result, after the last digital operation, is equal to zero. The = = 0 contact operation (CC0/CC1 equivalent) also combines with status bits OV and OS to indicate specific error results—see Table 13-3.

In LAD, examine for '1' using the = = 0 normally-open contact; examine for '0' using the = = 0 normally-closed contact. In FBD, examine for '1' using the = = 0 binary input; examine for '0' using the = = 0 negated binary input.

In the following example, the bit pattern in the input double word ID12 is combined with the binary pattern in the DWORD variable #Pattern_2, using WXOR_DW operation. The BOOL variable #CaseF4 is set to '1' if the XOR'ed result in MD130 is equal to zero AND the OV contact is '0'.

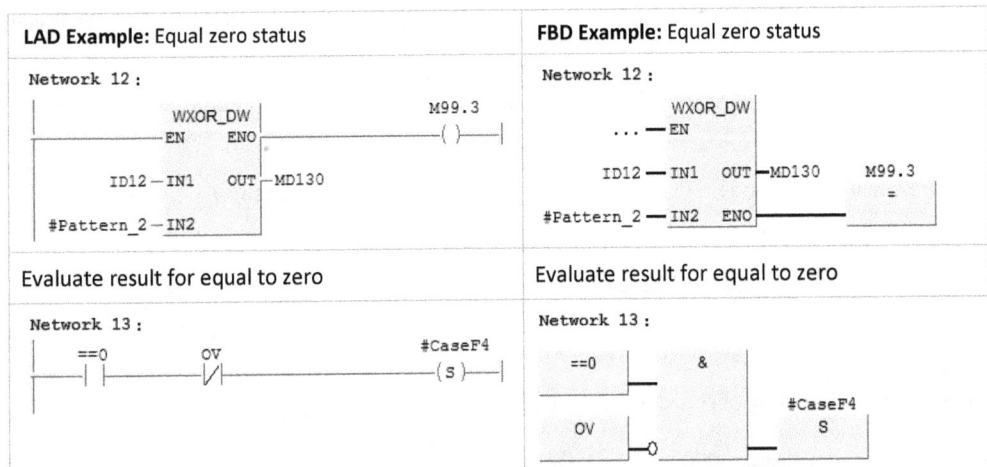

LAD Example: Equal zero status	**FBD Example:** Equal zero status
Network 12 :	Network 12 :
Evaluate result for equal to zero	Evaluate result for equal to zero
Network 13 :	Network 13 :

Figure 13-8. Illustration of checking for equal to zero result after a word logic operation

Greater Than or Equal Zero: >=0

The *Greater Than or Equal Zero* examine operations are used to determine if the accumulator 1 result, after the last digital operation, is greater than or equal zero. The > =0 contact operation (CC0/CC1 equivalent) also combines with status bits OV and OS to indicate other error results—see Table 13-3.

In LAD, examine for '1' using the > = 0 normally-open contact; examine for '0' using the > = 0 normally-closed contact. In FBD, examine for '1' using the > = 0 binary input; examine for '0' using the > = 0 negated binary input.

In the following example, the binary flag #CaseF2 is set to '1' if the bit memory address M70.1 is '1' **AND** the greater than or equal zero contact is '1' and the OV contact is '0'.

LAD Example: Greater than or equal zero status	**FBD Example:** Greater than or equal zero status
Network 10 :	Network 10 :

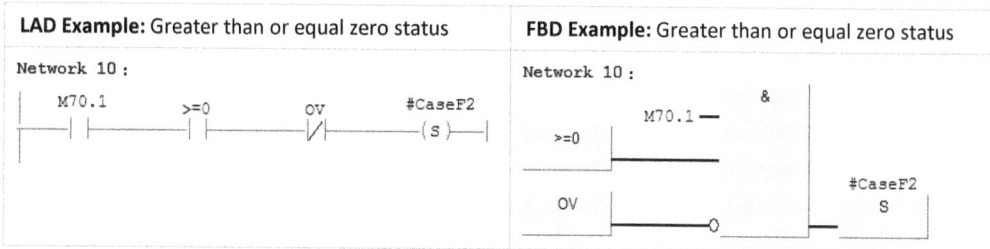

Figure 13-9. Illustration of checking for greater than or equal to zero result

Not Equal Zero: <> 0

The *Not Equal Zero* examine operations are used to determine if the accumulator 1 result, after the last digital operation, is not equal to zero. The < > 0 contact operation (CC0/CC1 equivalent) also combines with status bits OV and OS to indicate specific error results—see Table 13-3.

In LAD, examine for '1' using the < > 0 normally-open contact; examine for '0' using the < > 0 normally-closed contact. In FBD, examine for '1' using the < > 0 binary input; examine for '0' using the < > 0 negated binary input.

In the following example, the bit pattern in the input double word ID12 is combined with the binary pattern in the DWORD variable #Pattern_2, using the WXOR_DW operation. The BOOL variable #CaseF5 is set to '1' if the XOR'ed result in MD130 is not equal to zero AND the OV contact is '0'.

LAD Example: Not equal zero status	**FBD Example:** Not equal zero status
Network 12 :	Network 12 :
Evaluate result for not equal to zero	Evaluate result for not equal to zero
Network 14 :	Network 14 :

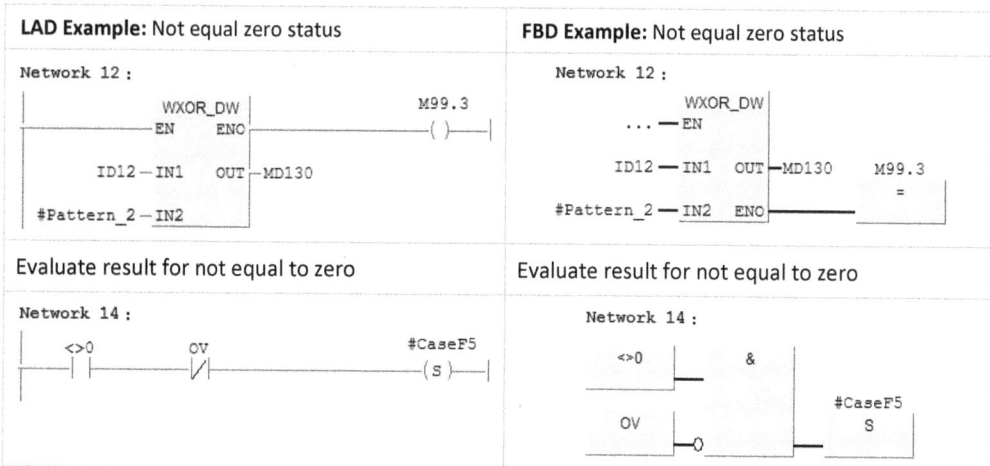

Figure 13-10. Illustration of checking for not equal to zero result after a word logic operation

Binary Result: BR

The *Binary Result status bit (BR)* contacts are used by LAD/FBD in implementing the EN/ENO mechanism in box operations. If an error occurs while **EN** is at logic 1, **ENO** is set to logic 0; otherwise the status of **ENO** follows the status of **EN**. The BR status bit is set based on the result of ENO.

In LAD, examine for '1' using the **BR** the normally-open contact; examine for '0' using the **BR** normally-closed contact. In FBD, examine for '1' using the **BR** binary input for; examine for '0' using the **BR** negated binary input.

In the network below, output Q 44.7 is Set (latched) if both M 101.1 AND M 101.2 are logic '1' AND the BR status bit is '0'.

LAD Example: Binary result status—BR	**FBD Example:** Binary result status—BR

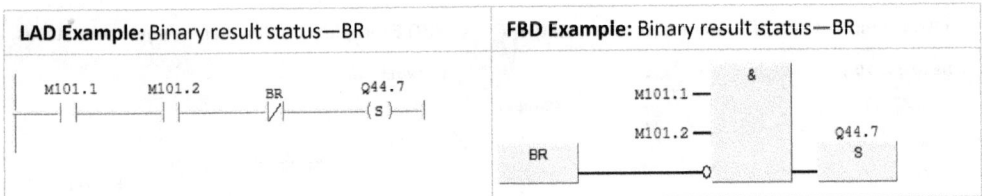

Figure 13-11. Illustration of the Binary Result (BR) status bit instruction

EVALUATING LAD/FBD DIGITAL OPERATIONS RESULTS

In LAD/FBD, *status bit operations* are implemented as contacts/binary inputs that may be combined with other operations as the basis of control logic.

As previously discussed, each status bit can be checked using either a normally-open or normally-closed contact. The NO contact lets you check for a status of '1'; the NC contact lets you check the results bit for a status of '0'. The following discussion summarizes the digital operations errors and results for which you may evaluate in your program.

Evaluating Conversion Results

When it comes to conversion operations, each operation differs in what is reported, and how it is reported—some operations may not report errors at all. In the topic *Conversion Descriptions* of Chapter 8 *LAD/FBD Conversion Operations*, you can determine what results or errors are reported for each operation.

If the conversion is to BCD from integer or double integer, you can evaluate status bits OV and OS, which are both set to '1' if the source value at parameter IN is too large to convert to 3-digit or 7-digit BCD. When the conversion is from BCD to integer or BCD to double integer, a parameter assignment error is reported by the CPU if an invalid BCD value is presented at the source IN. In such a case, synchronous error OB121 is called—if OB121 is not installed the CPU will switch to STOP.

In each rounding operation, the source value (value to be converted) is a REAL number and the converted result is a DINT value. When the operation is processed, if the source at IN is an invalid REAL number or is outside of the range expected to convert to a number within the permissible range of DINT values, then status bits OV and OS are set to '1' and the operation is aborted.

With negate operations NEG_I and NEG_DI, you can check to see whether the converted result is positive, zero, negative, or extreme negative. Extreme negative is an error where the converted result is -32768 (INT) or -2,147,483,648 (DINT)—this result is an error since there is no corresponding positive number from which this result could be obtained. See Table 13-3.

Evaluating Arithmetic and Math Functions

After each arithmetic or math function operation, you may use status bit operations to evaluate for a result within the normal range or a result with errors. For example, you may check for an overflow (permissible range exceeded) in the result by examining the **OV** status bit for '1', immediately following the operation; or you may perform a series of arithmetic operations and then check for an overflow result using the **OS** status bit (See Figure 13-3).

You can also check whether a numeric result is less than zero, equal to zero, or greater than zero. If the result is less than zero, then status bits CC0/CC1 are set to 1/0; if the result is equal to zero, then status bits CC0/CC1 are set to 0/0; if the result is greater than zero, then status bits CC0/CC1 are set to 0/1. As seen in Table 13-2, the LAD/FBD equivalent operations for CC0/CC1 indications of less than zero, equal to zero, and greater than zero are respectively the **< 0** contact, the **= = 0** contact, and the **> 0** contact. In each of these cases, where the result is within the numeric range, flagged by the indicator, the **OV** status bit is '0'.

Additional error indications, for which you may evaluate after arithmetic and math functions, are listed in Table 13-3. Each of these indications will affect a LAD/FBD equivalent operation for CC0/CC1, along with status bits OV and OS, which are both set to '1'. For example, in any operation with REAL numbers, status bit **UO** is set to '1' if an invalid source value is used. These additional error indications, grouped as INT/DINT operations, or REAL operations, are listed in the unshaded rows of Table 13-3.

Evaluating Compare Results

When it comes to compare operations, you may evaluate after those operations involving REAL numbers. For each of these operations, status bits CC0, CC1, OV, and OS all are all set to '1', and the operation is aborted if either source value (IN1 or IN2) is found to be an invalid REAL number.

You can also check whether a compare test for less than, equality, or greater than is affirmed. If a test for less than is True (1), then status bits CC0/CC1 are set to 1/0; if a test for equality is True (1), status bits CC0/CC1 are set to 0/0; if a test for greater than is True (1), status bits CC0/CC1 are set to 0/1.

As shown in Table 13-2, the LAD/FBD equivalent operations for CC0/CC1 indications for less than, equality, and greater than are the **< 0** contact, the **= = 0** contact, and the **> 0** contact respectively. In each of these cases, where the result is within the permissible range, the **OV** status bit is also '0'.

Evaluating Word Logic Results

After each word logic operation is completed, you may examine specific status bit instructions to determine if the bitwise logic result at **OUT** is 'equal to 0' or 'not = zero'. The instruction = = **0** (CC0= 0, CC1 = 0) is used to examine for zero; the instruction <> **0** (CC0= 0, CC1 = 1) is used to examine for a not equal to zero result.

Evaluating Shift Results

After each shift operation, the status of the last bit shifted out is loaded to the CC1 bit of the status word. In other words, the status bit CC1 is assigned the '1' or '0' value to reflect the shifted out bit position. In your program, you may examine the CC1 status bit using the LAD/FBD equivalent operation for the CC1 status bits. These equivalent operations are listed in Table 13-2. Based on this table, you can check shifted out bit for '0' using the = = **0** contact, and for '1' using the > **0** contact. See Table 13-3.

STEP 7 Programming Made Easy in LAD, FBD, and STL

Table 13-3. Digital operations results with status bits OV, OS, UO and CC0/CC1 equivalents

Result conditions for which to evaluate	LAD/FBD Status Bit Equivalents to Evaluate								
	UO	OV	OS	==0	<>0	<0	>0	<=0	>=0
INT/DINT Result Overflow (ADD_I, SUB_I)		1	1				1		
INT/DINT Result Overflow (MUL_I)		1	1					1	
INT/DINT Result Underflow (ADD_I, SUB_I)		1	1					1	
INT/DINT Result Underflow (MUL_I)		1	1				1		
INT Result = 32,768 for (DIV_I)		1	1					1	
INT/DINT Division by Zero	1	1	1						
INT/DINT Result = 0		0		1					
INT/DINT Result > 0 (valid positive)		0					1		
INT/DINT Result < 0 (valid negative)		0				1			
DINT Result = 2,147,483,647 for (DIV_D)		1	1				1		
REAL Overflow: + infinite (division by zero)		1	1				1		
REAL Underflow:- infinite (division by zero)		1	1			1			
Invalid REAL Value	1	1	1						
REAL Result ± (De-normalized)		1	1	1					
REAL Result = 0		0		1					
REAL Result > 0 (normalized)		0					1		
REAL Result < 0 (normalized)		0				1			
Word Logic Result = 0		0		1					
Word Logic Result <> 0		0			1				
Shift-Rotate—Shift Out Bit = 1		0					1		
Shift-Rotate—Shift Out Bit = 0		0		1					
Conversion Result = Extreme Negative, NEG_I		1	1			1			
Conversion Result = Extreme Negative, NEG_DI		1	1			1			
Conversion Result = 0		0		1					
Conversion Result > 0		0					1		
Conversion Result < 0		0				1			
Comparison with Invalid REAL Value (s)	1	1	1						
Compare Result Equal		0		1					
Compare Result Greater Than		0					1		
Compare Result Less Than		0				1			

Note: 1) Check for status equal '1' with the NO contact; check for status equal '0' with the NC contact. 2) CC0/CC1 result indications are the same for INT/DINT operations. The numeric range for INT is -32768 to +32,767; the range for DINT is -2147483648 to +2,147,483,647.

Copyright © 2013   234

Chapter 14

LAD/FBD
Program Flow Control
and Block Operations

This chapter introduces LAD/FBD Program Flow Control and Block Operations. With program control operations, you can manage how and when certain portions of your S7 program are processed. In short, with these operations you are able to alter how the CPU processes the program so that processing needs are serviced on an as-needed basis only.

Overall, these instructions help to minimize processor scan times and create a more efficient and better organized program. The discussions in this chapter include calls to blocks, internal jumps to labels, block termination, and definition of MCR zones.

Main Topics

- Program Flow Control and Block Operations
- Jump, Label, and Return Descriptions
- MCR Operations Descriptions
- Call and Open DB Operations Descriptions
- Programming Jumps, Labels, and Returns
- Programming MCR Operations
- Programming CALL and DB Operations

PROGRAM FLOW CONTROL OPERATIONS

For purposes of this discussion, operations covered in this chapter, and listed in Table 14-1, are grouped under 'Program Flow Control' operations. In the STEP 7 instruction browser, Jump and Label operations are contained in the 'Jumps' folder; Master Control Relay, Call, and Return operations are all in the 'Program Control' operations folder; and Open DB, is under 'DB Call'.

Table 14-1. LAD/FBD Program Flow Control Operations Summary

LAD	FBD	Brief Description
???	???	**LABEL.** Four-character label defining a network to which a **JMP** or **JMPN** operation directs program execution. Using jumps and labels, portions of logic can be skipped.
??? —(JMP)—	??? JMP	**Jump.** When the preceding RLO = 1, jump to the network identified by the label address specified above the **JMP** operation.
??? —(JMPN)—	??? JMPN	**Jump NOT.** When the preceding RLO = 0, jump to the network identified by the **label** address specified above the **JMPN** operation.
—(RET)—	RET	**RETURN.** When the preceding RLO = 1, terminate execution of the current block and return control to the calling block; otherwise continue with the following network.
—(MCRA)—	MCRA	**MCR Activate.** This unconditional operation, when encountered, enables MCR zones up to the point of the next **MCRD** (de-activate).
—(MCRD)—	MCRD	**MCR De-Activate.** This unconditional operation, when encountered, disables the use of MCR zones up to the point of the next **MCRA** (activate) operation.
—(MCR<)—	MCR<	**MCR Zone Start.** When the preceding RLO =1, this operation, which begins an MCR zone, enables the logic enclosed within the zone. Normally paired with the **MCR>** (End) operation.
—(MCR>)—	MCR>	**MCR Zone End.** This unconditional operation, when encountered, ends an MCR zone. Normally paired with the **MCR<** (Start) operation.
??? —(CALL)—	??? CALL	**CALL Block.** When the preceding RLO = 1, the referenced Function (FC), or System Function (SFC), is called for processing.
??? —(OPN)—	OPN	**OPEN DB.** When this unconditional operation is encountered, the data block (DB nnn) referenced above the operation is opened; data can then be accessed in subsequent networks.

JUMP, LABEL, AND RETURN DESCRIPTIONS

Jump and Label operations work together to facilitate jumps that are internal to a block, and are based on conditions that support either the execution of a section of code, or the skipping of that section. The Return operation, which is generally placed at the end of a section of code, provides a means by which a section of code is terminated, and program control is passed back to the calling block.

Label: <abcd>

The *label* element identifies a target network to which program execution may be passed by a jump instruction—for example **JMP** or **JMPN**. Each label is a unique four-character alphanumeric name that must start with a letter. Upper and lower case letters, for example 'CAS1' and 'cas1', are interpreted as being different. When inserted, the label is placed just above the network. You may insert the associated network immediately following the label.

In the following example, the network following the CAS1 label is processed when control is passed from a network whose binary coil/output is a **JMP** or **JMPN** instruction that references the label CAS1.

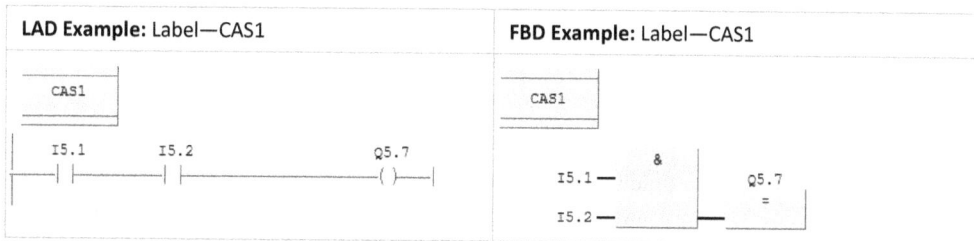

LAD Example: Label—CAS1	FBD Example: Label—CAS1

Figure 14-1. LAD/FBD illustration of Label instruction

Jump: JMP

The **JMP** instruction is a conditional jump, used to skip over one or more networks if the RLO=1 for the driving logic. When processed, program execution jumps to the network label that matches the four characters referenced above the **JMP** coil/output instruction. All networks between the **JMP** and the referenced label are skipped, and not processed. If the RLO = 0, the jump is not executed and the network immediately following the **JMP** network is processed.

The **JMP** instruction may also be programmed without driving logic conditions, which makes it unconditional. In such as case, program execution is always passed to the referenced label when the network is processed.

In the following example, if I 4.1 is True (RLO =1), then program processing is passed (jumps) to the network whose label is CAS1. If RLO = 0, then processing continues with the following network.

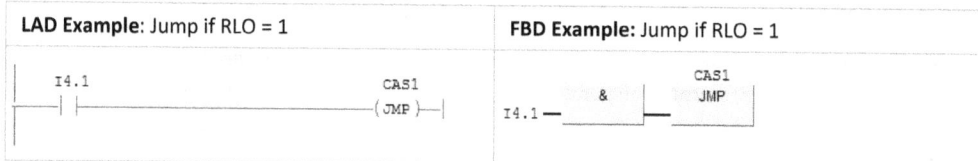

LAD Example: Jump if RLO = 1	FBD Example: Jump if RLO = 1

Figure 14-2. LAD/FBD illustration of Jump-to-Label instruction

Jump Not: JMPN

The **JMPN** instruction is a conditional jump, used to skip over one or more networks if the RLO = 0 for the driving logic. When executed, program processing jumps to the network LABEL that matches the four characters referenced above the **JMPN** instruction. All networks between the **JMPN** network and the referenced label are skipped, and not processed. If the RLO = 1, the jump is not executed and the network immediately following the **JMPN** network is processed.

Unlike the **JMP** operation, an unconditional jump is not possible when using the **JMPN** operation.

In the following example, if the address with the symbol name 'READY' is False (RLO =0), then program processing is passed (jumps) to the network whose label is CAS1. If the RLO = 1, then processing continues with the following network.

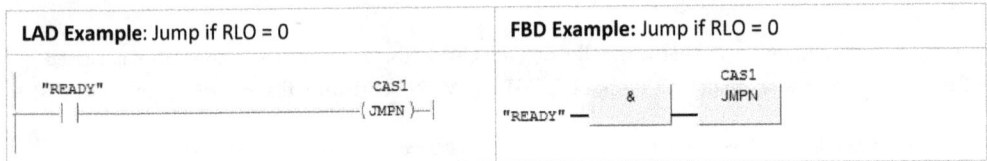

LAD Example: Jump if RLO = 0	**FBD Example**: Jump if RLO = 0
"READY" CAS1 ——| |————————————————(JMPN)—|	CAS1 & JMPN "READY" —

Figure 14-3. LAD/FBD illustration of Jump-to-Label instruction

Return: RET

The **RET** instruction is used to conditionally terminate the processing of a block before the normal end of the block is reached. You must always program conditional logic to drive the return coil/output. When executed, program execution of the current block is terminated, and control returns to the previous block from which the current block was called.

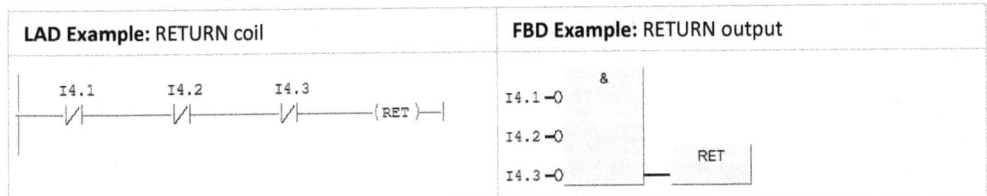

The **RET** instruction must be the only coil (LAD) or output (FBD) in the network.

In the following example if I 14.1 AND I 14.2 AND I 14.3 are all logic '0', then the RLO = 1 and the current block is terminated; control returns to the block from which the current block was called.

LAD Example: RETURN coil	**FBD Example**: RETURN output
I4.1 I4.2 I4.3 ——|/|————|/|————|/|————(RET)—|	& I4.1 —○ I4.2 —○ RET I4.3 —○

Figure 14-4. LAD/FBD illustration of Return instruction

PROGRAMMING JUMPS, LABELS, AND RETURNS

As stated previously, jump operations support the ability to direct the flow of processing in your program by passing control from one point in a code block to another point within the same block. Using these operations, it is possible to create sections of code within a block that perform specific operations when processed. Within a block, for example, one code section might be labeled **Auto**; another labeled **Manual**; and another labeled **OFF**.

Jump to Label

Jump operations allow jumps that take place internal to a block. When a jump operation is executed, statements between the jump instruction and the statement label to which it points are skipped. Processing of the current block resumes at the labeled statement. Although each label is unique, program execution may be passed to a single label from one or more **JMP** or **JMPN** instructions.

Conditional and Unconditional Jumps

Generally speaking, a jump is either unconditional or conditional. An unconditional jump is processed without regard to any conditions. Conditional jumps are always based upon the RLO status—in the case of the JMP operation, the jump is executed if the RLO=1; in the case of the JMPN operation, the jump is executed if the RLO=0.

The Return Operation

The Return operation, as previously described, allows the execution of a block to be terminated at an intermediate point in the block, prior to the normal block end. When the Return is processed, the current block is ended and processing returns to the block from which the terminated block was called.

By using both the label and the return operation, several partitioned code sections can be created within a block. Generally, each section starts with a Label operation and ends with a Return operation. When the specific label is the target of a jump, the operations between the label and the return are processed. When the Return operation is processed, control is passed back to the calling block.

MCR OPERATIONS DESCRIPTIONS

MCR operations allow you to create logic zones within your S7 program. Each zone, and the enclosed control logic, is enabled whenever the zone is enabled and is disabled whenever the zone is disabled.

There are four basic MCR operations. Each instruction, when programmed, must be physically or logically paired with the appropriate operation. The *MCR Activate* (**MCRA**) is paired with the *MCR De-Activate* (**MCRD**) to control activating and deactivating the S7 MCR functionality in your program. The *MCR Zone Start* (**MCR<**), and the *MCR Zone End* (**MCR>**), are paired to create a zone in which logic networks are enclosed.

MCR Activate: MCRA

The **MCR Activate** instruction enables the master control relay functionality in the S7 CPU and in your program. This coil/output instruction is programmed without any preceding logic conditions. When encountered, the **MCRA** instruction enables all MCR partitioned zones following this network, and up to the point where an MCR De-Activate (**MCRD**) instruction is encountered.

LAD Example: Activate MCR function	FBD Example: Activate MCR function
———————————(MCRA)—|	 MCRA |

Figure 14-5. LAD/FBD illustration of MCR Activate instruction

MCR De-Activate: MCRD

The **MCR De-Activate** instruction is used to control the disabling of the S7 master control relay functionality. This coil/output is always unconditional and programmed without any preceding logic. Following this instruction, no MCR zones will be enable for operation until after the S7 MCR function has been again activated with an **MCRA** instruction.

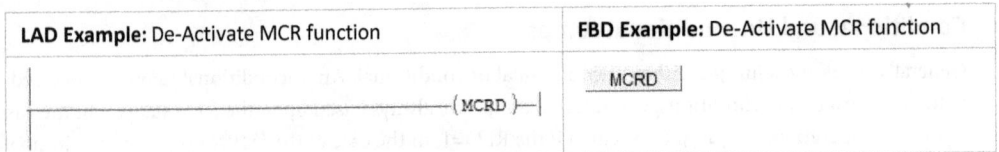

LAD Example: De-Activate MCR function	FBD Example: De-Activate MCR function
————————————————(MCRD)—|	MCRD

Figure 14-6. LAD/FBD illustration of MCR De-Activate instruction

MCR Zone Start: MCR<

The **MCR Zone Start** instruction is used to control the enabling or disabling of a partitioned zone of LAD/FBD networks. As illustrated in the figure below, the **MCR<** instruction marks the beginning of a partitioned zone and is paired with an MCR Zone End (**MCR>**) instruction, which marks the end of a partitioned zone. Up to eight zones may be opened (nested) before closing the inner-most zone.

When the preceding logic circuit to the **MCR<** coil/output has an RLO=1, the logic networks enclosed by the zone are enabled to be processed as normal; if the RLO = 0, then the logic networks enclosed by the zone are disabled and not processed.

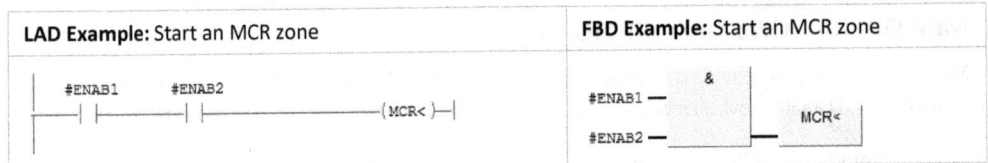

LAD Example: Start an MCR zone	FBD Example: Start an MCR zone
#ENAB1 #ENAB2 ——| |————| |——————(MCR<)—|	#ENAB1 — & #ENAB2 — MCR<

Figure 14-7. LAD/FBD illustration of MCR Zone Start instruction

MCR Zone End: MCR>

The **MCR Zone End** instruction marks the end of the nearest open MCR zone. As illustrated below, the **MCR>** instruction marks the end or close of a partitioned zone and is paired with an **MCR<** instruction, which marks the start of the partitioned zone. No logic conditions precede the MCR Zone End coil/output.

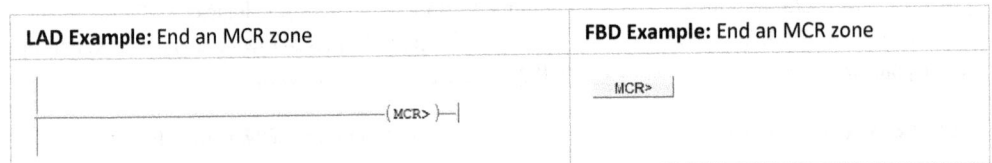

LAD Example: End an MCR zone	FBD Example: End an MCR zone
————————————————(MCR>)—|	MCR>

Figure 14-8. LAD/FBD illustration of MCR Zone End instruction

PROGRAMMING MCR OPERATIONS

As mentioned earlier, MCR operations supply the means for creating logic zones in the S7 program. The control logic within a zone is enabled when the zone is enabled, and is disabled when the zone is disabled. Similar to the behavior of circuits governed by a hardwired master control relay, when the MCR Start coil is de-energized, non-retentive coils/outputs inside the partitioned zone are switched off; retentive coils/outputs, for example Set/Reset, inside the partitioned zone retain their last state.

Activating and De-Activating MCR Functionality

The MCR Activate (**MCRA**) is paired with the MCR De-Activate (**MCRD**) to control activating and deactivating S7 MCR functionality in your program. The MCR-A operation must be used in each block in which an MCR zone will be opened or continued from a previous block. In other words, if a new block is called from within an open MCR zone, then that zone remains open in the called block, but must be activated within the new block. The MCR-A from the called block is then logically paired with the MCR-D of the calling block. This pairing is illustrated in Figure 14-9.

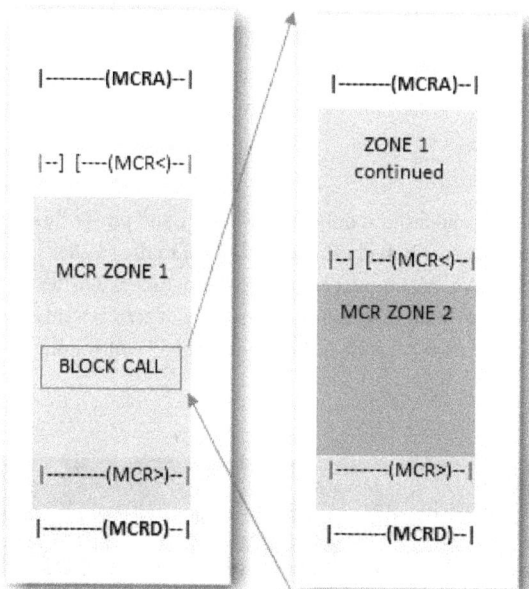

Figure 14-9. MCRA/MCRD behavior on block change

MCR Control Zones

The STEP 7 program may include up to a total of eight zones that may be individually isolated, or may overlap with zones that are nested inside of another zone. Since up to eight zones may be nested, it is important to understand how inner zones are affected by outer zones and how all zones are affected by the outer most MCR zone.

If MCR zones are nested, the first or outer-most zone controls the MCR dependency of all of the nested zones. If the outer zone is disabled, then the inner zones are also disabled regardless of whether their controlling MCR< operations is enabled or not. All inner zones are disabled when the outer most zone is disabled. And when the outer most zone is enabled, an inner zone that is also enabled will only have no MCR dependency if all zones in which that zone is nested are also enabled. MCR dependency behavior is outlined in Table 14-2.

```
|--------(MCRA)--|

|--] [----(MCR<)--|

   MCR ZONE 1

   I 2.2
|--] [----(MCR<)--|

   MCR ZONE 2



|--------(MCR>)--|

|--------(MCR>)--|

|--------(MCRD)--|
```

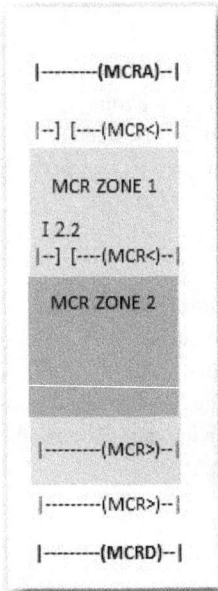

Figure 14-9. Nested MCR zones

MCR Dependency

Networks inside an MCR zone are processed in a normal fashion only if the RLO = 1 and power flows to the MCR< instruction. In this case, MCR dependency is switched off. If the RLO = 0 to the MCR< operation, then MCR zone dependency is switched ON. All outputs are reset to zero, except Set/Reset outputs and SR and RS flip-flop operations, which maintain their last states. A value of zero is written to all digital values or variables in box instructions. Table 14-2 describes the behavior of S7 coil/output instructions that are governed by an MCR zone.

Table 14-2. MCR Operations Dependency

Instruction Type	Response Description	
	MCR Dependency OFF Zone Enabled (RLO = 1)	MCR Zone Dependency ON Zone Disabled (RLO = 0)
Standard Coil	Assume normal program behavior	Assume signal state of '0'
Midline Coil	Assume normal program behavior	Assume signal state of '0'
Set Coil	Assume normal program behavior	Maintain last signal state
Reset Coil	Assume normal program behavior	Maintain last signal state
SR/RS Flip-Flop	Assume normal program behavior	Maintain last signal state
Box Binary Outputs	Assume normal program behavior	Write zero to OUT parameters/variables
Box Digital Outputs	Assume normal program behavior	Write zero to OUT parameters/variables
Move Box Outputs	Assume normal program behavior	Write zero to OUT parameters/variables

CALL OPERATIONS DESCRIPTIONS

With the exception of organization blocks, your program must have a call to each block (FC, FB, SFC, and SFB) in order that the block is processed. A block is called by referencing its absolute or symbolic address, for example FC10 (Unit_2_CTRL), with the appropriate call instruction.

In LAD/FBD, there is only one operation for making a block call—the CALL operation. There are, however, two basic ways in which a block is called. Both of these methods will be shown later in Programming Block Operations. As is described in the following discussions, there are different call requirements for FBs and FCs.

Call Coil/Call Output: CALL

In LAD/FBD, the **CALL** instruction is used to make a *conditional call* or an *unconditional call* from the current block to the specified user function (FC) or system function (SFC). Only FC and SFC blocks that do not have parameters to be passed may be called using the **CALL** coil/output instruction.

An unconditional call to a block is always executed; a conditional call is made only if the driving logic has continuity (RLO=1). If the RLO = 0, then the call is not executed, and the network following the **CALL** network is processed.

In the following example, if I 20.0 is logic '0', the call to FC1 is not made and program execution continues with the following network. If I 20.0 is logic '1', then the FC1 is called for processing. When FC1 is terminated, processing returns to the calling block and continues with the network immediately following the CALL network (Network 2). If I 20.0 is still at logic '1', then FC2 is called for processing.

In the last network, FC2 is called unconditionally—each time the current block is processed, FC2 is called for processing.

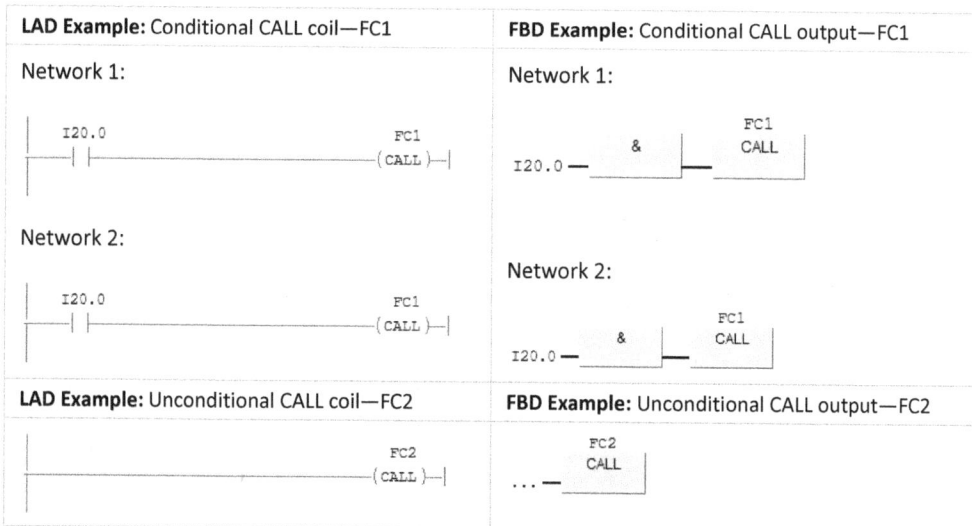

LAD Example: Conditional CALL coil—FC1	**FBD Example:** Conditional CALL output—FC1
Network 1:	Network 1:
Network 2:	Network 2:
LAD Example: Unconditional CALL coil—FC2	**FBD Example:** Unconditional CALL output—FC2

Figure 14-10. LAD/FBD illustration of CALL instruction

DB OPERATIONS DESCRIPTIONS

The data block (DB) resource supports the structuring and storing of data used in the user program. The following discussion presents the single LAD/FBD data block operation for opening a data block in order to access its contents in the control program. If you are not already familiar with data block concepts, including global and instance DBs and how they are used, you may wish to briefly review Chapter 2 *Data Blocks*.

Further operations for working with data blocks are presented in the *Data Block Operations* section of Chapter 26 *STL Program Flow Control and Block Operations*.

Open DB: OPN

The *OPN DB* instruction is used to open the referenced global data block for data access. The open DB coil/output is programmed without any preceding logic conditions. Since the open DB coil/output operation is unconditional, the instruction is executed each time the network is processed.

In networks following the **OPN** instruction, you can access bit, byte, word, and double word data locations in the open data block. The data block remains open and can be accessed until a new data block is opened.

In the following example, access to DB10 is opened in Network 1. In Network 2, data word 20 (DBDW20) is moved to bit memory word 20 (MW20).

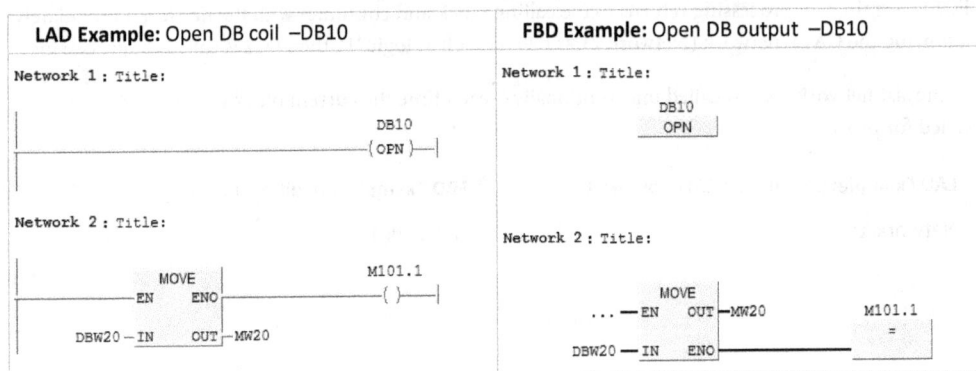

LAD Example: Open DB coil –DB10	**FBD Example:** Open DB output –DB10
Network 1 : Title: DB10 —(OPN)— Network 2 : Title: MOVE EN ENO M101.1 —( )— DBW20 – IN OUT – MW20	Network 1 : Title: DB10 OPN Network 2 : Title: MOVE ... — EN OUT – MW20 M101.1 = DBW20 — IN ENO

Figure 14-11. LAD/FBD illustration of Open Data Block instruction

PROGRAMMING BLOCK CALL OPERATIONS

A call to a block may be unconditional—without preceding logic conditions; or conditional, with preceding logic conditions. A call programmed without preceding logic is executed whenever it is processed (unconditional call). If logic conditions precede the call (conditional call), then the block is called for processing whenever the conditions are met; if the conditions are not met, the instruction immediately following the conditional call is processed.

Calling an FC in LAD/FBD

An FC/SFC can be called from any other block in LAD or FBD—including an OB, FB, or another FC. If the FC was written without formal parameters, then the **CALL** coil/output may be used. The block address (FC number) is specified above the **CALL** coil/output instruction.

In Figure 14-12, FC1 and FC2 are both called using the CALL coil (CALL output in FBD). Just above the coil, you may insert the absolute address or symbolic address for the FC. Also notice that the FC may be called with preceding logic (conditional), or without preceding logic (unconditional).

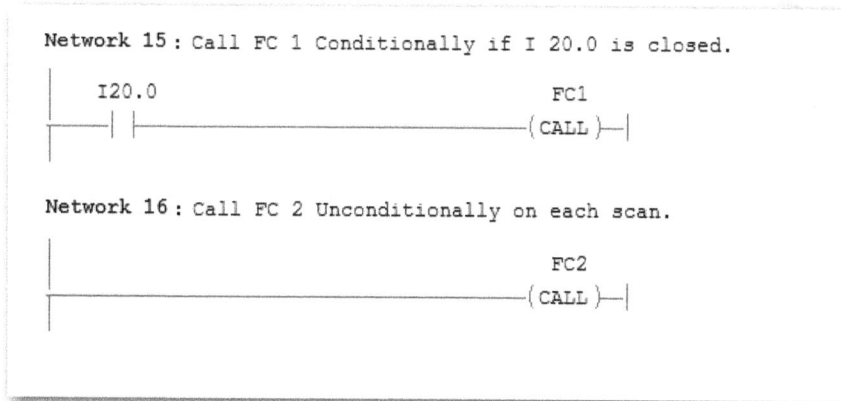

```
Network 15 : Call FC 1 Conditionally if I 20.0 is closed.

      I20.0                                    FC1
  ┌────┤ ├──────────────────────────────────(CALL)─┤
  │

Network 16 : Call FC 2 Unconditionally on each scan.

                                               FC2
  ┌──────────────────────────────────────────(CALL)─┤
  │
```

Figure 14-12. LAD/FBD conditional and unconditional use of CALL instruction

You may also drag the block from the *FC Blocks* folder of the Program Elements tab, and drop it onto the current network.

Notice in Figure 14-13, that both FC3 and FC30 appear as boxes with the EN/ENO mechanism. These Functions appear in this box fashion because of the method by which they were placed in the network. An FC appears as a box operation when you drag and drop it onto the network. Like with the CALL coil, you may insert the box call with or without preceding logic conditions.

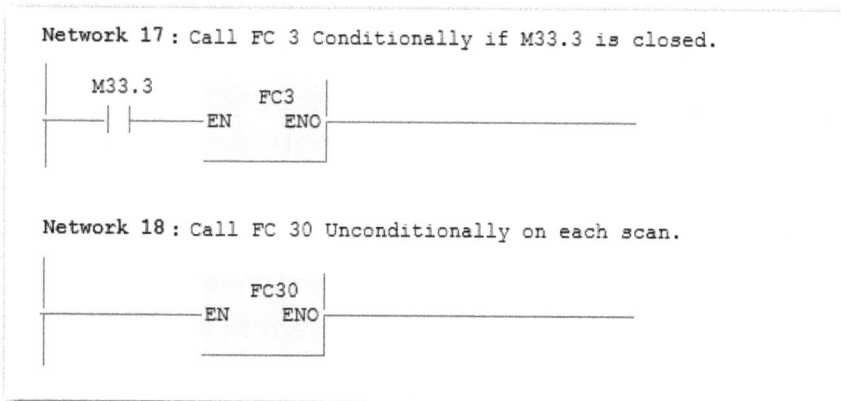

```
Network 17 : Call FC 3 Conditionally if M33.3 is closed.

      M33.3          FC3
  ┌────┤ ├──────┬── EN      ENO ──────────────────────
  │             │
              └──────────────┘

Network 18 : Call FC 30 Unconditionally on each scan.

                   FC30
  ┌────────────┬── EN      ENO ──────────────────────
  │            │
             └──────────────┘
```

Figure 14-13. LAD/FBD illustration of calling an FC as a box operation

If the FC has formal parameters, then you must drag and drop the block onto the network as described above, then you must specify a valid address or value at each FC parameter. See Chapter 28 *Developing Functions and Function Blocks* for examples of calling an FC with and without parameters.

Calling and FB in LAD/FBD

An FB/SFB, like an FC, can be called from any other code block in LAD or FBD—including an OB, FC, or another FB. With the FB, the CALL coil/output is never used. You must locate the FB you wish to call, in the FB Blocks folder of the Program Elements tab, and drag the block onto the current network. Then, you may specify the related instance DB just above the FB.

In Figure 14-14, FB100 is called without preceding logic (unconditional). FB100 is actually called twice—once with DB100 as the instance DB, and again with DB101 as the instance DB.

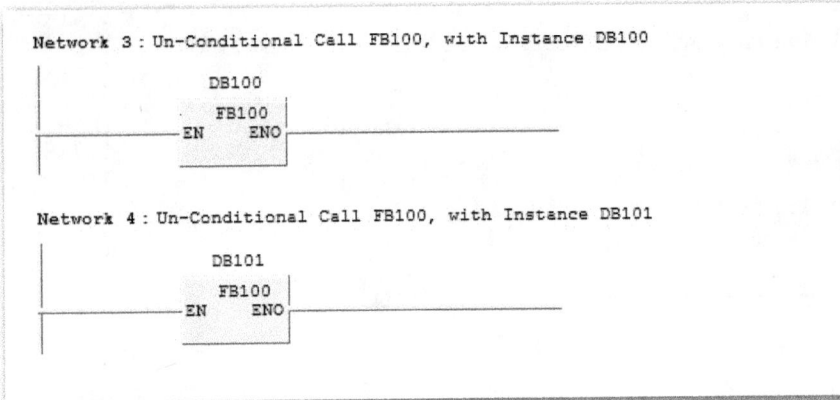

```
Network 3 : Un-Conditional Call FB100, with Instance DB100

                       DB100
                       FB100
                  ─EN      ENO─


Network 4 : Un-Conditional Call FB100, with Instance DB101

                       DB101
                       FB100
                  ─EN      ENO─
```

Figure 14-14 LAD/FBD illustration of unconditional calls to an FB

If the FB has formal parameters, then you must appropriately specify a valid address or value at each parameter, after the block has been inserted into the network. You must then specify the associated instance DB just above the FB. See Chapter 28 *Developing Functions and Function Blocks* for examples of an FC with and without parameters.

In the Figure 14-15, FB51 and FB60 are both called with preceding logic (conditional. The associated instance DB for each of these blocks is inserted just above the block.

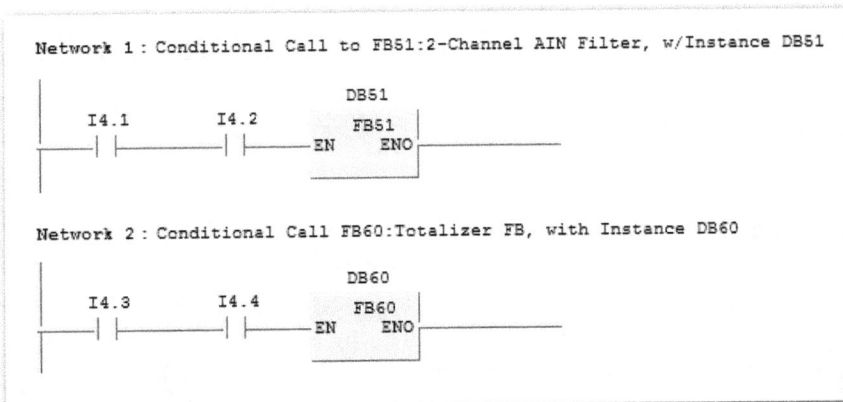

```
Network 1 : Conditional Call to FB51:2-Channel AIN Filter, w/Instance DB51

                              DB51
      I4.1        I4.2        FB51
     ─┤ ├────────┤ ├─────EN       ENO─


Network 2 : Conditional Call FB60:Totalizer FB, with Instance DB60

                              DB60
      I4.3        I4.4        FB60
     ─┤ ├────────┤ ├─────EN       ENO─
```

Figure 14-15. LAD/FBD illustration of conditional calls to an FB

Part III

The Statement List Language

Chapter 15

Introduction to Statement List

This chapter presents a brief introduction of the Statement List (STL) language. First we'll look at the basic instruction set categories and how each is used as a fundamental program building block. The basic elements of STL, for example, include binary examine operations and assignment operations.

Presented next is a brief overview of writing code blocks in STL. Since STL is a line-oriented statement-based language, we look at the structure of a statement and the construction of an STL network. Next, we examine how basic STL program elements combine to form a logic string and to complete a network. Like with LAD and FBD, an STL code block is developed incrementally by creating a number of networks.

Following this chapter, and for the next eleven chapters, the Statement List instruction categories are presented. These categories, as with LAD/FBD, include Binary Logic, Move (Load/Transfer), Timer, Counter, Conversion, Comparison, Integer and Real, Shift-Rotate, Word Logic, Status Bit, Program Flow Control and Block Operations.

Main Topics

- Statement List—An Overview
- Writing Code Blocks in STL

STATEMENT LIST—AN OVERVIEW

STL is one of three basic languages used to develop the STEP 7 program. STL, short for *Statement List*, corresponds to the IEC 1131-3 standard for *Instruction List*. When programming in STL you have full access to the S7 CPU instruction set. The basic elements of the STL language and the STL programming editor are described in this section.

To develop code in STL, you will need to understand the important STEP 7 concepts of data types and formats, as well as memory areas and addressing, as described in Chapter 1 *Memory and Addressing*. Also, if you are unfamiliar with the STEP 7 way of writing a PLC program in modular segments, referred to as 'blocks', then you may want to review these topics as well as other S7 programming principles introduced in Chapter 27 *Designing and Implementing the STEP 7 Program*.

Instruction Set Categories

The instruction categories of which the STL operation set is comprised are shown in Table 15-1. Generally speaking, these categories are the same as those of LAD and FBD. In fact, LAD and FBD graphic instructions are fundamentally based upon specific STL operations. For example, each of the box instructions of LAD and FBD are comprised of two or more STL operations. Starting with Chapter 16 *STL Binary Logic Operations*, STL instruction categories are described in the subsequent chapters.

As we begin to review STL operations in the next chapter, we'll take a look at how certain LAD/FBD box instructions translate to STL. This understanding will prove useful as you learn to develop in STL.

Table 15-1. Statement List Instruction Categories Summary

Instruction Categories		
Binary Logic Operations	Conversion Operations	Word Logic Operations
Load/Transfer Operations (Move)	Comparison Operations	Status Bit Operations
Timer Operations	Integer and REAL Operations	Program Flow Control
Counter Operations	Shift-Rotate Operations	Block Operations

Basic STL Program Elements

The basic elements of which the STL instruction set is comprised are shown in Table 15-2. These basic operations that include binary examine, assignment, logic construction, data transfer and accumulator, block management, and arithmetic instruction elements represent fundamental building blocks of the STL language.

Table 15-2. Some basic STL program elements

Binary Examine		Assignments		Block Management		
Standard	Special	Standard	Special	Jumps	Block Calls	Block Terminations
A, AN O, ON X, XN	FP FN OV, OS	=	S, R SET, CLR NOT	JMP, JMPN	UC, CC CALL	BE, BEC BEU, RET

Binary Logic Construction			Data Handling Operations	
Combine w/AND	Combine w/OR	Combine w/XOR	Load/Transfer	
A(, AN(	O(, ON(, O	X(, XN(	L, LC	T

Binary Examine Elements

Binary Examine Operations are used to examine the status of bit-addressable memory locations for '1' or '0', and to combine them to form simple series (AND) and parallel (OR) logic combinations. The examined bit addresses may represent the signals of externally connected input devices, or any bit variable or address in S7 memory. From this control logic, you can determine how internal or external outputs are controlled, whether or not a box instruction is executed, or if a specific code block should be called, or if a jump should be made, to process a section of code. Binary examine operations include both the positive and negated forms of AND, OR, and XOR, as shown in Table 15-2.

When you program a series circuit, the binary operands are combined according to the logical AND function, since all of the series conditions must be TRUE in order to yield a logic 1 result. The parallel circuit allows binary operands to be combined according to the logical OR function, since any one of multiple binary conditions must be TRUE in order to yield a logic 1 result.

Assignment Elements

STL *assignment operations* control the '1' or '0' (on/off) status of bit-addressable locations as required. These addresses may represent internal outputs, externally connected binary output devices, assignments with special functions, such as set/reset, or indicators for positive or negative edge transitions. The on/off state of these binary output variables is determined by the control logic inserted to control their operation—a basic operation described later in this section, as the *STL network*. In STEP 7, the completed binary logic result, as well as the intermediate result after each evaluation, is referred to as the *result of logic operations (RLO)*.

Binary Logic Construction Elements

In STL, creation of simple series and parallel circuits involves use of the positive and negated forms of the basic operations AND (A, AN), OR (O, ON), and XOR (X, XN). More complex circuits, however, may be developed by combining series circuits with parallel circuits. In STL, this combination is done using the binary logic construction operations A(, AN(, O(, ON(, O, X(, XN(as shown in Table 15-2. Each of the basic logic functions that has a form that includes an 'open parentheses', that when paired with a 'close parentheses', allows for logic combinations that considers the order of evaluation.

Block Management Elements

In STL, block management operations, as shown in Table 15-2, support the development of a program structure that involves block calls, from one block to another; internal block jumps to labels that accommodate a structure in which a block has several internal routines to which control may be passed for processing; and block termination operations, which support the previously mentioned structure, by allowing returns to a calling block, or the intermediate block terminations that are based on unconditional or conditional logic.

Data Handling Operations

Load and Transfer operations, as you'll learn, are all important in the development of an STL program and in the use of the digital operations. Digital operations, contrasted to binary operations, are those STEP 7 operations that involve byte, word, and double word variables or memory locations. Arithmetic, comparison, conversion, shift-rotate, and word logic are all digital operations.

In addition to Load and Transfer operations, STEP 7 supports several standard System Functions that support the handling of multiple locations or variables as a contiguous block. These callable functions that include SFC 20 (BLKMOV—Copy Area), SFC 81 (UBLKMOV—Copy Area without Gaps), and SFC 21 (FILL—Pre-Fill Area) are covered in Chapter 28 *Developing Functions and Function Blocks*.

WRITING CODE BLOCKS IN STL

Like with LAD and FBD, developing or editing a code block in STL generally starts by opening the block container from the Simatic Manager. From the Simatic Manager, you can double-click on a block or right-click on the block and select **Open** ➢ **Object** to open the block in the LAD/STL/FBD editor. Once the editor is opened, select **View** ➢ **STL** to edit the block in *Statement List* format.

Each block is opened in its own window. Each block window consists of a variable declaration pane, where variables are declared; and a code pane, where the program is written. If the block is new, you begin writing code by clicking inside the pane at the first network. To edit an existing block, find and click inside the network and specific line that you wish to edit. You may open as many blocks as needed. While a block is open, you may copy any block element—for example Network Title, Network Comment, and Network Logic, and paste it to the same or to a different network or block.

The STL Network

An *STL network* is comprised of one or more statement operations. Unlike LAD, an STL network does not have to be terminated. The following example illustrates a basic STL network—it consists of a set of *logic conditions* represented by the AND operation, arranged to form a series circuit that controls a binary output—in this case, a bit memory location. A code block, written in STL, will likely consist of several networks. When the CPU is placed in the Run Mode, and the block is called, each programmed network is processed and evaluated according to the operation of each instruction, and until a block termination operation is encountered.

Example A: A basic STL network using logical AND operation to develop series circuit

A	I	4.1	// Check I 4.1 for '1'
A	I	4.2	// Check I 4.2 for '1' and combine with previous RLO according to AND function
A	I	4.3	// Check I 4.3 for '1' and combine with previous RLO according to AND function
A	I	4.4	// Check I 4.2 for '1' and combine with previous RLO according to AND function
=		M101.1	// Assign RLO to bit memory address M101.1
A	I	5.1	// New RLO development starts with the next binary check operation

Statement Structure

In STL, the statement is the basic code component. The basic structure of a statement, shown in several examples in Table 15-3, includes four elements. From left-to-right, these elements include a *labe* that identifies the statement line; an *instruction,* also called *operation,* tells the CPU what to do; an *address,* also called *operand,* is the location upon which the CPU performs the operation; and finally, there is a *comment.* Both the label and comment are optional. Some statements may not include an explicit address upon which the instruction operates—in Example 7, below, the BEC instruction operates based on the current '1' or '0' RLO status—which is a bit status from the CPU status word.

In the remainder of this book, STL examples are represented in a table format as shown in Table 15-3, with four columns in which the four structural components of the statement will be placed. The optional label or comment may not always be included, although the column may be present.

Table 15-3. Examples of STL statement formats

Examples	STL Statement Structure			
	Label	Instruction	Address	Comment
Ex. 1		L	IW 4	// load contents of input word 4 to accumulator 1
Ex. 2	M001	A	I 6.7	// check I6.7 for '1' and combine using AND
Ex. 3		OW	#MASK	// combine accumulators using word OR function
Ex. 4	M002	JMP	M003	//jump unconditional to statement label M003
Ex. 5		INC	+2	//increment accumulator by +2
Ex. 6	M003	SLW	3	//multiply low word of accumulator by eight
Ex. 7		BEC		// terminate block if RLO = 1

Developing a Binary Logic String in STL

In STL, a *binary logic string* or *sequence* is developed using one or more of the Boolean logic functions and the assignment, or other conditional, instruction. A conditional instruction is one whose status is determined by the result of logic operations. After inserting the conditional logic, the string is terminated using a condition operation such as the assignment operation.

In the following example, the first logic string ends when the result of logic operations is assigned to bit memory address M101.1. Although the current RLO is then assigned to bit memory address M101.2, the new logic string is started with the following statement.

Example B: A basic STL network using logical AND operation to develop series circuit

A	I	4.1	// Check I 4.1 for '1'
A	I	4.2	// Check I 4.2 for '1' and combine with previous RLO according to AND function
A	I	4.3	// Check I 4.3 for '1' and combine with previous RLO according to AND function
=		M101.1	// Assign RLO to bit memory address M101.1
=		M101.2	
A	I	4.1	// Check I 4.1 for '1'
A	I	4.2	// Check I 4.2 for '1' and combine with previous RLO according to AND function
A	I	4.3	// Check I 4.3 for '1' and combine with previous RLO according to AND function
=		M101.1	// Assign RLO to bit memory address M101.1

Entering and Editing an STL Network

The most common method for entering and creating blocks in STL is the so-called incremental method. Using this method, each statement is entered on a single line. As previously described, the statement is comprised of an optional label, an instruction, an address (absolute or symbolic), and an optional comment. Generally, up to 999 networks may be inserted in a block; and each network can have up to 2000 statements. As you enter the statements of a code block, the order in which statements are entered is very important—it is the exact order in which each statement is processed.

With each block you create in STL, you may enter 1) **Variable Comments** of up to 80-characters for each variable you declare in the declaration section; 2) a **Block Title** to briefly title the code block; 3) a **Block Comment** to describe the purpose or operation of the block; 4) a **Network Title** to title each network you enter; and 5) a **Network Comment** that describes the purpose or operation of each network. In STL, each line may be commented, by inserting a Line Comment on the same line, after inserting two forward slash marks as a delimeter following the statement.

If the comment fields are not enabled (hidden), you may enable commenting from the editor menu. From the menu, select **View ➤ Display with ➤ Comments**. This selection toggles between enabling and disabling comment entry and display. Steps for displaying, entering, and editing program comments are described in Chapter 27 *Designing and Implementing the STEP 7 Program*.

Evaluating Binary Logic in an STL Network

A key to understanding STL is Boolean logic functions AND, OR, and XOR. Each of these operations is governed by a rule that is generally described by a *truth table*. When logic conditions are met according to the logic function and its truth table, the operation is said to be 'satisfied' or 'TRUE'. A TRUE value is normally represented by logic '1'; a FALSE value is usually represented by logic '0'. This TRUE/FALSE value is assigned to an output as the *result of logic operations (RLO)*.

In STEP 7, this logic result is called the result of logic operations. If a logic circuit has a TRUE result then it is said that the RLO = 1; otherwise, the RLO = 0. The logic state at any node in the logic network may also be referred to as the result of logic operation. The RLO may be evaluated after an binary examine operation, after any combination of logic, or at the output of the logic circuit.

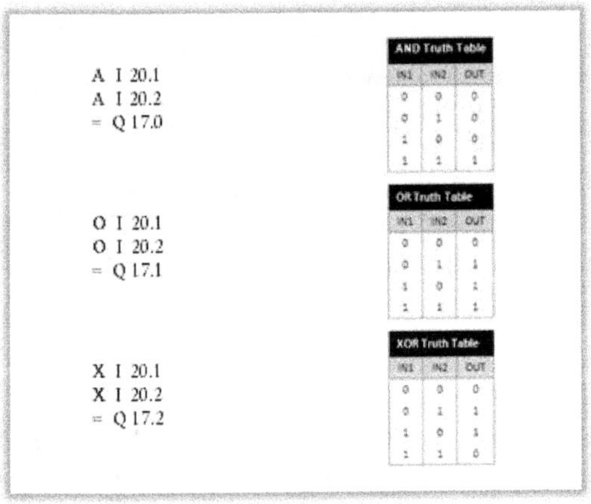

Figure 3-16. Basic evaluation of binary inputs in an STL

When you examine discrete input devices, remember that inputs are wired as normally-open or normally-closed. A device that is wired normally-open presents a logic '1' input to the gate when the device is activated; and presents a logic '0' to the gate when de-activated. This type of device is said to be 'Active HIGH'. On the other hand, a device wired normally-closed presents a logic '1' when not activated, and presents logic '0' when activated. This type of device is said to be 'Active LOW'.

How a device is wired and the effect it is expected to have in your control logic will determine which STL binary operand you should use to examine and combine inputs. In general, a AND operation 'A' is used when a logic '1' input is expected to produce a logic '1' as the RLO; and the AND NOT operation '**AN**' is used when a logic '0' is expected to produce a logic '1' as the RLO. In STL, the **AN** operation serves the same purpose as does the normally-closed contact in LAD.

Chapter 16

STL Binary Logic Operations

This chapter introduces STL Binary Logic Operations. In your S7 program development, these instructions are used primarily to create relay-equivalent control logic for on/off switching of externally connected devices. These operations also serve to create internal decision logic that will drive other STEP 7 operations—for example whether an arithmetic operation is carried out, or whether a code block is called. Decisions such as these are generally the result of examining binary conditions and applying control logic that has been constructed from Boolean functions such as AND, OR, and Exclusive OR. In this chapter, Binary Logic instructions have been divided into Basic Logic Operations and Memory Operations.

Before setting out on your review of this and subsequent STL operations chapters, if you have not already done so, you may find it beneficial to first review the material of Chapter 1 *Memory and Addressing*—there you will learn the important concepts of STEP 7 data types, memory areas, and addressing. These topics will aid you when it comes to specifying the appropriate address operand with each STL instruction.

Main Topics

- Binary Logic—Basic Operations
- Programming Basic Operations in STL
- Binary Logic—Memory Operations
- Programming Memory Operations in STL

BINARY LOGIC—BASIC OPERATIONS

In your STL program, Binary Logic Basic Operations, listed in the table below, will be used primarily to create relay-equivalent control logic. These operations also serve to create internal decision logic that drives other STEP 7 operations—including the logic that determines when and how often the various code blocks (FBs, FCs, SFBs, SFCs) of your program are called.

Table 16-1. STL Binary Logic—Basic Operations Summary

STL Operation	Brief Description
= <bit addr>	**Assign RLO.** Assign the result of logic operation (RLO) to the referenced bit address.
A <bit addr>	**AND Function.** Examine the referenced bit address or variable for logic 1, and combine with the RLO according to the AND function.
AN <bit addr>	**AND NOT Function.** Examine the referenced bit address or variable for logic 0, and combine with the RLO according to the AND function.
O <bit addr>	**OR Function.** Examine the referenced bit address or variable for logic 1, and combine with the RLO according to the OR function.
ON <bit addr>	**OR Not Function.** Examine the referenced bit address or variable for logic 0, and combine with the RLO according to the OR function.
X <bit addr>	**Exclusive OR Function.** Examine the referenced bit address or variable for logic 1, and combine with the RLO according to Exclusive OR function.
XN <bit addr>	**Exclusive OR NOT Function.** Examine the referenced bit address or variable for logic 0, and combine with the RLO according to the Exclusive OR NOT function.
A(	**AND Open Bracket.** Open nested AND operation.
AN(	**AND NOT Open Bracket.** Open nested AND NOT operation.
O(	**OR Open Bracket.** Open nested OR operation.
ON(	**OR NOT Open Bracket.** Open nested OR NOT operation.
X(	**Exclusive OR Open Bracket.** Open nested Exclusive OR operation.
XN(	**Exclusive OR NOT Open Bracket.** Open nested Exclusive OR NOT operation.
)	**Close Bracket.** Close Nested Operation.
O	**OR.** Combine AND operations using OR function.
NOT	**Negate RLO.** Invert Previous Result of Logic Operation.

PROGRAMMING BASIC OPERATIONS IN STL

Basic operations provide the fundamental elements for developing conditional logic to control ON/OFF signals to devices like motors and valves, as well internal signals used in program interlocking and other such binary signals. Such signals are used to enable or disable operations like timers, counters, program jumps, comparisons, and other parts of your program. Programming for these basic logic operations in STL is described in the following discussions.

Examining Bit Operands Using A, O, and X

With basic operations A, O, and X, you may examine bit addressable variables for '1' (a positive check) and combine the check result using AND, OR, and Exclusive OR functions. Recall that a positive binary check allows power flow in your logic circuit based on a '1' status. As such, a status of '1' yields a check result of '1', and thereby supports power flow; a status of '0' yields a check result of '0', and thereby interrupts power flow. You may examine binary addresses that include inputs (**I**), outputs (**Q**), bit memory (**M**), global data bits (**DBX**), instance data bits (**DIX**), local bits (**L**), and BOOL variables.

AND: A <bit addr.>

The **A** *operation* examines the referenced bit variable or address for '1', and combines the check result with the current RLO according to the AND function. When two binary operands are combined according to AND, the RLO = 1 only when both operands are '1'. If either operand is '0', then the RLO will be '0'. The AND function is like the series circuit in that all check results must be '1' to produce an RLO = 1. You may combine as many binary inputs using AND as required.

Example: Examine bit address for '1' and combine with the RLO using the AND operation

A	I 4.1	// Check I 4.1 for '1' and combine with RLO according to logical AND
A	I 4.2	// Check I 4.2 for '1' and combine with RLO according to logical AND
A	I 4.3	// Check I 4.3 for '1' and combine with RLO according to logical AND
		// If all results are '1' then the RLO = 1; if any result is '0' then the RLO = 0

OR: O <bit addr.>

The **OR** operation examines the referenced bit variable or address for '1', and combines the check result with the current RLO according to the OR function. When two binary operands are combined according to OR, the RLO = 1 only when both at least one operand is '1'. If both operands are '0', then the RLO will be '1'. The OR function is like the parallel circuit in that at least one check result must be '1' to produce an RLO = 1. You may combine as many binary inputs using OR as required.

Example: Examine bit address for '1' and combine with the RLO using the OR operation

O	M 2.1	// Check M 2.1 for '1' and combine with RLO according to logical OR
O	M 2.2	// Check M 2.2 for '1' and combine with RLO according to logical OR
O	M 2.3	// Check M 2.3 for '1' and combine with RLO according to logical OR
		// If at least one result is '1' then the RLO = 1; if all results are '0' the RLO = 0

Exclusive OR: X <bit addr>

The **X** operation examines the referenced bit variable or address for '1' and combines the check result with the current RLO according to the logical XOR function. When two binary operands are combined according to XOR, the RLO = 1 only if one of the two operands is '1', but not both. If both operands are '1', or if both operands are '0', then the RLO will be '0'.

Example: Examine bit address for '1' and combine with the RLO using Exclusive-OR

X	I 3.1	// Check I 3.1 for '1' and combine with RLO according to logical XOR
X	I 3.2	// Check I 3.2 for '1' and combine with RLO according to logical XOR
		// If the XOR operands are different, then the RLO = 1; otherwise RLO = 0

Examining Bit Operands Using AN, ON, and XN

With basic operations AN, ON, and XN, you may perform negated status checks and combine the check results according to AND, OR, and Exclusive OR. Recall that a negated check allows power flow based on a '0' status as opposed to a '1' status. As such, when negated operations are used, a status of '0' yields a check result of '1' (RLO = 1); a status of '1' yields a check result of '0' (RLO = 0). Like with operations A, O, and X, you may examine input (**I**) and output (**Q**) bits, bit memory (**M**), global (**DBX**) and instance (**DIX**) data bits, local (**L**) bits, and BOOL variables.

AND NOT: AN <bit addr>

The **AN** *operation* examines the referenced bit variable or address for '0' and combines the check result with the current RLO according to the AND function. When a bit variable is examined with the **AN** operation, a status of '0' yields a check result of '1'; a status of '1' yields a check result of '0'. When the check results are combined according to the logical AND function, a single check result of '0' produces an RLO = 0; if all check results are '1', then an RLO = 1 is produced.

Example: Examine bit address for '0' and combine with the RLO using the AND NOT operation

AN	I 4.4	// Check I 4.4 for '0'
AN	I 4.5	// Check I 4.5 for '0' and combine with RLO according to logical AND
A	I 4.6	// Check I 4.6 for '1' and combine with RLO according to logical AND
		// If all results are '1' then the RLO = 1; if any result is '0' then the RLO = 0

OR NOT: ON <bit addr>

The **ON** operation examines the referenced bit variable or address for '0' and combines the check result with the current RLO according to the logical OR function. If when checked, both the status and the current RLO = 1, the new RLO is equal to '1'. According to OR NOT, if, when checked, either the status is '0' or the current RLO = 0, then the new RLO is equal to '0'.

Example: Examine bit address for '0' and combine with the RLO using OR NOT operation

ON	I 4.5	// Check I 4.5 for '0'
O	I 4.6	// Check I 4.6 for '1' and combine with RLO according to logical OR
ON	I 4.7	// Check I 4.7 for '0' and combine with RLO according to logical OR
		// If at least one result is '1' then the RLO = 1; if all result are '0' then the RLO = 0

Exclusive OR NOT: XN <bit addr>

The *XN* operation examines the referenced bit variable or address for '0' and combines the check result with the current RLO according to the logical XOR function. If when checked, both the status and the current RLO = 1, the new RLO is equal to '1'. According to AND ,if, when checked, either the status is '0' or the current RLO = 0, then the new RLO is equal to '0'.

Example: Examine bit address for '0' and combine with the RLO using Exclusive-OR

X	I 3.4	// Check I 3.2 for '1' and combine with RLO according to the logical XOR
XN	I 3.3	// Check I 3.1 for '0'
		// If XOR (NOT) operands are the same, then the RLO = 1; otherwise the RLO = 0

Completing a Binary Logic String

In the Statement List language, the completion of a binary logic string or sequence is similar to the termination of a LAD/FBD network with a binary coil or output instruction. In STL, a binary logic string is completed with an assignment type of operation—for example, after the evaluation of a logic combination of binary examination operations. Assignment operations are those instructions that affect a memory operand or variable based on the result of logic operations. Specifically, an STL logic string is terminated with the instructions *assign RLO, set,* and *reset.* Set/reset operations are discussed in the following section, *Binary logic—Memory Operations.*

A major difference in the STL logic string is that a single logic network may include multiple logic strings. After one logic string is terminated, another may be started within the same network. When a logic string is evaluated, if the RLO = 1, then the following RLO dependent operation is executed; if the RLO = 0, then the following RLO dependent operation is not executed.

Assign RLO: = <bit addr>

The *assign RLO* operation performs the same function as the standard output coil in LAD. The result of logic operations (RLO) of the preceding logic is assigned to the address referenced by the assign output. If the preceding logic has an RLO = 1, then a value of '1' is assigned to the referenced address or variable; if the preceding logic has an RLO = 0, then a value of '0' is assigned to the referenced address or variable.

In the network below, output address Q 22.3 is energized if the input I 4.1 AND input I 4.2 is True (1).

Example: Assign '0' or '1' RLO value to the bit address specified in the assign statement

A	I 4.1	// Check I 4.1 for '1'
A	I 4.2	// Check I 4.2 for '1' and combine with RLO according to logical AND
=	Q 22.3	// Assign RLO = 1 or RLO = 0 to output address Q 22.3

Once the RLO is developed, you may use the assign operation as often as required, as seen below.

Example: Assign '0' or '1' RLO value to the bit address specified in the assign statement

A	I 4.1	// Check I 4.1 for '1'
=	Q 22.3	// Assign RLO = 1 or RLO = 0 to output address Q 22.3
=	Q 22.4	// Assign RLO = 1 or RLO = 0 to output address Q 22.4

Combining Logic Using Bracket Operations

Boolean AND, OR, Exclusive OR, and NOT functions facilitate development of various control logic combinations—for example AND, OR, XOR, NOT, AND-before-OR, and OR-before-AND logic. Basic Logic Operations are described in the following discussions.

AND-before-OR

The *and-before-or* combination in STL is the equivalent of two or more series circuits in parallel. In such a combination, a parallel branch has logic continuity (RLO = 1) if the AND combination is True. The network has logic continuity if the RLO = 1 for either of the AND combinations.

In the following example, the output address Q30.2 is activated (logic 1), when the AND combination of inputs I20.1 AND I20.2 result in RLO = 1 OR when the AND combination of inputs I20.3 AND I 20.4 result in RLO = 1. The output is de-activated whenever the RLO = 0 in both AND combinations.

Example A: AND-before-OR combination with two AND combinations

A	I	20.1	// Check I 20.1 for '1'
A	I	20.2	// Check I 20.2 for '1' and combine with RLO according to logical AND
O			// Combine previous RLO and following RLO according to logical OR
A	I	20.3	// Check I 20.3 for '1' and combine with RLO
A	I	20.4	// Check I 20.4 for '1' and combine with RLO according to logical AND
=		Q 30.2	// Assign RLO = 1 or RLO = 0 to output address Q 30.2

The following example of AND-before-OR involves three AND gates that are combined using the OR function; this combination is equivalent of three series circuits of two elements each that are combined as three parallel branches. You may add as many AND combinations in parallel as required.

In the following example, the output image address Q30.2 is energized when the RLO = 1 for any one of the AND combinations. Address Q30.2 is de-energized only when the RLO = 0 for all of the AND combinations.

Example B: AND-before-OR combination with three AND combinations

A	I	20.1	// Check I 20.1 for '1'
A	I	20.2	// Check I 20.2 for '1' and combine with RLO according to logical AND
O			// Combine previous RLO and following RLO according to logical OR
A	I	20.3	// Check I 20.3 for '1' and combine with RLO
A	I	20.4	// Check I 20.4 for '1' and combine with RLO according to logical AND
O			// Combine previous RLO and following RLO according to logical OR
A	I	20.5	// Check I 20.5 for '1' and combine with RLO
A	I	20.6	// Check I 20.6 for '1' and combine with RLO according to logical AND
=		Q 30.2	// Assign RLO = 1 or RLO = 0 to output address Q 30.2

OR-before-AND

The *or-before-and* combination in STL is the equivalent of two or more parallel circuits in series. In such a combination, all of the OR operations that are in series must have an RLO = 1 for the network to have complete logic continuity (RLO = 1).

In the following example, the output address Q30.6 is activated (logic 1) when the combination of I22.1 OR I22.3 is '1' AND the combination of I22.2 OR I22.4 is '1'. For example, Q30.6 will activate when I22.3 is closed AND I22.2 is closed. The output is de-activated whenever the RLO = 0 in any one of the OR combinations.

Example A: OR-before-AND combination

A(		// Combine according to AND; nested AND
O	I 22.1	// Check I 22.1 for '1'
O	I 22.3	// Check I 22.3 for '1' and combine with RLO according to logical OR
)		// Close nested AND
A(		// Combine the OR result according to AND
O	I 22.2	// Check I 22.2 for '1'
O	I 22.4	// Check I 22.4 for '1' and combine with RLO according to logical OR
)		// Close nested AND
=	Q 30.6	// Assign RLO = 1 or RLO = 0 to output address Q 30.6

The following example of OR-before-AND involves three OR gates that are combined using the AND function; this combination is equivalent to the series combination of three parallel circuits of two elements each. You may add as many OR combinations in series as required.

In the following example, the output image address Q30.6 is energized when the RLO = 1 for all three of the OR combinations. Address Q30.6 is de-energized when the RLO = 0 for any one of the OR combinations.

Example A: OR-before-AND combination

A(		// Combine according to AND; nested AND
O	I 22.1	// Check I 22.1 for '1'
O	I 22.2	// Check I 22.2 for '1' and combine with RLO according to logical OR
)		// Close nested AND
A(		// Combine the OR result according to AND
O	I 22.3	// Check I 22.3 for '1'
O	I 22.4	// Check I 22.4 for '1' and combine with RLO according to logical OR
)		// Close nested AND
A(		// Combine the OR result according to AND
O	I 22.5	// Check I 22.5 for '1'
O	I 22.6	// Check I 22.6 for '1' and combine with RLO according to logical OR
)		// Close nested AND
=	Q 30.6	// Assign RLO = 1 or RLO = 0 to output address Q 30.6

Inverting the Result of Logic Operations

In some cases, inversion of the RLO is required to achieve the desired logic. In LAD, the NOT contact can be placed between any two network nodes—for example between two contacts. In STL, the NOT operation may be placed at any point at which you wish to invert the RLO. The NOT in effect inverts the RLO at the point of insertion—an RLO = 1 becomes RLO = 0; and RLO = 0 becomes RLO =1.

Negate RLO: NOT

The **NOT** operation inverts or negates the state of the RLO at the point of insertion. If the RLO = 1 prior to the operation, then following the operation the RLO is inverted (RLO = 0). If the RLO = 0 prior to the operation, then after inversion the RLO = 1. The NOT operation does not reference any bit address, and may be placed at any point in the logic string.

In the example below, if I 4.7 is True (1), the RLO = 1, the output address Q22.4 is activated. The RLO = 1 following Q22.4, and is inverted after the NOT operation. If output Q22.4 = '1', then output Q22.5 = '0'. If output Q22.4 = '0', then output Q22.5 = '1'.

Example: Negated result of logic operations (RLO)

A	I 4.7	// Check I 4.7 for '1'
=	Q 22.4	// Assign the RLO value of '1' or '0' to output address Q 22.4
NOT		// Invert the RLO
=	Q 22.5	// Assign RLO = 1 or RLO = 0 to output address Q 22.5

In the example below, the output address Q 27.0 is energized if one or both of the input addresses I 44.5 and I 44.6 is open—that is when the RLO = 0 prior to the NOT operation. When both check results are '1' prior to the NOT operation, output address Q 27.0 is de-energized after the RLO is inverted. This operation is the negated AND function or NAND.

Example: Negated AND function—NAND function

A	I 44..5	// Check I 4.7 for '1'
A	I 44.6	// Check I 2.1 for '1' and combine with RLO according to logical AND
NOT		// Invert the RLO
=	Q 27.0	// Assign RLO = 1 or RLO = 0 to output address Q 22.4

In the example below, the output address Q27.1 is energized only when both bit memory operands M37.4 and M37.5 are '0'—that is when the RLO = 0 prior to the NOT operation. If either or both of the check results is '1', then the RLO = 1 prior to the NOT operation; after the inversion, Q27.1 is de-energized. This is the negated OR function or NOR.

Example: Negated OR function—NOR function

A(		// Combine according to AND; nested AND
O	M 37.4	// Check M 37.4 for '1'
O	M 37.5	// Check M 37.5 for '1' and combine with RLO
)		//
NOT		// Invert the RLO
=	Q 27.1	// Assign RLO = 1 or RLO = 0 to output address Q 27.1

BINARY LOGIC—MEMORY OPERATIONS

Binary logic memory operations are those operations that affect the '1' or '0' status of S7 bit addresses and variables in memory. These operations include *set/reset operations*, the equivalent of latch/unlatch operations of some controllers; and *edge-detection operations*, whose function is to detect and indicate the occurrence of rising-edge or falling-edge transitions in binary signals, or in the result of logic operations. Binary logic memory operations are summarized in Table 16-2.

Table 16-2. STL Binary Logic—Memory Operations Summary

STL Operation	Brief Description
S <bit addr>	**Set.** If the preceding RLO = 1, set the referenced bit address or variable to '1'. An RLO of '0' has no effect on the referenced bit address.
R <bit addr>	**Reset.** If the preceding RLO = 1, reset the referenced bit address or variable to '0'. An RLO of '0' has no effect on the referenced bit address.
FP <bit addr>	**Flag Positive RLO Edge.** If a 0-to-1 (rising edge) transition is detected in the RLO, set the new result of logic operation to '1' for one cycle (one-shot output).
FN <bit addr>	**Flag Negative RLO Edge.** If a 1-to-0 (falling edge) transition is detected in the RLO, set the new result of logic operation to '1' for one cycle (one-shot output).

Note: Valid bit addresses include Input (I), Output (Q), Bit Memory (M), Global Data (DBX), and Instance Data (DIX).

PROGRAMMING MEMORY OPERATIONS IN STL

As with basic binary logic operations, the code developed using binary logic memory operations helps to form the driving logic for externally connected discrete devices, as well as for triggering various internal operations. Like binary check operations, edge evaluation instructions may be inserted anywhere in the logic string. Set and Reset coils/outputs, on the other hand, are like the standard assign output operation and are driven by some combination of series/parallel logic. Also, as with the assign output operation set/reset, they may be used to terminate a binary logic string.

Edge detection operations are especially useful in that they supplement the standard binary examination operations, which simply check whether a discrete signal is ON or OFF—these operations detect,as well as indicate, both positive and negative edge transitions. Programming these basic logic memory operations in STL is described in the following discussions.

Set/Reset Operations

Set /reset operations **S** and **R** are individual operations that are generally programmed to operate as a pair, where both operations reference the same bit address or variable. When programmed as a pair, the order in which the two operations are inserted in the code will affect a characteristic of the set/reset operations referred to as dominance. Dominance determines which of the set/reset functions will override the other if the RLO preceding both operations transition from 0-to-1 during the same cycle.

With implementations of set/reset instructions, including the flip-flop implementation, dominance is determined by which of the two operations is programmed last; this is true since, if both operands are activated simultaneously during the same CPU cycle, the one processed last overrides the status previously determined by the other.

Set: S <bit addr>

When the **set** operation **S** is processed, the bit address or variable that it references is set to '1' if the preceding RLO transitions from 0-to-1. Because the referenced bit operand remains set regardless of a change in the RLO, the operand is said to be 'latched'. The referenced bit variable remains latched at '1' until reset (unlatched) by a **Reset** operation that references the same address.

In the following network, if the combination of input I 2.3 **AND** I 2.4 produces RLO = 1, then output address Q 33.7 is set (latched). The referenced bit operand remains set even if the RLO returns to '0'.

Example: Latching a bit address using the Set operation

A	I 2.3	// Examine I 2.3 for logic '1'
A	I 2.4	// Examine I 2.4 for logic '1', combine with RLO using logical AND
S	Q 33.7	// If RLO = 1 **Set** (latch) output address 33.7

Reset: R <bit addr>

When the **reset** operation **R** is processed, the bit address or variable that it references is set to '1' if the preceding RLO transitions from 0-to-1. Because the referenced bit operand remains reset regardless of a change in the RLO, the operand is said to be 'unlatched'.

In the following network, if the combination of input I 3.3 AND I 3.4 produces an RLO = 1, then output address Q 33.7 is reset (unlatched).

Example: Unlatching a bit address using the Reset operation

O	I 3.3	// Examine I 3.3 for logic '1'
A	I 3.4	// Examine I 3.4 for logic '1', combine with RLO using logical AND
R	Q 33.7	// If RLO = 1 **Reset** (unlatch) output address 33.7

Implementing Set/Reset Flip-Flop Functions

The **S** and **R** operations may be used in conjunction to form an *RS flip-flop* or an *SR flip-flop*. These operations are used to control the latching and unlatching of the same bit address or variable. Depending on your application, you would use the RS flip-flop if you want the set operation to have priority; and you would program the SR flip-flop if you need the reset operation to have priority.

Priority or dominance, as previously described, determines which of the two operations will override the other if both are processed with RLO=1 in the same cycle of OB1. If you want the **S** operation to be dominant, then it must be programmed after the **R** instruction; and conversely, the **R** operation to have dominance must be programmed after the **S** instruction.

Coding an RS Flip-Flop

The RS flip-flop performs both the latch and the unlatch functions, and is used when the set function must have priority. Both the Reset (**R**) and the Set (**S**) operations are executed only when the preceding RLO = 1. There is no effect on the **R** or **S** operation when the RLO = 0.

If the RLO prior to the **R** operation is '1', the referenced bit address is reset to '0' and remains reset unless set again. If the RLO prior to the **S** operation is '1', then the referenced bit address is set to '1'. If the RLO prior to the **R** and **S** operations both are '1' in the same program cycle, then the referenced bit address is first reset and then subsequently set (set priority) upon evaluation of the **S** operation.

Example: Implementing an RS-Flip-Flop with Set Priority

O	I 3.3	// Examine I 3.3 for logic '1'
A	I 3.4	// Examine I 3.4 for logic '1', combine with RLO using logical AND
R	Q 33.7	// If RLO = 1 **Reset** (unlatch) output address 33.7
A	I 2.3	// Examine I 2.3 for logic '1'
A	I 2.4	// Examine I 2.4 for logic '1', combine with RLO using logical AND
S	Q 33.7	// If RLO = 1 **Set** (latch) output address 33.7

Coding an SR Flip-Flop

The SR flip-flop performs both the latch and unlatch functions, and is used when the reset function must have priority. Both the Set (**S**) and the Reset (**R**) operations are executed only when the preceding RLO = 1. There is no effect on the **S** or **R** operation when the RLO = 0.

If the RLO prior to the **S** is '1', then the referenced bit address is set to '1' and remains set until reset. If the RLO prior to the **R** operation is '1', then the referenced bit address is reset to '0'. If the RLO prior to the **S** and **R** operations both are '1' in the same cycle, then the referenced bit address is first set and then subsequently reset (reset priority) upon evaluation of the **R** operation.

Example: Implementing an SR-Flip-Flop with Reset Priority

A	I 2.3	// Examine I 2.3 for logic '1'
A	I 2.4	// Examine I 2.4 for logic '1', combine with RLO using logical AND
S	Q 33.7	// If RLO = 1 **Set** (latch) output address 33.7
O	I 3.3	// Examine I 3.3 for logic '1'
A	I 3.4	// Examine I 3.4 for logic '1', combine with RLO using logical AND
R	Q 33.7	// If RLO = 1 **Reset** (unlatch) output address 33.7

Edge Detection Operations

You will mainly use *edge detection operations* to detect and indicate when a discrete transition occurs—for example, when a motor is switched off, an alarm is removed, or when a photo eye is first blocked. You might also use these operations to ensure that certain operations like an arithmetic calculation or move operation is triggered only once when its trigger event occurs.

In STL, the *Flag Positive Edge* operation **FP** detects and indicates when a positive-edge transition from 0-to-1 occurs—for example when a valve closes; the *Flag Negative Edge* operation **FN** detects and indicates when a negative transition from 1-to-0 occurs—for example when a motor is switched off.

STEP 7 Programming Made Easy in LAD, FBD, and STL

Flag Positive Edge RLO: FP <bit addr>

The **FP** operation detects a rising edge (0-to-1) transition in a specific address or in the result of logic operations at the point of insertion, and generates a one-shot indication using the edge-memory bit operand that you specify in the FP statement. The bit address you specify as the edge-memory bit saves the previous state of the RLO, and compares it to each new RLO. The edge-trigger may be developed by a specific address, or by the result of logic operations at a specific point in a logic string.

When the RLO preceding the edge trigger makes a 0-to-1 transition, the RLO after the FP operation is set to '1' for one program scan (OB1 cycle) and then returns to logic 0. You may insert the FP operation, as required, at any point in a logic string. If there is a difference between the current and previous RLO '0' state (detection of rising edge), the RLO bit will be '1' after the FP operation.

In the following network, the output Q 22.4 is energized for one cycle of OB1 when the logic combination of I 4.7 AND I 2.1 makes a 0-to-1 transition.

Example: Positive RLO Edge Detection

A	I	4.7	// Check I 4.7 for '1'
A	I	2.1	// Check I 4.7 for '1' and combine with RLO
FP	M	2.0	// On RLO = 1, set RLO = 1 for one OB1 cycle
=	Q	22.4	// If RLO = 1 **Set** (latch) output address 22.4

Flag Negative Edge RLO: FN <bit addr>

The **FN** operation detects a falling edge (1-to-0) transition in a specific address, or in the result of logic operations at the point of insertion, and generates a one-shot indication using the edge-memory bit operand that you specify in the FP statement. The bit address you specify as the edge-memory bit saves the previous state of the RLO, and compares it to each new RLO. The edge-trigger may be developed by a specific address, or by the result of logic operations at a specific point in a logic string.

When the RLO preceding the edge trigger makes a 1-to-0 transition, the RLO following the FN operation is set to '1' for one program scan (OB1 cycle), and then returns to logic 0. You may insert the FN operation, as required, at any point in a logic string. If there is a difference between the current and previous RLO '1' state (detection of falling edge), the RLO bit will be '1' after the FN operation.

In the following network, the output Q 22.4 is reset when the logic combination of I 4.7 AND I 2.1 makes a 1-to-0 transition.

Example: Negative RLO Edge Detection

A	I	4.7	// Check I 4.7 for '1'
A	I	2.1	// Check I 2.1 for '1' and combine with RLO
=	Q	22.4	
FN	M	2.0	// On RLO = 1, set RLO = 1 for one OB1 cycle
R	Q	22.5	// If RLO = 1 **Reset** (unlatch) output address 22.5

Copyright © 2013   266

Chapter 17

STL Load and Transfer Operations

In this chapter, you'll be introduced to STL Load and Transfer operations. In STL, load and transfer operations are used to copy data from one memory location or variable to another for various purposes. In this sense, the load and transfer operations combine to form the equivalent of the MOVE operation of LAD and FBD. Load/transfer operations sometimes simply store or protect data from being overwritten, and other times prepare data for subsequent processing and use in other operations.

Load and Transfer operations, as you'll learn, are all important in the development of a Statement List program and in the use of the STL digital operations. Digital operations, contrasted to binary operations, are those STEP 7 instructions that involve byte, word, and double word variables or memory locations. In load/transfer operations data movement is generally via the S7 accumulators; however, this chapter also introduces a subset of load/transfer operations that involve the S7 address registers AR1 and AR2.

In addition to STEP 7 Load and Transfer operations, which only move one value at a time, standard System Functions that support the handling of multiple locations or variables as a contiguous block are covered in Chapter 28 *Developing Functions and Function Blocks*. These callable Functions that include SFC 20 (BLKMOV—Copy Data Area), SFC 81 (UBLKMOV—Copy Data Area without Gaps), and SFC 21 (FILL—Pre-Fill Data Area Values), are available in the STEP 7 library, and are integrated in some S7 CPUs.

Main Topics

- Load and Transfer Operations
- Programming Load and Transfer Operations
- Accumulator Operations
- Programming Accumulator Operations
- System Functions for Moving Data

LOAD AND TRANSFER OPERATIONS

Load and Transfer instructions are essential to writing code that involves digital operations. Digital operations are those STEP 7 instructions that are performed using byte, word, and double word variables or memory locations. Arithmetic, comparison, conversion, shift-rotate, and word logic are all digital operations. In performing these operations, data is exchanged among S7 memory areas, input/output image areas, as well as peripheral I/O areas. In all of these cases, the data exchange is processed through the CPU's accumulators—specifically accumulator 1 (**ACCU 1**).

An example of data interchange via the accumulator is with an arithmetic calculation, where you first get two values, then perform the operation, and finally store the result. In this operation, the Load operation places the first value in ACCU 1. A subsequent load to get the second value causes the first value to be moved to accumulator 2 (**ACCU 2**). The operation may then be performed using the two accumulators—the result is placed in ACCU 1. You may then use the Transfer operation to move the result to the desired location or variable. The transfer operation always moves data from ACCU 1.

Data Interchange via S7 Accumulators

In STL, the *load operation* copies the contents of the source address you specify to the CPU's accumulator 1 for further processing. With each load operation, the value previously held in ACCU 1 is copied to ACCU 2, overwriting the previous contents of ACCU 2. The source address you specify in the Load statement may reference a constant, as shown in Table 17-2; a memory location (or its assigned symbol name), as listed in Table 17-1; or a declared block variable.

The *transfer operation* copies the contents of accumulator 1 to the destination address that you specify. The destination you specify in the Transfer statement may be an S7 memory location (or its symbol name), as listed in Table 17-3, or a declared block variable. Later we'll look at STL load operations that deal specifically with S7 address registers AR1 and AR2.

Loading—to Accumulator 1

In a load operation, the source location is always moved to the 32-bits of accumulator 1 right-justified. If the source is less than 32-bits—for instance a 16-bit word, then the 16-bits are copied right-justified to accumulator 1; the remaining higher bits of accumulator 1 are zeroed.

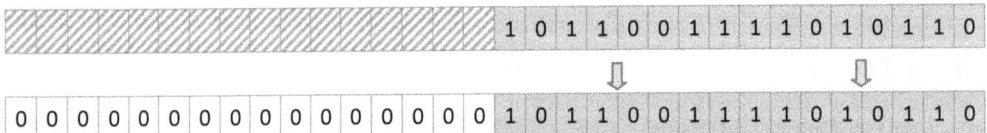

Figure 17-1. Load source to 32-bit accumulator 1, where source is less is less than 32-bits

Transferring—from Accumulator 1

In a transfer operation, the current 32-bit content of accumulator-1 is always moved to the specified destination right-justified. If the destination is less than 32-bits—for instance a 16-bit word, then the right-most word of accumulator 1 is copied to the destination. After a transfer, accumulator 1 is unchanged—thereby making subsequent transfers of the original content of the accumulator possible.

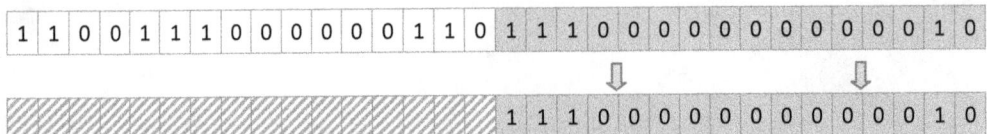

Figure 17-2. Transfer of 32-bit accumulator 1 to a destination that is less than 32-bits—a byte or word

PROGRAMMING LOAD AND TRANSFER OPERATIONS

Often data is moved prior to performing an operation that uses the data, or just after performing an operation, in order to preserve the data from overwriting. Programmatically changing a timer or counter preset, a process set point, or getting data for transfer to another device are all cases where Load and Transfer are used. Moving a contiguous block of data may call for system functions.

Loading S7 Memory Areas

With STL load operations, you may access any S7 memory area byte (B), word (W), or double word (DW), as shown in the table. These areas are accessable using absolute, symbolic, or indirect addresses.

Load: L <memory operand>

When processed, the load operation *L* copies the contents of a memory operand or variable to ACCU1, irrespective of the RLO. Using the statement format, as shown above, the L PIW 256 statement reads analog data from the analog input assigned this peripheral input address. It is important to note that with certain CPUs, to load an input (I) or output (Q) address, the module must be installed.

Table 17-1. Load Memory Operands Summary; Assigned symbol addresses may also be used

Memory Area	Operand	Example	Brief Explanation
Input Image	IB	L IB12	Load to ACCU 1 input image byte 12
	IW	L IW14	Load to ACCU 1 input image word 14
	ID	L ID16	Load to ACCU 1 input image double word 16
Output Image	QB	L QB20	Load to ACCU 1 output image byte 20
	QW	L QW22	Load to ACCU 1 output image word 22
	QD	L QD24	Load to ACCU 1 output image double word 24
Bit Memory	MB	L MB32	Load to ACCU 1 bit memory byte 32
	MW	L MW34	Load to ACCU 1 bit memory word 34
	MD	L MD36	Load to ACCU 1 bit memory double word 36
Peripheral Input	PIB	L PIB 17	Load to ACCU 1 peripheral input byte 17
	PIW	L PIW 256	Load to ACCU 1 peripheral input word 256
	PID	L PID42	Load to ACCU 1 peripheral input double word 42
Local	LB	L LB 8	Load to ACCU 1 local memory byte 8
	LW	L LW14	Load to ACCU 1 local memory word 14
	LD	L LD16	Load to ACCU 1 local memory double word 16
Data Block (global DB)	DBB	L DBB 12	Load to ACCU 1 global data byte 12
	DBW	L DBW 14	Load to ACCU 1 global data word 14
	DBD	L DBD 16	Load to ACCU 1 global data double word 16
Data Block (instance DB)	DIB	L DIB 12	Load to ACCU 1 instance data byte 12
	DIW	L DIW 14	Load to ACCU 1 instance data word 14
	DID	L DID 16	Load to ACCU 1 instance data double word 16
Timer	Tn	L T5	Load timer value as raw binary (decimal value)
	Tn	LC T5	Load timer value as binary coded decimal (BCD)
Counter	Cn	L C5	Load counter value as raw binary (decimal value)
	Cn	LC C5	Load counter value as binary coded decimal (BCD)
CPU Status Word	STW	L STW	Load to ACCU 1 the contents of the CPU status word

Loading Constant Values

With the load operation, you may also move a constant to accumulator 1, which can then be used in a subsequent digital operation. For example, you may want to perform a compare or an arithmetic operation where one of the values is a constant and the other a local variable or a value in a memory location. The constant value might be loaded as the first value, or as the second value, depending on your operation. Once both operands are loaded, the intended operation can be performed.

Load: L <constant>

The load instruction **L** is used to read constant values perhaps as one of two values that will be used in a word logic, arithmetic, or compare operation. The constant value is entered as a load statement parameter, using the appropriate constant notation for the value you wish to load. When processed, the load operation copies the specified constant to ACCU 1—irrespective of the RLO.

The statement **L** S5T#2H30M, for example, places a timer preset value of 2 hours and 30 minutes into accumulator 1. The next timer started will use the ACCU 1 value as its preset. Examples of loading a constant are shown in the following table.

Table 17-2. Load Constants Summary

Operand	Load Constant Example	Brief Description
B#16#	L B#16#FF	Load 8-bit hex value of FF
2#	L 2#1111_1111_1111_1111	Load 16-bit binary value
W#16#	L W#16#FAFA	Load 16-bit hex value of FAFA
C#	L C#750	Load BCD counter value of 750
B#	L B#(10,25)	Load 2 bytes with 10 and 25
2#	L 2#1111_1111_1111_1111_0000_0000_0000_0000	Load 32-bit binary value
W#16#	L DW#16#FFFF_2FEF	Load hex value of FFFF 2FEF
B#	L B#(25,75,1,100)	Load 4 bytes with 25, 75, 1, 100
INT	L +10	Load integer of +10
L#	L L#+45000	Load long integer of +45000
REAL	L 4.5E+4	Load floating-point +45000.0
CHAR	L 'NO'	Load ASCII characters 'NO'
	L 'DONE'	Load ASCII characters 'DONE'
S5T#/S5TIME#	L S5T#2H30M	Load Timer preset of 2hrs 30min.
D#/DATE#	L D#2009-06-03	Load Date value of June 03, 2009
T#/TIME#	L T#24D_20H_31M_23S_647_MS	Load Time as 24days,20hrs.,31 min., 23 sec., 647 milliseconds
TOD#	L TOD#2:32:59	Load Time of day as 02:32:59
P#byte.bit	L P#27.4	Load area-internal pointer
P#Abyte.bit	L P#M44.4	Load area-crossing pointer
P#<name>	L P#<name>	Load address of a local variable

Transferring Memory Areas

With the transfer operation, you may move the current contents of accumulator 1 to any byte (B), word (W), or double word (DW) of the S7 memory areas listed in Table 17-3. You may address the locations in these areas using absolute, symbolic, or indirect addressing. The *transfer* operation always copies the contents of accumulator 1 (ACCU 1) to the specified destination address or variable.

Transfer: T <memory operand>

The transfer operation *T* is used to write data from accumulator 1 to the S7 memory location you specify. When processed, the transfer operation moves the contents of ACCU 1 right-justified to the specified memory operand or variable, irrespective of the RLO. The original contents of accumulator 1 are unaffected. For example, after a speed calculation, the result in accumulator 1 is transferred to an analog output channel, directly via the peripheral output memory—the value remains in ACCU 1.

Using the statement format, as shown above, the statement **T** PQW 256 transfers the value currently in ACCU 1 to the analog output channel assigned the address of PQW 256. It is important to note that with certain CPUs, to transfer to an input (I) or output (Q) address, the module must be installed.

Table 17-3. Transfer Memory Operands Summary

Memory Area	Operand	Example	Brief Description
Input Image	IB	**T** IB12	Transfer ACCU 1 to input byte 12
	IW	**T** IW14	Transfer ACCU 1 to input word 14
	ID	**T** ID16	Transfer ACCU 1 to input double word 16
Output Image	QB	**T** QB20	Transfer ACCU 1 to output byte 20
	QW	**T** QW22	Transfer ACCU 1 to output word 22
	QD	**T** QD24	Transfer ACCU 1 to output double word 24
Bit Memory	MB	**T** MB32	Transfer ACCU 1 to memory byte 32
	MW	**T** MW34	Transfer from ACCU 1 to memory word 34
	MD	**T** MD36	Transfer ACCU 1 to memory double word 36
Peripheral Input	PQB	**T** PQB27	Transfer ACCU 1 to peripheral output byte 27
	PQW	**T** PQW256	Transfer ACCU 1 to peripheral output word 256
	PQD	**T** PQD42	Transfer ACCU 1 to peripheral output double word 42
Local	LB	**T** LB 8	Transfer from ACCU 1 to local memory byte 8
	LW	**T** LW14	Transfer ACCU 1 to local memory word 14
	LD	**T** LD16	Transfer ACCU 1 to local memory double word 16
Data Block (global DB)	DBB	**T** DBB12	Transfer ACCU 1 to global data byte 12
	DBW	**T** DBW14	Transfer ACCU 1 to global data word 14
	DBD	**T** DBD16	Transfer ACCU 1 to global data double word 16
Data Block (instance DB)	DIB	**T** DIB12	Transfer ACCU 1 to instance data byte 12
	DIW	**T** DIW14	Transfer ACCU 1 to instance data word 14
	DID	**T** DID16	Transfer ACCU 1 to instance data double word 16
CPU Status Word	STW	**T** STW	Transfer ACCU1 to the CPU status word

Loading and Transferring with Address Registers

In STL, a special subset of load and transfer operations supports the moving of pointers between an S7 address register and accumulator 1, and between each other. Recall, that the address register holds a 32-bit area pointer that can be loaded directly from a pointer constant, or from a pointer contained in ACCU 1. Area pointer were shown in Chapter 1 *Memory and Addressing*, Figure 1-20 and Figure 1-21.

Table 17-4. Load and Transfer Operations for Address Registers

Operation	Brief Description
LAR1	Load address register AR1 with the contents of accumulator 1
LAR1 <parameter>	Load address register AR1 with the specified pointer parameter (P#byte.bit, P#Abyte.bit or double word address MD, LD, DBD, or DID)
LAR1 AR2	Load address register AR1 with the contents of address register AR2
LAR2	Load address register AR2 with the contents of accumulator 1
LAR2 <parameter>	Load address register AR2 with the specified pointer parameter (P#byte.bit, P#Abyte.bit or double word address MD, LD, DBD, or DID)
TAR1	Transfer the contents of address register AR1 to accumulator 1
TAR1 <parameter>	Transfer the contents of address register AR1 to the specified address parameter
TAR1 AR2	Transfer address register AR1 contents to address register AR2
TAR2	Transfer the contents of address register AR2 to accumulator 1
TAR2 <parameter>	Transfer the contents of address register AR2 to the specified address parameter
CAR	Exchange (swap) the contents of address register AR1 and address register AR2
+AR1	Add to Address Register 1, the pointer offset of the low-order word of ACCU 1
+AR1 < P#Byte.Bit> +AR1 < P#AByte.Bit>	Add to Address Register 1, the specified area-internal/area-crossing pointer offset
+AR2	Add to Address Register 2, the pointer offset of the low-order word of ACCU 1
+AR2 < P#Byte.Bit> +AR2 < P#AByte.Bit>	Add to Address Register 2, the specified area-internal/area-crossing pointer offset

Note: Load/transfer operations for AR2 operate exactly as operations described for AR1. The bit address range of the pointer constant in this operation is 0.0 to 4095.7—the byte range when accessing areas I, Q, M, D, DX, and L.

Load Address Register 1: LAR1

When the *LAR1* operation is coded without a parameter, as shown in the following example, the 32-bit pointer currently contained in ACCU 1 is loaded to address register AR1. Like the standard load operation, the instruction is executed irrespective of the RLO and without affecting the status bits.

Example: Load address register 1 with the contents of accumulator 1

L	P#6.7	// Load to accumulator 1 a pointer value to the address location 6.7
LAR1		// Load AR1 with the pointer value (P#6.7) previously loaded to accumulator 1

Load Address Register 1: LAR1

The operation **LAR1**, as shown above, supports a parameter entry of either a pointer constant or a double word location that contains a 32-bit pointer. When processed, the statement loads address register AR1 with the pointer constant referenced directly in the statement, or with a 32-bit pointer contained in a specified double word location. Like the standard load operation, the instruction is executed irrespective of the RLO and without affecting the status bits.

Example: Load address register 1 with the specified pointer constant: P#Abyte.bit or P#byte.bit

LAR1	P#14.2	// Load AR1 with the area-internal pointer
LAR1	P#M100.0	// Load AR1 with the area-crossing pointer
LAR1	P##SRCE	// Load AR1 with the area-crossing pointer to the local variable **SRCE**

In the following example, address register 1 is loaded from bit memory double word MD42. Valid memory locations from which the pointer can be stored include bit memory, data block memory, and local memory—for example, DBD32, DID32, MD42, and LD44.

Example: Load address register 1 with the contents of the specified DWORD variable

LAR1	MD42	// Load AR1 with the pointer contained in memory double word MD42

Load Address Register 1 with AR2: LAR1 AR2

When processed, the statement **LAR1 AR2**, as shown above, loads address register AR1 with the 32-bit pointer currently contained in address register 2. Like the standard load operation, the instruction is executed irrespective of the RLO and without affecting the status bits.

Example: Load address register 1 with the contents of the address register 2

LAR1	AR2	// Load address register 1 with the contents of address register 2
+AR1	P#27.4	// Add to address register 1, the specified area-internal

Transfer address register 1: TAR1

When processed, the statement **TAR1**, as shown below, transfers the address register AR1 contents to ACCU 1. Like the standard transfer operation, the instruction is executed irrespective of the RLO and without affecting the status bits.

Example: Transfer the contents of address register 1 to accumulator 1

TAR1		// Transfer AR1 contents to accumulator 1
T	MD42	// MD42 copy pointer from accumulator 1 to bit memory address MD42

Transfer Address Register 1: TAR1

The operation **TAR1**, with parameter, supports a double word location to which a 32-bit pointer can be moved. Valid memory areas include bit memory (MD), data block memory (DBD and DID), and local memory (LD). Examples include MD 42, DBD12, DID14, and LD44.

When processed, the content of address register AR1 is transferred to the specified double word memory location. In the following example the contents of AR1 are moved to bit memory double word MD42. Like the standard transfer operation, the transfer AR1 operation is executed irrespective of the RLO and without affecting the status bits. ACCU 1 and ACCU 2 remain unchanged

Example: Transfer the contents of address register 1 to a double word location

TAR1	MD42	// Transfer the contents of AR1 to memory double word MD42

Transfer Address Register 1 to AR2: TAR1 AR2

When processed, the statement **TAR1 AR2** transfers the 32-bit pointer of address register AR1 to address register AR2. Like the standard load operation, the instruction is executed irrespective of the RLO and without affecting the status bits.

Example: Transfer the contents of address register 1 to address register 2

TAR1	AR2	// Transfer the contents of AR1 to address register AR2

Exchange AR1 and AR2: CAR

When processed, the statement **CAR**, as shown above, swaps the contents of AR1 and AR2. The contents of address register AR1 are moved to address register AR2 and the contents of address register AR2 are moved to address register AR1. Like other digital operations, the CAR operation is executed irrespective of the RLO and without affecting the S7 status bits.

Example: Swap the contents of address register AR1 and address register AR2

CAR	// Exchange (swap) address register AR1 with address register AR2

Add offset to Address Register 1: +AR1 <P#byte.bit>; +AR1 <P#Abyte.bit>; +AR1

The +**AR1** operation adds to the pointer value of AR1, a byte.bit offset value taken from the low-order word of accumulator 1, or specified as a statement parameter—as in the example, +AR1 **P#byte.bit**.

31	30	29	28	27	26	25	24	23	22	21	20	19	18	17	16	15	14	13	12	11	10	9	8	7	6	5	4	3	2	1	0
1	0	0	0	0	0	0	0	0	0	0	1	0	0	0	0	0	0	0	0	0	0	0	0	0	0	1	0	0	0	0	0
Bits 24, 25, 26 = memory area ID								(16-bit byte offset)																		Bit 0-7					

Figure 17-3. Double word pointer format of register indirect addressing—with area-crossing

Although the byte range of the pointer parameter P#byte.bit is 0-4095, the 16-bit unsigned integer byte offset value of Figure 17-3, has a valid byte range of 0-65535. The 16-bit integer is initially expanded to 24 bits with its correct sign, and then added to the right-most 24 bits of AR1. The memory area ID, encoded in bits 24 through 26 of AR1, is unchanged.

In the following example, the 16-bit integer offset value is taken from the low-order word of ACCU1.

Example: Add byte offset to the pointer value contents of AR1, where the offset value is in ACCU1

| OFST | L | +16 | // Load the integer value +16 to the low-order word of ACCU1 |
| | +AR1 | | // Add the current value from the low-order word of ACCU1 to AR1 |

In the following example, the byte offset value is taken from the specified pointer constant.

Example: Add byte offset to the pointer value of AR1, where the offset is stated as a parameter.

| OFST | +AR1 | P#16.0 | // Add the pointer offset value of 16.0 to AR1 |

Add offset to Address Register 2: +AR2 <P#byte.bit>; +AR2 <P#Abyte.bit>; +AR2

The **+AR2** operation adds to the pointer value of AR2, a byte offset value taken from the low-order word of accumulator 1 or specified as a statement parameter—as in the example, +AR2 **P#byte.bit**.

Although the byte range of the pointer parameter P#byte.bit is 0-4095, the 16-bit unsigned integer byte offset value of Figure 17-3, has a valid range of 0-65535. The 16-bit integer is initially expanded to 24 bits with its correct sign, and then added to the right-most 24 bits of AR2. The memory area ID, encoded in bits 24 through 26 of AR2, is unchanged.

In the following example, the 16-bit integer offset value is taken from the low-order word of ACCU1.

Example: Add byte offset to the pointer value of AR2, where the offset value is in ACCU1

| OFST | L | +32 | // Load the integer value +32 to the low-order word of ACCU1 |
| | +AR2 | | // Add the current value from the low-order word of ACCU1 to AR2 |

In the following example, the 16-bit integer offset value is taken from the low-order word of ACCU1.

Example Add byte offset to the pointer value of AR2, where the offset is stated as a parameter.

| OFST | +AR2 | P#32.0 | // Add the pointer offset value of 32.0 to AR2 |

ACCUMULATOR OPERATIONS

STL accumulator instructions are those operations that may be performed on the values contained in the accumulators once they have been loaded using load operations. Depending on the S7 CPU, there may be two accumulators or three accumulators in use. The load operation always places a value in accumulator 1—and a subsequent load moves the contents of accumulator 1 to accumulator 2. In this latter case, the original contents of accumulator 2 are lost. The following operations allow data to be moved between ACCU 1 and ACCU2 or among all 4 accumulators when available—as in the S7-400.

Table 17-5. STL Accumulator Operations Summary

Operation	Brief Description
POP	**POP 2-Accumulators.** In a 2-accumulator CPU, the entire content of ACCU 2 is copied to ACCU 1. ACCU 2 remains unchanged, and status bits are not affected.
POP	**POP 4-Accumulators.** In a 4-accumulator CPU, the entire content of ACCU 2 is copied to ACCU 1, the contents of ACCU 3 to ACCU 2, and the contents of ACCU 4 to ACCU 3. ACCU 4 is left unchanged, and the CPU status bits are unaffected.
PUSH	**Push ACCU 1 to ACCU 2.** In a 2-accumulator CPU, the entire content of ACCU 1 is copied to ACCU 2. ACCU 1 is left unchanged, and the CPU status bits are unaffected.
PUSH	**Push ACCU 1 through ACCU 4.** In a 4-accumulator CPU, the content of ACCU 3 is copied to ACCU 4, the contents of ACCU 2 to ACCU 3, and the contents of ACCU 1 to ACCU 2. ACCU 1 is left unchanged, and the CPU status bits are unaffected.
TAK	**Toggle Accumulators.** Interchange or swap the contents of ACCU 1 and ACCU 2.
ENT	**Enter Accumulator Stack.** Copy the contents of ACCU 3 to ACCU 4 and the contents of ACCU 2 to ACCU 3. If the ENT operation is placed directly before a load instruction, the previous result is saved in ACCU 3. The saved result is retrieved back to ACCU 2 after the next arithmetic operation.
LEAVE	**Leave Accumulator Stack.** Copy the contents of ACCU 3 into ACCU 2 and the contents of ACCU 4 into ACCU 3. The contents of ACCU 1 and ACCU 4 are left unchanged.
CAW	**Exchange Accumulator Word.** Reverse the two bytes (n and n+1) in ACCU 1.
CAD	**Exchange Accumulator Double Word.** Reverse the four bytes of ACCU 1.
NOP 0	**No Operation.** No operation is performed when this instruction is executed. The subsequent operation is processed.
NOP 1	**No Operation.** No operation is performed when this instruction is executed. The subsequent operation is processed.

PROGRAMMING ACCUMULATOR OPERATIONS

Accumulator operations will likely prove useful when you need to shuffle several data elements in a data processing transaction, or to retain intermediate results in a complex arithmetic calculation.

Processing Accumulator Operations

Accumulator operations, like other digital operations, are executed irrespective or without regard to the result of logic operations. Since these operations will always be processed, unless circumvented, you will have to employ jumps or other mechanisms if, in a given block, these operations should not be processed under certain conditions.

POP (2-Accumulator CPUs): POP

When the **POP** operation is processed in 2-accumulator CPUs, the entire content of ACCU 2 is copied to ACCU 1. ACCU 2 is unaltered, and the CPU status bits are not affected.

The POP instruction is illustrated below, for CPU's with two accumulators.

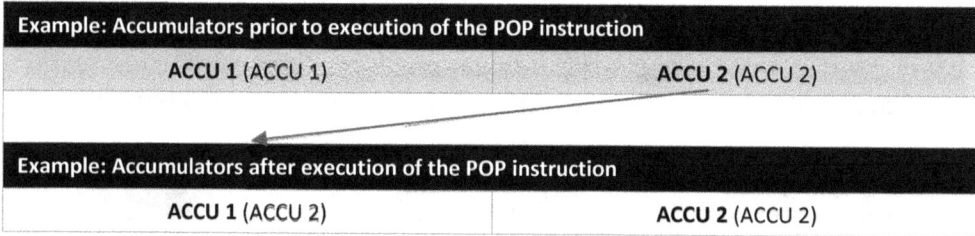

Example: Accumulators prior to execution of the POP instruction	
ACCU 1 (ACCU 1)	ACCU 2 (ACCU 2)

Example: Accumulators after execution of the POP instruction	
ACCU 1 (ACCU 2)	ACCU 2 (ACCU 2)

Figure 17-4. Illustration of POP instruction (2-accumulator CPU)

POP (4-Accumulator CPUs): POP

When the **POP** operation is processed in 4-accumulator CPUs, the entire content of ACCU 2 is copied to ACCU 1, the contents of ACCU 3 is copied to ACCU 2, and the contents of ACCU 4 is copied to ACCU 3. ACCU 4 is unaltered, and the CPU status bits are not affected.

The POP instruction is illustrated below, for CPU's with four accumulators.

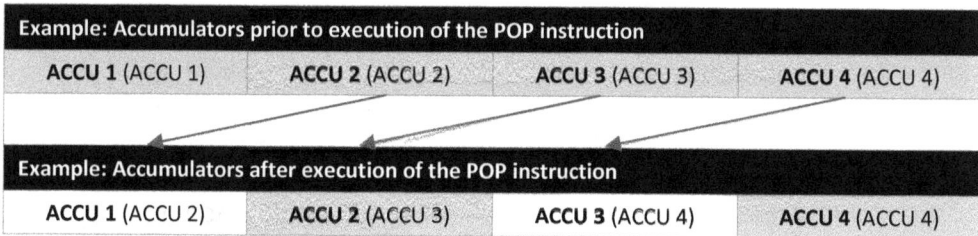

Example: Accumulators prior to execution of the POP instruction			
ACCU 1 (ACCU 1)	ACCU 2 (ACCU 2)	ACCU 3 (ACCU 3)	ACCU 4 (ACCU 4)

Example: Accumulators after execution of the POP instruction			
ACCU 1 (ACCU 2)	ACCU 2 (ACCU 3)	ACCU 3 (ACCU 4)	ACCU 4 (ACCU 4)

Figure 17-5. Illustration of POP instruction (4-accumulator CPU)

Push (2-Accumulator CPUs): PUSH

When the **PUSH** operation is processed in 2-accumulator CPUs, the entire contents of ACCU 1 is copied to ACCU 2. ACCU 1 is unaltered, and the CPU status bits are not affected.

The PUSH instruction is illustrated below, for CPU's with two accumulators.

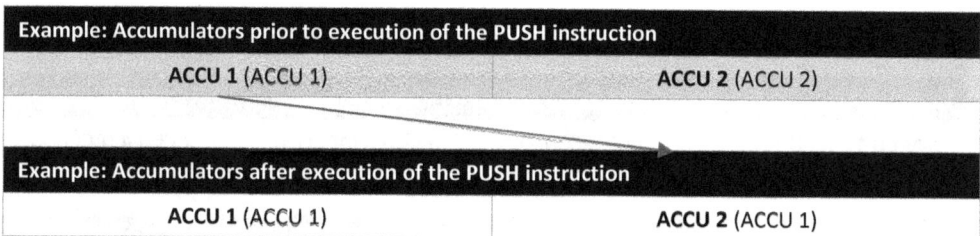

Example: Accumulators prior to execution of the PUSH instruction	
ACCU 1 (ACCU 1)	ACCU 2 (ACCU 2)

Example: Accumulators after execution of the PUSH instruction	
ACCU 1 (ACCU 1)	ACCU 2 (ACCU 1)

Figure 17-6. Illustration of PUSH instruction (2-accumulator CPU)

Push (4-Accumulator CPUs): PUSH

When the **PUSH** operation is processed in 4-accumulator CPUs, the entire contents of ACCU 3 is copied to ACCU 4, the contents of ACCU 2 is copied to ACCU 3, and the contents of ACCU 1 is copied to ACCU 2. ACCU 1 is unaltered, and the CPU status bits are not affected.

The PUSH instruction is illustrated below, for CPU's with four accumulators.

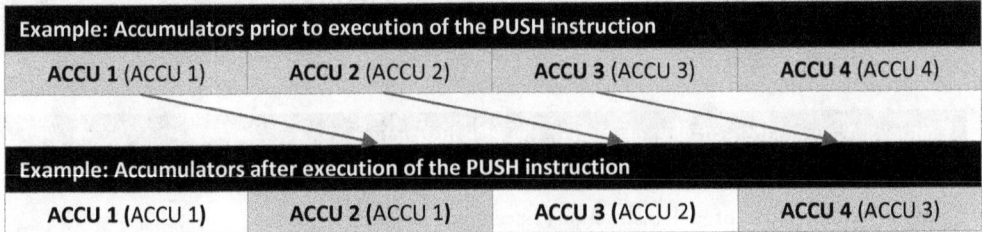

Example: Accumulators prior to execution of the PUSH instruction			
ACCU 1 (ACCU 1)	**ACCU 2** (ACCU 2)	**ACCU 3** (ACCU 3)	**ACCU 4** (ACCU 4)

Example: Accumulators after execution of the PUSH instruction			
ACCU 1 (ACCU 1)	**ACCU 2** (ACCU 1)	**ACCU 3** (ACCU 2)	**ACCU 4** (ACCU 3)

Figure 17-7. Illustration of PUSH instruction (4-accumulator CPU)

Toggle Accumulators: TAK

When the **TAK** operation is processed, the 32-bits of accumulator 1 are exchanged with the 32-bits of accumulator 2. Other CPU accumulators are unaltered, and the status bits are not affected.

The TAK instruction, illustrated below, always operates on accumulator 1 and accumulator 2.

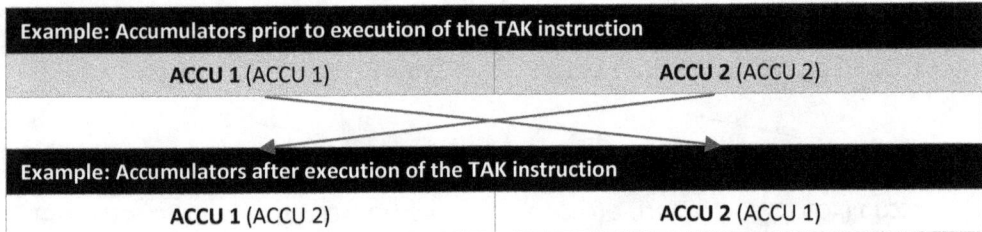

Example: Accumulators prior to execution of the TAK instruction	
ACCU 1 (ACCU 1)	**ACCU 2** (ACCU 2)

Example: Accumulators after execution of the TAK instruction	
ACCU 1 (ACCU 2)	**ACCU 2** (ACCU 1)

Figure 17-8. Illustration of TAK instruction

Enter Accumulator Stack: ENT

When the **ENT** operation is processed, the content of ACCU 3 is copied to ACCU 4 and the content of ACCU 2 is copied to ACCU 3—the contents of ACCU 1 and ACCU 2 are unaltered. When ENT is inserted prior to loading the next value in a calculation, the previous result is saved in accumulator 3. After the next arithmetic operation, ACCU 3 is moved back to ACCU2.

The ENT instruction, which operates on four accumulators, is illustrated below.

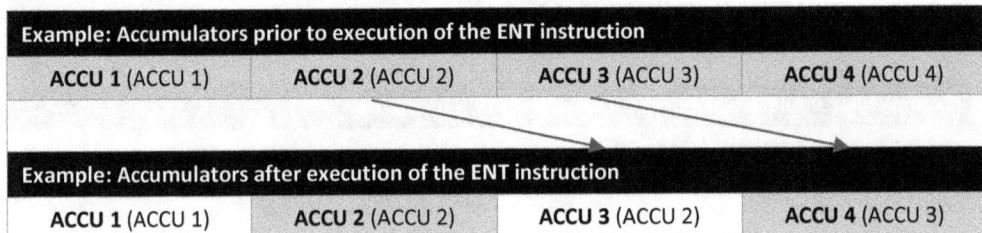

Example: Accumulators prior to execution of the ENT instruction			
ACCU 1 (ACCU 1)	**ACCU 2** (ACCU 2)	**ACCU 3** (ACCU 3)	**ACCU 4** (ACCU 4)

Example: Accumulators after execution of the ENT instruction			
ACCU 1 (ACCU 1)	**ACCU 2** (ACCU 2)	**ACCU 3** (ACCU 2)	**ACCU 4** (ACCU 3)

Figure 17-9. Illustration of ENT instruction (4-accumulator CPU)

Leave Accumulator Stack: LEAVE

When the **LEAVE** operation is processed, the contents of ACCU 3 is copied into ACCU 2 and the contents of ACCU 4 is copied into ACCU 3.

If you program the LEAVE instruction directly in front of a shift or rotate instruction, and combine the accumulators, then the leave instruction functions like an arithmetic instruction. The contents of ACCU 1 and ACCU 4 are unaltered.

The LEAVE instruction is shown below, for CPU's with four accumulators (ACCU 1 thru ACCU 4).

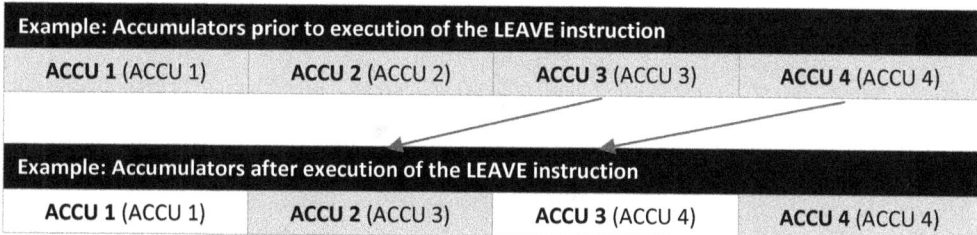

Example: Accumulators prior to execution of the LEAVE instruction			
ACCU 1 (ACCU 1)	**ACCU 2** (ACCU 2)	**ACCU 3** (ACCU 3)	**ACCU 4** (ACCU 4)

Example: Accumulators after execution of the LEAVE instruction			
ACCU 1 (ACCU 1)	**ACCU 2** (ACCU 3)	**ACCU 3** (ACCU 4)	**ACCU 4** (ACCU 4)

Figure 17-10. Illustration of LEAVE instruction (4-accumulator CPU)

Change Accumulator Word: CAW

When the **CAW** operation is processed, the 2-bytes of the low-order word of accumulator 1 are swapped. Byte-1 and byte-2 are swapped. The high-order bytes of accumulator 1 and the bytes of the remaining accumulators are not affected.

The CAW instruction is illustrated below, for the two low-order bytes of accumulator 1.

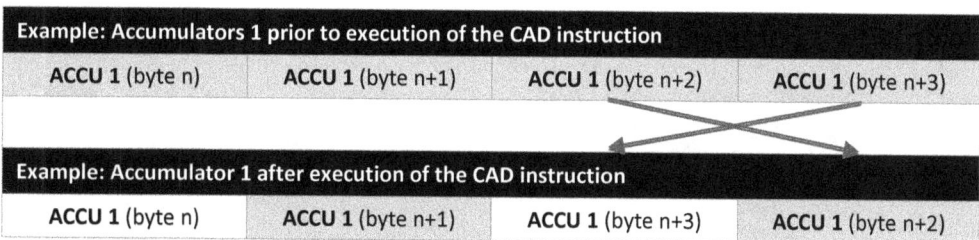

Example: Accumulators 1 prior to execution of the CAD instruction			
ACCU 1 (byte n)	**ACCU 1** (byte n+1)	**ACCU 1** (byte n+2)	**ACCU 1** (byte n+3)

Example: Accumulator 1 after execution of the CAD instruction			
ACCU 1 (byte n)	**ACCU 1** (byte n+1)	**ACCU 1** (byte n+3)	**ACCU 1** (byte n+2)

Figure 17-11. Illustration of CAW instruction

Change Accumulator Double Word: CAD

When the **CAD** operation is processed, the 4-bytes of accumulator 1 are interchanged. The low-order byte (byte n+3) is swapped with the high-order byte (byte n); byte-2 is swapped with byte-3. The bytes of the remaining accumulators are not affected.

The CAD instruction is illustrated below, for the 4-bytes of accumulator 1.

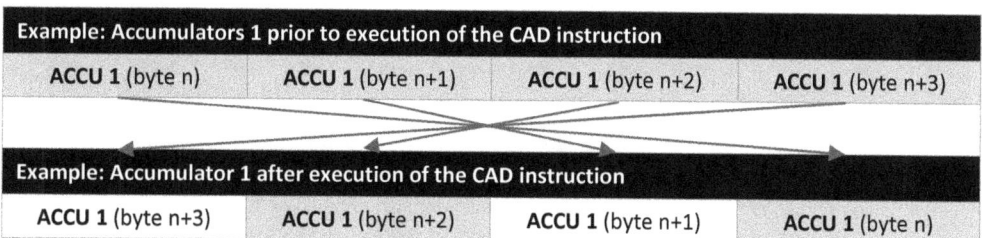

Example: Accumulators 1 prior to execution of the CAD instruction			
ACCU 1 (byte n)	**ACCU 1** (byte n+1)	**ACCU 1** (byte n+2)	**ACCU 1** (byte n+3)

Example: Accumulator 1 after execution of the CAD instruction			
ACCU 1 (byte n+3)	**ACCU 1** (byte n+2)	**ACCU 1** (byte n+1)	**ACCU 1** (byte n)

Figure 17-12. Illustration of CAD instruction

Chapter 18

STL Timer Operations

In this chapter, you'll be introduced to STL Timer Operations. As with LAD and FBD, STL supports the five S5 timer types.

Although timers have many applications, they are often used to enable or disable devices or process events after a timed delay. Timers might also be used to determine whether a device or process fault has occurred based on whether the timer has exceeded the preset delay, or if the timer has expired prematurely.

In addition to the STEP 7 timer instructions covered in this chapter, standard IEC timer functions are reviewed in Chapter 28 *Developing Functions and Function Blocks*. These callable Function Block operations, which include SFB 3 (TP—Pulse Generation), SFB 4 (TON—Timer On-Delay), and SFB 5 (TOF—Timer Off-Delay), are available in the STEP 7 library, and are integrated in some S7 CPUs.

Main Topics

- STL Timers Operations Overview
- Programming Timers in STL

STL TIMER OPERATIONS OVERVIEW

Statement List timer instructions for STEP 7 are listed in the table below. As in LAD and FBD, in STL there are five timer types including *on-delay (SD)*, *off-delay (SF)*, *stored on-delay (SS)*, *pulse (SP)*, and the *extended pulse (SE)*. In STL, a unique operation is used for starting each of these timer types. Like with the simple timers of LAD and FBD, each STL timer that you program is loaded with a preset, started, monitored, and reset using separate operations.

Although there are many applications for timers, they are generally used to enable or disable a device or process after a timed delay. Timers are also used to generate periodic pulses, as well as to determine whether a device or process fault has occurred. This latter application of timers is generally based on a timer that has exceeded the preset delay or that has expired prematurely.

Table 18-1. STL Timer Operations Summary

Operation	Brief Description
SD <Tn>	**Start ON Delay Timer:** When the preceding RLO transitions from 0-to-1, start the specified ON-Delay Timer <Tn> using the timer preset value previously loaded to ACCU 1.
SF <Tn>	**Start OFF Delay Timer:** When the preceding RLO transitions from 1-to-0, start the specified OFF-Delay Timer <Tn> using the timer preset value previously loaded to ACCU 1.
SS <Tn>	**Start Stored ON Delay Timer:** When the RLO transitions from 0-to-1, start the specified Stored ON-Delay Timer <Tn> using the timer preset value previously loaded to ACCU 1.
SP <Tn>	**Start Pulse Timer:** When the preceding RLO transitions from 0-to-1, start the specified Pulse Timer <Tn> using the timer preset value previously loaded to ACCU 1.
SE <Tn>	**Start Extended Pulse Timer:** When the preceding RLO transitions from 0-to-1, start the Extended Pulse Timer <Tn> using the timer preset value previously loaded to ACCU 1.
L <Tn>	**Load Timer Value as Decimal:** Load the current timer value (remaining time) of the specified timer as a decimal value. The value is in ACCU 1 for further processing.
LC <Tn>	**Load Timer Value as BCD:** Load the current timer value (remaining time) of the specified timer <Tn> as binary coded decimal (BCD). The value is in ACCU 1 for further processing.
R <Tn>	**Reset Timer:** When the RLO =1, reset the specified timer <Tn>. Reset the timer value and timer status.
FR <Tn>	**Enable Timer:** When the preceding RLO transitions from 0-to-1, enable the specified timer <Tn> to be restarted (re-triggered), provided the RLO = 1 for the associated timer inputs.

Note: Tn = Timer address—for example T0, T1, T2, and T3. ACCU1 = accumulator 1.

PROGRAMMING TIMERS IN STL

In STL, there are several operations associated with implementing the functions of a timer. These operations include loading a timer preset value or duration, starting the timer, monitoring the remaining time, checking the timer status, and resetting the timer—all of which are handled with separate operations. The following list shows the preferred sequence of entering these operations.

- Specify timer preset value

- Start a timer of specified type

- Reset the timer (R)

- Check timer status (binary check)

- Check remaining time value (in decimal)

- Check remaining time (in BCD)

Presetting and Starting a Timer

In STL, a timer is started based on the type of timer and on the logic you insert to start the timer. All timer types, except for the off-delay timer, start timing when the RLO immediately preceding the timer transitions from 0-to-1. The off-delay starts timing on a 1-to-0 transition of the preceding RLO.

The duration for which a timer will time, the timer preset value, is taken from accumulator 1 when the timer is started. How and when the timer preset is loaded to accumulator 1 is up to you, but must be loaded prior to starting the timer. The value may be loaded as a constant using the S5TIME constant format, as a declared variable of type S5TIME, or as a word-width memory location whose contents will be interpreted as the timer preset value. The preset value is loaded using the Load operation.

You may specify the timer's preset value as a constant, using the following S5T or S5TIME format where the range is S5T#0h_0m_0s_10ms to S5T#2h_46m_30s_0ms. The preset units of hour (h), minutes (m), seconds (s), and milliseconds (ms) are only specified if required. For example, the preset constant S5T#1m30s is 1 minute and 30 seconds.

Start Timer: SD <Tn>, SF <Tn>, SS <Tn>, SP <Tn>, SE <Tn>

Each of the five timers operates slightly different and is identified by a unique instruction. The start instruction for the *on-delay timer* is **SD**; the *off-delay timer* is **SF**; the *stored on-delay timer*, also called *retentive on-delay*, is **SS**; the *pulse timer* is **SP**; and the *extended pulse timer* is **SE**.

In STL, a timer is started based on the type of timer and on the logic you insert to trigger the timer. All timer types, except for the off-delay timer, start timing when the timer start impulse (RLO preceding the timer) transitions from 0-to-1. The off-delay starts on a 1-to-0 transition of the start impulse.

In the example below, the ON-delay timer T24 starts timing with a preset value of 500 milliseconds, when the combination of I 5.2 **OR** input I 5.3 transitions from '0' to '1'; the timer continues to time until the programmed delay expires, unless interrupted before the duration expires. A timer start function will always require an edge transition (0-to-1 or 1-to-0) at the RLO of the start input.

Example: Starting an ON-Delay timer

LTPV	L	S5T#500MS	// load timer preset of 500 milliseconds
	A	I 5.2	// timer start logic combines I 5.2 OR I 5.3
	O	I 5.3	// start on 0-to-1 transition of RLO
	SD	T 24	// start T24 as an ON-delay timer, using 500ms as the preset

Resetting a Timer

All timers, regardless of the type, are reset in the same fashion. Furthermore, the outcome of a reset is also the same in each of the timer types. In STL, a timer is reset based on the t logic you insert just prior to the reset operation **R**.

Reset Timer: R <Tn>

Whenever the RLO = 1, preceding the *reset timer* operation **R <Tn>**, the referenced timer will be reset. The remaining time value and the timer status are both reset to zero.

In the example below, the ON-delay timer T24 is reset on the rising edge of M133.7. Whenever I 5.4 transitions from '0' to'1', M133.7 goes high (1) for one scan and thereby resets timer T24.

Example: Resetting the current timer value

	A	I 5.4	// timer reset input
	FP	M133.7	// reset input on leading edge of I 5.4
RT24	R	T24	// reset timer T 24 on RLO = 1

Checking the Timer Status

The status of a timer is a binary operand that may be checked to determine whether a timer is enabled and still timing, or if the delay has expired. The result of the check may be combined with other binary operations such as AND, OR, and Exclusive OR functions to produce required control logic. Since each timer behaves differently, an intermediate check of the timer status, and whether it will yield '1' or '0' when checked, will depend on the specific timer type and at what point the check is made.

Later in this chapter, you can view the timing diagrams of each timer type—from these diagrams you may determine the response of the status throughout the timer's operating cycle.

Check Timer Status: A <Tn>, AN <Tn>, O <Tn>, ON <Tn>, X <Tn>, and XN <Tn>

The status of a timer is checked using binary check operations **A**, **AN**, **O**, **ON**, **X**, and **XN**. Using the required binary check instruction, the status of a specific timer is accessed by referencing the timer address as the operand address—for example AN T33. You may check the timer status for '0' using AN, ON, or XN; the status is checked for '1' using operations A, O, and X.

In the example below, if the on-delay timer T24 is expired, and the input image address I 5.4 (timer start input) is '1', then the output image address Q27.5 is activated. If the input image address I 5.4 (timer start input) is '1', and T24 is not yet expired (timer running), then the output image address Q27.6 is activated; Q 27.6 is de-activated in the cycle immediately following the timer expiration.

Example: Checking the timer status

	A	I 5.4	// check if the start input for T24 is enabled
	A	T 24	// check if the programmed delay for T24 is expired
	=	Q 27.5	// signal timer T24 is timed out
	AN	T 24	// check if status of T24 is '0'
	A	I 5.4	// check if the start input for T24 is enabled
	=	Q 27.6	// indicator for timer T24 running

Checking the Remaining Time

When you write code to check the timing progress while the timer is running, you will access the timer word directly, using one of two Load instructions—each operation references the timer address. Since S5 timers time down, the reported value will be the remaining time. The following two cases illustrate how the remaining time is accessed in STL as either a decimal value or as a BCD value.

Load Timer Value as Decimal: L <Tn>

Using the *load timer value as decimal* operation **L <Tn>**, and referencing the timer address as shown in the example below, you may access the remaining time as an integer (decimal) value. After the load, the remaining time is in accumulator 1—available for further processing.

In the example, the remaining time is transferred to bit memory word MW44. An intermediate check of the remaining time would involve comparing the value of MW44 to an integer variable or constant.

Example: Checking the remaining time as a decimal value

L	T 24	// load the remaining time of timer T24 as a decimal value
T	MW44	// store the remaining time of T24 to bit memory location MW44

Load Timer Value as Binary Coded Decimal: LC <Tn>

Using the *load timer value as BCD* operation (**LC**), and referencing the timer address, as shown in the example below, you may access the remaining time as a binary coded decimal (BCD) value. After the LC operation, the remaining time is a binary coded decimal value in accumulator 1—available for further processing.

In the example, the remaining time is transferred to output image word QW54. QW54 is output to a 3-digit BCD display.

Example: Checking the remaining time as a binary coded decimal value

LC	T 24	// load the remaining time of timer T33 as a BCD value
T	QW 54	// store the remaining time of T24 to output word QW 54 (3-digit BCD)

Enabling a Timer

The *enable operation* **FR** is not required for normal timer operation; however in certain applications, you may find it useful for the modified behavior that it provides. When triggered, this operation supports the restarting of a timer, without the normal edge transition required to start a timer. Depending on the specific timer and the point of the enable transition, the timer is started/restarted with the programmed delay or with the remaining time.

Enable Timer: FR <Tn>

Whenever the RLO preceding the *enable timer* operation **FR <Tn>** transitions from '0' to '1', the enable is active and the start operation of the specified timer is enabled for restart during the current program cycle. While the enable is active, normal timer start operations of SD, SF, SS, SP, and SE will execute without a positive edge—the only requisite is that the preceding RLO = 1 for these operations.

If the enable operation is used, it, along with the associated trigger logic, is placed prior to any other of the timer operations. An example of the enable timer logic is shown on the following page.

Example: Enabling a timer for retriggering

A	I 2.1	// on 0-to-1 transition, activate the enable operation
FR	T24	// enable timer T24 for restart
L	S5T#10s	// load timer preset as 10 seconds
A	I 5.2	//start input for timer T24 when enable is active
SD	T24	// Start/restart timer T24 as an ON-delay timer

Full Implementation of Timers in STL

Full implementation of timer functionality refers to use of STL operations to develop the LAD/FBD box equivalent of an S5 timer. The full implementation is best described in terms of the functional parameters of the LAD/FBD box timer, since each timer function is based upon an STL operation.

As with other LAD/FBD box operations implemented in STL, the sequence of STL operations is visually derived by taking each box parameter from top to bottom and left to right. Box timer parameters, as shown in order below, include the timer address (**???**); a start input (**S**), to start the timer operation; a preset value (**TV**), which defines the timing delay; a reset input (**R**), which resets the timer status and current value; a binary status flag (**Q**), which indicates the response of the timer; and the remaining time as a decimal (**BI**) value and binary coded decimal (**BCD**) value.

Table 18-2. LAD/FBD Timer box parameters. **Note:** Entry of all timer parameters is not required.

	Parameter	Type	Description	
❶	???		Timer Address < Tn > to Start	
❷	S		Start Input	
❸	TV	S5TIME	Timer Preset Value	
❹	R	BOOL	Reset Input	
❺	Q	BOOL	Timer Status Flag	
❻	BI	WORD	Remaining Time in Binary (Decimal)	
❼	BCD	WORD	Remaining Time in Binary Coded Decimal (BCD)	

Implementing the Basic Timer Functions

Before moving to the description and full programming of each timer type, we need to show how the STL timer operations correspond directly to the functions of a LAD/FBD box timer. This correlation is the same for each timer type. First, it's important to note that the previously described STL enable operation that supports the retriggering of a timer is not required nor implemented in LAD/FBD.

Excluding the timer enable function, the basic functions of the box timer can be summarized as four operations—❶ given the appropriate start input, start a specific timer address with a specific programmed duration; ❷ given a specific reset input, reset the timer; ❸ check the binary status of the timer to determine if it is enabled, still timing, or if the programmed delay has expired; and ❹ get the remaining time value and store it for further use as a decimal value or as a binary coded decimal value. Each of these functions is found in the full programming of each timer type. The functions are shown highlighted on the following page in the STL implementation of the on-delay timer.

Full Programming—ON-Delay Timer

If the RLO preceding the start input ❶ transitions from 0-to-1, then the *on-delay timer* starts timing, with the preset value loaded from accumulator 1. The timer will run while the start-input remains at '1', and continue until the preset time value expires. When the delay expires, the timer status is activated (on-delay) and will remain activated while the RLO at start input remains '1'. If the RLO at the start input transitions from 1-to-0, while the timer is running, the timer will stop.

Whenever the RLO = 1 preceding the reset input❷, both the timer status and the remaining time are reset to zero. When reset, a check of the timer status ❸ will yield '0' if checked for '1' and will yield '1' if checked for '0'. After a reset, the remaining time value is also reset to zero.

Prior to the time delay expiring, a check of the on-delay timer status ❸ for '1' will always yield '0' and a check for '0' yields '1'; when the timer has expired, a check for '1' will yield '1' while the RLO = 1 at the start-input; a check for '0' will yield '0'.

While the on-delay timer is timing, the remaining time ❹ may be loaded as a decimal value using the statement **L** T<n>, or as a binary coded decimal value, using the statement **LC** T<n>.

The on-delay timer is programmed below, and illustrated in the timing diagram of Figure 18-1.

Example: Full programming of the on-delay (SD) timer

Network 1:				**Box Timer—SD**
❶	A	I 2.1	// start input **(S)**	???
	L	S5T#10s	// load timer preset **(TV)** of 10 seconds	S_ODT
	SD	T1	// start timer T1 **(???)** as an **on-delay** timer	—S Q—
❷	A	I 3.1	// reset input **(R)**	—TV BI—
	R	T1	// reset timer T1	—R BCD—
❸	A	T1	// check status of timer T1 for '1' (expired)	
	=	Q 4.1	// timer status signal **(Q)**	
❹	L	T1	// load remaining time as binary (decimal) value	
	T	MW10	// store decimal time value **(BI)**	
	LC	T1	// load remaining time as binary coded decimal	
	T	MW12	// store BCD time value **(BCD)**	

The behavior of the on-delay timer is illustrated in the following timing diagram.

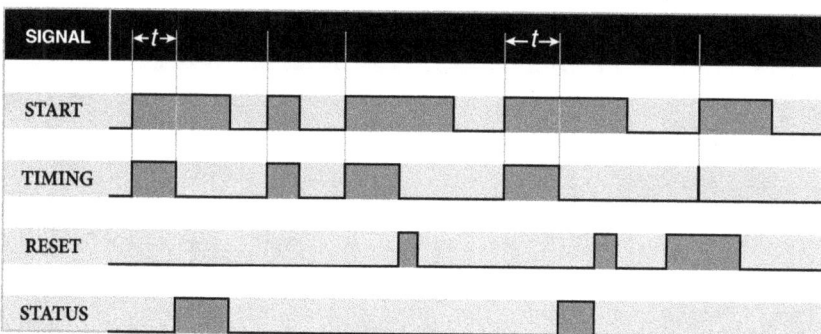

Figure 18-1. Timing diagram for the S5 ON-delay timer

Full Programming—OFF-Delay Timer

If the RLO preceding the start input ❶ transitions from 0-to-1, then the *off-delay timer* status is activated (1). If the start input transitions from '1' to '0', the timer starts with the preset value loaded from accumulator 1. While the start input remains at '0', the timer will run and continue with the timer status activated until the preset time expires. When the preset delay time expires, the off-delay timer status is de-activated (0). If the RLO preceding the start-input transitions from '0' to '1', while the timer is timing, the timer is reset and the status is de-activated.

Whenever the RLO = 1 preceding the reset input❷, both the timer status and the remaining time are reset to zero. When reset, a check of the timer status ❸ will yield '0' if checked for '1', and yields '1' if checked for '0'. After a reset, the remaining time value is also reset to zero.

Prior to the time delay expiring, a check of the off-delay timer status ❸ for '1' will yield '1', while the start input is '1' or while the timer is timing; and a check for '0' yields '1'; when the timer has expired, a check for '1' yields '0'; a check for '0' yields '1'.

While the off-delay is timing, the remaining time ❹ may be loaded as a decimal value using the statement **L** T<n>, or as a binary coded decimal value, using the statement **LC** T<n>. The off-delay timer is programmed below, and illustrated in the timing diagram of Figure 18-2.

Example: Full programming of the off-delay (SF) timer

Network 1:				Box Timer—SF
				???
	A	I 2.2	// start input (**S**)	S_OFFDT
❶	L	S5T#10s	// load timer preset (**TV**) of 10 seconds	—S Q—
	SF	T2	// start timer T2 (**???**) as an **off-delay** timer	—TV BI—
❷	A	I 3.2	// reset input (**R**)	—R BCD—
	R	T2	// reset timer T2	
❸	A	T2	// check status of timer T2 for '1'	
	=	Q 4.2	// timer status signal (**Q**)	
❹	L	T2	// load remaining time as binary (decimal) value	
	T	MW20	// store decimal time value (**BI**)	
	LC	T2	// load remaining time as binary coded decimal	
	T	MW22	// store BCD time value (**BCD**)	

The behavior of the off-delay timer is illustrated in the following timing diagram.

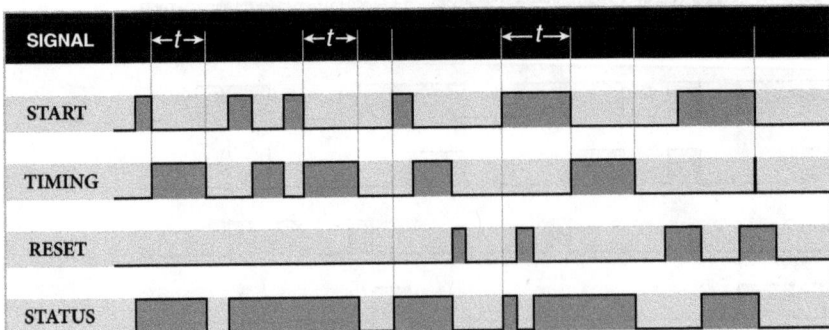

Figure 18-2. Timing diagram for the S5 OFF-delay timer

Full Programming—Stored (Retentive) ON-Delay Timer

If the RLO preceding the start input ❶ transitions from 0-to-1, then the *retentive on-delay timer* starts timing, with the preset value loaded from accumulator 1. The timer will run and continue until the preset time value expires, even if the start-input transitions from '1' to '0' before the timer expires. When the preset delay expires, the retentive on-delay timer status is activated and will remain activated until the timer is reset. If the start-input transitions from '1' to '0' while the timer is timing, and again from '0' to '1', the timer restarts and continues until the programmed delay expires.

Whenever the RLO = 1 preceding the reset input❷, both the timer status and the remaining time are reset to zero. When reset, a check of the timer status ❸ will yield '0' if checked for '1', and will yield '1' if checked for '0'. After a reset, the remaining time value is also reset to zero.

Prior to the time delay expiring, a check of the retentive on-delay timer status ❸ for '1' will yield '0', and a check for '0' yields '1'; when the timer has expired, a check for '1' yields '1' while the RLO = 0 at the reset input; a check for '0' yields '0'.

While the stored on-delay timer is timing, the remaining time ❹ may be loaded as a decimal value using the statement **L** T<n>, or as a binary coded decimal value, using the statement **LC** T<n>. The retentive on-delay timer is programmed below, and illustrated in the timing diagram of Figure 18-3.

Example: Full programming of the retentive (SS) on-delay timer

				Box Timer—SS
Network 1:				???
❶	A	I 2.3	// start input (S)	S_ODTS
	L	S5T#10s	// load timer preset (TV) of 10 seconds	—S Q—
	SS	T3	// start timer T3 (???) as a **retentive on-delay** timer	—TV BI—
❷	A	I 3.3	// reset input (R)	—R BCD—
	R	T3	// reset timer T3	
❸	A	T3	// check status of timer T3 for '1'	
	=	Q 4.3	// timer status signal (Q)	
❹	L	T3	// load remaining time as binary (decimal) value	
	T	MW30	// store decimal time value (BI)	
	LC	T3	// load remaining time as binary coded decimal	
	T	MW32	// store BCD time value (BCD)	

The behavior of the stored on-delay (retentive) timer is illustrated in the following timing diagram.

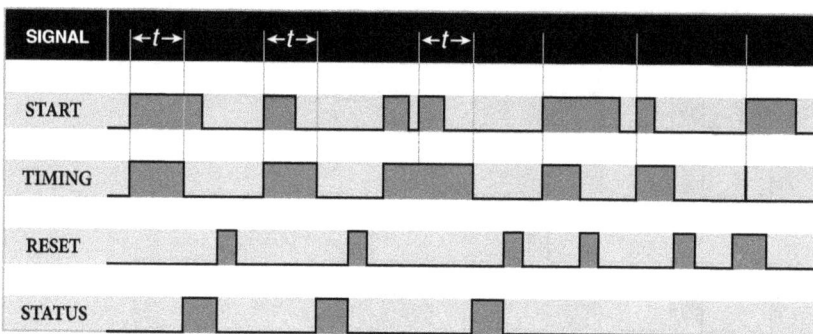

Figure 18-3. Timing diagram for the S5 Stored ON-delay timer

Full Programming—Pulse Timer

If the RLO preceding the start input ❶ transitions from 0-to-1, then the *pulse timer* starts timing, with the preset value loaded from accumulator 1. The timer will run while the start-input remains at '1', and continue until the preset time value expires. The pulse timer status is activated (1) for as long as RLO = 1 at the start-input and the time delay has not expired. If the start-input returns to '0', while the timer is timing, the timer stops and the timer status is de-activated (pulse—follows start input).

Whenever the RLO = 1 preceding the reset input❷, both the timer status and the remaining time are reset to zero. When reset, a check of the timer status ❸ will yield '0' if checked for '1', and will yield '1' if checked for '0'. After a reset, the remaining time value is also reset to zero.

Prior to the time delay expiring, a check of the pulse timer status ❸ for '1' will yield '1' while the RLO = 1 at the start-input, and a check for '0' will yield '0'; when the timer has expired, or the RLO = 0 at the start input, a check for '1' will yield '0'; a check for '0' will yield '1'.

While the pulse timer is timing, the remaining time ❹ may be loaded as a decimal value using the statement **L** T<n>, or as a binary coded decimal value, using the statement **LC** T<n>.

The pulse timer is programmed below, and illustrated in the timing diagram of Figure 18-4.

Example: Full programming of the pulse (SP) timer

				Box Timer—SP
Network 1:				???
	A	I 2.4	// start input (**S**)	S_PULSE
❶	L	S5T#10s	// load timer preset (**TV**) of 10 seconds	−S Q−
	SP	T4	// start timer T4 (**???**) as a **pulse** timer	−TV BI−
❷	A	I 3.4	// reset input (**R**)	−R BCD−
	R	T4	// reset timer T4	
❸	A	T4	// check status of timer T4 for '1'	
	=	Q 4.4	// timer status signal (**Q**)	
❹	L	T4	// load remaining time as binary (decimal) value	
	T	MW40	// store decimal time value (**BI**)	
	LC	T4	// load remaining time as binary coded decimal	
	T	MW42	// store BCD time value (**BCD**)	

The behavior of the pulse timer is illustrated in the following timing diagram.

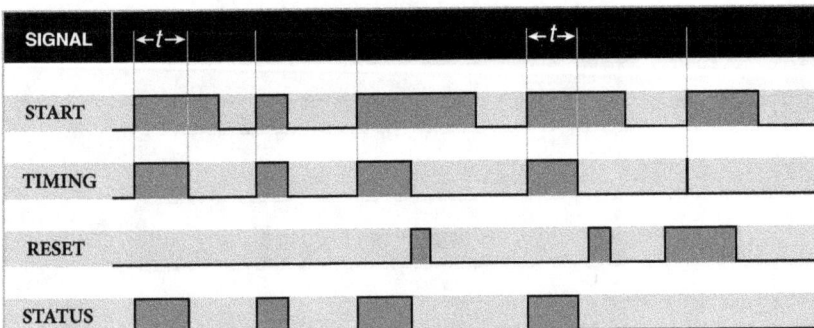

Figure 18-4. Timing diagram for the S5 PULSE timer

Full Programming—Extended Pulse Timer

If the RLO preceding the start input ❶ transitions from 0-to-1, then the *extended pulse timer* starts timing, with the preset value loaded from accumulator 1. The timer will run regardless of any signal change at the start-input and continue until the preset time value expires. From the time the start-input transitions from 0-to-1, the extended pulse timer status is activated (1) until the time delay expires (extended pulse). If while the timer is running, the start-input again transitions from 0-to-1, the timer will restart and continue until the preset delay expires.

Whenever the RLO = 1 preceding the reset input❷, both the timer status and the remaining time are reset to zero. When reset, a check of the timer status ❸ will yield '0' if checked for '1', and will yield '1' if checked for '0'. After a reset, the remaining time value is also reset to zero.

Prior to the time delay expiring, a check of the extended pulse timer status ❸ for '1' will yield '1', and a check for '0' yields '0'; when the timer has expired, a check for '1' will yield '0'; a check for '0' yields '1'.

While the extended pulse timer is timing, the remaining time ❹ may be loaded as a decimal value using the statement **L** T<n>, or as a binary coded decimal value, using the statement **LC** T<n>.

The extended pulse timer is programmed below, and illustrated in the timing diagram of Figure 18-5.

Example: Full Programming of the extended pulse (SE) timer

❶	A	I 2.5	// start input (S)	
	L	S5T#10s	// load timer preset (TV) of 10 seconds	
	SE	T5	// start timer T5 (???) as an **extended pulse** timer	
❷	A	I 3.5	// reset input (R)	
	R	T5	// reset timer T5	
❸	A	T5	// check status of timer T5 for '1'	
	=	Q 4.5	// timer status signal (Q)	
❹	L	T5	// load remaining time as binary (decimal) value	
	T	MW50	// store decimal time value (BI)	
	LC	T5	// load remaining time as binary coded decimal	
	T	MW52	// store BCD time value (BCD)	

Box Timer—SE ??? S_PEXT S Q TV BI R BCD

Network 1:

The behavior of the extended pulse timer is illustrated in the following timing diagram.

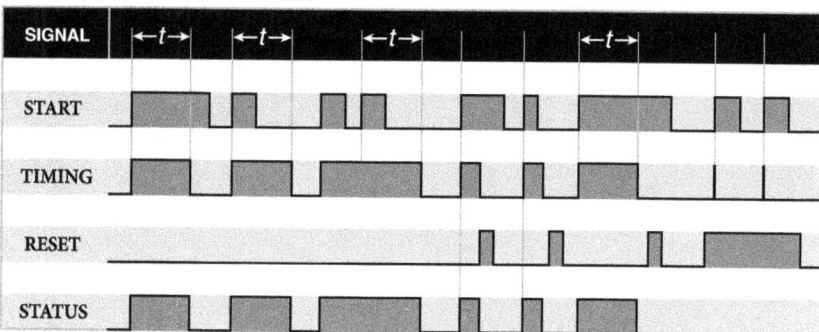

Figure 18-5. Timing diagram for the S5 EXTENDED PULSE timer

Excluding Timer Functions

When it comes to programming timer functions—not all of the operations used in the full timer implementation is required. The only operations that are absolutely needed are those instructions that load the timer preset and that start a specific timer type. You may decide which of the functions you actually need. For example, you may determine that explicit resetting of a timer is not always necessary. You determine, for example, whether you want to restart the timer on a new transition of the start input or after the timed duration expires. You may also decide to monitor the remaining time only in BCD.

In any case, if you always want timers to display in LAD/FBD, certain measures must be taken in your code. First, each timer must be placed in a separate network; also, you must insert a NOP for each timer function that you do not intend to implement. For example, to omit the reset operation and the remaining time in BCD, a NOP is substituted in place of these operations.

In the following implementation of an on-delay timer, the reset is omitted and the operations that load/transfer the remaining time as a BCD value are omitted. In each of these cases, a NOP operation replaces the omitted instruction. If the remaining operations of the timer are programmed as they would be in the full implementation, and the timer is the only code in the network, then the timer box will display in LAD/FBD.

Example: Omitting timer operations using NOP statement—Operations ❷ and ❹ omitted

				Box Timer—SD
Network 1:				
❶	A	I 2.1	// start input (**S**)	???
	L	S5T#10s	// load timer preset (**TV**) of 10 seconds	S_ODT
	SD	T1	// start timer T1 (**???**) as an on-delay timer	−S Q−
❷	NOP		// reset input (**R**)—**omitted**	−TV BI−
			// reset timer operation—**omitted**	−R BCD−
❸	A	T1	// check status of timer T1 for '1'	
	=	Q 4.1	// timer status signal (**Q**)	
❹	L	T1	// load remaining time as binary (decimal) value	
	T	MW10	// store decimal time value (**BI**)	
	NOP		// load remaining time as BCD—**omitted**	
			// store BCD time value—**omitted**	

Chapter 19

STL Counter Operations

In this chapter, you'll be introduced to STL Counter operations. As with LAD and FBD, UP, Down and UP/Down counters may be developed in STL.

Although counters have many program applications, they are generally used to enable or disable a control circuit or process event after a predetermined count is reached; or in production accounting and other event tracking applications.

In addition to standard STEP 7 counter operations covered in this chapter, standard IEC timer functions are reviewed in Chapter 28 *Developing Functions and Function Blocks*. These callable Function Block operations, which include SFB 0 (CTU—Up Counter), SFB 1 (CTD—Down Counter), and SFB 2 (CTD—Up/Down Counter), are available in the STEP 7 library, and are integrated in some S7 CPUs.

Main Topics

- STL Counter Operations Overview
- Programming Counters in STL

STL COUNTER OPERATIONS OVERVIEW

Statement List counter instructions are listed in the table below. In STL, there is the *up counter* and the *down counter*. The two operations may be used in conjunction to meet up/down counter requirements. Like with simple counters of LAD and FBD, each STL counter is loaded with a preset, incremented or decremented, monitored, and reset using separate operations.

Although counters have many process applications, they are generally used to enable or disable logic circuits or process events after reaching a preset count. Counters are also used in production accounting and other forms of event tracking.

Table 19-1. STL Counter Operations Summary

Operation	Brief Description
CU <Cn>	**Count UP.** When the RLO transitions from 0-to-1, increment the specified counter <Cn> by one count. A 0-to-1 transition has no effect after the count reaches a count of 999.
CD <Cn>	**Count Down.** When the RLO transitions from 0-to-1, decrement the specified counter <Cn> by one count. A 0-to-1 transition has no effect after the count reaches a count of 0.
S <Cn>	**Set Counter:** When the RLO transitions from 0-to-1, set the specified counter <Cn> using the count value from ACCU 1. The count value must be a BCD value between 0 and 999.
R <Cn>	**Reset Counter.** When the RLO =1, reset the specified counter <Cn> with a value of 0.
L <Cn>	**Load Counter Value as Decimal.** Load the current count value of the specified counter <Cn> as decimal. The value is left in ACCU 1 for further processing.
LC <Cn>	**Load Counter Value as BCD.** Load the current count value of the specified counter <Cn> as binary coded decimal (BCD). The value is left in ACCU 1 for further processing.
FR <Cn>	**Enable Counter.** When the RLO transitions from 0-to-1, enable operations of the specified counter <Cn> to be re-triggered, provided the RLO = 1 for the associated counter inputs.

Note: Cn = Counter address—for example C0, C1, C2, and C3.

PROGRAMMING COUNTERS IN STL

In STL, several operations are associated with implementing the functions of a counter. These operations include count up or down, load an initial preset value, check the current count, check the counter status, and reset a counter—all of which are handled with separate operations.

The preferred sequence of entering these operations is as shown below:

- Set Counter by specifying a counter preset value

- Count Up (CU) or Count Down (CD)

- Reset counter (R)

- Check counter status (binary check)

- Check counter value (in decimal)

- Check counter value (in BCD)

Setting a Counter Preset Value

A counter preset value is generally required with the down counter, but is used with the up counter if the counter must be re-initialized, or set to a specific count value. The counter preset value is always taken from accumulator 1. The preset value is loaded to accumulator 1 using the Load operation.

How or when the preset is loaded to the accumulator is irrelevant, but may be loaded from a word-width location, from a declared WORD variable, or directly loaded as a BCD constant (C#000-C#999), or as a hexadecimal constant (W#16#0000 –to- W#16#0999). If a word location—for example MW 100, or a WORD variable is used, then you may programmatically manipulate the preset value by writing new values to the specified word location or variable. In either case, for correct operation, a valid counter preset should be loaded to the accumulator prior to encountering a counter operation.

Set Counter: S <Cn>

The *Set Counter* operation **S <Cn>** is used to load a count value into the counter word address you specify. When the RLO immediately preceding the Set operation makes a 0-to-1 transition, the current contents of accumulator 1 is loaded as the new counter value. Valid values are from 000-999.

In the example below, when the logic combination of I5.3 OR M101.4 transitions from logic 0 to logic 1, then the counter C5 is loaded with a preset value of 100.

Example: Setting the current counter value

L	C#100	// load BCD value of 100 as a counter preset
A	I 5.3	// check I 5.3 for status of '1' ;
O	M 101.4	// check M 101.4 for '1' and combine with the RLO according to OR
S	C 5	// set counter value of C5 to 100.

Counting UP or Down

In STL, there are two operations that perform counting—*count up* (**CU**) and *count down* (**CD**) to the control logic that precedes the counter coil. The up counter will count up until reaching the maximum count value of 999; the down counter will count down until reaching the minimum count value of 000.

UP Counter: CU <Cn>

The *UP Counter* instruction CU will increment the counter value by +1 count whenever the RLO preceding the instruction makes a 0-to-1 transition. When the counter value has reached the maximum count of 999, the counter will not increment on subsequent 0-to-1 transitions of the RLO.

In the example below, whenever the logic combination of I5.2 AND NOT M3.7 transitions from logic 0 to logic 1, the up counter C5 is incremented by +1.

Example: Counting up

A	I 5.2	// check I 5.2 for status of '1';
AN	M 3.7	// check M 3.7 for '0' and combine with the RLO according to AND
CU	C 5	// increment counter C5 by +1 on positive edge transitions of RLO

DOWN Counter: CD <Cn>

The *Down Counter* operation **CD** will decrement the counter value by 1 whenever the RLO preceding the instruction makes a 0-to-1 transition. When the counter value has reached the minimum count of 000, the counter will not decrement on subsequent 0-to-1 transition of the RLO.

In the example below, whenever the logic combination of I5.2 AND NOT M3.7 transitions from logic 0 to logic 1, the down counter C5 is decremented by +1.

Example: Counting down

A	I 5.2	// check I 5.2 for status of '1';
AN	M 3.7	// check M 3.7 for '0' and combine with the RLO according to AND
CD	C 5	// decrement counter C5 by +1 on positive edge transitions of RLO

Resetting a Counter

A counter is reset using the standard reset instruction **R** and by referencing the counter address—for example C25. When a counter is reset, the current counter value and status are set to zero.

Reset Counter: R <Cn>

The reset operation is executed whenever the preceding RLO = 1. In the example, **C5** is reset whenever the RLO= 1 for the logic combination of **I** 5.4 OR M 101.4.

Example: Resetting the current counter value using the Reset operation

A	I 5.4	// check I 5.4 for status of '1' ;
O	M 101.4	// check M 101.4 for '1' and combine with the RLO according to OR
R	C 5	// reset counter value of C5 to 0.

Checking the Current Count

When writing code to check the current counter value, perhaps to compare to intermediate setpoints, you will need to access the counter word directly, using one of two Load instructions—each operation references the counter address as the source location. The following two operations illustrate how the current count is accessed in STL as either a decimal value or as a binary coded decimal value.

Load Count as Decimal Value: L <Cn>

Using the *load count as decimal value* operation **L** <Cn> and referencing the counter address, as shown above, you can access the current counter value as an integer (decimal) value. The current count after this load operation is a decimal value in accumulator 1. The value is available for further processing—for example to transfer the current count to bit memory word MW44.

In the following example, an intermediate check of the current count would mean comparing the counter value to an integer value.

Example: Checking the current count as a decimal value

L	C5	// get the current count of counter C5 as a decimal value
T	MW44	// store the current counter value to memory location MW44

Load Count as BCD Value: LC <Cn>

Using the *load count as BCD value* operation **LC <Cn>** and referencing the counter address, as shown, you can access the current count as a binary coded decimal value. The current count after this load operation is a BCD value in accumulator 1. The value is available for further processing—for example to transfer the current count to output image word QW54. QW54 will be output to a BCD display.

In the example below, an intermediate check of the remaining time would mean comparing the counter value to a BCD value.

Example: Checking the current count as a binary coded decimal value

LC	C5	// get the current count of counter C5 as a BCD value
T	QW54	// store the current counter value to output image location QW54

Checking the Counter Status

The status of a counter is a binary operand that you may check to determine whether a counter value is zero or greater than zero. The result of the check may be combined with other binary operations such as AND, OR, and Exclusive OR functions to produce the required control logic.

The status of the counter **C5,** for example, is '1' whenever the counter value is other than zero—otherwise the counter status is '0'.

Check Counter Status: A <Cn>, AN <Cn>, O <Cn>, ON <Cn>, X <Cn>, and XN <Cn>

The status of a counter is checked using binary check operations **A, AN, O, ON, X,** and **XN**. Using the appropriate operation, the status of a specific counter is accessed by referencing the counter address as the operand address—for example AN C33. You may check the counter status for '0' using AN, ON, or XN; the status is checked for '1' using operations A, O, and X. The counter status is '1' when the current count is greater than zero; the counter status is '0' when the current count is equal zero.

In the example below, the down counter C33 is decremented on each 0-to-1 transition of the input image address I 4.4. When the counter C33 reaches zero, the status of the counter is set to '0'. While the counter value is counting down and greater than zero, the check for '0' yields '0' and Q17.1 will be de-activated; when the counter reaches zero, the check for '0' yields '1' and Q17.1 will be activated.

Example: Checking the status of a counter using the binary check operators

A	I 4.4	// Down counter input
CD	C33	// Decrement counter C33
	.	//
	.	//
AN	C33	// Check if the status of C33 = '0' (if counter has reached zero)
=	Q17.1	// Activate output when the counter reaches zero

Enabling a Counter

The *enable counter* operation **FR** is not required for normal counter operation; however, you may find it useful for the modified counter behavior it offers. When triggered, the FR operation supports retriggering of the counter operations **CU**, **CD**, and **S** without a positive edge at these inputs.

Enable Counter: FR <Cn>

Whenever the RLO preceding the enable operation **FR** transitions from 0-to-1, the subsequent inputs of the specified counter are enabled for retriggering, during the current program cycle, provided that the RLO = 1 preceding the input. In other words, while the enable is active, the normal operations of **CU**, **CD**, and **S** (load preset) will execute without a positive edge, if the preceding RLO = 1.

If the enable operation is used, it, along with the associated trigger logic, is placed prior to any other of the timer operations. An example of the enable timer logic is shown below.

Example: Enabling a counter for retriggering

A	I 5.1	// enable input—check I 5.1 for status of '1'
FR	C5	//enable counter C5
A	I 5.2	// set counter input—check I 5.2 for status of '1'
L	#16#100	// load BCD value of 100 as a counter preset
S	C5	// set preset value for counter C5
A	I 5.3	// up count input—check I 5.3 for positive edge transition
CU	C5	// increment counter C5 by +1

Full Implementation of Counters in STL

The basic counter functions, and the operations used to make a full implementation in STL, are best described in terms of the functional parameters of the LAD/FBD box counter. This is true since it is the STL operations upon which the box counter is based.

As with other LAD/FBD box operations implemented in STL, the sequence of STL counter operations is visually derived by taking each box parameter from top to bottom and left to right. Box counter parameters, as shown in order below, include the counter address (**???**); a count input (**CU or CD**); a set input (**S**), which causes the preset value to be loaded; a preset value (**PV**), which defines the counter preset or override value; a reset input (**R**), which resets the counter status and current count; a binary status flag (**Q**), which is '1' if the counter value is > 0 and '0' otherwise; and the current count as a decimal value (**CV**) and as a binary coded decimal (**CV_BCD**) value.

Table 19-2. LAD/FBD Box counter parameters. **Note:** Entry of all parameters is not required.

	Parameter	Type	Description	Box Counter
❶			Counter Address < Cn >	???
❷	CU/CD		Counter Count Input	S_CD
❸	S	BOOL	Load Counter Preset Input Signal	—CD Q—
❹	PV	WORD	Counter Preset Value	—S CV—
❺	R	BOOL	Reset Counter Input	—PV CV_BCD—
❻	Q	BOOL	Counter Status Flag	—R
❼	CV	WORD	Current Counter Value in Decimal	
❽	CV_BCD	WORD	Current Counter Value in Binary Coded Decimal (BCD)	

Implementing the Basic Counter Functions

Before moving on to the description and full programming of each counter type, we need to show how the STL counter operations correspond directly to the functions of a LAD/FBD box counter. This correlation is the same for each counter type. First, it's important to note that the previously described STL enable operation that supports retriggering of a counter is not required nor implemented in the LAD/FBD box counter.

Excluding the counter enable function, the basic functions of the box counter can be summarized as four operations—❶ given the appropriate count input, count using a specific counter address with a specific preset value; ❷ given a specific reset input, reset the counter; ❸ check the binary status of the counter to determine if it is enabled, still timing, or if the preset value has been reached; and ❹ get the current count and store it for further use as a decimal value or as a binary coded decimal value. These same functions are found in the full programming of each counter type, as shown highlighted in the following example of the UP counter implementation.

Full Programming—UP Counter

For each transition from '0' to '1' at the **CU** input, the *up counter* increments the specified counter by +1, except for when the count has reached the maximum count of 999. A '0' to '1' transition at the set (**S**) input causes the count to be preset with the value you specify at **PV**. The preset value may be specified as a BCD (C#000 to C#999) or Hex (W#16#0000 to W#16#0999) constant, or a word location containing a valid count value.

The current count is output as a decimal value in the word location specified at **CV**, and in BCD in the location specified at **CV_BCD**. Whenever the RLO = 1 on the reset line **R**, the counter value is cleared to zero and the status signal **Q** is reset.

In the following example, the Up Counter increments by +1 when the count input I 5.3 transitions from '0' to '1'. Whenever the set count input I 5.2 is '1', the counter preset value is set to 100. The counter is reset whenever the RLO = 1 preceding the reset input of I 5.4.

Example: Full programming for Up (CU) counter

Network 1:				Box Counter—UP
❶	A	I 5.2	// Set counter input	
	L	#16#100	// Load counter preset value of 100 BCD	
	S	C5	// Set C5 with preset value from accumulator 1	
❷	A	I 5.3	// Count up input	
	CU	C5	// Increment counter C5 by +1 (on positive edge)	
❸	A	I 5.4	// Reset input	
	R	C5	// Reset counter C5	
❹	A	C5	// Check binary status of C5	
	=	M 104.1	// Counter status flag	
❺	L	C5	// Load current count of C5 as binary (decimal)	
	T	MW42	// Store current count as decimal in MW42	
	LC	C5	// Load current count of C5 as BCD	
	T	MW44	// Store current count as BCD in MW44	

Box Counter—UP diagram:
```
       ???
      s_CU
 —CU       Q—
 —S       CV—
 —PV  CV_BCD—
 —R
```

Full Programming—DOWN Counter

For each transition from '0' to '1' at the **CD** input, the *down counter* decrements the specified counter by +1, except for when the count has reached the minimum count of 000. A '0' to '1' transition at the set (**S**) input causes the count to be preset with the value you specify at **PV**. The preset may be specified as a BCD (C#000 to C#999) or Hex (W#16#0000 to W#16#0999) constant, or a word location containing a valid count value.

The current count is output as a decimal value in the word location specified at **CV**, and in BCD at the location specified at **CV_BCD**. Whenever the RLO = 1 on the reset line **R**, the counter value is cleared to zero and the status signal **Q** is reset.

In the following example the Down Counter decrements by +1 when the count input I 5.3 transitions from '0' to '1'. Whenever the set count input I 5.2 is '1', the counter preset value is set to 100. The counter is reset whenever the RLO = 1 preceding the reset input of I 5.4.

Example: Full programming for Down (CD) counter

Network 2:				Box Counter—UP
❶	A	I 5.2	// Set counter input	
	L	#16#100	// Load counter preset value of 100 BCD	
	S	C6	// Set C6 with preset value from accumulator 1	
❷	A	I 5.3	// Count down input	
	CD	C6	// Decrement counter C6 by +1 (on positive edge)	
❸	A	I 5.4	// Reset input	
	R	C6	// Reset counter C6	
❹	A	C6	// Check binary status of C6	
	=	M 104.1	// Counter status flag	
❺	L	C6	// Load current count of C6 as binary (decimal)	
	T	MW42	// Store current count as decimal in MW42	
	LC	C6	// Load current count of C6 as BCD	
	T	MW44	// Store current count as BCD in MW42	

Box Counter—UP:
```
        ???
       S_CD
  —CD        Q—
  —S        CV—
  —PV  CV_BCD—
  —R
```

Full Programming—UP/DOWN Counter

The *up-down counter* increments the specified counter by +1 on each '0' to '1' transition, at the **CU** input, except for when the count has reached the maximum count of 999. The counter decrements by +1 on each transition from '0' to '1' at the **CD** input, except for when at the minimum count of 000. A transition from '0' to '1' at the set (**S**) input causes the count to be preset with the value specified at **PV**. The preset may be specified as a BCD (C#000 to C#999) or Hex constant (W#16#0000 to W#16#0999), or as a word location containing a valid count value.

The current count is output in decimal in the word location specified at **CV**, and in BCD at the location specified at **CV_BCD**. Whenever the RLO = 1 on the reset line **R**, the counter value is cleared to zero and the status signal **Q** is reset.

In the following example, the Up-Down Counter increments by +1 when the logic combination of I 5.1 OR M 7.1 transitions from logic 0 to logic 1; and decrements by +1 when the logic combination of I 5.2 OR M 10.7 transitions from logic 0 to logic 1. Whenever the logic combination of I 5.3 AND M 101.4 transitions from logic 0 to logic 1, the counter preset value is set to 550. The counter is reset whenever the logic combination of I 5.4 AND NOT M 10.0 is True (1).

Example: Full programming for Up/Down (CD) counter

				Box Counter—UP
Network 3:				???
❶	A	I 5.2	// Set counter input	S_CUD
	L	#16#100	// Load counter preset value of 100 BCD	—CU Q—
	S	C7	// Set C7 with preset value from accumulator 1	—CD CV—
❷	A	I 5.3	// Count up input	—S CV_BCD—
	CU	C7	// Increment counter C7 by +1 (on positive edge)	—PV
❸	A	I 5.3	// Count down input	—R
	CD	C7	// Decrement counter C7 by +1 (on positive edge)	
❹	A	I 5.4	// Reset input	
	R	C7	// Reset counter C7	
❺	A	C7	// Check binary status of C7	
	=	M 104.1	// Counter status flag	
❻	L	C7	// Load current count of C7 as binary (decimal)	
	T	MW42	// Store current count as decimal in MW42	
	LC	C7	// Load current count of C7 as BCD	
	T	MW44	// Store current count as BCD in MW42	

Excluding Counter Functions

When it comes to programming counter functions, not all of the operations used in the full counter implementation are required. The only operations that are absolutely needed are those that count up or count down, and the load the counter preset, in the case of the down counter. You may decide which of the functions you actually need. For example, you may determine that explicit resetting of the counter is not necessary. You may decide to restart the counter after re-initializing the counter value.

In any case, if you always want counters to display in LAD/FBD, certain measures are required in your code. First, each counter must be placed in a separate network; also, you must insert a NOP for each counter function that you do not intend to implement. For example, to omit the reset operation and the current count in BCD, a NOP is substituted in place of these operations.

In the following counter implementation, both the reset and the operations that loads/and transfers the current count as a BCD value have been omitted. In each of these cases, a NOP operation replaces the omitted instructions. If the remaining operations of the timer are programmed as they would be in the full implementation, and the counter is the only code in the network, the counter box displays in LAD/FBD.

Example: Excluding counter functions

				Box Counter—UP
Network 4:				???
❶	A	I 5.2	// Set counter input	S_CU
	L	#16#100	// Load counter preset value of 100 BCD	—CU Q—
	S	C5	// Set C6 with preset value from accumulator 1	—S CV—
❷	A	I 5.3	// Count down input	—PV CV_BCD—
	CU	C5	// Decrement counter C5 by +1 (on positive edge)	—R
❸	NOP		// Reset input (R)—**omitted**	
			// Reset counter operation—**omitted**	
❹	A	C5	// Check binary status of C5	
	=	M 104.1	// Counter status flag	
❺	L	C5	// Load current count of C5 as binary (decimal)	
	T	MW42	// Store current count as decimal in MW42	
	NOP		// load current count as BCD—**omitted**	
			// store BCD count value—**omitted**	

Chapter 20

STL Conversion Operations

This chapter introduces STL Conversion Operations. Conversions are necessary when it comes to changing from one data type or format to another. Often a conversion is required before a value contained in a memory location or variable may be used in a compare or an arithmetic operation. For example, the conversion of an integer variable would be required prior to outputting it to a BCD device. The various S7 data types and formats are covered in Chapter 1 *Memory and Addressing*.

Main Topics

- STL Conversion Operations Overview
- Programming Conversions in STL

STL CONVERSION OPERATIONS OVERVIEW

Statement List *conversion operations* are listed in Table 20-1. Like with LAD and FBD, STL conversion operations include instructions for variables of type INT (integer) and DINT (double integer), REAL (floating-point) numbers, and BCD numbers. There are also four operations that convert a REAL number to DINT format by rounding. These rounding operations differ in how the number is rounded with respect to the fractional part of the floating-point number.

Using conversion instructions, you can change the format or data type of a declared variable or the value loaded from a specific memory location. For instance, a value input as BCD must be converted before use in an integer arithmetic operation. Conversely, if the integer result must be output to a BCD display, then the value would first require conversion to BCD. Since STEP 7 arithmetic operations are only performed on integer, double integer and real numbers, a variable not of these data types would first require the appropriate conversion.

Table 20-1. STL Conversion Operations Summary

Operation	Brief Description
BTI	**BCD to INT.** Convert 3-digit BCD value in ACCU1 to a 16-bit integer value.
ITB	**INT to BCD.** Convert 16-bit integer in ACCU1 to a 3-digit BCD value.
ITD	**INT to DINT.** Convert 16-bit integer in ACCU1 to a 32-bit double integer value.
BTD	**BCD to DINT.** Convert 7-digit BCD value in ACCU1 to a 32-bit double integer value.
DTB	**DINT to BCD.** Convert 32-bit double integer value in ACCU1 to a 7-digit BCD value.
DTR	**DINT to REAL.** When the RLO= logic 1, convert the 32-bit double integer value in ACCU1 to a REAL value (floating-point).
INV	**Invert Integer.** Complement or invert the integer value in ACCU1.
INVD	**Invert Double Integer.** Complement or invert the double integer value in ACCU1.
NEGI	**Negate INT.** Negate the integer value in ACCU1.
NEGD	**Negate DINT.** Negate the double integer value in ACCU1.
NEGR	**Negate REAL.** Negate the REAL value in ACCU1.
RND	**Round to Nearest Integer.** Convert the REAL value in ACCU1 by rounding to the nearest double integer (DINT) value. If the value lies halfway between the lower and upper integer values, then after conversion, the result is the even integer value (e.g. before = 2.5, after = 2).
TRUNC	**Truncate.** Convert the REAL value in ACCU1 to Integer by truncating the fractional part of the REAL value (e.g. before = 2.35, after = 2).
RND+	**Round to Next Higher Integer.** Convert the REAL value in ACCU1 by rounding to the next highest double integer (DINT) value that is greater than or equal to the original value (e.g. before = 2.35, after = 3).
RND-	**Round to Next Lower Integer.** Convert the REAL value in ACCU1 by rounding to the next lowest double integer (DINT) value that is less than or equal to the original value (e.g. before = 2.95, after = 2).

Note: The converted result is always right-justified in accumulator 1 (ACCU1) after the operation.

PROGRAMMING CONVERSIONS IN STL

Using conversion operations, you can alter the format or data type of a variable or a value loaded from a memory location. For instance, a value input as BCD must be converted before use in an integer arithmetic operation. Conversely, if the integer result must be output to a BCD display, then it would first require conversion to BCD. Since STEP 7 arithmetic operations are only performed on integer (INT), double integer (DINT), and REAL, variables not of these types would first require conversion.

Conversion Operations Sequence

Conversion operations, like other digital operations, are executed regardless of the RLO—without appropriate consideration a conversion will always be processed. If, in a given block, conversion operations should be processed only on certain conditions, then based on those conditions you will need to use jumps or some other mechanism to determine when the operations should be processed.

The general sequence for carrying out a conversion operation, as shown below, first involves a load operation that places the value to be converted in accumulator 1. STEP 7 will interpret the data format of the value in accumulator 1 according to the operation. The next statement is the conversion to be performed. After the conversion, the result, which remains in accumulator 1, may be stored or used in a follow-on operation.

Example: Standard format of a conversion operation sequence

CONV	L	MD134	// Load value to be converted	LAD/FBD Box Equivalent
	DTR		// Perform conversion operation	DI_R EN ENO
	T	MD234	// Store converted result	MD134 — IN OUT — MD234

Later in Chapter 25 *STL Status Bit Operations*, you can learn about how the execution of most conversion operations influences the CPU status bits, and may be evaluated in your program.

Converting Integer and BCD Numbers

If you are working with a 16-bit BCD value as a value to be converted (source) or as a result, then the value is either loaded from or stored to a word-width memory location or to a WORD variable. When working with a 16-bit integer as the source or as the result, then the value is either loaded from or stored to a word-width memory location or to an INT variable. Similarly, if you are dealing with a 32-bit BCD value as the source or as the result, then the value is either loaded from or stored to a double word location or a DWORD variable. Finally, when a 32-bit integer value is the source or the result, then the value is either loaded from or stored to a double word memory location or to a DINT variable.

BCD to Integer: BTI

When executed, the *BCD to Integer* instruction **BTI** interprets the lower 16-bits of accumulator 1 as a 3-digit BCD value. The sign of the BCD value is represented in bits 12 through 15, where all ones (1111) is negative, and all zeros (0000) is positive. Only bit-15 is considered in the conversion. The 3-digit BCD value, represented in bits 0 through 11, is converted to a 16-bit integer in the range of 0 to ± 999. The converted result is returned to ACCU1, and is available for further processing.

If the original ACCU1 content is interpreted as an invalid BCD number (A-F_{16}), then a parameter assignment error is reported in the CPU. This error causes the synchronous error OB 121 to be called. The CPU is interrupted and will transition to STOP if OB 121 has not been programmed.

In the following example, the WORD variable #BCD3_In is interpreted as a 3-digit BCD value that is converted to an INT value. The result is transferred to memory word MW70.

Example STL Code: Convert from BCD to Integer using BTI operation

CNV1	L	#BCD3_In	// Load the WORD variable #BCD3_In to ACCU1
	BTI		// Convert 3-digit BCD value to 16-bit integer
	T	MW70	// Store the result in bit memory word MW70

The previous example is illustrated below:

Example: Convert a 3-digit BCD value in declared variable #BCD3_In to Integer value		
0000 0101 0111 0110	WORD—#BCD3_In (ACCU1)	BCD +576
BTI		
0000 0010 0100 0000	INT RESULT (ACCU1)	INT +576

Figure 20-1. Illustration of BCD to Integer operation.

BCD to Double Integer: BTD

When executed, the *BCD to Double Integer* instruction **BTD** interprets the 32-bits of accumulator 1 as a 7-digit BCD value. The sign of the BCD value is represented in bits 12 through 15, where all ones (1111) is negative, and all zeros (0000) is positive. Only bit-15 is considered in the conversion. The 7-digit BCD value is converted to a double integer in the range of 0 to ± 9 999 999. The converted result is returned to ACCU1, and is available for further processing.

If the original ACCU1 content is interpreted as an invalid BCD number (A-F$_{16}$), then a parameter assignment error is reported in the CPU. This error causes the synchronous error OB 121 to be called. The CPU is interrupted and will transition to STOP if OB 121 has not been programmed.

In the following example, the WORD variable #BCD7_In is interpreted as a 7-digit BCD value that is converted, to a DINT (32-bit integer) value. The result is transferred to memory double word MD130.

Example: Convert from BCD to Double Integer using BTD operation

CNV2	L	#BCD7_In	// Load the WORD variable #BCD7_In to ACCU1
	BTD		// Convert 7-digit BCD value to 32-bit integer
	T	MD130	// Store the result in bit memory double word MD130

The previous example is illustrated below:

Example: Convert a 7-digit BCD value in declared variable #BCD7_In to Double Integer value		
1111 0000 0000 0101 0111 0011 0100 0100	DWORD—#BCD7_In (ACCU1)	- 57344 BCD
BTD		
1111 1111 1111 1111 0010 0000 0000 0000	DINT RESULT (ACCU1)	-57344 DINT

Figure 20-2. Illustration of BCD to Double Integer operation

Integer to BCD: ITB

When executed, the *Integer to BCD* instruction **ITB** interprets the lower 16-bits of accumulator 1 as a signed integer. The 16-bit integer value is converted to a 3-digit BCD value in the range of 0 to ± 999. The sign bit of the BCD result is represented in bits 12 through 15; all ones (1111) is negative, all zeros (0000) is positive. The converted result is returned to ACCU1. If the original integer value is too large for conversion (greater than 999), then status bits OV and OS are set and the conversion is aborted.

In the following example, the INT (16-bit integer) value in the bit memory word location MW72 is converted to a 3-digit BCD value. The result is transferred to the WORD variable #BCD3_Out.

Example: Convert from Integer to BCD using ITB operation

CNV3	L	MW72	// Load the memory word MW72 to ACCU1
	ITB		// Convert the 16-bit INT value to a 3-digit BCD value
	T	#BCD3_Out	// Store the result in the WORD variable #BCD3_Out

The previous example is illustrated below:

Example: Convert Integer value in memory word MW72 to a 3-digit BCD value			
0000 0000 1100 1011	INT Value—MW72 (ACCU1)	INT	+203
ITB			
0000 0010 0000 0011	DWORD RESULT (ACCU1)	BCD	+203

Figure 20-3. Illustration of Integer to BCD operation

Double Integer to BCD: DTB

When executed, the *Double Integer to BCD* instruction **DTB** interprets the 32-bits of accumulator 1 as a signed integer. The 32-bit integer value is converted to a 7-digit BCD value in the range of 0 to ±9,999,999. The sign bit of the BCD result is represented in bits 28 through 31; all ones (1111) is negative, all zeros (0000) is positive. The converted result is returned in ACCU1. If the original integer value is too large for conversion (greater than 9,999,999), then status bits OV and OS are set and the conversion is aborted.

In the following example, the DINT (32-bit integer) value in memory double word MD74 is converted to a 7-digit BCD value. The result is transferred to the DWORD variable #BCD7_Out.

Example: Convert from Double Integer to BCD using DTB operation

CNV4	L	MD74	// Load the memory double word MD74 to ACCU1
	DTB		// Convert the 32-bit INT value to a 7-digit BCD value
	T	#BCD7_Out	// Store the result in the DWORD variable #BCD7_Out

The previous example is illustrated below:

Example: Convert Double Integer value in memory double word MD74 to a 7-digit BCD value			
0000 0000 0000 0000 1110 0000 0000 0000	INT Value—MW72 (ACCU1)	DINT	+57344
DTB			
0000 0000 0000 0101 0111 0011 0100 0100	WORD—RESULT (ACCU1)	BCD	+57344

Figure 20-4. Illustration of Double Integer to BCD operation

Integer to Double Integer: ITD

When executed, the *Integer to Double Integer* instruction **ITD** interprets the lower 16-bits of accumulator 1 as a 16-bit integer. In the execution, the 16-bit integer is moved right-justified to ACCU1, with the sign bit (bit-15) placed in each of the high order bits (16-31) of ACCU1. The result is returned to ACCU1, and is available for further processing. No errors are signaled with this operation.

In the following example, the INT (16-bit integer) value of peripheral input word PIW 256 is converted to a DINT (32-bit integer) value. The result is transferred to memory double word MD134.

Example: Convert from Integer to Double Integer using ITD operation

CNV5	L	PIW256	// Load the analog input at peripheral input address PIW256
	ITD		// Convert 16-bit INT value to 32-bit DINT
	T	MD134	// Store the result in bit memory double word MD134

The previous example is illustrated below:

Example: Convert Integer value in memory word MW78 to a Double Integer value			
0000 0000 0000 0000 1111 1111 1111 0110	INT Value—MW78 (ACCU1)	-10	INT
ITD			
1111 1111 1111 1111 1111 1111 1111 0110	DINT RESULT (ACCU1)	-10	DINT

Figure 20-5. Illustration of Integer to Double Integer operation

Double Integer to REAL: DTR

When executed, the *Double Integer to Real* instruction **DTR** interprets the value in accumulator 1 as a 32-bit signed integer that is converted to a floating-point value in REAL format. The converted result is placed in ACCU1, and is available for further processing. For example, the result may be transferred to a double word memory location or variable of type REAL.

Because of its greater accuracy, the double integer value may be rounded to the next highest whole number when converted to REAL. No status bits are set with this conversion.

In the following example, the 16-bit INT value read from peripheral input word PIW256 is first converted to a double integer value, and then immediately converted to a 32-bit floating-point value. The result of the second conversion is transferred to memory double word MD234.

Example: Convert from Double Integer to REAL using DTR operation

CNV6	L	PIW256	// Load the analog input at peripheral input address PIW256
	ITD		// Convert 16-bit INT value to 32-bit DINT
	DTR		// Convert 32-bit DINT value to floating-point
	T	MD234	// Store the result in bit memory double word MD234

Converting REAL Numbers

The operations in this section include conversions from REAL numbers to 32-bit DINT values.

When the value to be converted is a REAL number, you may load the accumulator from a REAL variable, or a double word location that contains a floating-point value. With the following operations, the result is always a double integer value, which may be transferred to a double word memory location or to a DINT variable.

Round to Nearest Integer: RND

When executed, the *Round* instruction **RND** interprets the accumulator 1 contents as a 32-bit floating-point value. The result is the nearest whole number that is greater than or equal to the original number. If the input value lies halfway between the lower and upper integer values, then the result returns the even integer value. For example, if the source value = 2.5, then the result = 2; if the source = 2.59 then the result = 3. The result is returned in ACCU1, and is available for further processing.

If the value in ACCU1 is an invalid REAL number, or is outside of the permissible range of a double integer value, then the conversion is aborted and status bits OV and OS are set to '1'.

In the following example, the floating-point value in memory double word MD110 is rounded to the nearest whole DINT (32-bit integer) value. The result is transferred to memory double word MD210.

Example: Round ACCU1 contents to nearest double integer using RND operation

CNV7	L	MD110	// Load the bit memory double word MD110 to ACCU1
	RND		// Round the REAL value to nearest integer; leave DINT result in ACCU1
	T	MD210	// Store the result in bit memory double word MD210

Truncate: TRUNC

When executed, the *Truncate* instruction **TRUNC** interprets the accumulator 1 contents as a 32-bit floating-point value. The value is converted by truncating or dropping the fractional part of the original floating-point number. For example, if the value to be converted is equal 2.95, then the result will be 2. The converted result is returned in ACCU1, and is available for further processing.

If the value in ACCU1 is an invalid REAL number or is outside of the permissible range of a double integer value, then the conversion is aborted and status bits OV and OS are set to '1'.

In the following example, the REAL value in memory double word MD110 is truncated and the double integer result is returned to ACCU1. The result is then transferred to memory double word MD214.

Example: Truncate ACCU1 contents to double integer using TRUNC operation

CNVA	L	MD110	// Load the bit memory double word MD110 to ACCU1
	TRUNC		// Truncate the REAL value in ACCU1; leave DINT result in ACCU1
	T	MD214	// Store the result to bit memory double word MD214

Round to Next Higher Integer: RND+

When executed, the *Round to Next Higher Integer* instruction **RND+** interprets the accumulator 1 contents as a 32-bit floating-point value. The value is converted by rounding to the next higher integer value that is greater than or equal to the original number. For example, if the value to be converted is equal 2.3, then the result = 3. The result is returned in ACCU1, and is available for further processing.

If the value in ACCU1 is an invalid REAL number or is outside of the permissible range of a double integer value, then the conversion is aborted and status bits OV and OS are set to '1'.

In the following example, the REAL value in memory double word MD110 is rounded to the next higher whole DINT (32-bit integer) value. The result is transferred to memory double word MD218.

Example: Round ACCU1 contents to next higher double integer using RND+ operation

CNV8	L	MD110	// Load the memory double word MD110 to ACCU1
	RND+		// Round REAL value in ACCU1 to next higher integer; leave result in ACCU1
	T	MD218	// Store the result to bit memory double word MD218

Round to Next Lower Integer: RND-

When executed, the *Round to Next Lower Integer* instruction **RND-** interprets the accumulator 1 contents as a 32-bit floating-point value. The value is converted by rounding to the next lower integer value that is less than or equal to the original number. For example, if the value to be converted is equal 2.95 then the result is equal 2. The result is returned in ACCU1.

If the value in ACCU1 is an invalid REAL number or is outside of the permissible range of a double integer value, then the conversion is aborted and status bits OV and OS are set to '1'.

In the following example, the REAL value in memory double word MD110 is rounded to the next lower whole DINT (32-bit integer) value. The result is transferred to memory double word MD222.

Example: Round ACCU1 contents to next lower double integer using RND- operation

CNV9	L	MD110	// Load the memory double word MD134to ACCU1
	RND-		// Round REAL value in ACCU1 to next lower integer; leave result in ACCU1
	T	MD222	// Store the result to bit memory double word MD222

Negation Operations

The operations in this section include inversions and negations using one's and two's complement methods. These operations involve 16-bit integer (INT), 32-bit integer (DINT), and REAL values.

If you are dealing with a 16-bit integer value as the value as the source or as the result, then the value is either loaded from, or stored to, a word-width memory location, or to a declared INT variable. Similarly, if you are dealing with a 32-bit integer value as the source or as the result, then the value is either loaded from or stored to a double word memory location, or to a declared DINT variable. Finally, if you are converting a REAL number, the value is either loaded from, or stored to, a double word memory location or a REAL variable.

Invert Integer: INV

When executed, the *Invert Integer* instruction **INV** interprets the accumulator 1 contents as a 16-bit integer value. The INT value is inverted using the one's complement operation. This operation simply inverts each bit according to the NOT truth table. Upon execution, each '1' is changed to '0' and each '0' is changed to '1'. The result is returned in ACCU1 and is available for further processing. For example, you may transfer the result to an integer variable or word-width memory location.

In the example below, the INT value of input image word IW22 is inverted using one's complement. The result is transferred to memory word MW22. No errors are signaled with this operation.

Example: Invert ACCU1 integer using ones complement—INV operation

L	IW22	// Load the bit memory word MW44 to ACCU1
INV		// Invert INT value in ACCU1 using 1's complement; leave result in ACCU1
T	MW22	// Store the result to bit memory word MW22

The previous example is illustrated below:

Example: Invert Integer value in input word IW22	
1010 0101 1111 1100	INT Value—IW22 (ACCU1)
INV	
0101 1010 0000 0011	INT RESULT (ACCU1)

Figure 20-6. Illustration of Invert Integer operation

Invert Double Integer: INVD

When executed, the *Invert Double Integer* instruction **INVD** interprets the accumulator 1 contents as a 32-bit double integer value. The DINT value is inverted using the one's complement operation. This operation inverts each bit according to the NOT truth table. Upon execution, all 1s are changed to 0s and all 0s are changed to 1s. The result is returned in ACCU1 and is available for further processing. For example, you may transfer the result to a double integer variable or double word memory location.

In the example below, the DINT value of ID42 is inverted using ones complement. The result is transferred to memory double word MD142. No errors are signaled with this conversion operation.

Example: Invert ACCU1 double integer using ones complement—INVD operation

L	ID42	// Load input image double word ID42 to ACCU1
INVD		// Invert DINT value in ACCU1 using 1's complement; leave result in ACCU1
T	MD142	// Store the result to bit memory double word MD142

The previous example is illustrated below:

Example: Invert Double Integer value in input double word ID42	
1010 0101 1111 1100 1010 0101 1111 1100	DINT—ID42 (ACCU1)
INVD	
0101 1010 0000 0011 0101 1010 0000 0011	DINT RESULT (ACCU1)

Figure 20-7. Illustration of Invert Double Integer operation

Negate Integer: NEG

The negate instruction is useful in calculations where you simply need to change the sign of a number in a string of arithmetic operations, or to change the sign of the result.

When executed, the *Negate Integer* instruction **NEG** changes the sign of the 16-bit integer value in accumulator 1, by applying a two's complement operation. This operation changes the sign of the value, whereby positive values are changed to negative and negative values are changed to positive. The operation is equivalent to multiplying by -1, which is the same as first applying the ones complement and then adding +1. The converted result is placed in ACCU1.

If the converted result is outside of the permissible range for INT values (-32768 to +32767), then execution of the operation will result in an error—and CPU status bits set, as shown in Table 25-3.

In the following example, the INT value in memory word MW230 is negated. The result is transferred to from ACCU1 to memory word MW232.

Example: Negate ACCU1 integer using twos complement—NEG operation

L	MW230	// Load the bit memory word MW230 to ACCU1
NEG		// Negate INT value in ACCU1 using 2's complement; leave result in ACCU1
T	MW232	// Store the result to bit memory word MW232

The previous example is illustrated below:

Example Negate		Negate Integer value in memory word MW232 using 2's Complement	
MW24	**ACCU1**		
+255	-255	0000 0000 1111 1111	**ACCU1 = + 255**
		1111 1111 0000 0000	Complement
		0000 0000 0000 0001	Add +1
		1111 1111 0000 0001	**RESULT = - 255**

Figure 20-8. Illustration of Negate Integer operation

Negate Double Integer: NEGD

The negate double integer instruction **NEGD** is useful in calculations where you simply need to change the sign of a number in a string of arithmetic operations, or to change the sign of the result.

When executed, the *Negate Double Integer* instruction **NEGD** changes the sign of the 32-bit double integer value in accumulator 1 by applying a two's complement operation. This operation changes the sign of the value, whereby positive values are changed to negative and negative values are changed to positive. The operation is equivalent to multiplying by -1, which is the same as first applying the ones complement and then adding +1. The result is placed in accumulator 1 and is available for further processing. For example, the value can be transferred to a double word location or variable.

If the result is outside of the permissible range for DINT values -2,147,483,648 to +2,147,483,647), then execution of the operation will result in an error—and CPU status bits set, as shown in Table 25-3.

In the following example, the DINT value in memory double word MD234 is negated using the two's complement operation. The result is transferred from ACCU1 to memory word MD238.

Example: Negate ACCU1 double integer using twos complement—NEGD operation

L	MD234	// Load the bit memory double word MD234 to ACCU1
NEGD		// Negate DINT value in ACCU1 using 2's complement; leave result in ACCU1
T	MD238	// Store the result in bit memory double word MD238

The previous example is illustrated below:

Example Negate		Negate Double Integer value in double word MD234 using 2's Complement	
MD24	ACCU1		
-5	+5	0000 0000 0000 0000 0000 0000 0000 0101	ACCU1 = + 5
		1111 1111 1111 1111 1111 1111 1111 1010	Complement
		0000 0000 0000 0000 0000 0000 0000 0001	Add +1
		1111 1111 1111 1111 1111 1111 1111 1011	RESULT = - 5

Figure 20-9. Illustration of Negate Double Integer operation

Negate REAL: NEGR

When executed, the *Negate Real* instruction **NEGR** negates the floating-point value in accumulator 1 by changing the sign of the number. Positive values are changed to negative and negative values are changed to positive. The NEGR operation is equivalent to multiplying by -1. The Negate Real changes the sign of the mantissa of the original value. The negated result is placed in ACCU1 and is available for further processing. No errors are signaled with this conversion.

In the following example, the REAL value in memory double word MD234 is negated using the two's complement operation. The result is transferred from ACCU1 to memory double word MD242.

Example: Negate ACCU1 REAL value using two's complement—NEGR operation

L	MD234	// Load the bit memory double word MD234 to ACCU 1
NEGR		// Negate REAL value by inverting the sign bit; leave result in ACCU1
T	MD242	// Store the result in memory double word MD242

Chapter 21

STL Compare Operations

This chapter introduces STL Compare Operations. The various compare operations compare two word or double word locations, or variables. The outcome of the compare test is indicated by setting the result of logic operations to '1' if the result is True, and to '0' if the result is False.

In your control program, the compare result may be the basis of many other decisions. For example, the compare operation might be used to determine if a timer preset, counter value, arithmetic calculation, or other process value has reached a setpoint or is within range. Depending on the outcome of the compare test, the program may activate or de-activate a device, or enable or disable a process event. The compare result might also determine whether a block is called, or if an iterative process should be terminated.

Main Topics

- STL Compare Operations Overview
- Programming Comparisons in STL

STL COMPARE OPERATIONS OVERVIEW

STEP 7 compare operations are listed in Table 21-1. Generally speaking, there are six compare tests that include test for *equal, not equal, less than, greater than, less than-or-equal,* and *greater-than-or-equal.* Each of the six compare tests are supported for comparison of integer (**I**), double integer (**D**), and Real (**R**) values. For example, to compare two integer values for non-equality, you would use < > **I**; you would use < > **D** to compare two double integer values; and < > **R** for two Real values.

Table 21-1. STL Compare Operations Summary

Operation	Brief Description
= = I	Compare the Integer values in ACCU2 and ACCU1 for equal
<> I	Compare the Integer values in ACCU2 and ACCU1 for not equal
< I	Compare the Integer value in ACCU2 for less than ACCU1
> I	Compare the Integer value in ACCU2 for greater than ACCU1
<= I	Compare the Integer value in ACCU2 for less than or equal ACCU1
>= I	Compare the Integer value in ACCU2 for greater than or equal ACCU1
= = D	Compare the Double Integer values in ACCU2 and ACCU1 for equal
<> D	Compare the Double Integer values in ACCU2 and ACCU1 for not equal
< D	Compare the Double Integer value in ACCU2 for less than ACCU1
> D	Compare the Double Integer value in ACCU2 for greater than ACCU1
<= D	Compare the Double Integer value in ACCU2 for less than or equal ACCU1
>= D	Compare the Double Integer value in ACCU2 for greater than or equal ACCU1
== R	Compare the REAL values in ACCU2 and ACCU1for equal
<> R	Compare the REAL values in ACCU2 and ACCU1for not equal
< R	Compare the REAL value in ACCU2 for less than ACCU1
> R	Compare the REAL value in ACCU2 for greater than ACCU1
<= R	Compare the REAL value in ACCU2 for less than or equal ACCU1
>= R	Compare the REAL value in ACCU2 for greater than or equal ACCU1

Note: Each compare operation compares the contents of ACCU2 and ACCU1, right justified.

PROGRAMMING COMPARE OPERATIONS IN STL

Typical uses of compare operations include range checking, setpoint control, and calculation limit checks. Compare operations might be used, for instance, to check the remaining time of a timer or if a counter has reached a certain intermediate or final value. In operations that involve a loop or an iterative calculation, like the averaging of analog readings, then the compare is used to check if the process is complete or if the loop should be terminated.

The permissible number range for comparisons of 16-bit integers (INT), 32-bit double integers (DINT) and for floating-point values (REAL) are shown in the following table.

316

Table 21-2. Valid constants, memory operands, and variables used in compare operations

IN1/IN2	INT	DINT	REAL
Constants (in permissible range)	-32,768 to +32,767	-2,147,483,648 to +2,147,483,647	± 1.75495e-38 to ± 3.402823+e38
S7 Memory Locations (Absolute or Symbolic)	IW 28 QW 42 MW 54 DB6.DBW12	ID 28 QD 42 MD 54 DB6.DBD12	ID 28 QD 42 MD 54 DB6.DBD12
Declared Variables (of correct data type)	#Value_1 #Value_2	#Value_1 #Value_2	#Value_1 #Value_2

Note: Symbolic addresses may be substituted for absolute addresses.

Compare Operations Sequence

Compare operations, like other digital operations, are executed regardless of the RLO—and without appropriate consideration a conversion will always be processed. If, in a given block, a compare operation should be processed only on certain conditions, then based on those conditions, you will need to use jumps or some other mechanism to determine when the operation should be processed.

The general sequence for carrying out a compare operation, as shown below, first involves subsequent load operations that place the values to be compared in accumulators 1 and 2. Next, the compare operation to be performed is selected based on the type of values being compared. Finally, you may flag the outcome of the compare, using a binary operand that indicates the 'True' or 'False' result. The binary indicator that you specify may be further combined with other binary operands.

Example: Standard format of a compare operations sequence

				LAD/FBD Box Equivalent
CMP3	L	#Speed_1	// Load first compare value	CMP <I M101.4
	L	+1500	// Load second compare value	—( )—
	<I		// Perform compare operation	#Speed_1 — IN1
	=	M101.4	// Indicate compare outcome	1500 — IN2

The result of each compare operation influences CPU status bits CC0 and CC1, based on the relationship between the values in accumulator 1 and accumulator 2. Later in Chapter 25 STL *Status Bit Operations*, and in Chapter 26 *STL Program Flow Control and Block Operations*, you can learn how to evaluate these status bits and make decisions in your program.

Comparing INT, DINT, or REAL Numbers

When comparing integers, you may use a constant in the range of -32768 to +32767, variables of data type INT, or you may reference word-width memory locations. If you are comparing double integers, you may use constants in the range of -2,147,483,648 to +2,147,483,647, variables of data type DINT, or you may reference double word memory locations. When comparing REAL values, you may use REAL constants, variables of data type REAL, or you may reference double word memory locations.

The following discussions describe the six integer compare operations. The discussions would be much the same for DINT or REAL compare operations; the operations would change respectively to **D** or **R**.

Compare for Equal: ==I; ==D; ==R

When processed, the *Compare Equal* instruction interprets the values in ACCU1 and ACCU2 as integer, double integer, or REAL according to the instruction you specify. The operation determines if the contents of ACCU2 *equal to* ACCU1. If the test result is TRUE, then '1' is assigned to the result of the logic operation; if the test result is FALSE, then '0' is assigned to the RLO.

In the example, if the integer value of +1000 is **equal to** the integer value contents of bit memory word MW62, then the compare result is TRUE and the RLO=1. RLO = 1 is assigned to bit address M101.1.

Example: Comparing 16-bit integer values for equal

CMP1	L	+1000	// Load integer constant +1000
	L	MW62	// Load bit memory word MW62
	==I		// Check if +1000 (ACCU2) is **equal to** contents of MW62 (ACCU1)
	=	M101.1	// Assign true (1)/false (0) binary outcome of compare test to M101.1

Compare for Not Equal: <>I; <>D; <>R

When processed, the *Compare Not Equal* instruction interprets the values in ACCU1 and ACCU2 as integer, double integer, or REAL according to the instruction you specify. The operation checks for the contents of ACCU2 *not equal to* ACCU1. If the test result is TRUE, then '1' is assigned to the result of the logic operation; if the test result is FALSE, then '0' is assigned to the RLO.

In the following example, if the integer value in DBW10 is **not equal to** the integer value in DBW12 then the compare result is TRUE and the RLO=1. The RLO = 1 is assigned to the bit address M101.2.

Example: Comparing 16-bit integer values for not equal

CMP2	L	DBW10	// Load contents of DBW10
	L	DBW12	// Load contents of data word DBW12
	<>I		// Check if DBW10 (ACCU2) is **not equal to** DBW12 (ACCU1)
	=	M101.2	// Assign true (1)/false (0) binary outcome of compare test to M101.2

Compare for Greater Than: >I; >D; >R

When processed, the *Compare Greater Than* instruction interprets the values in ACCU1 and ACCU2 as integer, double integer, or REAL, according to the instruction you specify. The operation checks for the contents of ACCU2 *greater than* ACCU1. If the test result is TRUE, then '1' is assigned to the result of the logic operation; if the test result is FALSE, then '0' is assigned to the RLO.

In the following example, if the integer variable #Speed_1 is **greater than** the integer value +1500, then the compare result is TRUE and the RLO=1. The RLO = 1 is assigned to the bit address M101.3.

Example: Comparing 16-bit integer values for greater than

CMP3	L	#Speed_1	// Load integer variable #Speed_1
	L	+1500	// Load integer constant +1500
	>I		// Check if #Speed_1 (ACCU2) is **greater than** +1500 (ACCU1)
	=	M101.3	// Assign true (1)/false (0) binary outcome of compare test to M101.3

Compare for Less Than: <I; <D; <R

When processed the *Compare Less Than* instruction interprets the values in ACCU1 and ACCU2 as integer, double integer, or REAL according to the instruction you specify. The operation checks for the contents of ACCU2 *less than* ACCU1. If the test result is TRUE, then '1' is assigned to the result of the logic operation; if the test result is FALSE, then '0' is assigned to the RLO.

In the following example, if the integer variable #Speed_1 is **less than** the integer variable #Speed_2, then the compare result is TRUE and the RLO=1. The RLO = 1 is assigned to the bit address M101.4.

Example: Comparing 16-bit integer values for less than

CMP4	L	#Speed_1	// Load INT variable #Speed_1
	L	#Speed_2	// Load INT variable #Speed_2
	<I		// Check if #Speed_1 (ACCU2) is **less than** #Speed_2 (ACCU1)
	=	M101.4	// Assign true (1)/false (0) binary outcome of compare test to M101.4

Compare for Greater Than or Equal: >=I; >=D; >=R

When processed the *Compare Greater Than or Equal* instruction interprets the values in ACCU1 and ACCU2 as integer, double integer, or REAL according to the instruction you specify. The operation checks for the contents of ACCU2 *greater than or equal* ACCU1. If the test result is TRUE, then '1' is assigned to the result of the logic operation; if the test result is FALSE, then '0' is assigned to the RLO.

In the following example, if the integer variable #Speed_4 is **greater than or equal** to the integer value of +3500, then the result is TRUE and the RLO=1. RLO = 1 is assigned to the bit address M101.5.

Example: Comparing 16-bit integer values for greater than or equal

CMP4	L	#Speed_4	// Load integer variable #Speed_4
	L	+3500	// Load integer constant +3500
	>=I		// Check if #Speed_4 (ACCU2) is **greater than or equal** +3500 (ACCU1)
	=	M101.5	// Assign true (1)/false (0) binary outcome of compare test to M101.5

Compare for Less Than or Equal: <=I; <=D; <=R

When processed the *Compare Less Than or Equal* instruction interprets the values in ACCU1 and ACCU2 as integer, double integer, or REAL according to the instruction you specify. The operation checks for the contents of ACCU2 *less than or equal* ACCU1. If the test result is TRUE, then '1' is assigned to the RLO; if the test result is FALSE, then '0' is assigned to the RLO.

In the following example, if the integer variable #Speed_5 is **less than or equal** the integer value of zero, then the compare result is TRUE and the RLO=1. RLO = 1 is assigned to the bit address M101.6.

Example: Comparing 16-bit integer values for less than or equal

CMP5	L	#Speed_5	// Load integer variable #Speed_5
	L	0	// Load integer constant +0
	<=I		// Check if #Speed_5 (ACCU2) is **less than or equal** 0 (ACCU1)
	=	M101.6	// Assign true (1)/false (0) binary outcome of compare test to M101.6

Combining Compare Operations

As shown in the example below, compare operations may also be placed in series. If in series, all of the compare tests must be satisfied to complete logic continuity. In the following example, M33.6 is activated when both of the compare tests are satisfied.

Example: Compare operations in series

CMP4	A(		
	L	#Speed_1	// Load INT variable #Speed_1 to ACCU1
	L	1100	// Load INT constant 1100.
	= = I		// Check if #Speed_1 is **equal to** +1100
	)		
	A(		
	L	#Speed_2	// Load INT variable #Speed_2 to ACCU1
	L	1500	// Load INT constant 1500 to ACCU1.
	= = I		// Check if #Speed_2 is **equal to** +1500
	)		
	=	M33.6	// If both tests are TRUE (1) assign RLO = 1 to M33.6; else RLO= 0

Compare operations may also be logically OR'ed (paralleled). If the result of either of the compare operations is satisfied or True (1), then the RLO = 1. In the following example, M33.7 is activated if either of the compare tests is satisfied.

Example: Compare operations in parallel

CMP4	L	#Speed_2	// Load INT variable #Speed_2 into ACCU1
	L	+1200	Load INT constant +1200 to ACCU1.
	<=I		// Check if #Speed_2 is **less than or equal to** +1200
	O		
	L	#Speed_2	// Load INT variable #Speed_2 to ACCU1
	L	+0	// Load INT constant +0 to ACCU1.
	> = I		// Check if #Speed_2 is **greater than or equal to** '0'
	=	M33.7	// If either test is True (1), assign RLO = 1 to M33.7; else RLO= 0

Chapter 22

STL Integer and Real Operations

This chapter introduces STL Integer and REAL operations. These operations support simple to complex calculations using 16-bit integer (INT), 32-bit double integer (DINT), and 32-bit floating-point (REAL) values. In STEP 7, these operations represent basic arithmetic operations, as well as standard trigonometric and mathematical functions. In STL, these operations may be combined in series and in parallel to form basic calculations, control algorithms, or other numerical solutions.

An important part of Integer and REAL operations is that information pertaining to each result, whether valid or in error, is available for evaluation in S7 status bits CC0, CC1, OV, and OS. How Integer and REAL operations influence these status bits is described in Chapter 25 *STL Status Bit Operations*.

Main Topics

- STL Integer and REAL Arithmetic Operations
- Programming Integer and REAL Arithmetic in STL
- Standard Math Functions Overview
- Programming Standard Math Functions in STL

STL INTEGER AND REAL ARITHMETIC OPERATIONS

Statement List basic arithmetic operations include *addition*, *subtraction*, *multiplication*, and *division* on numbers of type INT (integer), DINT (double-integer), and REAL (floating-point). With these operations, you can implement solutions for production accounting, control algorithms, diagnostic routines, and other computing functions. While the DIV_DI instruction produces the normal quotient result, the remainder part of the quotient is the result in the MOD_DI operation.

In addition to these basic arithmetic operations, which operate on ACCU1 and ACCU2, the last three operations of Table 22-1 perform their operation using the value currently in ACCU1. After the operation is performed the result is available in ACCU1 for further processing.

Table 22-1. STL Integer and REAL Arithmetic Operations Summary

Operation	Brief Description
+ I	**INT Addition.** Add the INT value of ACCU2 to the INT value of ACCU1.
- I	**INT Subtraction.** Subtract from the INT value of ACCU2, the INT value of ACCU1.
* I	**INT Multiplication.** Multiply the INT value of ACCU2 times the INT value of ACCU1.
/ I	**INT Division.** Divide the INT value of ACCU2 by the INT value of ACCU1.
+D	**DINT Addition.** Add the DINT value of ACCU2 to the DINT value of ACCU1.
- D	**DINT Subtraction.** Subtract from the DINT value of ACCU2, the DINT value of ACCU1.
*D	**DINT Multiplication.** Multiply the DINT value of ACCU2 times the DINT value of ACCU1.
/D	**DINT Division.** Divide the DINT value of ACCU2 by the DINT value of ACCU1. The quotient result remains in ACCU1.
MOD	**DINT Division.** Divide the DINT value of ACCU2 by the DINT value of ACCU1. The quotient remainder remains in ACCU1.
+ R	**REAL Addition.** Add the REAL value of ACCU2 to the REAL value of ACCU1.
- R	**REAL Subtraction.** Subtract from the REAL value of ACCU2, the REAL value of ACCU1.
* R	**REAL Multiplication.** Multiply the REAL value of ACCU2 times the REAL value of ACCU1.
/ R	**REAL Division.** Divide the REAL value of ACCU2 by the REAL value of ACCU1.
INC <n>	**Increment Accumulator 1.** Add the parameter value +n (0-255) to the value of ACCU1. The result remains in ACCU 1.
DEC <n>	**Decrement Accumulator 1.** Subtract the parameter value +n (0-255) from the value of ACCU1. The result remains in ACCU 1.
+ <const>	**Add Constant to Accumulator 1.** Add the specified constant to current value of ACCU1.

PROGRAMMING INTEGER AND REAL ARITHMETIC IN STL

The permissible number range for arithmetic operations of 16-bit integers (INT), 32-bit double integers (DINT) and for floating-point values (REAL) are shown below.

Table 22-2. Valid constants, memory operands, and variables used in arithmetic operations

IN1/IN2/OUT	INT	DINT	REAL
Constants (in permissible range)	- 32768 to +32767	-2,147,483,648 to +2,147,483,647	± 1.75495e-38 to ± 3.402823+e38
S7 Memory Locations (Absolute or Symbolic)	IW 28 , QW 42 MW 54, DB6.DBW12	ID 28, QD 42 MD 54, DB6.DBD12	ID 28, QD 42 MD 54, DB6.DBD12
Declared Variables (of correct data type)	#Value_A #Value_B #Total	#Value_A #Value_B #Total	#Value_A #Value_B #Total

Arithmetic Operations Sequence

Arithmetic operations, like other digital operations, are executed regardless of the RLO—without appropriate consideration these operations are always processed. If, in a given block, an arithmetic operation should be processed only on certain conditions, then based on those conditions you will need to use jumps or some other mechanism to determine when the operation should be processed.

The general sequence for carrying out an arithmetic operation, as shown below, first involves subsequent load operations that place the values to be operated upon in accumulators 1 and 2. Next, the arithmetic operation to be performed is selected based on the type of the previously loaded values. Finally, the result, which remains in accumulator 1, may be stored or used in a follow-on calculation. Later in Chapter 25 *STL Status Bit Operations* and in Chapter 26 *STL Program Flow Control and Block Operations*, you can learn how to evaluate status bits and make decisions in your program.

Example: Standard format of an arithmetic operations sequence

LABL	L	#VALUE_1	// Load first value
	L	#VALUE_2	// Load second value
	+I		// Perform arithmetic operation
	T	#Sum_1	// Store resultant value

LAD/FBD Box Equivalent

```
              ADD_I
         ─── EN    ENO ─
#Value_1 ─ IN1   OUT ├─ #Sum_1
#Value_2 ─ IN2
```

Integer Arithmetic Operations

In STL, integer arithmetic operations are denoted by **I** for INT, which represents 16-bit integer operations. The four basic operations are **+I** for add integers, **-I** for subtract integers, ***I** for multiply integers, and **/I** for divide integers. Generally, you will use integer operations when whole number results in the range of -32,768 to + 32,767 are acceptable or required.

As previously stated, arithmetic operations are performed on the values currently in accumulator 1 and accumulator 2. How and when these values are loaded is irrelevant, but in the case of integer operations the values may be loaded from a word-width memory location, from declared INT variables, or directly as an integer constant. The accumulator 1 result is available for you to transfer to a new location as required. The location to which you move the result may be a declared variable of type INT or a word-width memory location—for example MW 44, or DB10.DW26.

Add Integer: +I

When the *Add Integer* operation **+I** is executed, the integer addition of the values in accumulator 1 and accumulator 2 is processed as ACCU2 **plus** ACCU1. The result is placed in ACCU1, and is available for further processing.

In the following example, the add integer operation is processed as the integer value in MW12 **plus** the integer constant +100. The ACCU1 result is transferred to bit memory location MW 14.

Example: Addition using integer values

LABL	L	MW12	// Load bit memory word MW12
	L	+100	// Load integer constant +100
	+I		// Integer add MW12 (ACCU2) **plus** 100 (ACCU1)
	T	MW14	// Store the sum to bit memory word MW14

Subtract Integer: -I

When the *Subtract Integer* operation **-I** is executed, the integer subtraction of the values in accumulator 1 and accumulator 2 is processed as ACCU2 **minus** ACCU1. The result is placed in ACCU1.

In the following example, the subtract integer operation is processed as integer variable #VALUE_1 **minus** integer variable #VALUE_2. The ACCU1 result is transferred to the integer variable #Total_1.

Example: Subtraction using integer values

LABL	L	#VALUE_1	// Load INT variable #VALUE_1
	L	#VALUE_2	// Load INT variable #VALUE_2
	-I		// Integer subtract #VALUE_1 (ACCU2) **minus** #VALUE_2 (ACCU1)
	T	#Total_1	// Store the result to INT variable #Total_1

Multiply Integer: *I

When the *Multiply Integer* operation *** I** is executed, the integer multiplication of the values in accumulator 1 and accumulator 2 is processed as ACCU2 **times** ACCU1. The result is placed in ACCU1, and is available for further processing.

In the following example, the multiply integer operation is processed as the integer contents of data block DB11, word DBW12 **times** the integer value of +2. The ACCU1 result is transferred to data block DB21, word DBW12.

Example: Multiplication using integer values

LABL	L	DB11.DBW12	// Load DB11 data word DBW12
	L	+2	// Load integer constant +2
	*I		// Integer multiply DBW12 (ACCU2) **times** +2 (ACCU1).
	T	DB21.DBW12	// Store the product in DB21 data word DBW12

Divide Integer: /I

When the *Divide Integer* operation / **I** is executed, the integer division of the values in accumulator 1 and accumulator 2 is processed as ACCU2 **divided by** ACCU1. The result is placed in ACCU1, and is available for further processing.

In the following example, the divide integer operation is processed as the integer contents of data block DB21, word DBW12 **divided by** the integer constant +10. The ACCU1 result is transferred to the bit memory word location MW14.

Example: Division using integer values

LABL	L	DB21.DBW12	// Load DB11 data word DBW12
	L	+10	// Load integer constant +10
	/I		// Integer divide DBW12 (ACCU2) **divided by** +10 (ACCU1)
	T	MW14	// Store the quotient result to bit memory word MW14

Double Integer Arithmetic Operations

In STL, double integer arithmetic operations are denoted by **D** for DINT, which represents 32-bit integer operations. The four basic operations are **+D** for add integer, **-D** for subtract integer, ***D** for multiply double integers, and **/D** for divide double integers. Generally, you will use double integer instructions when whole number results in the range of -2,147,483,648 to +2,147,483,647 are required.

As previously stated, arithmetic operations are performed using the values currently in accumulator 1 and accumulator 2. How and when these values are loaded to the accumulators is irrelevant, but in the case of double integer operations, the values may be loaded from a double word memory location, a declared DINT variable, or directly as a double integer constant. The accumulator 1 result is available for you to transfer to a new location as required. The location to which you move the result may be a declared variable of type DINT or a double word location—for example MD 44, or DB22.DD24.

Add Double Integer: +D

When the *Add Double Integer* operation **+D** is executed, the double integer addition of the values in accumulator 1 and accumulator 2 is processed as ACCU2 **plus** ACCU1. The result is placed in ACCU1, and is available for further processing.

In the following example, the double integer add operation is processed as the double integer contents of bit memory double word MD72 **plus** the long integer constant +100. The ACCU1 result is transferred to bit memory double word location MD74.

Example: Addition using double integer values

LABL	L	MD72	// Load bit memory double word MD72
	L	L# +100	// Load long integer constant +100
	+D		// Double integer add MD72 (ACCU2) **plus** 100 (ACCU1)
	T	MD74	// Store the sum to bit memory double word MD74

Subtract Double Integer: -D

When the *Subtract Double Integer* operation - **D** is executed, the double integer subtraction of the values in accumulator 1 and accumulator 2 is processed as ACCU2 **minus** ACCU1. The result is placed in ACCU1, and is available for further processing.

In the following example, the double integer subtract operation is processed as double integer variable #VALU_3 **minus** double integer variable #VALU_4. The ACCU1 result is transferred to the double integer variable #Total_2.

Example: Subtraction using double integer values

LABL	L	#VALU_3	// Load DINT variable #VALU_3
	L	#VALU_4	// Load DINT variable #VALU_4
	- D		// Double integer subtract #VALU_3 (ACCU2) **minus** #VALU_4 (ACCU1)
	T	#Total_2	// Store the result to DINT variable #Total_2

Multiply Double Integer: *D

When the *Multiply Double Integer* operation * **D** is executed, the double integer multiplication of the values in accumulator 1 and accumulator 2 is processed as ACCU2 **times** ACCU1. The result is placed in ACCU1, and is available for further processing.

In the following example, the double integer multiply operation is processed as the double integer contents of data block DB11, data double word DBD22 **times** the long integer value of +2. The ACCU1 result is transferred to data block DB21, data double word DBD22.

Example: Multiplication using double integer values

LABL	L	DB11.DBD22	// Load DB11 data double word DBD22
	L	L# +2	// Load long integer constant +2
	*D		// Double integer multiply DBD22 (ACCU2) **times** +2 (ACCU1)
	T	DB21.DBD22	// Store the product result to DB21 double data word DBD22

Divide Double Integer: /D

When the *Divide Double Integer* operation /D is executed, the double integer division of the values in accumulator 1 and accumulator 2 is processed as ACCU2 **divided by** ACCU1. The result is in ACCU1, and is available for further processing.

In the following example, the double integer divide operation is processed as the double integer contents of data block DB21, data double word DBD26 **divided by** the long integer value of +10. The ACCU1 result is transferred to the bit memory double word location MD74.

Example: Division with quotient result using double integer values

LABL	L	DB21.DBD26	// Load DB21 data double word DBD26
	L	L# +10	// Load long integer constant +10
	/D		// Double integer divide DBD26 (ACCU2) **divided by** +10 (ACCU1)
	T	MD74	// Store the quotient result to memory double word MD74

Modulus Divide Integer: MOD

When the *Modulus Divide* operation **MOD** is executed, the double integer division of the values in accumulator 1 and accumulator 2 is processed as ACCU2 **divided by** ACCU1. The quotient remainder (fractional part of the result) is placed in ACCU1, and is available for further processing.

In the following example, the modulus divide operation is processed as the double integer contents of data block DB21, data double word DBD26 **divided by** the long integer value of +10. The ACCU1 result is transferred to the bit memory double word location MD78.

Example: Modulus division with remainder result using double integer values

LABL	L	DB21.DBD26	// Load DB21 data double word DBD26
	L	L# +10	// Load long integer constant +10
	MOD		// Modulus divide DBD26 (ACCU2) **divided by** +10 (ACCU1)
	T	MD78	// Store the quotient remainder to double word MD78

Accumulator Integer Operations

Accumulator 1 integer operations provide simple addition and subtraction operations whereby a parameter constant is directly added to or subtracted from the value in accumulator 1. These simple operations do not provide status bit error and result information, nor do they affect CPU accumulators 2, 3, and 4. For the most part, these operations are used either to index a count in the range is 0-255, or to manipulate an address. For calculations, you should use the 16-bit and 32-bit arithmetic operations.

Increment: INC <+n>

When executed, the *Increment* operation **INC** causes the low-order byte of accumulator 1 to be increased (incremented) by the parameter value <**n**>. The value **n** has a valid range of 0 to 255; meaning that the low-byte of accumulator 1 may be incremented from 0 to 255.

If the result exceeds 255, then the low-order byte of accumulator 1 resets to zero. The INC operation does not affect the upper bytes of ACCU 1 or ACCU 2. The CPU status bits are also unaffected.

In the following example, the value contained in memory byte MB40 is incremented by +4 on each pass through of the code, starting with the statement labeled INCR. On each pass, +4 is added to the previous value, and the result is in the low-byte of accumulator 1. The new result is transferred back to memory byte MB40. In this fashion the code is indexing the counter byte (MB40) by +4 on each pass of this section of code.

Example: Add (increment) the current contents of accumulator 1 by the specified constant

	L	+0	//
	T	MB40	// Initialize counter byte
INCR	L	MB40	// Load the value in memory byte MB40 to the low-byte of ACCU 1
	INC	4	// Add +4 to the value in the low-byte of ACCU 1 (MB40)
	T	MB40	// Store the incremented result to memory byte MB40
	L	4	
	== I		
	=	M101.1	

Decrement: DEC <+n>

When executed, the *Decrement* operation **DEC** causes the low-order byte of accumulator 1 to be reduced (decremented) by the parameter value <**n**>. The value **n** has a valid range of 0 to 255, meaning that the low-byte of accumulator 1 may be decremented from 255 down to 0.

If the result falls below zero, then the low-order byte of accumulator 1 resets to 255. The DEC operation does not affect the upper bytes of ACCU 1 or ACCU 2. The status bits are also unaffected.

In the following example, the value contained in memory byte MB40 is decremented by +1 on each pass through of the code, starting with the statement labeled **DECR**. On each pass, +1 is subtracted from the previous value, and the result is returned to the low-byte of accumulator 1. The new result is transferred back to MB40. In this fashion the original MB40 value is decreasing by +1, until it is zero.

Example: Subtract (decrement) the current contents of accumulator 1 by the specified constant

DECR	L	MB40	// Load value from MB40 to low-byte of ACCU1
	DEC	1	// Subtract +1 from the low-byte of ACCU1
	T	MB40	// Transfer the result (low-byte of ACCU1) to byte MB40
	L	0	// Load integer constant of 0 to the low-byte of ACCU1
	== I		// If ACCU2 = ACCU1; or (0) is equal to (MB40)
	=	M101.1	// then assign '1' to M101.1; otherwise assign '0' to M101.1

Add Constant: + <const>

When the *Add Constant* operation is executed, the constant value specified as a parameter is added to the current contents of accumulator 1. If you specify a negative constant, the value is subtracted from the accumulator.

The constant you specify may be a hexadecimal BYTE, in the range of B#16#00 to B#16#FF; a decimal (INT) value, in the permissible range of -32,768 to +32,767; or a double integer (DINT) constant, in the permissible range of L#-2,147,483,648 to L#+2,147,483,647. If you specify a positive or negative integer that exceeds the valid INT range, the value will be interpreted as a DINT value.

The add constant operation does not provide carry-over from the low byte of the low word of accumulator 1 to the high byte of the low word of accumulator 1, and therefore not equivalent to using standard 16- and 32-bit arithmetic operations.

Example: Add the specified integer constant to the current contents of accumulator 1

	L	MW12	// Load the value of MW12 into low-order word of ACCU1
	L	MW14	// Load the value of MW14 into low-order word of ACCU1
	+I		// Add low-order word of ACCU2 to low-order word of ACCU1
PLUS	+	50	// Add +50 to ACCU1; result is returned in low-order word of ACCU1
	T	MW18	// Store the result to memory word MW18

In the following example, a DINT (double integer) constant is subtracted from the accumulator:

Example: Add the specified long integer constant to the current contents of accumulator 1

	L	MD12	// Load the value of MD12 into ACCU1
	L	MD16	// Load the value of MD16 into ACCU1
PLUS	+	L#-40000	// Add DINT constant -40,000 to the integer value of ACCU1
	T	DB4.DBD2	// Store the result to data block DB4 double data word DBD2

In the following example, a hexadecimal BYTE constant is added to the accumulator:

Example: Add the specified Hex constant to the current contents of accumulator 1

	L	IB32	// Load the value of MW12 into the low-order of ACCU1
	L	IB33	// Load the value of MW14 into the low-order of ACCU1
PLUS	+	B#16#25	// Add hex constant 25 to the byte value of ACCU1
	T	MW20	// Store the result to bit memory word MW20

REAL Arithmetic Operations

In STL, REAL arithmetic operations are denoted by **R** for REAL, which represents 32-bit floating-point operations. The four basic operations are **+R** for add REAL, - **R** for subtract REAL, ***R** for multiply REAL, and **/R** for divide REAL. Generally, you will use REAL instructions when integer results or calculations are not acceptable, or when your process requires the accuracy of floating-point numbers.

As stated earlier, arithmetic operations are done using the values in accumulator 1 and accumulator 2. How and when these values are loaded to the accumulators is irrelevant, but in the case of REAL operations the values may be loaded from a double word location, from declared REAL variables, or directly as a floating-point constant. After the operation, the accumulator 1 result is available for transfer to a new location as required. The location to which you move the result may be a declared variable of type REAL or a double word memory location—for example DB22.DD12 or MD 42.

Add REAL: +R

When the *Add REAL* operation **+R** is executed, the floating-point addition of the values in accumulator 1 and accumulator 2 is processed as ACCU2 **plus** ACCU1. The result is placed in ACCU1, and is available for further processing.

In the following example, the Add REAL operation is processed as the floating-point value in MD12 **plus** the floating-point constant +100.0. The ACCU1 result is transferred to the double word MD18.

Example: Addition using REAL values

LABL	L	MD12	// Load bit memory double word MD72
	L	1.0E+02	// Load floating point constant +100.0
	+R		// REAL add MD12 (ACCU2) **plus** 100.0 (ACCU1)
	T	MD18	// Store the sum in bit memory double word MD18

Subtract REAL: -R

When the *Subtract REAL* operation **-R** is executed, the floating-point subtraction of the values in accumulator 1 and accumulator 2 is processed as ACCU2 **minus** ACCU1. The result is placed in ACCU1, and is available for further processing.

In the following example, the Subtract REAL operation is processed as the REAL variable #REAL_V1 **minus** REAL variable #REAL_V2. The ACCU1 result is transferred to the REAL variable #R_TOTAL1.

Example: Subtraction using REAL values

LABL	L	#REAL_V1	// Load REAL variable #REAL_V1
	L	#REAL_V2	// Load REAL variable #REAL_V2
	-R		// REAL subtract #REAL_V1 (ACCU2) **minus** #REAL_V2 (ACCU1)
	T	#R_TOTAL1	// Store the result in REAL variable #R_TOTAL1

Multiply REAL: *R

When the *Multiply REAL* operation ***R** is executed, the floating-point multiplication of the values in accumulator 1 and accumulator 2 is processed as ACCU2 **times** ACCU1. The result is placed in ACCU1, and is available for further processing.

In the following example, the Multiply REAL is processed as the floating-point contents of data block DB31, data double word DBD12 **times** the floating-point constant of +2.0. The ACCU1 result is transferred to data block DB41, data double word DBD12.

Example: Multiplication using REAL values

LABL	L	DB31.DBD12	// Load DB31 double data word DBD12
	L	2.0E0	// Load REAL constant +2.0
	*R		// REAL multiply DBD12 (ACCU2) **times** +2.0 (ACCU1)
	T	DB41.DBD12	// Store the result in DB41 data double word DBD12

Divide REAL: /R

When the *Divide REAL* operation **/R** is executed, the floating-point division of the values in accumulator 1 and accumulator 2 is processed as ACCU2 **divided by** ACCU1. The result is in ACCU1, and is available for further processing.

In the following example the Divide REAL is processed as the floating-point contents of data block DB41, data double word DBD12 **divided by** the floating-point constant +10.0. The ACCU1 result is transferred to the bit memory double word MD84.

Example: Division using REAL values

LABL	L	DB41.DBD12	// Load DB41 double data word DBD12
	L	1.0E01	// Load REAL constant +10.0
	/R		// REAL divide DBD12 (ACCU2) **divided by** +10.0 (ACCU1)
	T	MD84	// Store the quotient result to memory double word MD84

Combining Arithmetic Calculations

How you proceed with an extended calculation will depend on whether you are working with a 2-accumulator CPU, as in the S7-300, or with a 4-accumulator CPU, as in the S7-400. Generally, each load and transfer operation is always affecting accumulator 1—when two successive loads are made, the value previously held in accumulator 1 is moved to accumulator 2. If a value was already in accumulator 2, that value is lost in a 2-accumulator CPU.

In a CPU where there are only two accumulators, an initial or intermediate result remains in accumulator 1; to continue, you simply need to load a new value and then follow with the next calculation—of course recognizing that the operation will use ACCU 1 and ACCU 2.

Example: Extended arithmetic calculation in a 2-accumulator CPU

CLC1	L	#M	// Load INT variable #M
	L	#X	// Load INT variable #X
	*I		// Find product of #M times #X
	L	#B	// Load INT variable #B
	+I		// #MX + #B
	T	#Y	// Store value in INT variable #Y (Y = MX + B)

In a four-accumulator CPU, operations like ENT and TAK set the stage for extended, or carry-on, calculations by allowing you to manipulate data movement among the accumulators. For instance, with the ENT operation, the contents of accumulator 3 is copied to accumulator 4 and the contents of accumulator 2 is copied to accumulator 3; when the TAK operation is processed, the 32-bits of accumulator 1 are exchanged with the 32-bits of accumulator 2. These operations, and others like them, were covered earlier in the *Accumulator Operations* section of Chapter 17 *STL Load and Transfer Operations*.

In the following example, the ENT instruction is inserted prior to loading the next value in a follow-on calculation—the previous result of accumulator 2 is thereby pushed to accumulator 3. When the next arithmetic operation is carried out, the intermediate result, previously stored in accumulator 3, is copied back to accumulator 2 for use in the next calculation.

Example: Extended arithmetic calculation in a 4-accumulator CPU

CLC2	L	#Y2	// Load INT variable #Y2 (to ACCU 1)
	L	#Y1	// Load INT variable #Y1 (to ACCU1) after moving Y2 to ACCU 2
	-I		// Perform subtraction #Y2—#Y1
	L	#X2	// Load INT variable #X2 (to ACCU1) after moving Y2-Y1 result to ACCU 2
	ENT		//Copy ACCU 3 to ACCU 4; copy ACCU 2 (Y2-Y1 result) to ACCU 3.
	L	#X1	// Load INT variable #X1 (to ACCU1
	-I		// Perform subtraction ACCU 2—ACCU 1; leave (X2-X1) result in ACCU 1 and copy ACCU 3 (Y2-Y1 result) back to ACCU 2
	/I		// Divide ACCU 2 (Y2-Y1) by ACCU 1 (X2-X1)
	T	#M	// Store result in INT variable #M (#M = #Y2-#Y1 ÷ #X2-#X1)

STANDARD MATH FUNCTIONS OVERVIEW

In addition to the four basic arithmetic operations for REAL numbers, the STL instruction set includes the standard mathematical functions listed in the table below. These operations include functions for computing the angle or radian measurement, absolute value, square root, square of a number, natural log of a number, and for computing the power of a number with a radix of base e.

Table 22-3. STL Standard Mathematical Functions Summary

Operation	Brief Description
SIN	**Sine.** Determine the *sine* of the REAL angle given in radians in ACCU1.
COS	**Cosine.** Determine the *cosine* of the REAL angle given in radians in ACCU1.
TAN	**Tangent.** Determine the *tangent* of the REAL angle given in radians in ACCU1.
ASIN	**Arc Sine.** Determine the *arc sine* of the REAL source value (*-1 to +1*) in ACCU1.
ACOS	**Arc Cosine.** Determine the *arc cosine* of the REAL source value (*-1 to +1*) in ACCU1.
ATAN	**Arc Tangent.** Determine the *arc tangent* of the REAL source value (*full range*) in ACCU1.
ABS	**Absolute Value.** Determine and return the *absolute value* of REAL source value in ACCU1.
SQRT	**Square Root.** Determine and return the *square root* of the REAL source value in ACCU1.
SQR	**Square.** Determine and return the *square* of the REAL source value in ACCU1.
EXP	**Exponent.** Determine and return the *exponent* of the REAL source value in ACCU1.
LN	**Natural Log.** Determine and return the *natural log* of the REAL source value in ACCU1.

Note: The calculated result of each standard math function is returned in accumulator 1 (ACCU1).

PROGRAMMING STANDARD MATH FUNCTIONS IN STL

Standard functions are performed using floating-point (REAL) numbers. When working with these operations, you may use double word memory locations or declared REAL variables. Standard functions are generally combined with other REAL arithmetic operations to create computational algorithms and numerical control solutions.

Standard Math Functions Sequence

Like with other digital operations, standard mathematical operations are executed irrespective to the result of logic operations (RLO). Without appropriate consideration, these operations will always be processed. If standard math operations, in a given block, should be processed only under certain conditions, then based on those conditions, you will have to employ jumps or some other mechanism to circumvent execution of these operations.

In STL, standard math functions operate on the value last loaded to accumulator 1. After execution of each operation, the result remains in accumulator 1. Valid data loaded to accumulator 1 include variables of data type REAL, or the contents of double word locations like DB22.DD12 or MD44. The accumulator 1 result may be transferred to variable of data type REAL, or to a double word location.

As with arithmetic operations, the outcome of standard math functions will affect the CPU status bits. In the event the source value in accumulator 1, at execution time, is an invalid REAL number, then an invalid REAL number is returned by the function and status bits CC0, CC1, OV, and OS are set to '1'.

While the standard math functions each operate on real numbers, there are restrictions on the permissible range of input values in some of the operations, as shown in the following table.

Table 22-4. Valid Range at IN and OUT in Math Functions

IN and OUT Values of Trig Function			90º (Π/2)
Function	**IN Value**	**OUT Result**	
Sine	Full range	-1 to + 1	
Cosine	Full range	-1 to + 1	
Tangent	Full range	Full range	180° (Π) ··· 0° / 360° (2Π)
Arc Sine	-1 to + 1	-Π/2 to +Π/2	
Arc Cosine	-1 to + 1	0 to +Π	
Arc Tangent	Full range	-Π/2 to +Π/2	270° (3Π/2)

Note: Pi = 3.141592. For the trigonometric functions sine, cosine, and tangent, the greatest accuracy is achieved when the source value at **IN** is greater than -2Π and less than 2Π.

Calculate the Sine: SIN

When processed, the *sine* operation **SIN** interprets the contents of accumulator 1 as a floating-point number that represents an angle expressed in radians. The operation computes the sine of the angle and returns the result in accumulator 1. A valid result for this operation should always be greater than or equal to -1, and less than or equal to +1. The accumulator 1 result may be further operated upon or may be transferred to a double word memory location or REAL variable.

In the following example, the **SIN** function operates on the floating-point value loaded from memory double word MD42. The computed sine is transferred from ACCU1 to memory double word MD142.

Example: Find the sine of the angle given in radians

L	MD42	// Load the memory double word MD42 to ACCU1
SIN		
T	MD142	// Move the ACCU1 result to memory double word MD142

Calculate the Cosine: COS

When processed, the *cosine* operation **COS** interprets the contents of accumulator 1 as a floating-point number that represents an angle expressed in radians. The operation computes the cosine of the angle and returns the result in accumulator 1. A valid result for this operation should always be greater than or equal to -1, and less than or equal to +1. The accumulator 1 result may be further operated upon or may be transferred to a double word memory location or a REAL variable.

In the following example, the **COS** function operates on the floating-point value loaded from memory double word MD44. The computed cosine is transferred from ACCU1 to memory double word MD144.

Example: Find the cosine of the angle given in radians

L	MD44	// Load the memory double word MD44 to ACCU1
COS		
T	MD144	// Move the ACCU1 result to memory double word MD144

Calculate the Tangent: TAN

When processed, the *tangent* operation **TAN** interprets the contents of accumulator 1 as a floating-point number that represents an angle expressed in radians. The operation computes the tangent of the angle and returns a REAL value result in accumulator 1. The accumulator 1 result may be further operated upon, or may be transferred to a double word memory location or a REAL variable.

In the following example, the **TAN** function operates on the floating-point value loaded from memory double word MD46. The computed tangent is transferred from ACCU1 to memory double word MD146.

Example: Find the tangent of the angle given in radians

L	IM46	// Load the memory double word MD46 to ACCU1
TAN		
T	MD146	// Move the ACCU1 result to memory double word MD146

Calculate the Arc Sine: ASIN

When processed, the *arc sine* operation **ASIN** reads the floating-point value in accumulator 1 and computes its inverse—a, floating-point number that represents an angle in radians. A valid source value in accumulator 1 is expected in the range of -1 <= **source** <= +1. The computed arc sine, a floating-point value that represents an angle in radian measure, should always be greater than or equal to -Π/2 and less than or equal to + Π/2 (pi = 3.141593). The accumulator 1 result may be further operated upon, or may be transferred to a double word memory location or a REAL variable.

If the result exceeds the permissible range, status bits CC0, CC1, OV, and OS are set to '1'.

In the following example, the **ASIN** function operates on the floating-point value loaded from memory double word MD48. The computed result of ACCU1 is transferred to memory double word MD148.

Example: Find the arcsine of the angle given in radians

L	MD48	// Load the memory double word MD48 to ACCU1
ASIN		
T	MD148	// Move the ACCU1 result to memory double word MD148

Calculate the Arc Cosine: ACOS

When processed, the *arc cosine* operation **ACOS** reads the floating-point value in accumulator 1 and computes its inverse—a floating-point number that represents an angle in radians. A valid source value in accumulator 1 is expected in the range of -1 <= **source** <= +1. The computed arc cosine, a floating-point value that represents an angle in radian measure, should always be greater than or equal to zero, and less than or equal to + Π (pi = 3.141593). The accumulator 1 result may be further operated upon, or may be transferred to a double word memory location or a REAL variable.

If the result exceeds the permissible range, status bits CC0, CC1, OV, and OS are set to '1'.

In the example below, the **ACOS** function operates on the floating-point value loaded from memory double word MD50. The computed sine is transferred from ACCU1 to memory double word MD150.

Example: Find the arc cosine of the angle given in radians

L	MD50	// Load the memory double word MD50 to ACCU1
ACOS		
T	MD150	// Move the ACCU1 result to memory double word MD150

Calculate the Arc Tangent: ATAN

When processed, the *arc tangent* operation **ATAN** reads the floating-point value in accumulator 1 and computes its inverse—a, floating-point number that represents an angle in radians. The computed arc tangent, a floating-point value that represents an angle in radian measure, should always be greater than or equal to -Π/2, and less than or equal to + Π/2 (pi = 3.141593). The accumulator 1 result may be further operated upon, or may be transferred to a double word memory location or a REAL variable.

If the result exceeds the permissible range, status bits CC0, CC1, OV, and OS are set to '1'.

In the example below, the **ATAN** function operates on the floating-point value loaded from memory double word MD42. The computed sine is transferred from ACCU1 to memory double word MD142.

Example: Find the arc tangent of the angle given in radians

L	MD42	// Load the memory double word MD42 to ACCU1
ATAN		
T	MD142	// Move the ACCU1 result to memory double word MD142

Absolute Value: ABS

When executed, the *Absolute Value* operation **ABS** computes the absolute value of the floating-point number in accumulator 1, simply by changing the sign bit of the mantissa to '0'. The result, which is always a positive form of the original value, is returned in accumulator 1. If the source value is a positive number—for example +5, then the computed result is +5. If the source value is a negative number—for example -5, then the computed result is +5. The **ABS** operation outcome does not affect CPU status bits.

In the following example, the floating-point value in memory double word MD110 is multiplied by the floating-point value in MD114, after first ensuring that the absolute value MD110 is used. The result is transferred to memory double word MD210.

Example: Perform calculation after first computing the absolute value of the multiplier

CMPT	L	2.0E+00	// Load the constant value 2.0 to accumulator 1
	L	MD110	// Load the memory double word MD110 to ACCU1
	ABS		// Compute the absolute value of the value in ACCU 1 (MD110)
	*R		// Find the product of the 2.0 times the absolute value of (MD110)
	T	MD210	// Move the result to memory double word MD210

Calculate the Square Root: SQRT

When processed, the *square root* operation **SQRT** reads the floating-point value in accumulator 1 and computes its square root—a floating-point value returned to accumulator 1. The source value of accumulator 1 should be greater than or equal to zero. The square root operation always returns a positive result when the source value in accumulator 1 is greater than zero.

If the source value is minus zero, then minus zero is returned. If there is an attempt to extract the square root of a negative number other than zero, an invalid REAL number is returned in accumulator 1 and status bits CC0, CC1, OV, and OS are set to '1'.

In the example below, the **SQRT** function operates on the floating-point value loaded from memory double word MD42. The square root is transferred from ACCU1 to memory double word MD142.

Example: Calculate the square root

	L	MD42	// Load the memory double word MD42 to ACCU1
	SQRT		
	T	MD142	// Move the ACCU1 result to memory double word MD142

Calculate the Square: SQR

When processed, the *square* operation **SQR** reads the floating-point value in accumulator 1 and computes its square—a floating-point value that is returned in accumulator 1.

In the following example, the local variables #A and #B are each squared using the **SQR** function, and then added as two REAL values, whose sum is returned in accumulator 1. The **SQRT** functions computes the square root of accumulator 1 ($A^2 + B^{2)}$; the result is transferred to the local variable #C.

Example: Return the square of a number

L	#A	// Load the declared variable #A to ACCU 1
SQR		// Compute the square of #A
L	#B	// Load the declared variable #B to ACCU 1
SQR		// Compute the square of #B
+R		// Get the sum of $A^2 + B^2$
SQRT		// Compute the square root of $A^2 + B^2$
T	C#	// Move the ACCU1 result to the declared variable #C

Calculate the Exponent: EXP

When processed, the *exponent* operation **EXP** reads the floating-point value in ACCU 1 and computes the exponential power. The value is computed from base e (= 2.718282e+0) and the floating-point value in accumulator 1, using the expression (e^{IN} = **EXP**). The result is returned in ACCU 1.

In the following example, the **EXP** function operates on the floating-point value loaded from memory double word MD42. The computed exponent is transferred to memory double word MD142.

Example: Determine the exponential value from base e (2.718282e +00) and the value in ACCU 1

L	MD42	// Load the memory double word MD42 to ACCU1
EXP		
T	MD142	// Move the ACCU1 result to memory double word MD142

Calculate the Natural Log: LN

When processed, the *Natural Log* operation **LN** computes the natural logarithm (of base e, base e = 2.718282e+00) of the floating-point value in accumulator 1. The floating-point result is returned in accumulator 1. If the value in accumulator 1 is less than or equal to zero, an invalid REAL number is returned by the function, and status bits CC0, CC1, OV, and OS are set to '1'.

In the following example, the **LN** function operates on the floating-point value loaded from memory double word MD42. The computed natural log is transferred to memory double word MD142.

Example: Find the natural log of a number

L	MD42	// Load the memory double word MD42 to ACCU1
LN		
T	MD142	// Move the ACCU1 result to memory double word MD142

Chapter 23

STL Shift-Rotate Operations

In this chapter, you'll be introduced to STL Shift-Rotate operations. In your S7 program development, shift-rotate operations are used to manipulate the contents of a memory location or variable by moving the data by some number of bit positions to the left or to the right. The shift and rotate operations are often used in tracking applications. Shift operations specifically, may be used to perform quick multiplications or divisions by two, or by any multiple of two.

Main Topics

- STL Shift-Rotate Operations Overview
- Programming Shift-Rotate Operations in STL

STL SHIFT-ROTATE OPERATIONS OVERVIEW

STL *shift-rotate instructions* are listed below. These operations are used move the bit contents of a location to the left or right by a specified number of positions. These instructions are often used in materials conveying applications where information associated with discrete parts must be tracked, while in motion, for some distance and usually prior to performing a secondary control operation.

As seen in Table 23-1, shift operations are available for both 16-bit and 32-bit operations, whereas rotate operations are performed on the full 32-bits of accumulator 1. A closer look at the operations will also reveal that each operation has two forms—in the first form the number of bit positions (**n**) by which to shift or rotate is specified as a parameter with the instruction; in the second form, the number of bit positions to shift or rotate is not explicitly expressed, but instead is loaded from accumulator 2.

Table 23-1. STL Shift-Rotate Operations Summary

Operation	Brief Description
SLW <n>	**Shift Left Word.** Shift low-order 16-bits of ACCU1 by **n**-bits to the left.
SLW	**Shift Left Word.** Shift low-order 16-bits of ACCU1 by **N**-bits to the left—**N** is in ACCU2.
SRW <n>	**Shift Right Word.** Shift low-order 16-bits of ACCU1 by **n**-bits to the right.
SRW	**Shift Right Word.** Shift low-order 16-bits of ACCU1 by **N**-bits to the right—**N** is in ACCU2.
SLD <n>	**Shift Left Double Word.** Shift 32-bits of ACCU1 by **n**-bits to the left.
SLD	**Shift Left Double Word.** Shift 32-bits of ACCU1 by **N**-bits to the left—**N** is in ACCU2.
SRD <n>	**Shift Right Double Word.** Shift 32-bits of ACCU1 by **n**-bits to the right.
SRD	**Shift Right Double Word.** Shift 32-bits of ACCU1 by **N**-bits to the right—**N** is in ACCU2.
RLD <n>	**Rotate Left Double Word.** Rotate 32-bits of ACCU1 by **n**-bits to the left.
RLD	**Rotate Left Double Word.** Rotate 32-bits of ACCU1 by **N**-bits to the left—**N** is in ACCU2.
RRD <n>	**Rotate Right Double Word.** Rotate 32-bits of ACCU1 by **n**-bits to the right.
RRD	**Rotate Right Double Word.** Rotate 32-bits of ACCU1 by **N**-bits to the right—**N** is in ACCU2.
SSI <n>	**Shift Signed Integer.** Shift INT value of ACCU1 by **n**-bits to the right.
SSI	**Shift Signed Integer.** Shift INT value of ACCU1 by **N**-bits to the right—**N** is in ACCU2.
SSD <n>	**Shift Signed Double Integer.** Shift DINT value of ACCU1 by **n**-bits to the right.
SSD	**Shift Signed Double Integer.** Shift DINT value of ACCU1 by **N**-bits to the right—**N** is in ACCU2.
RLDA	**Rotate Left via CC1.** Rotate 32-bits of ACCU1 left by 1-bit through status bit CC1.
RRDA	**Rotate Right via CC1.** Rotate 32-bits of ACCU1 right by 1-bit through status bit CC1.

PROGRAMMING SHIFT-ROTATE OPERATIONS IN STL

In a shift operation, the bits vacated on the shift end are filled with zeros; bits shifted out on the opposite end are lost. In a rotate operation, bits pushed out on the opposite end are carried around to fill the vacated positions. *Shift Signed Integer* and *Shift Signed Double Integer* operations shift the bits of a 16-bit or 32-bit integer, thereby changing the integer value, but without altering the original sign. The bit positions vacated in a shift integer operation are always filled with the original sign bit.

Shift Operations Sequence

The shift operation sequence, like other digital operations, is executed regardless of the RLO—without appropriate consideration, they will always be processed. If, in a given block, shift-rotate operations should be processed only on certain conditions, then based on those conditions you will need to use jumps or some other mechanism to determine when the operations should be processed.

As previously mentioned, there are two general sequences for constructing the shift operation in STL. The first method, as shown directly below, involves two subsequent load operations, the first of which is the number of bit positions by which to shift. The second load operation is of the location or variable to be shifted. When the shift variable is loaded, the previously loaded value (**n**) is moved to accumulator 2. Next the shift operation is performed, using the value (**n**), in accumulator 2. The shifted result, which remains in accumulator 1, may be stored, or used in a follow-on operation—including a subsequent shift operation.

Example: Shift sequence 1—the positions by which to shift <N> is in accumulator 2

SHF4	L	MW4	// Load **N** as the number of shift positions	LAD/FBD Box
	L	MD40	// Load address to be shifted	SHR_DW
	SRD		// Shift the contents of accumulator 1 by n-positions	EN ENO
	T	MD140	// Store shifted result	MD40 — IN OUT —
				MW4 — N

The second sequence for coding the shift operation is shown below. In this shift method, the number of positions to shift (**n**) is specified as a statement parameter. First, the shift location or variable is loaded to accumulator 1. Next, the shift operation is inserted along with the parameter <**n**> to specify the number of bit positions by which to shift the value in accumulator 1. Finally the result, returned in accumulator 1, may be stored, or used in a follow-on operation.

Example: Shift sequence 2—the positions by which to shift is specified as a parameter <n>

SHF3	L	MW20	// Load address/variable to be shifted
	SRW	2	// Shift the contents of accumulator 1 by n-positions
	T	MW120	// Store shifted result

Shift WORD and DWORD Operations

How or when this value is loaded to the accumulator is irrelevant, but is up to you. When loading a value for one of the 16-bit operations, you must load a 16-bit constant, word-width memory locations, or an appropriate 16-bit variable. If loading values for a 32-bit operation, the value may be a 32-bit constant, a double word memory location, or a 32-bit variable. Execution of each shift operation will influence the CPU status bits, which you can learn how to evaluate in Chapter 25 *STL Status Bit Operations.*

A typical application of the shift operation is to multiply by a factor of two for each position shifted to the left; or to divide by a factor of two, for each position shifted to the right. Each shift represents a factor of 2^n, where n is the number of positions shifted. In other words $2^1 = 2$, $2^2 = 4$, $2^3 = 8$, $2^4 = 16$, $2^5 = 32$, and so on. So, a variable is multiplied by 2 simply by shifting to the left by one. A shift by three positions to the left multiplies by 2^3 or 8. Conversely, shifting by 1 position to the right, results in division by 2; shifting 2 places to the right is division by 2^2 or 4.

Shift Left Word: SLW <n>; SLW

When processed, the *Shift Left Word* operation **SLW** shifts the low-order 16-bits of accumulator 1 bitwise by **n**-positions to the left. The n-positions vacated on the right are filled with zeros. The shifted result, with bits 16-31 unaffected, remains in accumulator 1 and is available for further processing.

You may specify the number of shift positions (n= 0-15) as a statement parameter in the **SLW <n>** format; or as a positive integer (0-255) in the low-byte of accumulator 2. In either format, if **n** is within the permissible range, the last shifted bit value is assigned to status bit CC1; if **n** = 0, the operation is treated as a **NOP** (no operation); and if 0 < **n** ≤ 15 then status bits CC0 and OV are reset. If the value **n** of ACCU2 is greater than 16, then zero (W#16#0000) is returned in ACCU1, and status bits CC1, CC0, and OV are reset to '0'.

In the following example, the value contained in memory word MW20 (6) is shifted left by 3 positions—the 3-bits vacated on the right are filled with '0'. A shift to the left by 3 is equivalent to multiplying by 2^3 or +8. The result is equivalent to the multiplication of 6 times 8, or 48.

Example: Shift Left Word using SLW, where bit-positions <n> is explicitly specified

SHF1	L	MW20	// Load memory word MW20 to low-order word of ACCU1
	SLW	3	// Shift the low-order bits 0-15 of ACCU1 (MW20) left by 3-positions
	T	MW120	// Move the shifted ACCU1 result to memory word MW120

The previous example is illustrated below:

Example: Shift Left Word	
0000 0000 0000 0110	**MW20** = 6
	n = 3 positions
0000 0000 0011 0000	**MW120** = Shifted Result = 48

Figure 23-1. Illustration of Shift Left Word

Shift Left Double Word: SLD <n>; SLD

When processed, the *Shift Left Double Word* operation **SLD** shifts the 32-bit contents of accumulator 1 bitwise by n-positions to the left. The n-positions vacated on the right are filled with zeros. The shifted result remains in accumulator 1 and is available for further processing.

You may specify the number of shift positions (n= 0-32) as a statement parameter in the **SLD <n>** format; or as a positive integer (0-255) in the low-byte of accumulator 2. In either format, if **n** is within the permissible range, the last shifted bit value is assigned to status bit CC1; if n = 0, the operation is treated as a **NOP** (no operation); and if 0 < **n** ≤ 32 then status bits CC0 and OV are reset. If the value **n** of ACCU2 is greater than 32, then zero (DW#16#0000) is returned in ACCU1, and status bits CC1, CC0, and OV are reset to '0'.

In the following example, the value DW#16#38FB in double word MD44 is shifted left by 8 positions. The value n = 8, loaded first, is in accumulator 2. The 8-bits vacated on the right are filled with '0'.

Example: Shift Left Double Word using SLD, where bit-positions <N> is in accumulator 2

SHF2	L	+8	// Load integer constant +8 (**n**-positions) to ACCU1
	L	MD44	// Load MD44 to ACCU1. Move old ACCU1 value (n-positions) to ACCU2
	SLD		// Shift 32-bits of ACCU1 left by **8**-positions
	T	MD140	// Move shifted result from ACCU1 to memory double word MD140

The previous example is illustrated below:

Example: Shift Left Double Word	
0000 0000 0000 0000 0011 1000 1111 1011	**MD44** = DW#16#38FB
	ACCU 2 = N = 8 positions
0000 0000 0011 1000 1111 1011 0000 0000	**MD140 = Shifted Result**

Figure 23-2. Illustration of Shift Left Double Word

Shift Right Word: SRW <n>; SRW

When processed, the *Shift Right Word* operation **SRW** shifts the low-order 16-bits of accumulator 1 bitwise by **n**-positions to the right. The n-positions vacated on the left are filled with zeros. The shifted result, with bits 16-31 unaffected, remains in accumulator 1 and is available for further processing.

You may specify the number of shift positions (**n**= 0-15) as a statement parameter in the **SRW <n>** format; or as a positive integer (0-255) in the low-byte of accumulator 2. In either format, if **n** is within the permissible range, the last shifted bit is assigned to status bit CC1; if **n** = 0, the operation is treated as a **NOP** (no operation); and if $0 < n \leq 15$, then status bits CC0 and OV are reset. If the value **n** of ACCU2 is greater than 16, then zero (W#16#0000) is returned in ACCU1, and status bits CC1, CC0, and OV are reset to '0'.

In the following example, the value in memory word MW20 (128) is shifted right by 2 positions—the 2-bits vacated on the left are filled with '0'. A shift to the right by 1 is equivalent to dividing by 2^1 or +2. The result is equivalent to 128 divided by +2, or 64. The result is placed in MW120.

Example: Shift Right Word using SRW, where bit-positions <n> is explicitly specified

SHF3	L	MW20	// Load memory word MW20 contents to low-order word of ACCU1
	SRW	1	// Shift the low-order 16-bits of ACCU1 (MW20) right by 1-position
	T	MW120	// Move the shifted ACCU1 result to memory word MW120

The previous example is illustrated below:

Example: Shift Right Word	
0000 0000 1000 0000	**MW20 = 128**
	n = 1 position
0000 0000 0010 0000	**MW120** = Shifted Result = 64

Figure 23-3. Illustration of Shift Right Word

Shift Right Double Word: SRD <n>; SRD

When processed, the *Shift Right Double Word* operation **SRD** shifts the 32-bits of accumulator 1 bitwise by n-positions to the right. The n-positions vacated on the left are filled with zeros. The shifted result remains in accumulator 1 and is available for further processing.

You may specify the number of shift positions (n= 0-32) as a statement parameter in the **SRD <n>** format; or as a positive integer (0-255) in the low-byte of accumulator 2. In either format, if **n** is within the permissible range, the last shifted bit value is assigned to status bit CC1; if n = 0, the operation is treated as a **NOP** (no operation); and if 0 < **n** ≤ 32, then status bits CC0 and OV are reset. If the value **n** of ACCU2 is greater than 32, then zero (DW#16#0000) is returned in ACCU1, and status bits CC1, CC0, and OV are reset to '0'.

In the example below, the value in double word MD44 is shifted right by 16 positions. The value n=16, in MW40, is loaded from accumulator 2. The 16-bits vacated on the left are filled with '0'.

Example: Shift Right Double Word using SRD, where bit-positions <N> is in accumulator 2

SHF4	L	MW40	// Load memory word MW40 (**N** = 16) to ACCU1
	L	MD44	// Load MD44 to ACCU1. Move old ACCU1 value (**N=16**) to ACCU2
	SRD		// Shift 32-bits of ACCU1 right by **16** -positions; **N** is in ACCU2
	T	MD140	// Move shifted result from ACCU1 to memory double word MD140

The previous example is illustrated below:

Example: Shift Right Double Word	
1000 0000 0000 0000 0000 0000 0000 0000	**MD44**
	MW40 = N = 16 positions
0000 0000 0000 0000 1000 0000 0000 0000	**MD140** = Shifted Result

Figure 23-4. Illustration of Shift Right Double Word

Shift Signed Integer: SSI <n>; SSI

When processed, *the Shift Signed Integer* operation **SSI** shifts the low-order 16-bits of accumulator 1, including the sign bit, bitwise by n-positions to the right. Each position vacated on the left is filled with the value of the original sign bit (bit-15); '0' if the number is positive, '1' if the number is negative. The shifted result, with bits 16-31 unaffected, remains in accumulator 1.

You may specify the number of shift positions (n= 0-15) as a statement parameter in the **SSI <n>** format; or as a positive integer (0-255) in the low-byte of accumulator 2. In either format, if n is within the permissible range, the last shifted bit value is assigned to status bit CC1; if n = 0, the operation is treated as a NOP (no operation); and if 0 < n ≤ 15, then status bits CC0 and OV are reset. If the value n (ACCU2) is ≥ 16, then depending on the sign, the shift result (ACCU1) is W#16#0000 and CC1 = 0; the shift result is W#16#FFFF and CC1 = 1.

In the following example, the integer value (+64) in memory word MW20 is shifted right by 2 positions. A shift to the right by 2 is equivalent to dividing by 2^2 or +4. The result is equivalent to the division of +64 by +4, or +16. Notice that each position vacated on the left is filled with the original sign bit of '0'.

Example: Shift Signed Integer using SSI, where bit-positions <n> is explicitly specified

SHF5	L	MW20	// Load the signed 16-bit integer of memory word MW20 to ACCU1
	SSI	2	// Shift the 16-bit integer of ACCU1 (MW20) right by 2-positions
	T	MW120	// Move the shifted ACCU1 result to memory word MW120

The previous example is illustrated below:

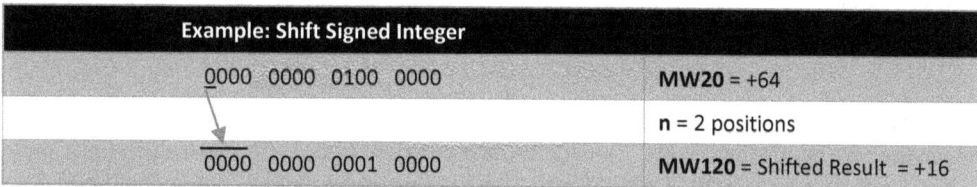

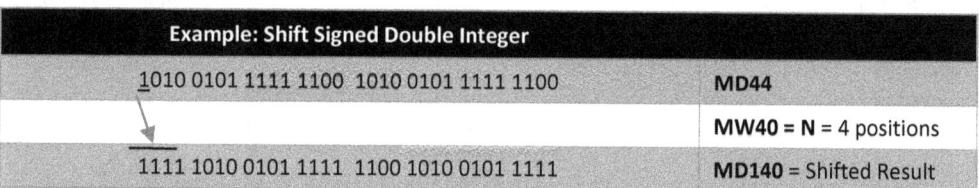

Example: Shift Signed Integer	
0000 0000 0100 0000	**MW20** = +64
	n = 2 positions
0000 0000 0001 0000	**MW120** = Shifted Result = +16

Figure 23-5. Illustration of Shift Right Integer

Shift Signed Double Integer: SSD <n>; SSD

When processed, the *Shift Signed Double Integer* operation **SSD** shifts the 32-bits of accumulator 1 bitwise by n-positions to the right. Each position vacated on the left is filled with the value of the original sign bit (bit-31); '0' if the number was positive, '1' if the number was negative. The shifted result remains in accumulator 1 and is available for further processing.

You may specify the number of shift positions (n= 0-32) as a statement parameter in the **SSD <n>** format; or as a positive integer (0-255) in the low-byte of accumulator 2. In either format, if n is within the permissible range, the last shifted bit value is assigned to status bit CC1; if n = 0, the operation is treated as a NOP (no operation); and if $0 < n \le 31$, then status bits CC0 and OV are reset. If the value n of ACCU2 is ≥ 32, then depending on the sign, the shift result (ACCU1) is DW#16#0000_0000 and CC1 = 0; the shift result is DW#16#FFFF_FFFF and CC1 = 1.

In the following example, the integer value in MD44 is shifted right by 4. The value n = 4, in MW40, is loaded from accumulator 2. The 4-bits vacated on the left are filled with the original sign bit of '1'.

Example: Shift Signed Double Integer using SSD, where bit-positions <N> is in accumulator 2

SHF5	L	MW40	// Load contents of memory word MW40 (**N** -value) to ACCU1
	L	MD44	// Load the contents of memory double word MD44 to ACCU1. Move old ACCU1 value (shift value **N**) to ACCU2
	SSD		// Shift 32-bit integer of ACCU1 right by **N**-positions; **N** is in ACCU2
	T	MD140	// Move shifted result from ACCU1 to memory double word MD140

The previous example is illustrated below:

Example: Shift Signed Double Integer	
1010 0101 1111 1100 1010 0101 1111 1100	**MD44**
	MW40 = N = 4 positions
1111 1010 0101 1111 1100 1010 0101 1111	**MD140** = Shifted Result

Figure 23-6. Illustration of Shift Right Double Integer

Rotate Double Word Operations

In a rotate operation, bits pushed out on the opposite end are carried around to fill the vacated positions. Unlike with shift operations, which support shifting 16-bit and 32-bit variables, rotate operations only involve 32-bit variables. All of the rotate operations are 32-bit operations—therefore you must load a 32-bit constant, double word location, or a 32-bit variable.

Rotate Left Double Word: RLD <n>; RLD

When processed, *the Rotate Left Double Word* operation **RLD** shifts the 32-bits of accumulator 1 bitwise by n-positions to the left. The n-positions vacated on the right are filled with the n-bit values shifted out on the left. The result remains in accumulator 1 and is available for further processing.

You may specify the number of rotate positions (n= 0-32) as a statement parameter in the **RLD <n>** format; or as a positive integer (0-255) in the low-byte of accumulator 2. If n = 0 in either format, the operation is treated as a **NOP** (no operation). When RLD is executed, if n > 32, the last bit shifted is moved to the CC1 status bit, and accumulator 1 is shifted based on modulo 32—for example the shift variable is shifted by 1 position, n = 33; shifted by 2, n = 34; shifted by 3 position, n = 35; and so on.

In the following example, the value in memory double word MD20 is rotated left by 4 positions—the 4-bits vacated on the right (1100) are filled with the 4-bits rotated out on the left (1010).

Example: Rotate Left Double Word using RLD, where bit-positions <n> is explicitly specified

SHF6	L	MD20	// Load the contents memory double word MD20 to ACCU1
	RLD	4	// Rotate the 32-bits of ACCU1 (MD20) left by 4-positions
	T	MD24	// Move the rotated result from ACCU1 to memory double word MD24

The previous example is illustrated below:

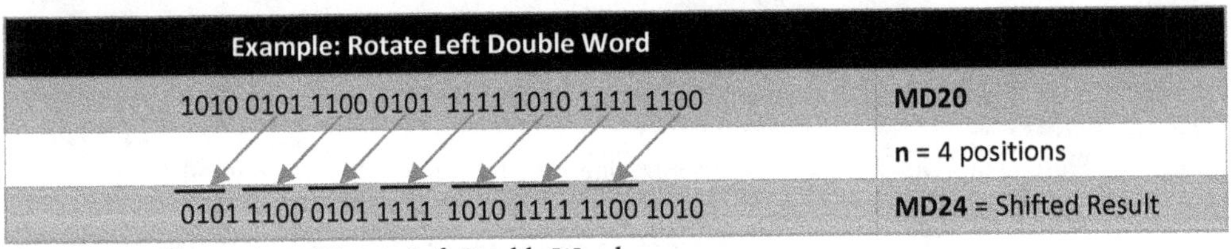

Figure 23-7. Illustration of Rotate Left Double Word

Rotate Right Double Word: RRD <n>; RRD

When processed, the *Rotate Right Double Word* operation **RRD** shifts the 32-bits of accumulator 1 bitwise by n-positions to the right. The n-positions vacated on the left are filled with the n-bit values shifted out on the right. The result remains in accumulator 1 and is available for further processing.

You may specify the number of rotate positions (n= 0-32) as a statement parameter in the **RRD <n>** format; or as a positive integer (0-255) in the low-byte of accumulator 2. If n = 0 in either case, the operation is treated as a **NOP** (no operation). When RRD is executed, if n > 32, the last bit shifted is moved to the CC1 status bit, and accumulator 1 is shifted based on modulo 32—for example the shift variable is shifted by 1 position, n = 33; shifted by 2, n = 34; shifted by 3 position, n = 35; and so on.

In the following example, the value contained in memory double word MD 44 is rotated right by 4 positions. The value n = 4, contained in MW40, is loaded from accumulator 2. The 4-bits vacated on the left (1010) are filled with the 4-bits rotated out on the right (1100).

Example: Rotate Right Double Word using RRD, where bit-positions <N> is in accumulator 2

SHF7	L	+4	// Load the integer constant +4 (**N**-positions) to ACCU1
	L	MD44	// Load contents of memory double word MD44 to ACCU1. Move old ACCU1 value (shift value **N**) to ACCU2
	RRD		// Rotate 32-bits of ACCU1 right by **N** -positions; **n** is in ACCU2
	T	MD140	// Move rotated result from ACCU1 to memory double word MD140

The previous example is illustrated below:

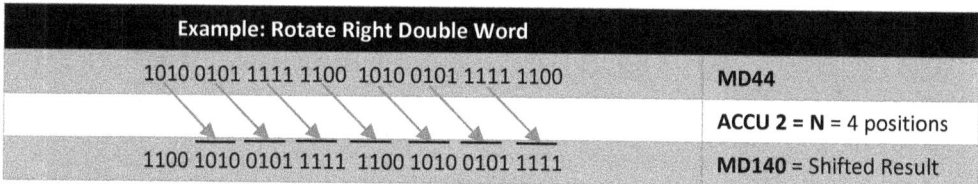

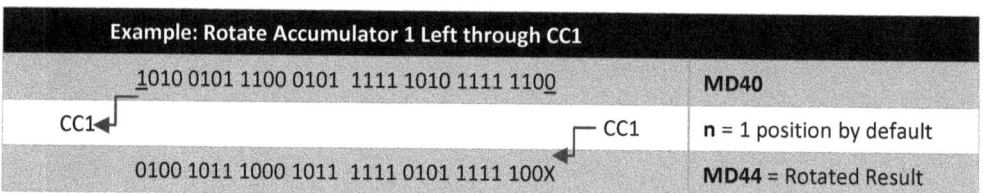

Example: Rotate Right Double Word	
1010 0101 1111 1100 1010 0101 1111 1100	**MD44**
	ACCU 2 = N = 4 positions
1100 1010 0101 1111 1100 1010 0101 1111	**MD140** = Shifted Result

Figure 23-8. Illustration of Rotate Right Double Word

Rotate Left through CC1: RLDA

When processed, the *Rotate Left through CC1* operation **RLDA** shifts the 32-bits of accumulator 1 bitwise by one position to the left. The number of rotate positions defined in this operation is implicitly one position, and is fixed. The 1-position vacated on the right (bit-0) assumes the current CC1 status; the status bit CC1 assumes the state of the bit rotated out on the left (bit-31); and status bit CC0 is set to '0'. The rotated result remains in accumulator 1 and is available for further processing.

In the example below, the 32-bits of accumulator 1 are rotated left by 1 position—the bit vacated on the right (bit-0) assumes the CC1 status; CC1 assumes the status of bit-31, which is '1' in this example.

Example: Rotate Left through CC1 status bit using RLDA

SHF9	L	MD40	// Load contents of memory word MD40 to ACCU1
	RLDA		// Rotate the 32-bits of ACCU1 (MD40) left by 1-position, through CC1
	T	MD44	// Move the shifted ACCU1 result to memory word MD44

The previous example is illustrated below:

Example: Rotate Accumulator 1 Left through CC1	
1010 0101 1100 0101 1111 1010 1111 1100	**MD40**
CC1 ◄┘ ┌ CC1	**n** = 1 position by default
0100 1011 1000 1011 1111 0101 1111 100X	**MD44** = Rotated Result

Figure 23-9. Illustration of Rotate Left through CC1

Rotate Right through CC1: RRDA

When processed, the *Rotate Right through CC1* operation **RRDA** shifts the 32-bits of accumulator 1 bitwise by one position to the right. The number of rotate positions defined in this operation is implicitly one position, and is fixed. The 1-position vacated on the left (bit-31) assumes the current CC1 status; the status bit CC1 assumes the state of the bit rotated out on the right (bit-0); and status bit CC0 is set to '0'. The rotated result remains in accumulator 1 and is available for further processing.

In the example, the 32-bits of accumulator 1 are rotated right by 1 position—the bit vacated on the left (bit-31) assumes the CC1 status; CC1 assumes the status of bit-0, which is '0' in this example.

Example: Rotate Right through CC1 status bit using RRDA

SHFA	L	MD40	// Load contents of memory word MD40 to ACCU1
	RRDA		// Rotate the 32-bits of ACCU1 (MD40) right by 1-position, through CC1
	T	MD44	// Move the shifted ACCU1 result to memory word MD44

The previous example is illustrated below:

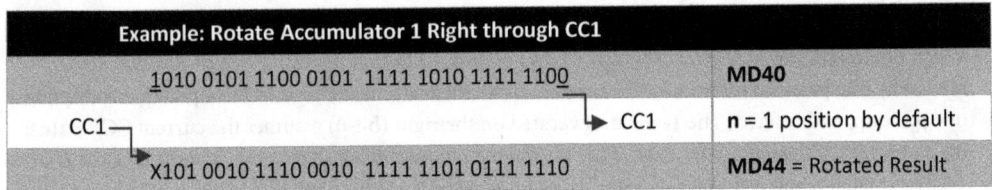

Example: Rotate Accumulator 1 Right through CC1		
1010 0101 1100 0101 1111 1010 1111 1100		MD40
CC1 ⟶ ⟶ CC1		n = 1 position by default
X101 0010 1110 0010 1111 1101 0111 1110		MD44 = Rotated Result

Figure 23-10. Illustration of Rotate Right through CC1

Chapter 24

STL Word Logic Operations

This chapter introduces STL Word Logic instructions. In S7 program development, these operations are used to perform bit-wise Boolean logic combinations of S7 memory locations or variables. Using Word Logic operations, you can perform standard Boolean operations of AND, OR and Exclusive OR on two 16-bit (word) or two 32-bit (double word) locations or variables.

Word Logic operations generally serve as a mechanism for applying various masking techniques for manipulating specific bit positions in word or double word patterns.

Main Topics

- STL Word Logic Operations Overview
- Programming Word Logic Operations in STL
- Word Logic Applications

STL WORD LOGIC OPERATIONS OVERVIEW

With STL *word logic operations,* you can perform standard Boolean operations of AND, OR, and Exclusive OR in a bit-wise manner using WORD (16-bit) or DWORD (32-bit) variables. As shown in the table, the Boolean functions are represented by the letters **A** for AND, **O** for OR, and **XO** for Exclusive OR. A WORD operation is represented by **W**, and a DWORD operation by **DW**. Therefore, the three 16-bit operations are **AW, OW, XOW**; the 32-bit operations are **AD, OD,** and **XOD**.

Table 24-1. STL Word Logic Operations Summary

Operation	Brief Description
AW	**AND WORD.** Use AND function to combine low-order 16-bits of ACCU1 and ACCU2.
OW	**OR WORD.** Use OR function to combine low-order 16-bits of ACCU1 and ACCU2.
XOW	**Exclusive OR WORD.** Use XOR to combine low-order 16-bits of ACCU1 and ACCU2.
AD	**AND DWORD.** Using AND function, combine 32-bits of ACCU1 and ACCU2.
OD	**OR DWORD.** Using OR function, combine 32-bits of ACCU1 and ACCU2.
XOD	**Exclusive OR DWORD.** Using XOR function, combine 32-bits of ACCU1 and ACCU2.

Note: Each operation is performed bitwise (bit-by-bit) according to its Boolean function and Truth Table.

PROGRAMMING WORD LOGIC OPERATIONS IN STL

In S7 program development, word logic operations perform bit-wise Boolean logic combinations of S7 word or double word locations or variables. With these instructions, and techniques referred to as *masking,* you are able to selectively set bit positions to '1' or '0'. It is also possible to selectively invert multiple bits of a location or variable and to determine what bits differ from an expected bit pattern.

Word Logic Operations Sequence

Word logic operations, like other digital operations, are executed regardless of the RLO—without appropriate consideration, these operations will always be processed. If, in a given block, word logic operations should be processed only on certain conditions, then based on those conditions you will need to use a jump or some other mechanism to determine when the operations should be processed.

Example: Standard format of a word logic operation sequence

				LAD/FBD Box Equivalent
LABEL	L	IW22	// Load first value to ACCU1	WOR_W
	L	MW22	// Load second value to ACCU1	EN ENO
	OW		// Perform word logic operation	IW22 — IN1 OUT — QW42
	T	QW42	// Store accumulator 1 result	MW22 — IN2

The general sequence for carrying out a word logic operation, as shown above, first involves two subsequent load operations that place the two values to be operated upon in accumulators 1 and 2. Next, there is the word logic operation to be performed. The result, which is always left in accumulator 1, may be stored, or may be used in a follow-on calculation after another value is loaded. Later in Chapter 25 *STL Status Bit Operations,* you can learn about how the execution of word logic operations influences the CPU status bits, and may be evaluated in your program.

16-bit Word Logic Operations

As previously stated, word logic operations are performed using the values currently in accumulator 1 and accumulator 2. How or when these values are loaded to the accumulators is irrelevant. When loading the accumulators for 16-bit operations, you may load WORD constants, variables of type WORD, or word-width locations like IW8, MW28, or DB22.DW12. The accumulator 1 result can be transferred to a new location as required. The location to which you move the result may be a variable of type WORD or a word-width memory location—for example MW 44, QW38, or DB10.DW26.

AND Word: AW

When executed, the *AND Word* operation **AW** applies a bit-wise logical AND according to the AND truth table, on the lower 16-bits of ACCU1 and ACCU2. The result is placed in the corresponding bit positions of the lower 16-bits of ACCU1, while the contents of ACCU2 are left unchanged. According to the AND truth table, the output is '1' only if both inputs are '1', and is '0' if either of the inputs is '0'.

In the example below, the AND Word serves as a 'mask-out' operation. To mask-out a bit from a source word—IW22 in this case, requires a '0' in the same bit position(s) of the mask variable. In this example, MW22 is the mask. Notice, that the upper eight bits of the result are all zeros or masked-out. Only the lower eight bits of IW22 passed through to the result. To leave any bit of the source word unaltered in the result, place a value of '1' in the same position of the mask word. This masking effect of the AND function, as described earlier, is based on the rules '0' **AND** 'X' = '0' and '1' **AND** 'X' = X.

AND Truth Table			Example: AND Word IW22 and MW22	
IN1	IN2	OUT	1010 0101 1011 1100	WORD—IW22 (ACCU2)
0	0	0	AND	
0	1	0	0000 0000 1111 1111	WORD—MW22 (ACCU1)
1	0	0	=	
1	1	1	0000 0000 1011 1100	WORD RESULT (ACCU1)

Figure 24-1. Illustration of AND Truth Table and AND Word operation

The STL code for this example is shown below:

Example: Combining 16-bit variables using the logic AND function AW

L	IW22	// Load Input Word IW22 into ACCU1
L	MW22	// Load bit memory Word MW22 into ACCU1. Shift old ACCU1 to ACCU2
AW		// Combine lower 16-bits of ACCU2 and ACCU1 according to logical AND
T	MW42	// Transfer ACCU1 results to bit memory word MW42

In addition to loading each value, as shown in the previous, you may specify a constant value as an operand in the word logic operation. In the code listing below, the mask value of Figure 24-1 is entered as a constant in the AND Word statement. The constant value is combined with the ACCU1 contents.

Example: Combining a WORD constant with ACCU1 using the logic AND function AW

L	IW22	// Load Input Word IW22 into ACCU1
AW	W#16#00FF	// Combine hex constant W#16#00FF bitwise with IW22 according to AND
T	MW42	// Transfer ACCU1 results to bit memory word MW 42

OR Word: OW

When executed, the *OR Word* operation **OW** applies a bit-wise logical OR on the lower 16-bits of ACCU1 and ACCU2, according to the OR truth table. The result is placed in the corresponding bit positions of the lower 16-bits of ACCU1, while the contents of ACCU2 are left unaltered. According to the OR truth table, the output is '1' if any one of the inputs is '1', and is '0' if all inputs are '0'.

In the example below, the OR Word serves as a 'mask-in' (Set to '1') operation. To mask-in a bit from a source word (IW22 in this case), even if the bit is '0', requires a '1' in the corresponding bit position of the mask—MW22 in this case. The upper eight bits of the result are masked in. To leave any bit of the source word unchanged in the result, a value of '0' is placed in the corresponding position of the mask. Notice how the lower eight bits of IW22 are passed through to the result. This masking effect of the OR function, as described earlier, is based on the rules '1' **OR** 'X' = 1 and '0' **OR** 'X' = X.

OR Truth Table			Example: OR Word IW22 and MW22	
IN1	**IN2**	**OUT**	1010 0101 1101 1001	**WORD—IW22** (ACCU2)
0	0	0	OR	
0	1	1	1111 1111 0000 0000	**WORD—MW22** (ACCU1)
1	0	1	=	
1	1	1	1111 1111 1101 1001	**WORD RESULT** (ACCU1)

Figure 24-2. Illustration of OR Truth Table and OR Word operation

The STL code for the previous example is shown below:

Example: Combining 16-bit variables using the logic OR function OW

L	IW22	// Load Input Word IW22 into ACCU1
L	MW22	// Load bit memory Word MW22 into ACCU1. Shift old ACCU1 to ACCU2
OW		// Combine the lower 16-bits of ACCU2 and ACCU1 according to logical OR
T	QW42	// Transfer ACCU1 results to output word QW42

In addition to loading each value, as shown in the example above, you may specify a constant value as an operand in the word logic operation. In the code listing below, the mask value of our example is entered as a constant in the OR Word statement. The constant is combined with the ACCU1 contents.

Example: Combining a WORD constant and a 16-bit variable using the logic OR function OW

L	IW22	// Load Input Word IW22 into ACCU1
OW	W#16#FF00	// Combine hex constant W#16# FF00 (mask) bit-wise according to OR
T	QW42	// Transfer ACCU1 results to output word QW42

Exclusive-OR Word: XOW

When executed, the *XOR Word* operation **XOW** applies a bit-wise logical XOR on the lower 16-bits of ACCU1 and ACCU2, according to the XOR truth table. The result is placed in the corresponding bit positions of the lower 16-bits of ACCU1, while the contents of ACCU2 are left unaltered. According to the XOR truth table, as seen below, the output is always '0' if both inputs are the same; and is '1' if either input is '1', but not both—that is to say that the result is '1' whenever the inputs are different.

In some masking situations, you will need to have the bit value of specific bit positions inverted—the XOR function is used in these situations. In the brief tutorial on applying the XOR for this purpose, the rules for setting the mask value was based on '1' **XOR** 'A' = Ā. The term Ā means **A** inverted. So then, the mask will contain a '1' in each bit that corresponds to a bit position in the source that needs to be inverted. The other rule for setting the mask for XOR was to leave other bits of the source unaffected—that rule is '0' **XOR** 'A' = A. So then, the mask will contain a '0' in each bit that corresponds to a bit position in the source that will pass through. In this example, the source is IW22; the mask is MW24.

XOR Truth Table			Example: XOR Word IW22 and IW24	
IN1	**IN2**	**OUT**	1110 0101 1011 1100	**WORD—IW22** (ACCU2)
0	0	0	XOR	
0	1	1	1111 1111 1111 0000	**WORD—MW24** (ACCU1)
1	0	1	=	
1	1	0	0001 1010 0100 1100	**WORD RESULT** (ACCU1)

Figure 24-3. Illustration of XOR Truth Table and XOR Word operation

The STL code for implementing the above example is shown below:

Example: Combining 16-bit variables using the logic Exclusive OR function XOW

L	IW22	// Load Input word IW22 into ACCU1
L	IW24	// Load bit memory word IW24 into ACCU1. Shift old ACCU1 to ACCU2
XOW		// Combine ACCU2 (IW22) with ACCU1 (IW24) according to XOR
T	MW42	// Transfer ACCU1 results to bit memory word MW42

In addition to loading each value from a memory location or variable, as shown in the example above, you may specify a constant value as one of the word logic operands. In the following example, the mask value of our example is entered as a constant in the XOR Word statement. A constant parameter might be used in a case such as comparing the source word IW22 (actual machine states) to the expected states (mask constant). The constant value, what the machine states (16-bits) should be, is combined with the ACCU1 contents.

Example: Combining a WORD constant and 16-bit variable using the logic XOR function XOW

L	IW22	// Load Input word IW22 into ACCU1
XOW	W#16#FFF0	// Combine hex W#16#FFF0 (mask) bit-wise according to XOR
T	MW42	// Transfer ACCU1 results to bit memory word MW42

32-bit Word Logic Operations

As previously stated, word logic operations are performed using the values currently in accumulator 1 and accumulator 2. How or when these values are loaded to the accumulators is irrelevant. When loading the accumulators for 32-bit operations, you may load DWORD constants, variables of data type DWORD, or double word locations like ID14, MD44, and DB22.DD12. You may transfer the accumulator 1 result to a new location as required. The location to which you move the result may be a variable of type DWORD or a double word location—for example MD44, QD62, or DB22.DD24.

AND Double Word: AD

When executed, the *AND Double Word* operation **AD** applies a bit-wise logical AND according to the AND truth table, on the 32-bits of ACCU1 and ACCU2. The result is placed in the corresponding bit positions of the 32-bits of ACCU1, while the contents of ACCU2 are left unchanged. According to the AND truth table, the output is '1' only if both inputs are '1', and is '0' if either of the inputs is '0'.

In the following example, the result contains a '1' only in those positions where a '1' occurs in the corresponding positions of both ACCU2 and ACCU1.

AND Truth Table			Example: AND Double Word ID22 and ID42	
IN1	**IN2**	**OUT**	1010 0101 1111 1100 1010 0111 0001 1110	**DWORD—ID22** (ACCU2)
0	0	0	AND	
0	1	0	1111 0000 0000 0011 1111 0110 0101 0011	**DWORD—ID42** (ACCU1)
1	0	0	=	
1	1	1	1010 0000 0000 0000 1010 0110 0001 0010	**DWORD RESULT** (ACCU1)

Figure 24-4. Illustration of AND Truth Table and AND DWORD operation

The STL code for the previous example is shown below:

Example: Combining 32-bit variables using the logic AND function AD

L	ID22	// Load Input double word ID22 into ACCU1
L	ID42	// Load Input double word ID42 into ACCU1. Shift old ACCU1 to ACCU2
AD		// Combine the 32-bits of ID22 with the 32-bits of ID42 according to AND
T	MD42	// Transfer the ACCU1 logic result to bit memory double word MD42

In addition to loading each value, as shown in the example above, you may specify a constant as an operand in the word logic operation. In the code listing below, the mask value of our example is entered as a constant in the AND Double Word statement—the constant is combined with the ACCU1 contents.

Example: Combining a DWORD constant with 32-bit variable using the logic AND function AD

L	ID22	// Load Input double word ID22 into ACCU1
AD	DW#16#F003_F653	// Combine hex DW#16#F003_F653 with ID22 according to AND
T	MD42	// Transfer the ACCU1 logic result to bit memory word MD42

OR Double Word: OD

When executed, the *OR Double word* operation **OD** applies a bit-wise logical OR according to the OR truth table, on the 32-bits of ACCU1 and ACCU2. The result is placed in the corresponding bit positions of the 32-bits of ACCU1, while the contents of ACCU2 are left unchanged. According to the OR truth table, the output is '1' if any one of the inputs is '1', and is '0' if all inputs are '0'.

In the following example, the result contains a '1' in those positions where a '1' occurs in the corresponding position in either ACCU2 or in ACCU1, or in both.

OR Truth Table			Example: OR Double Word ID22 and ID42	
IN1	**IN2**	**OUT**	1010 0101 1111 1100 1010 0101 0000 0011	DWORD—ID22 (ACCU2)
0	0	0	OR	
0	1	1	1111 1000 0010 0011 1111 0000 0000 0100	DWORD—ID42 (ACCU1)
1	0	1	=	
1	1	1	1111 1101 1111 1111 1111 0101 0000 0111	DWORD RESULT (ACCU1)

Figure 24-5. Illustration of OR Truth Table and OR DWORD operation

The STL code for the previous example is shown below:

Example: Combining 32-bit variables using the logic OR function OD

L	ID22	// Load Input double word ID22 into ACCU1
L	ID42	// Load Input double word ID42 into ACCU1. Shift old ACCU1 to ACCU2
OD		// Combine the 32-bits of ID22 with the 32-bits of ID42 according to OR
T	MD42	// Transfer the ACCU1 logic result to bit memory double word MD42

In addition to loading each value, as shown in the example above, you may specify a constant as an operand in the word logic operation. In the code listing below, the mask value of our example is entered as a constant in the OR Double Word statement—the constant is combined with the contents of ACCU1.

Example: Combining a DWORD constant with 32-bit variable using the logic OR function OD

L	ID22	// Load Input double word ID22 into ACCU1
OD	DW#16#F823_F004	// Combine hex DW#16# F823_F004 with ID22 according to OR
T	MD42	// Transfer the ACCU1 result to bit memory double word MD42

Exclusive OR Double Word: XOD

When executed, the *XOR Double Word* operation **XOD** applies a bit-wise logical Exclusive-OR according to the XOR truth table, on the 32-bits of ACCU1 and ACCU2. The result is placed in the corresponding bit positions of the 32-bits of ACCU1, while the contents of ACCU2 are left unaltered. According to the XOR truth table, as seen below, the output is always '0' if both inputs are the same; and is '1' if either input is '1', but not both—that is to say whenever the inputs are different.

In the following example, the result contains a '1' only in those positions where a '1' occurs in either ACCU2 or in ACCU1, but not in both.

XOR Truth Table			Example: XOR Double Word ID22 and ID42	
IN1	**IN2**	**OUT**	1010 0101 1111 1100 1010 0101 1111 1100	DWORD—ID22 (ACCU2)
0	0	0	XOR	
0	1	1	1111 0000 0000 0011 1000 1011 0000 0011	DWORD—ID42 (ACCU1)
1	0	1	=	
1	1	0	0101 0101 1111 1111 0000 1010 1111 1111	DWORD RESULT (ACCU1)

Figure 24-6. Illustration of XOR Truth Table and XOR DWORD operation

The STL code for the previous example is shown below:

Example: Combining 32-bit variables using the logic XOR function XOD

L	ID22	// Load Input double word ID22 into ACCU1
L	ID42	// Load Input double word ID42 into ACCU1. Shift old ACCU1 to ACCU2
XOD		// Combine the 32-bits of ID22 with the 32-bits of ID42 according to XOR
T	MD42	// Transfer the ACCU1 results to bit memory double word MD42

In addition to loading each value, as shown in the example above, you may specify a constant as an operand in the word logic operation. In the code listing below, the mask value of our example is entered as a constant in the XOR Double Word statement—the constant is combined with the contents of ACCU1.

Example: Combining a DWORD constant and a 32-bit variable using logic XOR function XOD

L	ID22	// Load Input double word ID22 into ACCU 1
XOD	DW#16#F003_8B03	// Combine hex DW#16#F003_8B03 with ID22 according to XOR
T	MD42	// Transfer ACCU1 results to bit memory double word MD42

WORD LOGIC APPLICATIONS

By applying a technique called *masking,* these Boolean operations let you selectively manipulate specific bits within a word or double word variable, while leaving other bits unaffected. *Masking* in general is setting certain bit values to achieve a desired outcome. This ability is likely the most common application of Word Logic operations. Using the AND operation, you can force selected bit values to '0'; with OR selected bit values are set to '1'; and with XOR, bits can be selectively inverted.

An example of masking may be illustrated in a situation where all zeros must be guaranteed in the upper 8-bits of a 16-bit variable 'X'. To achieve this result, you would AND the variable 'X' with a mask value in which the upper 8-bits are all '0'. The mask value would contain all '1s' in the lower 8-bits to cause these bits of the original variable to be unaffected (pass through). This bitwise AND, as shown below, forces the high-order 8-bits to '0' and passes the low-order 8-bits of 'X' unchanged. Masking principles for AND, OR, and XOR are illustrated below and described in the following discussions.

AND Function with Mask	OR Function with Mask	XOR Function with Mask
1111 1010 0101 1100 Var. 'X'	1111 1010 0101 1100 Var. 'X'	1111 1010 0101 1100 Var. 'X'
AND	OR	XOR
0000 0000 1111 1111 Mask	0000 0000 1111 1111 Mask	0000 0000 1111 1111 Mask
0000 0000 0101 1100 Result	0000 0000 1111 1111 Result	1111 1010 1010 0011 Result

Figure 24-7. Illustration of 'masking' with AND, OR, and XOR

Masking with AND

An important characteristic of the **AND** operation is that, with masking, you can selectively force a '0' result in multiple bit positions. Masking, as mentioned earlier, involves setting one of the operand values of the operation in a way that causes a known result. For example, with the AND, if one of the operands is '0', the result is always '0' regardless of the other operand. Conversely, if one operand is '1', the result is always the second operand value. Stated another way, '0' **AND** 'X' = 0, and '1' **AND** 'X' = X. You can apply these rules to set the bit positions of a mask variable to affect a desired outcome.

Querying the status of a bit—Masking with AND Example

A typical application of the bitwise AND operation is to determine the status of a specific bit position of a given source word. The technique is referred to as "querying the status of a bit."

It is possible to use a bit mask to check the state of specific bits regardless of the other bits. The technique involves a first step of selectively forcing all bits off except for the bits in question; the other bits are turned off using the bitwise mask with AND, based on the previously described rule—'0' **AND** 'X' = 0. After performing the bitwise AND, using the source and mask variable, the resultant value is compared to '0'. A result equal to '0' indicates that the bit in question was 'OFF'; if the result of the compare is any other value, then the bit in question was 'ON'. This technique is quite convenient since it is not necessary to determine the actual value of the result—just that the result is not equal to zero.

Example: Querying the status of one or more bit positions—with AND	
1010 0101 1111 1100 1010 0101 1111 1100	SOURCE VARIABLE
1000 0000 1000 0000 1000 0000 1000 0000	MASK VARIABLE
1000 0000 1000 0000 1000 0000 1000 0000	QUERIED RESULT

Figure 24-8. Illustration of masking with AND to query the status of one or more bit positions

Masking with OR

An important characteristic of the **OR** operation is that, with masking, you can selectively force a '1' result in multiple bit positions. Masking, as mentioned earlier, involves setting a bit position in a way that causes a known result, when combined. For example, with the OR, if one of the operands is '1', the result is always '1' regardless of the second operand's value. Conversely, if one operand is set to '0', the result is always the second operand value. Stated another way '1' **OR** 'X' = 1, and '0' **OR** 'X' = X. You can apply these rules to set the bit positions of a mask variable to affect the outcome.

Setting the status of a bit to '1'—Masking with OR Example

A typical application of the OR Word operation is to simply ensure that one or more specific bit positions are set to '1'. Ensuring that a bit position is '1' is based on the rule '1' **OR** 'X' = 1.

When you define the mask value, '1' needs to be inserted for each bit position you wish to ensure is '1'. In the example below, the mask variable is set to ensure that the low-order 16-bits of the source variable are set to '1'. Notice, that by placing '0' in the mask for bit positions, the high-order 16-bits are completely unaffected—this is true because of the rule with the bitwise OR, '0' **OR** 'X' = X.

Example: Inverting the status of one or more bit positions—with OR	
1010 0101 1111 1100 1010 0101 1011 1100	**SOURCE VARIABLE**
0000 0000 0000 0000 1111 1111 1111 1111	**MASK VARIABLE**
0101 0101 1111 1111 0000 1010 1111 1111	**SET BIT POSITIONS RESULT**

Figure 24-9. Illustration of masking with OR to set the status of a bit position(s) to '1'

Masking with XOR

An important characteristic of the **XOR** operation is that, with masking, you can selectively cause one or more bit positions to be inverted. Masking, as mentioned previously, involves setting the bit value of one of the operands (mask variable) in a way that produces a known result. For example, with the XOR, if one of the operands is '1', the XOR result is always the inverse of the other operand. If the other operand is '1', for example, the XOR result would be '0'; if the other operand is '0', the result would be '1'. In both cases the result is the second operand inverted. Stated as a rule, '1' **XOR** 'A' = Ā. Conversely, if one of the operands of an XOR is fixed at '0' (mask operand), then the XOR result is always equal to the second operand. Stated another way '0' **XOR** 'A' = A.

Inverting the status of a bit—Masking with XOR Example

A typical application of the Exclusive OR Word operation is to invert the status of one or more bit positions—if the current value is '0', it is toggled to '1'; if the current value is '1', it is toggled to '0'.

Bit inversion using XOR is possible based on the rule '1' **XOR** 'A'= Ā. Ā is the inverted status of the original value 'A'. When you define the mask value, '1' needs to be inserted for each bit position you wish to invert. In the example below, every bit position needs to be inverted—therefore every bit position of the mask has a value of '1'.

Example: Inverting the status of one or more bit positions—with XOR	
1010 0101 1111 1100 1010 0101 0011 0001	**SOURCE VARIABLE**
1111 1111 1111 1111 1111 1111 1111 1111	**MASK VARIABLE**
0101 1010 0000 0011 0101 1010 1100 1110	**INVERTED RESULT**

Figure 24-10. Illustration of masking with XOR to invert the status of a bit position(s)

Checking a Variable Against Itself—XOR Example

A typical application of the bitwise XOR is to verify that a specific bit pattern is what it is expected to be in a known situation. For example, imagine you need to verify whether the actual states of 32 inputs are match the 32 expected states at a specific step of a machine or process. If the process is operating correctly as it should, the actual states of the 32 inputs should always match the 32 bit values of the expected states.

If all is as it should be, applying the XOR using the actual states (operand variable-1) and the expected states (operand variable-2), is in essence combining the source variable with itself (mask variable). An examination of the XOR Truth table (Figure 24-6), shows another XOR characteristic—if both operands are the same, both are '1' or both are '0', then the XOR result is always '0'. Stated another way, 'A' **XOR** '**A**'= '0'.

In the following example, the actual states of 32 inputs are combined with the 32 expected states using the XOR function. In each bit position, a match of the actual and expected values is confirmed where the result of the XOR is '0'. Wherever a bit position has a result of '1', then the actual and expected bit values do not match.

Example: Using XOR to compare actual states (Source) with expected states (Mask)	
1010 0101 1111 1100 1010 0101 1011 0011	**OPERAND VARIABLE—1 (Actual states)**
1010 0101 1111 1100 1010 0101 0011 0001	**OPERAND VARIABLE—2 (Expected states)**
0000 0000 0000 0000 0000 0000 1000 0000	**RESULT**—differing state(s) include bit-7

Figure 24-11. Illustration of using with XOR to compare actual states to expected states

Chapter 25

STL Status Bit Operations

In this chapter, you'll be introduced to operations that provide access to the S7 CPU status word. The CPU status word contains control and indicator bits that are used and affected by the CPU as it processes your control program. Access to these status bits let you evaluate this relevant information and take actions in your S7 program. Generally, these operations are used to evaluate numerical results, or respond to errors, that may occur in digital operations including arithmetic, standard functions, comparison, conversion, shift-rotate, and word-logic operations.

In Chapter 26 *Program Flow Control and Block Operations*, you'll be introduced to a set of operations that facilitate internal block JUMP operations in response to the same status bit information presented in this chapter.

Main Topics

- Status Bit Operations Overview

- Programming Status Bit Operations

- Evaluating Digital Operations Results

STATUS BIT OPERATIONS OVERVIEW

S7 CPU's generate status bits as a result of executing each STEP 7 operation. Status bits are used by the CPU as it processes binary logic operations like AND/OR logic, and are set as it processes digital operations like compares, arithmetic, conversions, shift-rotate, and word logic. Relevant status bit information, derived from the so-called *CPU status word*, is accessible in your STL user program.

Access to status bit results, as seen in Table 25-1, is provided in two ways: the first method involves the use of *status bit operands* that can be checked using binary examination operations; the second method involves three operations that can directly affect status bits. Take a moment to review these operations.

Table 25-1. STL Status Bit Operations Summary. (See note following table)

Operation	Brief Description
STL Operand	Check for '1': Using positive binary operands: A, O, and X—Example: A **OV** Check for '0': Using negated binary operands: AN, ON, and XN—Example: A **OS**
OV	**Overflow.** Check for overflow in last digital operation.
OS	**Overflow Stored.** Check for overflow in last series of math operations.
BR	**Binary Result.** Check the binary result (BR) bit.
UO	**Unordered.** Check for invalid REAL number in last operation.
> 0	**Greater Than Zero.** Check result of last operation for greater than zero.
< 0	**Less Than Zero.** Check last operation for less than zero.
> = 0	**Greater Than or Equal Zero.** Check last operation for greater than or equal zero.
< = 0	**Less Than or Equal Zero.** Check last operation for less than or equal zero.
= = 0	**Equal Zero.** Check last operation for equal zero.
< > 0	**Not Equal Zero.** Check last operation for not equal zero.
STL Operation	The following operations affect the result of logic operations (RLO)
SET	**Set RLO.** Set the result of logic operation to '1'.
CLR	**Clear RLO.** Reset the result of logic operation to '0'.
SAVE	**Save RLO to BR.** Save the result of logic operations to the binary result status bit (BR).

Note: With exception of OV, OS, and BR, status bit operations in this table are derived from condition code status bits CC0 and CC1. Actual S7 status bits are described in the following discussion.

Status Word Description

The CPU status word is divided into two groups—*binary flags*, which include /FC, RLO, STA, OR, and BR, are used and affected in binary operations; and *digital flags*, which include OS, OV, CC0, and CC1, are used and affected in digital operations. The following discussion describes the S7 CPU status bits.

Bit	08	07	06	05	04	03	02	01	00
Name	BR	CC1	CC0	OV	OS	OR	STA	RLO	/FC

Figure 25-1. S7 CPU Status word

/FC—First Check

The *first check* (/FC) status bit flags the first step in a new logic string—it is a key mechanism that marks the start of a new logic string as well as the development of a new RLO. In other words, as one logic string ends, another is started. The new logic string begins with the first binary check (/FC=1)—for example an AND operation that follows a condition operation such as the assignment '='. Condition operations, such as the assignment '=', are dependent on the RLO, and mark the end of a logic string (/FC=0). Other such operations include Set 'S' and Reset 'R'. Although one or more of these condition operations may follow one another, a new logic string is started with the next binary check (/FC=1).

RLO—Result of Logic Operation

In STEP 7, the completed binary logic result, as well as the intermediate result after each evaluation, is referred to as the *result of logic operations (RLO)*. In binary logic operations, the CPU uses the RLO to reflect the logic state at any step in a logic string. The RLO, for example, is updated after each binary check, and after each logic combination to reflect whether the operation was fulfilled. The binary check yields '0' (RLO=0) if the operation is not fulfilled; and yields '1' (RLO=1) if the operation is fulfilled. The RLO in turn is used to affect condition operations such as the assignment '=', set 'S', and reset 'R'.

STA—Status

The *status bit (STA)* reflects the actual '0' or '1' state of an address that is examined by the CPU using any of the binary examination instructions. For example, a digital input may be checked and found to have a status of '1'. Once examined, for example during the evaluation of a logic string, the status bit STA is set to reflect the determined state of the address. If the status is '0', STA is set to '0'; if the status is '1', then STA is set to '1'. Where the referenced address involves a memory operation, for example S, R, or the standard assignment '=', then the '0' or '1' value assigned to STA is the same as the RLO.

OR—OR Status

In a logic string involving two or more AND-before-OR combinations, the *OR status bit* buffers the result of a completed AND combination, and indicates to a subsequent OR combination that a '1' result has already been determined. In such a case, the status bit OR is set to '1'. The status bit OR is reset to '0' by all other binary operations.

BR—Binary Result

The *binary result (BR) status bit* is used by STEP 7 in managing the graphic implementation of the EN/ENO mechanism, used in LAD and FBD box operations. The function of the BR status bit in box operations may be seen by viewing a LAD/FBD box operation displayed in STL. If the functions of a box were developed in STL, there would be no need to implement the BR functionality—unless you wished to view the box in LAD/FBD, or simply if you wanted to save and later check the RLO status immediately following a digital operation—for example following a compare or word logic operation.

OV—Overflow

The *overflow (OV) status bit* indicates that the last operation used or resulted in an invalid REAL number, or that the permissible numeric range of the operation was exceeded. Such an error might involve an arithmetic operation, standard function, conversion operation, or a compare involving REAL numbers. On overflow, the OV bit is set—and is only reset after the next error-free operation.

OS—Overflow Stored

The *overflow stored (OS) status bit* is set whenever the OV status bit is set—it actually stores the initial occurrence of the overflow. Whereas the OV bit is reset after the next error-free arithmetic operation, the OS status bit, on the other hand, remains set until a new block call. This characteristic of the OS status bit means that in a series of calculations it is not necessary to evaluate (using OV) for an error immediately after each operation—instead you may evaluate the entire calculation for a numeric overflow, or use of an invalid REAL number, by evaluating the OS status bit after the last operation.

CC0 and CC1—Condition Code Bits

Status bits *CC0* and *CC1* provide information as to the numeric result, relative to zero, after arithmetic, standard function, comparison, conversion, word logic, and shift operations. In STL, status bits CC0/CC1 are not accessed directly, but instead by using the equivalent STL status bit operands as listed in Table 25-2. For example, using these operand equivalents, you can determine whether a numeric result is less than zero (negative), equal to zero, or greater than zero (positive). In each of these cases where the numeric result is within the permissible range, status bit OV is '0'.

Status bits CC0 and CC1 also combine with status bits OV and OS to report specific errors when the result is outside of the permissible number range, or other error conditions exist, in a digital operation. These error results are listed in Table 25-3. In each of these cases where the numeric result is outside of the permissible range, status bit OV and OS are both set to '1'.

Table 25-2. Condition Code Status Bits CC0/CC1 Indications and STL Operands Equivalents

CC0 and CC1 Status Bit Indication	STL Operand Equivalent
Result Equal Zero: (CC0 = 0; CC1 = 0)	= = 0
Greater Than Zero: (CC0 = 0 AND CC1 = 1)	> 0
Result Less Than Zero: (CC0 = 1 AND CC1 = 0)	< 0
Result Greater Than or Equal Zero: (CC0 = 0)	> = 0
Result Less Than or Equal Zero: (CC1 = 0)	< = 0
Result Not Equal Zero: (CC0 = 0 AND CC1 = 1) OR (CC0 = 1 AND CC1 = 0)	< > 0
Unordered (invalid REAL): (CC0 = 1 AND CC1 = 1)	UO

PROGRAMMING STATUS BIT OPERATIONS

CPU status bits indicate specific results after digital operations compare, conversion, word logic, shift-rotate, arithmetic, and standard math functions. This information is available for use in your program using the status bit operations briefly described in Table 25-1. These operands, for example, let you determine if the result of an arithmetic operation or math function is equal to zero, not equal to zero, positive, or negative; or in shift operations, whether the last bit shifted out is '1' or '0'.

You may use status bit instructions alone to evaluate the result of a specific digital operation, or in conjunction with other binary operations as the basis of control logic—for example to determine if a user-written function that handles errors should be called.

Examining Status Bit Operands

You may use status bit operands alone to evaluate the result of a specific digital operation, or they may be combined with other binary operations as the basis of control logic. You may use binary check operations including A, O, and X to check status bit operands for '1'; and you may use binary check operations including AN, ON, and XN to check status bit operands for '0'.

Overflow: OV

If an invalid REAL number was used, or the permissible number range was exceeded, in a previous operation, the *overflow (OV)* status bit is set to '1' and is not reset until the next error-free operation.

In the example below, the integer variables #VAR1 and #VAR2 are multiplied. A check for overflow in the accumulator 1 result is made immediately after the integer multiply operation. You may always check the OV status immediately after an operation that may result in overflow, or at least prior to another such operation. If an overflow is detected, then the indicator variable #Flag1 is activated.

Example: Illustration of checking for an overflow, immediately after individual operation

M001	L	#VAR1	// Load the variable VAR1 to ACCU 1
	L	#VAR2	// Load the variable VAR2 to ACCU 1—move old value to ACCU 2
	*I		// Multiply ACCU 2 value (VAR1) by ACCU 1 value (VAR2)
	A	OV	// Check accumulator result for overflow after integer multiply
	==	#Flag1	// // Indicator flag1 will reflect the '0' or '1' status of the OV status bit

Overflow Stored: OS

Whenever the overflow bit OV is set, the *overflow stored (OS)* bit is also set. Whereas the OV bit is reset with the next error-free operation, the OS bit is not reset until a new block call, or the current block terminates. Leaving the OS bit set allows it to be used to check for an overflow after a series of operations instead of immediately after each operation. If the result of any one of the operations produces an overflow result, then the OS bit is set to '1'.

In the example below, the result of the addition is always subsequently multiplied in the following multiplication operation. The subsequent network evaluates for an overflow in the completed calculation. The OS status =1 if either the addition or multiplication operation results in an overflow.

Example: Illustration of checking for an overflow, after a series of calculations

M002	L	1.53750 e+001	//
	L	#VAR1	//
	+R		//
	T	MD66	//
	A	OV	// Check result for overflow in the previous REAL addition
	L	MD12	//
	*R		//
	T	MD16	//
	A	OS	// Check for overflow at end of calculation
	=	#Flag2	// Indicator flag2 will reflect the '0' or '1' status of the OS status bit

Unordered: UO

The *Unordered (UO)* status bit operand is used to determine whether an invalid REAL number was involved in the last operation. For example, if either of the source values used in a REAL arithmetic or REAL compare operation is invalid, the UO status bit is set to '1'; otherwise the bit is set to '0'. The UO status bit also combines with status bits OV and OS to indicate other error results—see Table 25-3.

In the following example, when processing is passed to label M003, the REAL compare is performed, and evaluation is made, to check if an invalid REAL number was used. The binary indicator #Flag3 is activated if an invalid REAL value is reported.

Example: Illustration of checking for use of an invalid REAL number

M003	L	MD16	
	L	3.2e+006	
	>R		
	A	UO	// Check for invalid REAL value in previous operation
	A	OV	
	A	OS	
	=	#Flag3	// Indicator flag3 is turned on if all three status bits are set

Greater Than Zero: > 0

The *Greater Than Zero* status bit operand equivalent is used to check if the accumulator 1 result is greater than zero after the last arithmetic or standard function. The > **0** operand (CC0/CC1 equivalent) also combines with status bits OV and OS to indicate specific error results—see Table 25-3.

In the example, when processing is passed to label M004, the DINT addition is processed. The result is evaluated for a positive number that is not overflowed. Flag4 is activated if the result in MD66 is within the normal positive range.

Example: Illustration of checking for a result that is greater than zero and within the valid range

M004	L	L#1575	
	L	MD62	
	+D		
	T	MD66	
	A	>0	// Check if accumulator result is a positive value
	AN	OV	// Check overflow bit for '0'
	=	#Flag4	// Overflow indicator flag4 is turned on for positive result and overflow bit is not set

Less Than Zero: < 0

The *Less Than Zero* status bit operand equivalent is used to determine whether the accumulator 1 result is less than zero after the last digital operation. The < **0** operand (CC0/CC1 equivalent) also combines with status bits OV and OS to indicate specific error results—see Table 25-3.

In the following example, the BOOL variable #CaseF1 is set if the result of a previous operation is within the normal negative range (no underflow), or the OV status is '0'.

Example: Illustration of checking for a result that is less than zero and within the valid range

A	<0	// Check if accumulator result is a negative value
AN	OV	// Check overflow bit for '0'
S	#CaseF1	// Set flag for case 1 if the previous accu1 result less than zero

Less Than or Equal Zero: <= 0

The *Less Than or Equal Zero* status bit operand equivalent is used to check if the accumulator 1 result is less than or equal zero after the last digital operation. The < = **0** operand (CC0/CC1 equivalent) also combines with status bits OV and OS to indicate specific error results—see Table 25-3.

In the following example, the BOOL variable #CaseF3 is set if the result of a previous operation is less than or equal zero, and the OV status is '0' (no underflow).

Example: Illustration of Results Bit Less Than or Equal Zero (<=0) instruction

A	<=0	// Check if accumulator result is a negative value or zero
AN	OV	// Check overflow bit for '0'
S	#CaseF3	// Set flag for case F3 if the previous accu1 result less than or equal to zero

Equal Zero: == 0

The *Equal Zero* status instructions are used to determine if the accumulator 1 result, after the last digital operation, is equal to zero. The = = **0** operand (CC0/CC1 equivalent) also combines with status bits OV and OS to indicate specific error results—see Table 25-3.

In the following example, when processing is passed to label M005, the bit pattern in the input double word ID12 is combined with the binary pattern in the DWORD variable #Pattern_2, using the XOD word logic operation. The BOOL variable #CaseF4 is set to '1' if the XOR'ed result in MD130 is equal to zero, and the OV status is '0'.

Example: Illustration of Results Bit Equal Zero (==0) instruction

M005	L	ID12	
	L	#Pattern_ 2	
	XOD		
	T	MD130	
	=	M99.3	
	A	==0	// Check if accumulator result is all zeros after the XOR
	AN	OV	// Check overflow bit for '0'
	S	#CaseF4	// Set the flag for case F4 if the previous accu1 result is equal to zero

Greater Than or Equal Zero: >= 0

The *Greater Than or Equal Zero* status instructions are used to determine if the accumulator 1 result, after the last digital operation, is greater than or equal zero. The > =0 operand (CC0/CC1 equivalent) also combines with status bits OV and OS to indicate specific error results—see Table 25-3.

In the following example, the BOOL variable #CaseF2 is set if the result of a previous operation is greater than or equal zero, and the OV status is '0' (no overflow).

Example: Illustration of Results Bit Greater Than or Equal Zero (>=0) instruction

A	>=0	// Check if accumulator result is a positive value or zero
AN	OV	// Check overflow bit for '0'
S	#CaseF2	// Set flag for case F2 if the accu1 result is greater than or equal to zero

Not Equal Zero: <> 0

The *Not Equal Zero* status instructions are used to determine if the accumulator 1 result, after the last digital operation, is not equal to zero. The < > 0 operand (CC0/CC1 equivalent) also combines with status bits OV and OS to indicate specific error results—see Table 25-3.

In the following example, when processing is passed to label M006, the bit pattern in the input double word ID12 is combined with the binary pattern in the DWORD variable #Pattern_2, using the XOD word logic operation. The BOOL variable #CaseF5 is set to '1' if the XOR'ed result in MD130 is not equal to zero, and the OV status is '0'.

Example: Illustration of Results Bit Not Equal Zero (<>0) instruction

M006	L	ID12	
	L	#Pattern_2	
	XOD		
	T	MD130	
	=	M99.3	
	A	<>0	// Check if accumulator result is not all zeros after the XOR
	AN	OV	// Check overflow bit for '0'
	S	#CaseF5	// Set the flag for case F5 if the previous accu1 result is not equal to zero

Setting, Clearing and Saving the RLO

SET, CLR, and SAVE, described in the following discussions, provide the ability to manipulate the binary status of the RLO in order to facilitate the convenient use of the RLO throughout the STL program. The SET operation, for example, eliminates the need to develop an 'always 1' bit to have available for immediate use throughout the program; conversely, the CLR operation eliminates the need to develop an 'always 0' bit for further use throughout the program. With the SAVE operation, you are able to store the RLO to the binary result (BR) status bit—also for further use in you program.

Set RLO: SET

When processed, the *SET* operation will unconditionally set the result of logic operations to equal '1'— that is without any preceding conditions. You may insert the SET operation any point in the logic string—the newly developed RLO can be used immediately if necessary.

In the network below, output Q 22.3 and Q 22.4 are switched ON (1), without any preceding logic.

Example: Setting the result of logic operations using the SET operation

SET		//Set the RLO to '1'
=	Q22.3	// Assign a value of RLO = 1 to Q22.3
=	Q 22.4	// Assign a value of RLO = 1 to Q22.4

Clear RLO: CLR

When processed, the *clear (CLR)* operation will unconditionally set the result of logic operations to equal '0'—that is without any preceding conditions. You may insert the CLR operation at any point in the logic string—the newly developed RLO can be used immediately if necessary.

In the network below, output Q 22.3 and Q 22.4 are switched OFF (1), without any preceding logic.

Example: Clearing the result of logic operations using the CLR operation

CLR		//Clear the RLO—or assign a value of '0' to RLO
=	Q22.3	// Assign a value of RLO = 0 to Q22.3
=	Q 22.4	// Assign a value of RLO = 0 to Q22.4

Binary Result: BR

In STL, the *binary result (BR)* status bit is examined for a status of '1' using the *BR* status bit operand, and examined for a status of '0' using the negated *BR* status bit operand. In the network below, output Q 44.7 is Set (latched) if both M 101.1 AND M 101.2 are '1' AND the BR status bit is '0'.

Example: Illustration of the Binary Result (BR) status bit instruction

A	M101.1	// Check M101.1 for '1'
A	M101.2	// Check M101.2 for '1' and combine with RLO using AND
AN	BR	// Check the BR status for '0' and combine with RLO using AND
S	Q44.7	//

Save RLO: SAVE

When processed, the *SAVE* operation unconditionally saves the current value of the RLO to the binary result (BR) status bit. You may insert the SAVE operation at any point in the logic string, and later use the saved status of the RLO by checking the BR status bit.

Example: Saving the result of logic operations to the BR status bit

SET		//Set the RLO to '1'
=	Q22.3	// Assign a value of RLO = 1 to Q22.3
=	Q 22.4	// Assign a value of RLO = 1 to Q22.4
SAVE		// Save the RLO to the BR status bit for later use

EVALUATING STL DIGITAL OPERATIONS RESULTS

In STL, *status bit operations* are implemented as binary operands that may be checked using any of the binary operand check operations, and combined with other operations as the basis of control logic.

As previously discussed, you can check each status bit operand for '1' using the binary check instructions A, O, and XOR; and you can check each status bit operand for '0' using binary check operations AN, ON, and XN. The following discussions summarize error and result statuses for which you can evaluate, following STL digital operations in your program.

Evaluating Conversion Results

When it comes to conversions, each operation differs in what is reported, and how it is reported—some operations may not report errors at all. In the descriptions of *conversion operations* in Chapter 20 *STL Conversion Operations*, you can determine what results or errors are reported for each operation.

If the conversion is to BCD from integer or double integer, you can evaluate status bits OV and OS, which are both set to '1', if the source value is too large to convert to 3-digit or 7-digit BCD. When the conversion is from BCD to integer or BCD to double integer, a parameter assignment error is reported by the CPU if a source value is an invalid BCD value. In such a case, synchronous error OB121 is called—if OB121 is not installed, the CPU will switch to STOP.

In each of the rounding operations, the source value to be converted is a REAL number, and the converted result is a DINT value. When the rounding operation is processed, if the source value is an invalid REAL number, or is outside of the range that will correctly convert to a number within the permissible DINT range, then status bits OV and OS are set to '1' and the operation is aborted.

With negate operations NEGI and NEGD, you can check to see whether the converted result is positive, zero, negative, or extreme negative. Extreme negative is an error where the converted result is -32768 (INT), or -2,147,483,648 (DINT)—this result is an error since there is no corresponding positive number from which this result could be obtained. See Table 25-3.

Evaluating Arithmetic and Standard Function Results

After each arithmetic or standard function operation, you may use status bit operations to evaluate for a result within the normal range, or a result with errors. For example, you may check for an overflow (permissible range exceeded) in the result by examining the **OV** status bit for '1', immediately following the operation; or you may perform a series of arithmetic operations, and then check for an overflow result using the **OS** status bit. See *Examining Status Bit Operands*, earlier in this chapter.

You can also check if a numeric result is less than zero, equal to zero, or greater than zero. If the result is less than zero, status bits CC0/CC1 are set to 1/0; if the result is equal to zero, status bits CC0/CC1 are set to 0/0; if the result is greater than zero, status bits CC0/CC1 are set to 0/1. As seen in Table 25-2, the STL operand equivalents of CC0/CC1 indications of less than zero, equal to zero, and greater than zero are respectively the < 0 operand, the = = 0 operand, and the > 0 operand. In each of these cases, where the result is within the numeric range flagged by the indicator, the **OV** status is also '0'.

Additional error indications, for which you may evaluate after arithmetic and math functions, are listed in Table 25-3. Each of these error indications will affect a LAD/FBD equivalent operation for CC0/CC1, along with status bits OV and OS, which are both set to '1'. For example, in any operation involving REAL numbers, status bit **UO** is set to '1' if an invalid source value is used. These additional error indications, grouped as INT/DINT operations or REAL operations, are listed in the unshaded rows of Table 25-3.

Evaluating Compare Results

When it comes to compare operations, you may evaluate after those operations involving REAL numbers. For each of these operations, status bits CC0, CC1, OV, and OS all are all set to '1', and the operation is aborted if either source value (ACCU1 or ACCU2) is found to be an invalid REAL value.

You can also check whether a compare test for less than, equality, or greater than is affirmed. If a test for less than is True (1), then status bits CC0/CC1 are set to 1/0; if a test for equality is True (1), status bits CC0/CC1 are set to 0/0; if a test for greater than is True (1), status bits CC0/CC1 are set to 0/1.

As shown in Table 25-2, the STL equivalent operations for CC0/CC1 indications of less than, equality, and greater than are respectively the **< 0** operand, the **= = 0** operand, and the **> 0** operand. In each of these cases, where the result is within the permissible range, the **OV** status bit is also '0'.

Evaluating Word Logic Results

After each word logic operation is completed, you may examine specific status bit instructions to determine if the bitwise logic result at OUT is 'equal to 0' or 'not = zero'. The status bit instruction = = **0** (CC0= 0, CC1 = 0) is used to examine for zero; the status bit instruction <> **0** (CC0= 0, CC1 = 1) is used to examine for a not equal to zero result.

Evaluating Shift Results

After each shift operation, the status of the last bit shifted out is loaded to the **CC1** bit of the status word. In other words, the status bit CC1 is assigned the '1' or '0' value to reflect the shifted out bit position. In your program, you may examine the CC1 status bit using the STL status bit operand equivalent to the CC1 status bits. These equivalent operands are listed in Table 25-2. Based on this table, you can check the shifted out bit for '0' using the = = **0** operand, and for '1' using the > **0** operand. See Table 25-3.

Table 25-3. Digital operations results with status bits OV, OS, UO and CC0/CC1 equivalent operands

Result conditions for which to evaluate	STL Status Bit Operand Equivalents to Evaluate								
	UO	OV	OS	==0	<>0	<0	>0	<=0	>=0
INT/DINT Result Overflow (ADD_I, SUB_I)		1	1				1		
INT/DINT Result Overflow (MUL_I)		1	1					1	
INT/DINT Result Underflow (ADD_I, SUB_I)		1	1					1	
INT/DINT Result Underflow (MUL_I)		1	1				1		
INT Result = 32,768 for (DIV_I)		1	1					1	
INT/DINT Division by Zero	1	1	1						
INT/DINT Result = 0		0		1					
INT/DINT Result > 0 (valid positive)		0						1	
INT/DINT Result < 0 (valid negative)		0				1			
DINT Result = 2,147,483,647 for (DIV_D)		1	1					1	
REAL Overflow: + infinite (division by zero)		1	1					1	
REAL Underflow: - infinite (division by zero)		1	1			1			
Invalid REAL Value	1	1	1						
REAL Result ± (De-normalized)		1	1	1					
REAL Result = 0		0		1					
REAL Result > 0 (normalized)		0						1	
REAL Result < 0 (normalized)		0				1			
Word Logic Result = 0		0		1					
Word Logic Result <> 0		0			1				
Shift-Rotate—Shift Out Bit = 1		0						1	
Shift-Rotate—Shift Out Bit = 0		0		1					
Conversion Result = Extreme Negative, NEG_I		1	1			1			
Conversion Result = Extreme Negative, NEG_DI		1	1			1			
Conversion Result = 0		0		1					
Conversion Result > 0		0						1	
Conversion Result < 0		0				1			
Comparison with Invalid REAL Value (s)	1	1	1						
Compare Result Equal		0		1					
Compare Result Greater Than		0						1	
Compare Result Less Than		0				1			

Note: 1) Check for status equal '1' using positive binary check operands; check for status equal '0' using negated binary check operands. 2) CC0/CC1 indications are the same for INT/DINT operations. The numeric range for INT is -32768 to +32,767; the range for DINT is -2147483648 to +2,147,483,647.

Chapter 26

STL Program Flow Control and Block Operations

This chapter introduces STL Program Control and Block Operations. With this group of operations, you can manage how and when certain portions of your S7 program are processed and terminated. In short, with these operations you are able to alter how the CPU processes the program so that processing needs are serviced on an as-needed basis only.

Overall, these instructions help to minimize processor scan times and create a more efficient and better organized program. The discussions of this chapter include calls to blocks, internal jumps to labels, block termination, and definition and operation of MCR zones.

Main Topics

- Program Flow Control—Jump Operations
- Programming Jump Operations in STL
- Program Flow Control—Block Operations
- Programming Block Operations in STL
- Program Flow Control—MCR Operations
- Programming MCR Operations in STL

PROGRAM FLOW CONTROL—JUMP OPERATIONS

STL Program Control has been divided into three operations categories that include *jump operations*, *block operations*, and *MCR operations*. Block jumps, the first of these three categories, provides the facility for making internal block jumps—that is the passing of control between sections of code within a block. This passing of control is accomplished using statements that include a jump operation and a label parameter to which the jump is targeted. These jumps, which include both conditional and unconditional operations, are based on status bit results as covered in the last chapter. Take a few moments to review the brief descriptions of the STL jump operations listed in Table 26-1.

Table 26-1. STL Program Flow Control—Jump Operations Summary

Operation	Brief Description
JU <label>	**Jump Unconditional**. Jump to specified label irrespective of the RLO.
JL <label>	**Jump to List**. Process jump function by passing control to one of several target labels.
LOOP <label>	**Loop Jump**. Jump to specified label if the value in the low-order word of ACCU 1 <> 0.
JC <label>	**Jump Conditional on RLO = 1**. Jump to specified label if RLO = 1.
JCN <label>	**Jump Conditional on RLO = 0**. Jump to specified label if RLO = 0.
JCB <label>	**Jump Conditional with BR on RLO = 1**. Jump to label if RLO = 1; save RLO to BR.
JNB <label>	**Jump Conditional with BR on RLO = 0**. Jump to label if RLO = 0; save RLO to BR.
JBI <label>	**Jump Conditional on BR = 1**. Jump to specified label if the Binary Result = 1.
JNBI <label>	**Jump Conditional on BR = 0**. Jump to specified label if the Binary Result = 0.
JO <label>	**Jump Conditional on Overflow**. Jump to specified label if the status bit OV = 1.
JOS <label>	**Jump Conditional on Overflow Stored**. Jump to specified label if the status bit OS = 1.
JUO <label>	**Jump Conditional on Invalid REAL**. Jump to specified label if the result bit UO = 1.
JZ <label>	**Jump if result is Zero**. Jump to specified label if status bit CC0 = 0 and CC1 = 0.
JN <label>	**Jump if result Not-zero**. Jump to specified label if status bit CC0 = 1 and CC1 = 0.
JP <label>	**Jump if result Positive**. Jump to specified label if status bit CC0 = 0 and CC1 = 1.
JM <label>	**Jump if result Minus**. Jump to specified label if status bit CC0 = 1 and CC1 = 0.
JPZ <label>	**Jump if result Positive or Zero**. Jump to specified label if status bit CC0 = 0 and CC1 = 0.
JMZ <label>	**Jump if result Minus or Zero**. Jump to specified label if status bit CC0 = 1 and CC1 = 0.

Note: See Chapter 25 *STL Status Bit Operations* for instructions for examining CPU Status Bits

PROGRAMMING JUMP OPERATIONS IN STL

STL jump operations support the ability to direct the flow of processing in your program, by passing control from one point in a code block to another point within the same block. Generally speaking, a jump will either be unconditional or conditional. An unconditional jump is processed without regard to any conditions. Conditional jumps are all based upon a specific S7 status bit—for example the RLO status, the OV status, the binary result bit BR, or status bits CC0/CC1, which are set after a digital operation. These same status bit operations, implemented using STL operands and described in the previous chapter, are implemented here using jump instructions. Finally, a jump may be based on a counter value held in accumulator 1, which acts as the index value of a loop operation.

Jump to Label

Jump to label operations facilitate jumps that take place internal to a block. When a jump operation is executed the statements between the jump instruction and the statement label to which it points are skipped. Processing of the current block resumes at the labeled statement.

A *statement label* identifies a target statement to which program execution may be passed, by a jump instruction—for example **JMP** <label> and **JMPN** <label>. Each statement label is a unique four-character alphanumeric name that must start with a letter. Upper and lower case letters are interpreted as being different—for example 'CAS1' and 'cas1' are different. Although each label must be unique, program execution may be passed to a single label from one or more jump instructions. Both forward and backward jumps are possible.

Unconditional Jump Operations

In the STL operations set, the **JU** and **JL** instructions are considered unconditional jump operations. These operations are executed irrespective of the result of logic operations (RLO), status bit results, or any other condition. These operations, when processed, are executed unconditionally.

Jump Unconditional: JU <label>

The **JU** operation is executed without any conditions—in other words, unconditional. When the JU operation is processed, program processing jumps to the statement whose label matches the label parameter referenced in the JU statement. All statements between the JU operation and the referenced label are skipped, and not processed.

In the following example, when each case of the JU operation is processed, program execution jumps to the statement labeled STRT. The On-delay timer T1 is started with a different preset value, based on whether program execution is passed from the section labeled RUN1, RUN2, or RUN3. Although there are three jumps to the label STRT, the label STRT is unique within the block.

Example: Unconditional Jump to specified label

RUN1	L	S5T#30M	// Set Run Timer Preset value to 30 minutes
	JU	STRT	// Go Start
			//
			//
RUN2	L	S5T#40M	// Set Run Timer Preset value to 40 minutes
	JU	STRT	// Go Start
			//
			//
RUN3	L	S5T#50M	// Set Run Timer Preset value to 50 minutes
	JU	STRT	// Go Start
			//
			//
STRT	AN	T1	// Scan start permissive for '1'—begin main block sequence
	SD	T1	// Start run timer

Jump to List: JL <label>

The **Jump to List** operation **JL**, also called **Jump Distributor**, provides a mechanism by which control is passed to one of several target labels. The target labels for this operation are implemented in the form of a list of up to 255 jump unconditional (**JU**) statements that immediately follow the JL statement. The list of JU statements is entered without blank line between each operation. The list is terminated by a labeled statement, whose label is the target parameter in the JL statement.

When the **JL** operation is executed, the label to which control is passed is determined by the value currently in the low-order byte of accumulator 1. The value may be a number in the range of 0-255. If a value of '0' is in ACCU 1, the first jump is executed; if a value of '1' is in ACCU 1, the second jump is executed; and so on. If the value in ACCU is greater than the length of the jump list, then control is passed to the end of the list—the statement immediately following the last **JU** operation.

The following example illustrates the structure and operation of the Jump to List operation, and how it offers at least one method of organizing the processing of multiple sections in a code block. The code starting with the label CAS0 is processed if the value in ACCU 1 is '0'; CAS1 code is processed if the value in ACCU 1 is '1'; CAS2 is processed if the value in ACCU 1 is '2'; and CAS3 code is processed if the value in ACCU 1 is '3'. You may use any valid 4-characater label names as the jump targets.

Example: Use jump distribution to a list of code sections based on current jump number

	L	MW10	//Load memory word MW100 to accumulator 1—total number of jumps.
	JL	END	// JL label parameter points to end of the jump list.
	JU	CAS0	// Jump to CAS0 if value in low-order byte of ACCU 1 is '0' target jump (1)
	JU	CAS1	// Jump to CAS1 if value in low-order byte of ACCU 1 is '1' target jump (2)
	JU	CAS2	// Jump to CAS2 if value in low-order byte of ACCU 1 is '2' target jump (3)
	JU	CAS3	// Jump to CAS3 if value in low-order byte of ACCU 1 is '3' target jump (4)
END	BEU		// End of jump distribution list.
CAS0			// Code section for Case (0)—Target position 1
			// End code for this case
	JU	MAIN	// Continue with the main block code.
CAS1			// Code section for Case (1)—Target position 2
			// End code for this case
	JU	MAIN	// Continue with the main block code.
CAS2			// Code section for Case (2)—Target position 3
			// End code for this case
	JU	MAIN	// Continue with the main block code
CAS3			// Code section for Case (3)—Target position 4
			// End code for this case
MAIN			// Start main block code
			//
			//
			//
			//
BEND			// Block end

Conditional Jumps based on RLO and BR Status

The following jump instructions are conditional operations that are based on either the result of logic operations (RLO), or on the binary result (BR) status bit. These status bits, along with other CPU status bits, are described in the Chapter 25 *STL Status Bit Operations* discussion on *S7 CPU Status Word*.

Jump to label if RLO = 1: JC <label>

The **JC** operation is based on the result of logic operations (RLO). When the JC operation is processed, if the preceding RLO=1, program processing jumps to the statement whose label matches the label parameter referenced in the JC statement. If the preceding RLO = 0, the jump is not executed and processing resumes immediately following the **JC** statement.

The following example illustrates how the JC operation can be used to pass control to defined (labeled) code sections when the associated permissive is active.

Example: Jump conditionally to specified label on RLO = 1

	A	I 2.0	// Check I 2.0 for '1'—Case 1 enabled.
	JC	CAS1	// If RLO = 1 jump to the section starting with the statement labeled CAS1
			// otherwise continue
	A	I 2.1	// Check I 2.1 for '1'—Case 2 enabled
	JC	CAS2	// If RLO = 1 jump to the section starting with the statement labeled CAS2
			// otherwise continue
CAS1			// Process code section for CAS1
			//
CAS2			// Process code section for CAS2

Jump to label if RLO = 0: JCN <label>

The **JCN** operation is based on the result of logic operations (RLO). When the **JCN** operation is processed, if the preceding RLO=0, program processing jumps to the statement whose label matches the label parameter referenced in the JCN statement. If the preceding RLO = 1, the jump is not executed and processing resumes with the statement immediately following the JCN statement.

In the following example, control is passed to the main section of the block code while the RLO = 0. When the step number reaches 10, two outputs are reset and the program is terminated.

Example: Jump conditionally to specified label on RLO = 0

	L	+10	//
	L	STEPNO	// step number
	= I		//
	JCN	MAIN	// If step no ≠ 10 (RLO = 0) jump to main; else reset outputs Q33.0 and Q33.1
	R	Q33.0	// Reset output image bit Q33.0
	R	Q33.1	// Reset output image bit Q33.1
	BEC		// terminate block on RLO = 1
MAIN			//
			//

Jump to label if RLO = 1, with BR: JCB <label>

When the JCB operation is processed, if the preceding RLO=1, program processing is passed to the statement whose label matches the label parameter referenced in the JCB statement. The binary result bit (BR) is set to '1' (= RLO) and is carried with the jump. If the RLO = 0, the jump is not executed, the BR bit is set to '0', and processing resumes with RLO = 0 at the following statement.

Example: Jump conditionally to specified label on RLO = 1; save RLO = 1 to BR status bit

	L	#Speed_4	// Load integer variable #Speed_4
	L	+3500	// Load integer constant +3500
	> I		// Check if #Speed_4 is greater than +3500
	JCB	NXT1	// If RLO = 1 jump to NXT1 (#Speed_4 > 3500); else continue , BR=0
	L	PIW 260	// Continue here if jump not executed
	T	MW40	
NXT1			// Code sections for CAS2

Jump to label if RLO = 0, with BR: JNB <label>

If the preceding RLO=0, when the **JNB** operation is processed, program processing is passed to the statement whose label matches the label parameter referenced in the JNB statement. The RLO is saved to the binary result bit (BR =0) and is carried with the jump. If the RLO = 1, the jump is not executed, and the RLO is saved to the binary result (BR=1), and processing continues with RLO = 1 at the following statement.

Example: Jump conditionally to specified label on RLO = 0; save RLO = 0 to BR status bit

	L	+3500	// Load integer constant +3500
	L	#Speed_4	// Load integer variable #Speed_4
	< I		// If #Speed_4 is greater than +3500 (RLO=0)
	JNB	OVRS	// BR = 0 jump to OVRS; else continue to next
OVRS			// Code sections for OVRS

Jump to label if BR = 1: JBI <label>

The **JBI** operation is based on the binary result status bit (BR). When the JBI operation is processed, if BR = 1, program processing jumps to the statement label that matches the label referenced in the JBI statement. If BR = 0, then the jump is not executed; processing resumes with the following statement.

Example: Jump conditionally to specified label on BR = 1

CMP4	L	#Speed_1	// Load INT variable #Speed_1
	L	#Speed_2	// Load INT variable #Speed_2
	> I		// Check if #Speed_1 is greater than #Speed_2
	SAVE		//
	JBI	CAS2	// If if #Speed_1 > #Speed_2 (BR= 1) jump to CAS2
			// else BR=0; continue here
CAS2			// Code sections for CAS2

Jump to label if BR = 0: JNBI <label>

The **JNBI** operation is based on the binary result status bit (BR). When the JNBI instruction is processed, if BR = 0, program processing jumps to the statement label that matches the label parameter referenced in the JNBI statement. If BR = 1, the jump is not executed and processing resumes with the following statement.

Example: Jump conditionally to specified label on BR = 0

CMP4	L	#Speed_1	// Load INT variable #Speed_1
	L	#Speed_2	// Load INT variable #Speed_2
	< I		// Check if #Speed_1 is less than #Speed_2
	SAVE		//
	JNBI	END1	// If if #Speed_1 > #Speed_2 (BR= 0) jump to END1
	L	0	// else BR=1; continue here
	T	PQW 260	//
END1			// End code section
			//

Jumps based on CC0 and CC1 Status

Condition code bits CC0 and CC1 are bits of the CPU status word that indicate specific results after executing arithmetic, standard functions, compare, word logic, and shift-rotate operations. Results indicated by CC0 and CC1 are not directly available, but are available in your STL program as Jump operations—as listed in Table 26-2, and as binary operands as were described in the previous chapter.

With the appropriate jump instructions, you can implement code to check for specific results. For example, after an arithmetic or math function you can implement a jump based on a resulting value of zero, not equal zero, positive, or negative. If the STL editor is set to display status bits, you can view the status of CC0 and CC1 in your STL program, as they are updated, in the online debug monitor mode.

Table 26-2. Condition Code Status Bits CC0/CC1 and Equivalent STL Jump/Status bit Operands

CC0	CC1	Status Bit Operand	Executed Jump Operations on CC0/CC1 combination	
0	0	= = 0	JZ	Jump if Zero
		< = 0	JMZ	Jump if minus (less than zero) or zero
		> = 0	JPZ	Jump if positive (greater than zero) or zero
1	0	< 0	JM	Jump if minus (less than zero)
		< = 0	JMZ	Jump if minus (less than zero) or zero
		<> 0	JN	Jump if not zero
0	1	> 0	JP	Jump if positive (greater than zero)
		> = 0	JPZ	Jump if positive (greater than zero) or zero
		<> 0	JN	Jump if not zero
1	1	UO	JUO	Jump if result is invalid (unordered)

Jump if Zero: JZ <label>

The **Jump if Zero** operation **JZ** is based on the condition code status bits CC0 and CC1. If CC0 = 0 and CC1 = 0, then program processing jumps to the statement whose label matches the label parameter referenced in the **JZ** statement. If the status of CC0/CC1 is other than 0/0, then the jump is not executed and processing resumes with the following statement. See Table 26-2 for the relationship of status bits CC0/CC1 to jump instructions.

With any one of the cases listed below, if CC0 = 0 and CC1 = 0, the **JZ** operation is executed.

1. The value in ACCU 1 is zero after an arithmetic or standard math function.

2. The value in ACCU 1 and ACCU 2 are equal after a compare operation.

3. The value in ACCU 1 is zero after a word logic operation.

4. The value of the last shifted bit is '0' after a shift operation.

Example: Jump if zero to specified label (C C0 = 0 and CC1 = 0)

	L	MW46	// Load input word IW44 to ACCU 1
	SRW	4	// Shift the bits of low-order word of ACCU 1 right by 4 bit positions
	JZ	TST1	// Jump to TST1 if bit-3 (the last shifted bit) of accumulator 1 is '0'
			//
TST1			// Process this code when the last shifted bit equal '0'

Jump if Not Zero: JN <label>

The **Jump if Not Zero** operation **JN** is based on the condition code status bits CC0 and CC1. If CC0 = 0 and CC1 = 1, or CC0 = 1 and CC1 = 0, then program processing jumps to the statement whose label matches the label parameter referenced in the **JN** statement. If the status of CC0/CC1 is 0/0 or 1/1, the jump is not executed and processing resumes with the following statement. See Table 26-2 for the relationship of status bits CC0/CC1 to jump instructions.

If in any of the following cases CC0 = 0 and CC1 = 1, or CC0 = 1 and CC1 = 0, the **JN** operation is executed:

1. The value in ACCU 1 is not zero after an arithmetic or standard math function.

2. The value in ACCU 1 and ACCU 2 are not equal after a compare operation.

3. The value in ACCU 1 is not zero after a word logic operation.

4. The value of the last shifted bit is '1' after a shift operation.

Example: Jump if not zero to specified label (CC0 = 0 and CC1 = 1, or CC0 = 1 and CC1 = 0)

	L	IW44	// Load input word IW44 to ACCU 1
	L	IW46	// Load input image word IW46 to ACCU 1—move old value to ACCU 2
	XOW		// Combine input image word IW44 with output word DBW44 using XOW
	JN	TST2	// Jump to TST2 if all bits are not zero in accumulator 1 after the XOW
			//
TST2			// Process this code when all bits are not zero in ACCU 1

Jump if Positive: JP <label>

The **Jump if Positive** operation **JP** is based on the condition code status bits CC0 and CC1. If CC0 = 0 and CC1 = 1, then program processing jumps to the statement whose label matches the label parameter referenced in the **JP** statement. If the status of CC0/CC1 is other than 0/1, then the jump is not executed and processing resumes with the following statement. See Table 26-2 for the relationship of status bits CC0/CC1 to jump instructions.

If in any of the following cases CC0 = 0 and CC1 = 1, the **JP** operation is executed:

1. The value in ACCU 1 is within the permissible positive range after an arithmetic/math function.
2. The value in ACCU 2 is greater than the value in ACCU 1 after a compare operation.
3. The value in ACCU 1 is not zero after a word logic operation.
4. The value of the last shifted bit is '1' after a shift operation.

Example: Jump if positive to specified label (CC0 = 0 and CC1 = 1)

	L	MW46	// Load input word MW46 to ACCU 1
	L	MW48	// Load input image word MW48 to ACCU 1
	- I		// Subtract ACCU 1 value (MW48) minus the ACCU 2 value (MW46)
	JP	TST3	// Jump to TST3 if the subtraction result is positive
			//
			//
TST3			// Process this code when the accumulator 1 result is positive
			//

Jump if Minus: JM <label>

The **Jump if Minus** operation **JM** is based on the condition code status bits CC0 and CC1. If CC0 = 1 and CC1 = 0, then program processing jumps to the statement label that matches the label parameter referenced in the **JM** statement. If the status of CC0/CC1 is other than 1/0, then the jump is not executed and processing resumes with the following statement. See Table 26-2 for the relationship of status bits CC0/CC1 to jump instructions.

If in any of the following cases CC0 = 1 and CC1 = 0, the **JM** operation is executed:

1. The value in ACCU 1 is within the permissible negative range after an arithmetic/math function.
2. The value in ACCU 2 is less than the value in ACCU 1 after a compare operation.

Example: Jump if minus to specified label (CC0 = 1 and CC1 = 0)

	L	MW46	// Load input word MW46 to ACCU 1
	L	MW48	// Load input image word MW48 to ACCU 1
	- I		// Subtract ACCU 1 value (MW48) minus the ACCU 2 value (MW46)
	JM	TST4	// Jump to TST4 if the subtraction result is minus (negative)
			//
			//
TST4			// Process this code when the accumulator 1 result is negative
			//

Jump if Positive or Zero: JPZ <label>

The *Jump if Positive or Zero (JPZ)* operation is based on the condition code status bit CC0. Only if CC0 = 0 does program processing jump to the statement label that matches the label parameter referenced in the **JPZ** statement. If CC0 = 1, the jump is not executed and processing resumes with the following statement. See Table 26-2 for the relationship of status bits CC0/CC1 to jump instructions.

If in any of the following cases CC0 = 0, the **JPZ** operation is executed:

1. The value in ACCU 1 is within the permissible positive range or equal to zero after an arithmetic/math function.

2. The value in ACCU 2 is greater than or equal to the value in ACCU 1 after a compare.

3. The value in ACCU 1 is zero or greater after execution of each word logic operation.

4. The value in ACCU 1 is zero or greater after execution of a shift operation.

Example: Jump if positive or zero to specified label (CC0 = 0)

	L	MW46	// Load input word MW46 to ACCU 1
	L	MW48	// Load input image word MW48 to ACCU 1—move old value to ACCU 2
	>I		// Compare if ACCU 2 value (MW46) is less than ACCU 1 value (MW48)
	JPZ	TST5	// Jump to TST5 if the ACCU 2 value (MW46) is > than ACCU 1 value (MW48)
			//
			//
TST5			// Process this code when MW46 is greater than MW48
			//

Jump if Minus or Zero: JMZ <label>

The *Jump if Zero (JMZ)* operation is based on the condition code status bit CC1. Only if CC1 = 0 does program processing jump to the statement label that matches the label parameter referenced in the **JMZ** statement. If CC1 = 1, then the jump is not executed and processing resumes with the following statement. See Table 26-2 for the relationship of status bits CC0/CC1 to jump instructions.

If in any of the following cases CC1 = 0, the **JMZ** operation is executed:

1. The value in ACCU 1 is within the permissible negative range or equal to zero after an arithmetic/math function.

2. The value in ACCU 2 is less than or equal to the value in ACCU 1 after a compare.

Example: Jump if minus or zero to specified label (CC1 = 0)

	L	MW46	// Load input word MW46 to ACCU 1
	L	MW48	// Load input image word MW48 to ACCU 1—move old value to ACCU 2
	<I		// Compare if ACCU 2 value (MW46) is less than ACCU 1 value (MW48)
	JMZ	TST6	// Jump to TST6 if the ACCU 2 value (MW46) is < than ACCU 1 value (MW48)
			//
			//
TST6			// Process this code when MW46 is less than MW48
			//

Jump if Unordered (UO): JUO <label>

The *Jump if Unordered (JUO)* operation is based on the condition code status bits CC0 and CC1. If CC0 = 1 and CC1 = 1, then program processing jumps to the statement label that matches the label parameter referenced in the statement. If the status of CC0/CC1 is other than 1/1, then the jump is not executed and processing resumes with the following statement. See Table 26-2 for the relationship of status bits CC0/CC1 to jump instructions.

If in any of the cases listed below CC0 = 1 and CC1 = 1, the **JUO** operation is executed

1. An attempt to divide by zero occurred in an arithmetic operation.

2. An invalid REAL value is specified as the source value or is the result in an arithmetic operation.

Example: Jump if unordered to specified label (CC0 = 1 and CC1 = 1)

	L	MD46	// Load memory double word MD46 to ACCU 1
	L	MD50	// Load memory double word MD50 to ACCU 1—move old value to ACCU 2
	/R		// Divide ACCU 2 value (MD46) by ACCU 1 value (MD50)
	JP	ERR1	// Jump to ERR1 if there was an attempt to divide by zero (MD50 = 0)
ERR1			// Process this code on attempt to divide by zero

Jumps based on OV and OS Status

The next two jump operations are conditional and are based on the status of the overflow (OV) and the overflow stored (OS) status bits. These status bits, along with other status bits of the CPU status word, are described in the Chapter 25 *STL Status Bit Operations* discussion on *S7 CPU Status Word*.

Jump if Overflow (OV): JO <label>

The *jump if overflow (JO)* operation is executed based on the overflow status bit. If the **OV** = **1**, program processing jumps to the label that matches the label parameter referenced in the **JO** statement. If **OV** = **0**, the jump is not executed and processing resumes with the following statement.

If in any of the following cases the **OV** status bit is set, the **JO** operation is executed:

1. After any arithmetic operation or standard math function.

2. After compare operations involving REAL numbers.

3. After conversion operations INT/DINT-to-BCD and REAL-to-DINT.

4. After conversion operation involving two's complement.

Example: Jump to specified label if the previous operations resulted in overflow

	L	MD46	// Load memory double word MW46 to ACCU 1
	L	MD50	// Load input image word MW48 to ACCU 1—move old value to ACCU 2
	*D		// Multiply ACCU 2 value (MW46) by ACCU 1 value (MW48)
	JO	OVR1	// Jump to OVR1 if the result exceeds the permissible DINT range
	T	MD54	// Transfer the result to memory double word MD54
			//
OVR1			// Process this code when the computed result has overflowed

Jump if Overflow Stored (OS): JOS <label>

The *jump if overflow stored (JOS)* operation is executed based on the overflow stored status bit. If the status bit **OS = 1**, then program processing jumps to the statement label that matches the label parameter referenced in the jump statement. If **OS = 0**, the jump is not executed and the statement immediately following the **JOS** instruction is processed.

In the following example, the result of the multiplication is subsequently divided in the following operation. If an overflow occurs in either calculation, the program jumps to the label OVR1.

Example: Jump to specified label if any of the previous operations resulted in overflow

	L	MD46	// Load memory double word MW46 to ACCU 1
	L	MD50	// Load input image word MW48 to ACCU 1—move old value to ACCU 2
	*D		// Multiply ACCU 2 value (MW46) by ACCU 1 value (MW48)
	L	L#2	// Load long constant 2 to ACCU 1—move old ACCU 1 result to ACCU 2
	/D		// Divide result (ACCU 2) by 2
	JOS	OVR1	// Jump to OVR1 if the ACCU 1 result exceeds the permissible DINT range
	T	MD54	// and overflow stored status bit is set; otherwise, store result in MD54
			//
			//
OVR1			// Process this code when the overflowed stored status is set

Loop Jump: LOOP <label>

When the *LOOP* operation is processed, the loop count value of accumulator 1 is decremented by +1 before passing control to the specified statement label. Program execution will resume at the labeled statement—processing the statements between the LOOP statement and the target label. The loop is processed in this fashion, while the loop counter value of accumulator 1 is not equal to zero. When the loop count is zero, the loop is terminated, and the processing continues with the operation following the loop instruction.

The **LOOP** operation supports both forward and backward jumps, where the maximum jump distance is -32768 or +32767. The jump distance, roughly equal to the number of jumped statements, depends on the mix of statements in the block, and the memory requirement of each statement that is jumped.

In the following example, the initial loop counter value of 10 is transferred to memory word MW10. The count is decremented by one on each pass of the LOOP operation. The statements between the **LOOP** operation and the loop label are processed on each iteration of the loop.

Example: Process defined loop until loop index counter = 0

	L	#LCNT	// Load maximum loop count into the low-order word of ACCU 1
	T	MW10	// MW10 is the loop counter that will count down to zero.
CONT:	L.	PIW260	//
			//
	L	MW10	// Load contents of loop counter into ACCU 1
	LOOP	CONT	// Decrement ACCU 1 value and jump to the specified label if MW10 <> 0
	A	I 5.5	// Program scan resumes here after loop is finished
MAIN			//

PROGRAM FLOW CONTROL—BLOCK OPERATIONS

Block operations is the second of the three categories into which STEP 7 Program Control operations are divided—these operations are divided further into four subsets: *block termination instructions*, *data block instructions*, *block call instructions* and *null operations*. Block call operations support conditional or unconditional calls to code blocks, whereas block termination instructions are used to terminate block execution either conditionally or unconditionally.

Data block instructions support the management of data block registers, and access to data in your program. Take a moment to review the block operations instructions of Table 26-3.

Table 26-3. STL Program Control—Block Operations

Operation Type	Operation	Brief Description
BLOCK TERMINATION	BEU	**Block End Unconditional.** Terminate block processing irrespective of the RLO and return to the calling block.
	BEC	**Block End Conditional.** Terminate block processing if RLO = 1.
	BE	**Block End.** Terminate block. Last operation of each block.
DATA BLOCK	OPN *<DBn>*	**Open Data Block.** Open Data Block for access using DB register.
	OPN *<DIn>*	**Open Data Block.** Open Data Block for access using DI register.
	CDB	**Exchange DB Registers.** Exchange Data Block Registers DI and DB.
	L DBNO	Load number of the DB opened with DB register to ACCU 1.
	L DINO	Load number of the DB opened with DI register to ACCU 1.
	L DBLG	Load length of data block opened with DB register to ACCU 1.
	L DILG	Load length of data block opened with DI register to ACCU 1.
BLOCK CALL	CALL *<FBn>*	**Call Function Block.** Call the designated FB for processing.
	CALL *<FCn>*	**Call Function.** Call the designated FC for processing.
	CALL *<SFBn>*	**Call System Function Block.** Call designated SFB for processing.
	CALL *<SFCn>*	**Call System Function.** Call the designated SFC for processing.
	UC *<FBn>*	**Unconditional Call to Function Block.** Call designated FB for processing, irrespective of the RLO.
	CC *<FBn>*	**Conditional Call to Fnction Block.** If RLO = 1 call designated FB for processing.
	UC *<FCn>*	**Un-Conditional Call to Function.** Call designated FC for processing irrespective of the RLO.
	CC *<FCn>*	**Conditional Call to Function.** If RLO = 1 call designated FC for processing.
NULL OPERATIONS	NOP 0	**No Operation 0.** Process statement without any affect.
	NOP 1	**No Operation 1.** Process statement without any affect.
	BLD *<nnn>*	**Build Operation.**

PROGRAMMING BLOCK OPERATIONS IN STL

Block operations in STL can be divided into three parts—*block termination operations*, *data block operations*, and *block call operations*. Block call operations are used to make conditional or unconditional calls to code blocks, whereas block termination instructions are used to terminate the execution of a block either conditionally or unconditionally. Data block operations support the management of data blocks and access to data in your program.

Block Termination

In STL, there are three block termination operations—**BEU**, **BEC**, and **BE**. These operations are used to conditionally, or unconditionally, end processing of a block. When a block is terminated, whether conditional or unconditional, control returns to the block from which the terminated block was called.

Block End Unconditional: BEU

The *block end unconditional (BEU)* operation is used to unconditionally terminate the processing of a block prior to the normal end of the block. When the BEU operation is executed, program execution exits the current block and returns to the previous block from which the current block was called. A typical example is shown below, where the block is unconditionally terminated at several intermediate points in the block. In this example, a BEU instruction is placed after each conditional jump test, as well as after the last statement in the each code section to which a jump is made.

In the following example, the BEU operation is used to unconditionally terminate defined sections of program code as well as the main line if no section has been enabled.

Example: Unconditional block termination at intermediate point in block

	A	I 4.1	// Scan I 4.1 for '1'—step #1 enabled
	JC	STP1	// If RLO = 1 go process STP1 code
	A	I 4.2	// Scan I 4.2 for '1'—step #2 enabled
	JC	STP2	// If RLO = 1 go process STP2 code
	BEU		// Terminate block unconditionally if no step is enabled
STP1	L	DBW0	// Load data word DBW0 (pattern #1) to accumulator 1
	T	QW10	// Transfer output pattern #1 for step #1 to output image word 10
	BEU		// Terminate block unconditionally when done with step #1 code
STP2	L	DBW2	// Load data word DBW2 (pattern #2) to accumulator 1
	T	QW10	// Transfer output pattern #2 for step #2 to output image word 10
	BEU		// Terminate block unconditionally when done with step #2 code

Block End Conditional: BEC

The *block end conditional (BEC)* operation is used to terminate processing of a block if the RLO = 1. When the BEC operation is executed, program execution exits the current block and returns to the previous block from which the current block was called—for example OB1. If the RLO =0, the operation is not executed, the CPU sets the RLO to '1', and resumes with the operation immediately following the BEC.

The BEC operation allows a block to be ended conditionally, at intermediate points in the block. A typical use of the instruction, as shown in the example, is to check at the very beginning of a block whether the required permissives are met; if not, then terminate execution of the block. This method supports the idea of placing all code associated with a block within that block, including its execution.

Example: Terminate block execution at intermediate point on condition RLO = 1

	L	+100	//
	L	DB10.DBW2	//
	<> I		//
	ON	I 2.1	// RLO = 1 indicates continuation permissives are not met
	BEC		// On RLO = 1, terminate the block, exit and return to the calling block
CONT	A	I 2.2	//
	A	I 2.3	//
	O		//
	A	I 2.4	//
	A	I 2.5	//
	=	Q4.7	//

Block End: BE

The *block end (BE)* instruction is always the last instruction of the last network of a block—it represents the normal end of the block. You may insert the BE operation as the last instruction, but even if you fail to do so, STEP 7 inserts BE when the block is saved. When the BE operation is processed, program execution is terminated in the current block, regardless of the RLO. Processing returns to the previous block from which the current block was called.

Example: Normal block end statement

			//
	AN	I 2.1	//
	BEC		// On RLO = 1, terminate the block, exit and return to the calling block
CONT	A	I 2.2	// If the BEC is not processed, this statement is processed as a new string
			//
			//
			//
			//
	BE		// Normal block end. The block is exited upon execution of this operation

Data Block Operations

The data block (DB) resource supports the structuring and storing of data used in the user program. If you are not already familiar with data block concepts, including global and instance DBs, and how they are used, you may wish to briefly review Chapter 2 *Data Blocks*.

The following discussions include operations for opening data blocks using either the DB register or the DI register; and operations for determining the data block number and data block length of the last data block opened in the DB register or in the DI register. When dealing with data blocks, a vital functionality is that at least one shared DB and one instance DB may be open at the same time.

Open Data Block via DB Register: OPN DB <n> or OPN DB <name>

When the *OPN DB* operation is executed, a data block, of the number or symbolic name that is specified as a parameter, is opened using the DB register. The data block is opened irrespective of the RLO, and does not affect the RLO, accumulators, status bits, or block nesting depth.

In the following example, data word 10 of DB40 is transferred to bit memory word MW10. The hexadecimal constant 00FF is copied to data word 0.

Example: Using the OPN DB instruction and partial addressing of data elements

OPN	DB40	// Open data block DB20 as a global data block
L	DBW10	// Load data word DBW10 of DB40 to accumulator 1
T	MW10	// Transfer the content of accumulator 1 (DBW10) into MW10
L	W#16#00FF	//
T	DBW0	// Transfer the content of accumulator 1 (00FF) to data word DBW0

Open Data Block via DI Register: OPN DI <n> or OPN DI <name>

When the *OPN DI* operation is executed, a data block, of the number or symbolic name that is specified as a parameter, is opened using the DI register. The data block is opened irrespective of the RLO and does not affect the RLO, accumulators, status bits, or block nesting depth.

In the following example, data word 10 of DB40 is copied first to bit memory word MW10; MW 10 is subsequently copied to data word 10 of instance data block DI 50.

Example: Using the OPN DI instruction and partial addressing of data elements

OPN	DB40	// Open data block DB40 using the DB register
L	DBW10	// Load data word DBW10 of DB40 to accumulator 1
T	MW10	// Transfer accumulator 1 contents (DBW10) to memory word MW10
OPN	DI50	// Open data block DB50 using the DI register
L	MW10	// Load MW10 (DBW10 of DB40) to accumulator 1
T	DIW10	// Transfer the content of accumulator 1 (MW10) into DIW10

Exchange Data Block Registers: CDB

When executed, the CDB operation exchanges the current contents of the data block registers (DB and DI). In effect, the contents of the registers are swapped—a shared data block becomes an instance data block and vice-versa. The operation is executed regardless of the RLO, and it does not affect the RLO.

In the example, with the CDB operation the DI register is exchanged with the DB register. This technique is necessary to access the data of an instance DB, which is otherwise not accessible outside of the associated FB. By swapping the DB and DI registers, the DB number passed to the block is opened via the DB register, and read/write access to the DB is then possible. After the desired data is accessed in the called block, the CDB operation restores the DI and DB registers to their previous content.

Example: Exchange data block register contents

PT1	CDB		//
	OPN	#DBNUM	// open data block whose DB number was passed to block as a parmeter
PT2	CDB		//
			//

Load DB Number from DB Register: L DBNO

When executed, the *L DBNO* loads to accumulator 1, the number of the data block that was previously opened, using the DB register. The data block is opened irrespective of the RLO, and does not affect the RLO, accumulators, status bits, or block nesting depth. See the following example.

Load DB Number from DI Register: L DINO

When executed, the *L DINO* operation loads to accumulator 1, the number of the data block that was previously opened, using the DI register. The data block is opened irrespective of the RLO, and does not affect the RLO, accumulators, status bits, or nesting depth.

Example: Using operations L DBNO or L DINO

	L	+100	// Load upper range of data block numbers from which to read
	L	DBNO	// Load to accumulator 1 the number of the currently open data block
	>I		// If the DB number is greater than +100 then terminate block; else continue
	JC	END1	//
CONT			//
			//
			//
END1	BEU		//

Load DB Length from DB Register: L DBLG

When executed, the *L DBLG* operation loads to accumulator 1 the length of the data block that was previously opened, using the DB register. The data block is opened irrespective of the RLO, and does not affect the RLO, accumulators, status bits, or block nesting depth. See the following example.

Load DB Length from DI Register: L DILG

When executed, the *L DILG* operation loads to accumulator 1 the length of the data block that was previously opened, using the DI register. The data block is opened irrespective of the RLO, and does not affect the RLO, accumulators, status bits, or block nesting depth.

Example: Using operations L DBLG or L DILG

	OPN	DB40	
	L	DBLG	// Load to accumulator 1 the length of the currently open data block
	L	+75	// Load integer value of +75 to accumulator 1
	>=I		// Is the accumulator 2 value (DB length) greater than or equal to +75
	JC	FLT	// If the DB length is greater or equal to +75 then jump to FLT1
	A	I 7.7	// Else continue
	=	M10.7	//
	BEU		//
FLT1	CALL	FC 30	// Fault response Function
			//

Block Calls

With the exception of organization blocks (OBs), your program must have a call to each FC and FB in order that the block is called for processing. A block is called by referencing its absolute or symbolic address—for example FC10 or Comm_Unit, using the appropriate call operation.

A call to an FC may be made without conditions (unconditional), or based on specific conditions (conditional)—the same is true for an FB. Based on the purpose of a block, you determine where it is called from, how often it is called, and what conditions, if any, should trigger the call. When processed, an unconditional call is always executed; a conditional call is made only if the driving logic has continuity (RLO=1). If the RLO = 0, the call is not executed, and the following network is processed.

In STL, there are three block call instructions—**CALL, CC,** and **UC.** The CC operation is always used to make a conditional call; the UC operation is always used to make an unconditional call. Whereas the CALL operation may be used to call an FC or an FB that has no parameters, it must be used whenever the called block—FC (SFC) or FB (SFB)—has block parameters.

Call to FC with Parameters: CALL FC <n>

You must use the *CALL* operation on each call of an FC with parameters, and supply an absolute or symbolic address at each parameter. The address that you enter at each parameter must appropriately match the data type of the parameter, and be uniquely assigned to the device or operation for which the block is called.

The CALL operation is always executed irrespective of the RLO—without appropriate consideration, the operation is always processed. To cause the operation to be processed only on certain conditions, and then based on those conditions, you will need to use a jump or some other mechanism to determine when the CALL operation should be processed.

In the following example, a call is made to the parameterized FC 40, which controls an AC motor. Such a block may be called as often as required. As you use such a block in the development of your program, you would supply different addresses at each parameter, on each call of the block.

Example: STL call to FC 10—with parameters, using CALL

M3	CALL	FC 40	// call FC 40 ac motor start function
		START: I 5.1	// start PB address
		STOP: I 5.2	// stop PB address
		ESTOP: I 5.3	// emergency stop PB address
		TIMER: T33	// on delay timer
		HCOIL: I 5.4	// auxiliary coil contact address
		STOPPED: Q10.1	// motor stopped indicator
		RUNNING:Q10.2	// motor running indicator
		MOTOR: Q10.3	// motor starter coil

Call to FB with Parameters: CALL FB<n>, DB<n>

You must use the **CALL** operation on each call of an FB with parameters, and supply an absolute or symbolic address at each parameter. The address that you enter at each parameter must appropriately match the data type of the parameter, and be uniquely assigned to the device or operation for which the block is called.

The CALL operation is always executed irrespective of the RLO—without appropriate consideration, the operation is always processed. To cause the operation to be processed only on certain conditions, then based on those conditions, you will need to use a jump or some other mechanism to determine when the CALL operation should be processed.

In the following example, a call is made to the parameterized FB 53, which samples two analog input channels and produces an average reading for each channel. The block allows the addresses for two analog input channels to be specified, an integer value for the number of sample readings, a location to store each sampled channel, and finally, a binary signal that indicates the process is complete. As you use such a block in the development of your program, you would supply different addresses at each parameter, on each call of the block.

Example: STL call to FB 53—with parameters, using CALL

	AN	I 4.4	//
	BEC		// terminate block if I 4.4 = 0
M1	CALL	FB53, DB53	// call FB and open instance DDB
	CH0_IN:	#Temp1	// read temp1 (analog input channel)
	CH1_IN:	#Temp2	// read temp2 (analog input channel)
	SAMPLE:	6	// number of times to sample
	CH0_AVG:	DB10.DBW0	// store average reading on #Temp1
	CH1_AVG:	DB10.DBW1	// store average reading on #Temp2
	DONE:	M10.7	// done averaging two channels

Unconditional Call to FB without Parameters: UC FB <n>, DB <n>

The STL *unconditional call (UC)* operation is used to make an unconditional call to an FB that has no parameters. The call may be from the main organization block (OB1), or from any other block—including an OB, FC, or another FB. The UC operation is followed by the absolute or symbolic address of an FB, and its associated instance DB, separated by a comma.

In the following example, both FB 40 and FB 50 are called unconditionally. As FB 40 is called, its associated instance data block DB 10 is also opened automatically, and the block is processed until it finishes or is terminated. Upon termination of FB 40, processing returns to the calling block at the statement immediately following the FB 40 call. Similarly, as FB 50 is called, its instance data block DB 11 is also opened automatically. Processing of FB 50 is as previously described with FB40.

FB 40 and FB 50, regardless of any logic preceding, are always called when the UC operation is processed.

Example: STL unconditional call to FB 40—without parameters, using UC

			//
M2	UC	FB40, DB10	// unconditional call to FB 40
			//
M3	UC	FB50, DB11	// unconditional call to FB 50
			//

Unconditional Call to Function without Parameters: UC FC <n>

The STL *unconditional call (UC)* operation is used to make an unconditional call to an FC that has no parameters. The call may be from the main organization block (OB1), or from any other block—including an OB, FB, or another FC. The UC operation is followed by the absolute or symbolic address of an FC.

In the following example, unconditional calls are made to FC10, FC 20, FC 30, and FC 40. First, FC 10 is called—the block is processed until it finishes or is terminated; upon termination of FC 10, processing returns to the calling block at the statement immediately following the FC 10 call—in this case that is FC 20. Similarly, FC 20 is processed, and when control returns to the calling block, processing continues with FC 30, and so on with FC 40 until the calling block is finished.

Example: STL unconditional call to FC 10—without parameters, using UC

		//
UC	FC 10	// unconditional CALL to FC 10—Manage Unit 1
UC	FC 20	// unconditional CALL to FC 20—Manage Unit 2
UC	FC 30	// unconditional CALL to FC 30—Manage Unit 3
UC	FC 40	// unconditional CALL to FC 40—Manage Unit 4
		//

Conditional Call to FB without Parameters: CC FB <n>, DB <n>

The STL *conditional call (CC)* operation is used to make a conditional call to an FB that has no parameters. The call may be from the main organization block (OB1), or from any other block—including an OB, FC, or another FB. The CC operation is followed by the absolute or symbolic address of an FB, and its associated instance DB, separated by a comma.

In the following example, FB 40 is called if the RLO = 1. As FB 40 is called, its associated instance data block DB 10 is also opened automatically, and the block is processed until it finishes or is terminated. Upon termination of FB 40, processing returns to the calling block at the statement immediately following the FB 40 call. Similarly, FB 50 is called if the RLO = 1. As FB 50 is called, its instance data block DB 11 is also opened automatically. Processing of FB 50 is as previously described with FB40.

Example: STL conditional call to FB 40—without parameters, using CC

A	I 7.1	// call FB 40 on RLO=1
CC	FB40, DB10	//
A	I 7.2	// call FB 50 on RLO=1
CC	FB50, DB11	//
		//
		//

Conditional Call to Function without Parameters: CC FC <n>

The STL *conditional call (CC)* operation is used to make a conditional call to an FC that has no parameters. The call may be from the main organization block (OB1), or from any other block—including an OB, FB, or another FC. The CC operation is followed by the absolute or symbolic address of an FC.

In the following example, FC 10, FC20, and FC 30 are each called on the condition that the preceding RLO = 1. First, if I 4.1 is '1', FC 10 is called—the block is processed until it finishes or is terminated. Upon termination of FC 10, processing returns to the calling block at the statement immediately following the FC 10 call—in this case I 4.2 is checked for '1'. If the RLO =1, FC 20 is called; otherwise, if the RLO = 0 the call is not made and I 4.3 is checked for '1'. If the RLO = 1, FC 30 is called, and when control returns to the calling block, processing continues until the calling block is finished.

Example: STL conditional call to FC 10—without parameters, using CC

		//
A	I 4.1	// check if Unit-1 enabled
CC	FC 10	// conditional CALL to FC 10—Manage Unit 1
A	I 4.2	//
CC	FC 20	// conditional CALL to FC 20—Manage Unit 2
A	I 4.3	//
CC	FC 30	// conditional CALL to FC 30—Manage Unit 3

Null Operations

The statements **NOP 0** and **NOP 1** are STL null operations. They have no effect upon execution. The STL operation **BLD** is also considered a null operation.

No Operation 0: NOP 0; NOP 1

The statements *NOP 0* and *NOP 1* are null operations—when processed, they have no effect. These operations are generally used wherever you need a statement that has no processing effect. For example, there may need for a jump operation, but only for the purpose of skipping certain other operations. Since label operations always require a statement, the NOP is often a choice if no other operation is required. Null operations do use memory and have an execution time.

Example: Null operation

	A	I 4.1	// check if Unit-1 enabled
	JC	NULL	
	L	0	//
	T	MW46	//
NULL	NOP	0	// Null operation

Build Operation: BLD <nnn>

The STEP 7 editor uses the display instruction *BLD nnn* to incorporate decompiled information into the program when switching between LAD/FBD to STL. The BLD statements are not displayed.

PROGRAM FLOW CONTROL—MCR OPERATIONS

MCR operations allow you to create logic zones in your S7 program. The entire zone, and the enclosed control logic, is enabled whenever the zone is enabled, and disabled whenever the zone is disabled. Similar to the behavior of the hardwired master control relay, when the MCR Start coil/output is de-energized, non-retentive outputs inside the partitioned zone are switched off; retentive outputs inside the partitioned zone retain their last state.

Table 26-4. STL Program Control—MCR Operations Summary

Operation	Brief Description
MCRA	MCR Activate. Activate MCR Function
MCRD	MCR Deactivate. Deactivate MCR Function
MCR(	MCR Zone Start. Start MCR Zone
)MCR	MCR Zone End. End MCR Zone

STL MCR OPERATIONS DESCRIPTIONS

There are four basic MCR operations. Each operation, when programmed, is paired with one of the others. The *MCR Activate* (**MCRA**) is paired with the *MCR De-Activate* (**MCRD**) to control activating and deactivating S7 MCR functionality in your program. The *MCR Zone Start* **MCR(** and the *MCR Zone End*)**MCR**, are paired to create a zone in which logic networks are enclosed.

MCR Activate: MCRA

The *MCRA* operation enables the master control relay functionality in the S7 CPU. This operation is always executed without regard to the RLO. When encountered, the **MCRA** operation enables all MCR partitioned zones following this statement, and up to the point where an *MCR de-activate* (**MCRD**) operation is encountered.

MCR Deactivate: MCRD

The *MCRD* operation disables the S7 master control relay functionality. This operation is always executed without regard to the RLO. No MCR zones may be placed following this operation, until after the S7 MCR functionality has been again enabled with an *MCR activate* (**MCRA**) operation.

MCR Zone Start: MCR(

The *MCR Zone Start* operation is used to control the enabling or disabling of zone dependence of a partitioned zone of operations. The MCR Start zone operation **MCR(** marks the beginning of a partitioned zone and is always paired with an MCR Zone End operation)**MCR**, which marks the end of a partitioned zone. Up to eight zones may be opened (or nested) before you must close the inner-most zone.

When the conditions driving the **MCR(** operation are True (RLO=1), the statements inside of the zone are enabled to be processed as normal. When the conditions are not True (RLO = 0), the statements inside of the zone are disabled and not processed as normal.

MCR Zone End:)MCR

The *MCR Zone End* operation marks the end of the nearest open MCR zone. As illustrated in the example on the following page, the)**MCR** operation marks the end or close of a partitioned zone, and is paired with an **MCR(** operation, which marks the start of the partitioned zone. The MCR zone end operation is executed unconditionally.

PROGRAMMING MCR OPERATIONS

As mentioned earlier, MCR operations supply the means for creating logic zones in the S7 program. The control logic within a zone is enabled when the zone is enabled, and is disabled when the zone is disabled. Similar to the behavior of circuits governed by a hardwired master control relay, when the MCR Start coil is de-energized, non-retentive coils/outputs inside the partitioned zone are switched off; retentive coils/outputs, for example Set/Reset, inside the partitioned zone retain their last state.

Activating and De-Activating MCR Functionality

The MCR Activate (**MCRA**) is paired with the MCR De-Activate (**MCRD**) to control activating and deactivating S7 MCR functionality in your program. The MCR-A operation must be used in each block in which an MCR zone will be opened, or continued from a previous block. In other words, if a new block is called from within an open MCR zone, then that zone remains open in the called block, but must be activated within the new block. The MCR-A from the called block is then logically paired with the MCR-D of the calling block.

MCR Control Zones

The STEP 7 program may include up to a total of eight zones that may be individually isolated or may overlap with zones that are nested inside of another zone. Since up to eight zones may be nested, it is important to understand how inner zones are affected by outer zones and how all zones are affected by the outer most MCR zone.

If MCR zones are nested, the first or outer-most zone controls the MCR dependency of all of the nested zones. If the outer zone is disabled, then the inner zones are also disabled regardless of whether the MCR zone start operation is enabled or not. All inner zones are disabled when the outer most zone is disabled. And when the outer most zone is enabled, an inner zone that is also enabled will have no MCR dependency, if all zones in which that zone is nested are also enabled. MCR dependency behavior is outlined in Table 26-5.

Example: Illustration of Nested MCR Zones with dependency controlled by outer zone

			//
	MCRA		// Activate MCR Function
ZON1	A	Input1	// Zone dependency check
	MCR(		// Start MCR Zone 1
			//
			//
			//
ZON2	A	Input2	// Zone dependency check
	MCR(		// Start MCR Zone 2
			//
			//
			//
	)MCR		End Zone 2
			//
	)MCR		End Zone 1
	MCRD		// Deactivate MCR Function

MCR Control Zones Dependency

Networks inside an MCR zone are processed in a normal fashion only if the RLO = 1 at the statement prior to the)**MCR** statement. In this case, MCR dependency is switched off and the operations inside the zone are processed normally. If the RLO = 0 in the statement just prior to the **MCR(** statement, then MCR zone dependency is switched ON for the enclosed zone. All outputs (assignments) are reset to zero, except outputs controlled by the Set/Reset operations, which maintain their last state. A value of zero is written to all digital values or variables in digital operations. Table 26-5 describes the behavior of S7 coil/output instructions that are governed by an MCR zone.

Table 26-5. MCR Operations Dependency

Instruction Type	Response Description	
	MCR Dependency OFF Zone Enabled (RLO = 1)	MCR Zone Dependency ON Zone Disabled (RLO = 0)
Assignment	Assume normal program behavior	Assume signal state of '0'
Set bit	Assume normal program behavior	Maintain last signal state
Reset bit	Assume normal program behavior	Maintain last signal state
Write Digital Value	Assume normal program behavior	Write zero to OUT parameters/variables
Transfer Digital Value	Assume normal program behavior	Write zero to OUT parameters/variables

Part IV

The STEP 7 Program

Chapter 27

Designing and Implementing the STEP 7 Program

This chapter presents a systematic approach to designing and implementing a STEP 7 program. First, we'll look at some basic principles of STEP 7 programming and a method of defining the controlled machine or process as partitioned units. This first step provides a basis for structuring the S7 program.

Next, we'll review the basic STEP 7 building blocks including Functions (FCs), Function Blocks (FBs), Organization Blocks (OBs), System Functions, and Data Blocks (DBs). We'll examine the intent for each of these block types, as well as how each is applied in developing a modular program structure—a structure that can accommodate the way in which the machine or process is partitioned. We'll revisit I/O addressing using both the absolute and symbolic methods; and, finally, we'll do a walkthrough of how each block type is created, and the essential considerations in determining which should be used.

After examining how blocks are developed, we'll take a look at the methods for calling those same blocks from within your program. Finally, you'll be introduced to OB1 and how, as the main program block, it is used to organize calls to the other blocks of your program.

Main Topics

- Program Development Overview
- STEP 7 Program Resources
- Addressing I/O Modules
- Assigning Symbolic Addresses
- Creating Code Blocks and Data Blocks
- Documenting the STEP 7 Program
- Organizing and Processing the Program
- Developing an OB1 Call Strategy

PROGRAM DEVELOPMENT OVERVIEW

The STEP 7 programming environment supports the structuring of control tasks, and the associated code, into functional units. In this approach, a machine or process is divided into major units based on equipment (technology) function, or process function. If necessary, each unit is further broken down into subunits. Once the machine/process is broken down into units, you can apply STEP 7 programming resources to develop the control code in modular sections that reflect the segmentation of the control tasks. This program development strategy is examined in the following discussions.

Program Design Strategy

While the traditional linear approach to PLC programming is possible, a structured design that uses modular blocks of code to perform different task is the basis of the STEP 7 program architecture. Whether you choose a linear or structured approach, the design strategy is best developed in three steps. The first step is to divide the overall control task by defining the machine or process units. The next step involves defining the tasks and operations of each machine/process unit. The final step is to develop the control logic for each of the operations of a task. STEP 7 offers three standard languages, any of which you may use separately, or in combination, to develop your control logic code.

Defining Machine or Process Units

Any machine or process can be partitioned into logical units—this is a first step in developing a well-organized STEP 7 program. These units, which reflect the main technological subsystems or control functions, each perform a functional part of the whole system. By dividing the system into units, development team members can quickly grasp the control system scope. Well planned partitioning allows for easier adaptation of the program for reuse and future expansion.

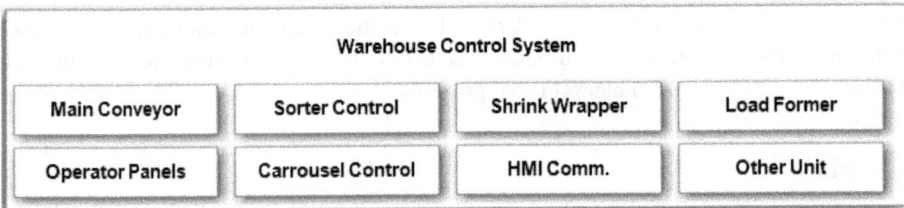

Figure 27-1. Typical division of machine/process into major functional or technological units

Defining Tasks and Operations

Machine or process units generally reflect the major systems involved in producing a finished product. Therefore, each unit can usually be further divided into functional *tasks* that must be accomplished to complete the entire unit function. Each task will normally consist of several operations that require control of individual devices, process loops, or other elements for which control logic must be defined. This concept, illustrated in the following diagram, is one method of conveying the major procedures, tasks, and operations for which a program must be developed. Other tasks like communications, alarm handing, error processing, operator controls, and production accounting must all be considered.

Defining the Control Logic

The third stage of the development strategy deals with creating control logic for the individual tasks of each process unit. In other words, how the operations for specific *devices, equipment components,* or *process loops* will be implemented, must be defined. A given subtasks, for instance, may require control of several digital and analog actuators. Completion of the tasks may also require data processing, error checking, and communications. One-by-one, the control logic for these operations must be defined and developed to complete the unit control.

400

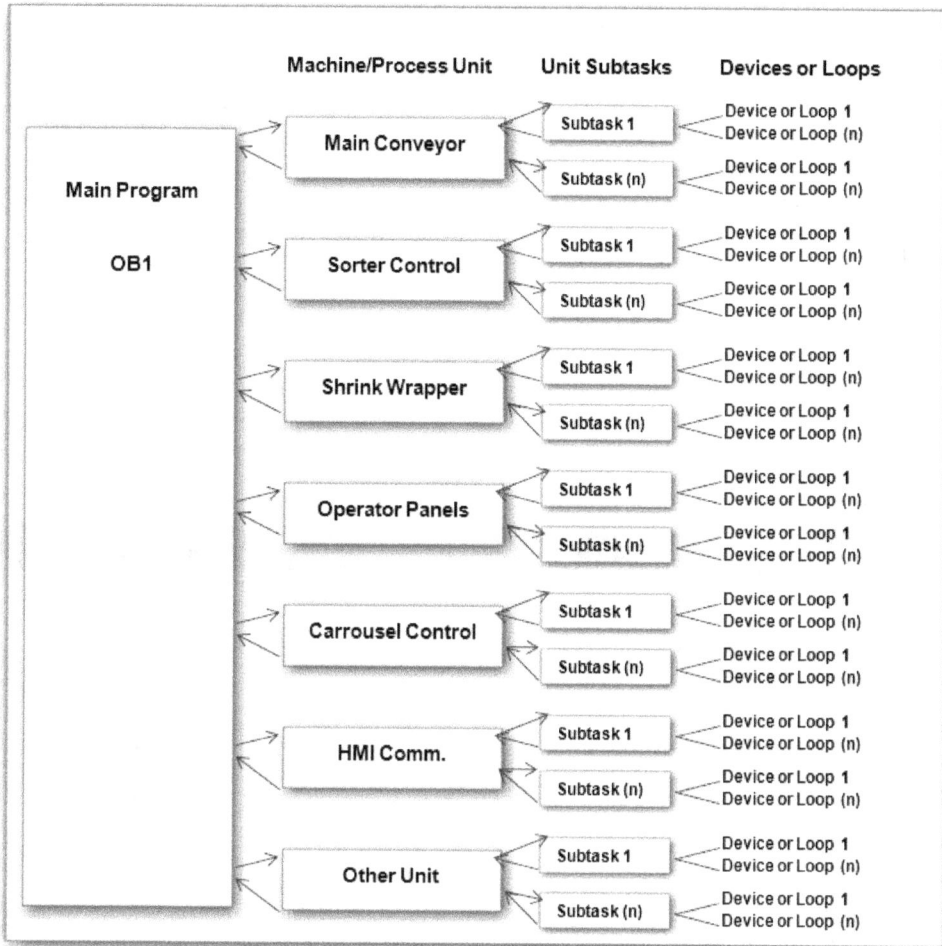

Figure 27-2. Program design based on division of control task into major functional or technological units, unit tasks, and into device or control loop operations.

Defining Modular Functionality with Blocks

Figures 27-1 and 27-2, together, illustrate how STEP 7 program development can be initiated by first defining units of functionality, each of which performs a part of the overall control task. This process results in a structure that is readily translated into individual modules of code, referred to as *blocks*.

In the diagram above, for example, a block might be assigned to each of the process units. A block might also be assigned to each of the unit subtasks. In practice, the size or level of complexity of the subtask will help you to determine if it should be coded in a single block or subdivided into multiple blocks. Each block assigned to a subtask would contain the control logic for the devices or operations that complete the functions of the subtask. Generally, the top level block, assigned to each of the process units, is responsible for coordinating the calls to the blocks that control each of the associated subtasks. This function simply involves enabling or disabling block calls based on required conditions.

As you develop the program, you may subdivide your code into as many sections or blocks as you like. To the degree that it is possible, each block should be self-contained and relatively independent of other blocks, requiring little or no interlocking. With this approach, each block can be easily tested and debugged, and later problems can be easily pinpointed to a specific block based on its function.

STEP 7 PROGRAM RESOURCES

In STEP 7, there are three categories of blocks from the user program is developed. *User-blocks* are that S7 block category in which you may write code or store data. *System blocks* are blocks integrated in the S7 CPU, but you may call from within the user program. *Standard blocks* are off-the-shelf user blocks that you may also use in your program. Standard blocks include common applications like PID control, and communications drivers. Several groups of standard blocks are supplied with STEP 7. As you begin to determine how to divide your code and to define what blocks need to be developed, be sure to consider system and standard blocks that may serve to reduce development time.

While you may use all of the resource types listed in Table 27-1, the actual number of each user block type that may be programmed, or system block that is integrated in a CPU, is dependent on the CPU. Furthermore, the length of a programmed block, in bytes, is also CPU dependent. You may view the performance characteristics for your CPU from the online window of the SIMATIC Manager, by selecting the CPU, then from the menu selecting **PLC** ➢ **Module Information**.

Table 27-1. STEP 7 Block Resources Overview

Block Type	Identifier	Range	Category	Comment
Data Blocks	DB	DB 1 to DB n	User Block	User Numbered
System Blocks	SFB/SFC/SDB	CPU Dependent	System Block	Pre-Numbered
Function Blocks	FB	FB 1 to FC n	User Block	User Numbered
Functions	FC	FC 1 to FC n	User Block	User Numbered
Organization Blocks	OB	OB 1 to OB n	User Block	Pre-Numbered

Data Blocks

The Data Block (DB) is of the user block category, and is the equivalent to the *data area* found in many PLCs. In STEP 7, the DB is a resource for organizing and storing constant and variable data used in your program. Typical constants might include analog set points, PID loop data, and recipe data values in a batch process. Variables might include analog input and output data, or various production totals. Like other STEP 7 resources, data blocks are modular. You may use as many DBs as required or supported by your CPU, include as many values as needed, and structure the data as you see fit. The application and programming of data blocks are covered, in detail, in Chapter 2 *Data Blocks*.

DB10 -- "Zone Data" -- New_prj\SIMATIC 300(1)\CPU 316\...\DB10

Address	Name	Type	Initial value	Comment
0.0		STRUCT		
+0.0	TMP_PRESET_1	INT	350	Temperature Preset Zone1
+2.0	TMP_PRESET_2	INT	375	Temperature Preset Zone2
+4.0	TMP_PRESET_3	INT	400	Temperature Preset Zone3
+6.0	OVR_LMT_T1	BOOL	FALSE	ZONE 1 OVER Limit
+6.1	OVR_LMT_T2	BOOL	FALSE	ZONE 2 OVER Limit
+6.2	OVR_LMT_T3	BOOL	FALSE	ZONE 3 OVER Limit
+7.0	OPER_RESP_1	CHAR	'N'	OPERATOR RESPONSE 'YES'
+8.0	OPER_RESP_2	CHAR	'N'	OPERATOR RESPONSE 'NO'
+10.0	NEW_CNT	WORD	W#16#900	COUNTER OVERRIDE
=12.0		END_STRUCT		

Figure 27-3. Each data block may be defined according to your data requirements

System Blocks (SFB, SFC, and SDB)

System function blocks (SFBs) and system functions (SFCs) are code blocks that are integrated in the S7 CPU. With these already developed and tested functions, you can reduce the overall development time of your program. There are SFC and SFB code blocks for managing distributed I/O, accessing module data, enabling and disabling interrupts, setting and reading the real-time clock, message-handling and data transfer operations, handling various S7 communications services, accessing services of the CPU in your user program, and many other functions. A small sample of system blocks are listed below. A more extensive list of SFCs and SFBs is provided in Appendix B *System Blocks*.

Table 27-2. System blocks for Copy Data Area and Block Operations

SFC	Abbreviation	Description
SFC 20	BLKMOV	Copy a contiguous group of variables to designated destination
SFC 21	FILL	Initialize a contiguous memory area with data
SFC 22	CREAT_DB	Create a Data Block during runtime
SFC 23	DEL_DB	Delete a Data Block during runtime
SFC 24	TEST_DB	Test a Data Block during runtime
SFC 25	COMPRESS	Compress CPU user memory
SFC 44	REPL_VAL	Transfer substitute value to accumulator 1 (ACCU1)
SFC 81	UBLKMOV	Un-Interruptible Block Move

Function Block (FB)

A *Function Block* is of the user block type—a block in which portions of your program may be written. The FB is generally used in developing control code or algorithms where data, that is needed or generated by the block, must be retained in memory from one call of the block to the next. To handle this requirement, a Data Block (DB) must be assigned to each FB. Stored static variables, and parameters passed to and from the FB, are retained in this DB. If, for example, you were an FB to handle a special function related to a closed-loop process, the changing data associated with the loop variables must be available on each call of the block. In cases such as this, you'll want to use an FB. FB application and programming are covered in Chapter 28 *Developing Functions and Functions Blocks*.

Function (FC)

A *Function*, like the Function Block, is a user block in which you may write parts of the control program. Unlike the FB, the FC only supports use of temporary variables, and does not have memory that is available from one call of the block to the next. Variables used in an FC are not retained after processing has ended. The fact that FC variables are not available after the block is ended makes the FC primarily suited to developing simple to complex control logic with no memory requirements. In many S7 applications, FCs account for as much as 60 percent of the program. The application and programming of FCs is covered in Chapter 28 *Developing Functions and Functions Blocks*.

Organization Block (OB)

An *Organization Block,* or OB, as it is often called, is an S7 resource for organizing your code to efficiently respond to events and conditions that occur in the controlled process, in your program, and in the PLC itself. As one of three types of blocks in which you may write your program, OBs provide the structured means by which the code blocks (FBs, FCs, SFBs, SFCs) of your program are called and processed under start-up, normal, background, periodic, interrupt, and error and fault conditions.

OBs are different from FCs and FBs in that their processing is predetermined by specific events that can occur in the process, in the control program, or in the control system. Typical events include timed or hardware interrupts, error or fault conditions in the PLC, or even the startup of the PLC. Table 27-3 introduces the basic categories of organization blocks and the types of S7 events to which each category is associated. An OB, once downloaded, is called by the operating system based upon the occurrence of any one of its associated triggering events.

Table 27-3. S7 Organization Block Categories—and the associated call event

OB Type	OB No.	Call Event and Application Description
Normal Processing	1	Called after each CPU start-up, and cyclically at the end of each CPU cycle (I/O update to process the main user program).
Time-of-Day Interrupts	10-17	Called at user-configured time-of-day interrupts.
Time-of-Delay Interrupts	20-23	Called at user-configured time-delay interrupts.
Cyclic Interrupts	30-38	Called cyclically at user defined intervals (e.g., 500 ms).
Hardware Interrupts	40-47	Called on occurrence of a process or module-generated interrupt.
Multi-Computing Interrupt	60	Called on interrupt by a CPU in a multi-CPU configuration.
Redundancy Interrupts	70-73	Called redundancy error in an S7-400 H-system.
Asynchronous Interrupts	80-87	Called on a system-related fault, like a power supply error or fault.
Background Processing	90	Called when current cycle time is less than the set minimum cycle.
Start-up Processing	100-102	Called at each CPU start-up, prior to calling OB1.
Synchronous Interrupts	121-122	Called on program-related runtime errors, like block not installed.

Examples of OBs include OB 100, OB 101, and OB 102—they are all start-up OBs associated with the different startup types. If these blocks are in the user program at the time of start-up, then the appropriate block is called based on the type of start-up that occurs. If you have code that needs to be processed at each start-up, then you might write this code in the appropriate start-up OB, or in a block called from the start-up OB. OB application and programming are covered in Chapter 29 *Developing Organization Blocks*.

ADDRESSING I/O MODULES

The following discussions describe I/O module addressing for both the S7-300 and S7-400 controllers. As we look at addressing for the S7-300, you will see that, by default, the addressing is slot-dependent. That is to say that the addresses for each installed module are determined by the slot in which it is placed. This slot-based addressing is the only addressing in most S7-300 CPUs, but not all. Some S7 300 CPUs allow free assignment of addresses, as with all S7-400 controllers.

In the S7 memory discussions of Chapter 1 *Memory and Addressing*, we examined the memory areas into which discrete and analog I/O modules are actually mapped. In the following discussions, we'll see how each module address is initially defined in the Hardware Configuration tool. Immediately after these discussions of absolute address assignment, we'll take a look at how symbolic addresses are assigned to I/O addresses, as well as to other S7 memory areas.

Addressing S7-300 Digital I/O

As you insert each digital module in an S7-300 hardware configuration, the addresses are assigned automatically to the module according to the module slot. From Rack-0/Slot-4, each slot reserves four-bytes in the S7 memory, starting with byte-0. The 4-bytes are reserved in either the Input (I) image area or Output (Q) image area, depending on whether the module is an input or output module. Reserving four bytes allows each slot to support modules of up to 32 inputs or 32 outputs.

A maximum expansion of the S7-300 is shown below, starting with Rack-0. Slot-4 reserves bytes 0-3; slot-5 reserves bytes 4-7; and so on. Installing an input module in a slot, reserves bytes in the input (**I**) area; if an output module is installed, the 4-bytes are reserved in the output (**Q**) area. Depending on the module type inserted, the address area of the opposite module type goes unused. Although most S7-300s use this fixed slot-dependent addressing, some S7-300 CPUs allow free assignment of the address range. In these CPUs, input and output modules may use the same starting byte addresses. For example, you might have addresses I 3.0 and Q 3.0—in this case no byte addresses go unused.

		I/Q	I/Q	I/Q	I/Q	I/Q	I/Q	I/Q	I/Q
Rack-3	**Slot-3**	**Slot-4**	**Slot-5**	**Slot-6**	**Slot-7**	**Slot-8**	**Slot-9**	**Slot-10**	**Slot-11**
Power Supply	IM 361	96.0 to 99.7	100.0 to 103.7	104.0 to 107.7	108.0 to 111.7	112.0 to 115.7	116.0 to 119.7	120.0 to 123.7	124.0 to 127.7
		I/Q	I/Q	I/Q	I/Q	I/Q	I/Q	I/Q	I/Q
Rack-2	**Slot-3**	**Slot-4**	**Slot-5**	**Slot-6**	**Slot-7**	**Slot-8**	**Slot-9**	**Slot-10**	**Slot-11**
Power Supply	IM 361	64.0 to 67.7	68.0 to 71.7	72.0 to 75.7	76.0 to 79.7	80.0 to 83.7	84.0 to 87.7	88.0 to 91.7	92.0 to 95.7
		I/Q	I/Q	I/Q	I/Q	I/Q	I/Q	I/Q	I/Q
Rack-1	**Slot-3**	**Slot-4**	**Slot-5**	**Slot-6**	**Slot-7**	**Slot-8**	**Slot-9**	**Slot-10**	**Slot-11**
Power Supply	IM 361	32.0 to 35.7	36.0 to 39.7	40.0 to 43.7	44.0 to 47.7	48.0 to 51.7	52.0 to 55.7	56.0 to 59.7	60.0 to 63.7
		I/Q	I/Q	I/Q	I/Q	I/Q	I/Q	I/Q	I/Q
Rack-0	**Slot-3**	**Slot-4**	**Slot-5**	**Slot-6**	**Slot-7**	**Slot-8**	**Slot-9**	**Slot-10**	**Slot-11**
CPU and Power Supply	IM 360	0.0 to 3.7	4.0 to 7.7	8.0 to 11.7	12.0 to 15.7	16.0 to 19.7	20.0 to 23.0	24.0 to 27.7	28.0 to 31.7

Figure 27-4. Default Digital Addressing of the S7-300, starting at byte-0. Each slot reserves 4-bytes

Both input and output addresses are referenced as **byte.bit**. Digital input addresses are prefixed with the input memory identifier 'I'; digital output addresses are prefixed with the output memory identifier 'Q' as a prefix to the byte address. A 32-point output module in Rack-0/Slot-4 would use addresses Q 0.0 through Q 3.7, where Q 0.0 corresponds to the first output circuit; an input module in the same slot would use I 0.0 through I 3.7, where I 0.0 corresponds to the first input circuit. Inputs and outputs may also be accessed by byte, word, or double word—for example IB 8, IW 8, and ID 8; QB 8, QW 8, and QD 8. For additional addressing details, refer to *Input Image* and *Output Image* in Chapter 1 *Memory and Addressing*, under *S7 Memory Areas*.

Addressing S7-300 Analog I/O

Default analog I/O addressing in the S7-300, like with digital I/O, is predetermined by the physical slot in which a module is inserted. Each slot, as shown in the figure below, reserves sixteen bytes of the peripheral memory (**P**) area, starting with byte 256 in Slot-4 of Rack-0. Since each analog input or output channel uses two bytes to handle its digital representation of the signal, an allocation of sixteen bytes per slot supports modules with up to 8 analog channels (inputs or outputs).

Maximum expansion of the S7-300, using analog I/O, is shown below. Rack-0/Slot-4 reserves bytes 256 - to- 271; slot-5 reserves bytes 272 -to- 287; and so on. If an input module is inserted, each input channel address is prefixed with **PIW** *(peripheral input word)*; if an output module is inserted, each output channel address is prefixed with **PQW** *(peripheral output word)*. The first analog input is PIW 256 (bytes 256-257); the second input is PIW 258. If an analog output module is placed in the first slot, the first output address is PQW 256 (bytes 256-257); the second output channel is PQW 258.

Whereas the default start byte-address, assigned to each module, is generally fixed in the S7-300, some CPUs support free assignment of addresses. In those CPUs with free assignment, input and output modules may use the same starting byte address. For example, you might have PIW 256 and PQW 256.

		PIW/PQW Slot-4	PIW/PQW Slot-5	PIW/PQW Slot-6	PIW/PQW Slot-7	PIW/PQW Slot-8	PIW/PQW Slot-9	PIW/PQW Slot-10	PIW/PQW Slot-11
Rack-3	Slot-3	Slot-4	Slot-5	Slot-6	Slot-7	Slot-8	Slot-9	Slot-10	Slot-11
Power Supply	IM 361	640 to 655	656 to 671	672 to 687	688 to 703	704 to 719	720 to 735	736 to 751	752 to 767
Rack-2	Slot-3	Slot-4	Slot-5	Slot-6	Slot-7	Slot-8	Slot-9	Slot-10	Slot-11
Power Supply	IM 361	512 to 527	528 to 543	544 to 559	560 to 575	576 to 591	592 to 607	608 to 623	624 to 639
Rack-1	Slot-3	Slot-4	Slot-5	Slot-6	Slot-7	Slot-8	Slot-9	Slot-10	Slot-11
Power Supply	IM 361	384 to 399	400 to 415	416 to 431	432 to 447	448 to 463	464 to 479	480 to 495	496 to 511
Rack-0	Slot-3	Slot-4	Slot-5	Slot-6	Slot-7	Slot-8	Slot-9	Slot-10	Slot-11
CPU and Power Supply	IM 360	256 to 271	272 to 287	288 to 303	304 to 319	320 to 335	336 to 351	352 to 367	368 to 383

Figure 27-5. Default S7-300 analog addressing, starting at byte-256 of peripheral memory

Unlike digital I/O where the input image is read, and the output image is written, cyclically by the CPU, you must manage analog I/O updates in your program. This is true since with peripheral memory, modules are read from or written to directly. In LAD/FBD, an analog input is read by specifying the analog input address as the source address in a Move operation; an output is written to by specifying the analog output address as the destination address of the Move. In STL, a Load operation reads an analog input by specifying the input address—for example in **L PIW 258**. A value is sent to an analog output with a Transfer operation that specifies the output—for example in **T PQW 260**.

Addressing S7-400 Digital I/O

As in the S7-300, each digital input module inserted in the S7-400 configuration reserves 4-bytes of the input image (**I**) memory. Each output module reserves 4-bytes of the output image (**Q**) memory. Reserving four bytes per slot allows for a module of up to 32 inputs or 32 outputs. Although default addressing in the S7-400 is similar to the S7-300, it is neither fixed nor slot-dependent. Furthermore, input and output modules may use the same starting byte address. This method of assignment means that no addresses in the input image (**I**) or output image (**Q**) areas need to go unused.

As modules are inserted in the hardware configuration, STEP 7 assigns the first input module with a start address of byte-0. This assignment is regardless of the slot you use. The first output module is also assigned a start address of byte-0. The start address of each successive input module is based on the next available byte address after the previously inserted input module (for inputs), or previously inserted output module (for outputs), even if a slot is left empty.

		I	I	I	I	Q	Q	Q	Q
		Slot-4	Slot-5	Slot-6	Slot-7	Slot-8	Slot-9	Slot-10	Slot-11
PS-407	S7-CPU	Byte-0	Byte-4	Byte-8	Byte-12	Byte-0	Byte-4	Byte-8	Byte-12
		0.0	4.0	8.0	12.0	0.0	4.0	8.0	12.0
		to	to	to	to	to	to	to	to
		0.7	4.7	8.7	12.7	0.7	4.7	8.7	12.7
		Byte-1	Byte-5	Byte-9	Byte-13	Byte-1	Byte-5	Byte-9	Byte-13
		1.0	5.0	9.0	13.0	1.0	5.0	9.0	13.0
		to	to	to	to	to	to	to	to
		1.7	5.7	9.7	13.7	1.7	5.7	9.7	13.7
		Byte-2	Byte-6	Byte-10	Byte-14	Byte-2	Byte-6	Byte-10	Byte-14
		2.0	6.0	10.0	14.0	2.0	6.0	10.0	14.0
		to	to	to	to	to	to	to	to
		2.7	6.7	10.7	14.7	2.7	6.7	10.7	14.7
		Byte-3	Byte-7	Byte-11	Byte-15	Byte-3	Byte-7	Byte-11	Byte-15
		3.0	7.0	11.0	15.0	3.0	7.0	11.0	15.0
		to	to	to	to	to	to	to	to
		3.7	7.7	11.7	15.7	3.7	7.7	11.7	15.7

Figure 27-6. Default digital I/O addresses in the S7-400 starts at byte-0, regardless of the slot, and reserves 4-bytes per slot. Input and output modules may use the same byte addresses.

If input modules are installed in the first four slots, as seen in the figure above, slot-4 would reserve bytes 0-3; slot-5 would reserve bytes 4-7; slot-6 would reserve bytes 8-11; and so on. If output modules are installed in the following four slots, slot-8 would reserve bytes 0-3; slot-9 would reserve bytes 4-7; slot-10 would reserve bytes 8-11; and so on. In the S7-400, the automatically assigned starting byte-addresses may be modified under the address properties of the module. STEP 7 will notify you if you attempt to assign a start byte address that is already assigned .

Digital inputs are accessed using the input memory identifier 'I' as a prefix to byte addresses; digital outputs are accessed using the output memory identifier 'Q' as a prefix to byte addresses. Both inputs and outputs are referenced as '**byte.bit**'. A 32-point output module in slot-4 would use addresses Q 0.0 through Q 3.7, where Q 0.0 corresponds to the first output circuit; an input module would use I 0.0 through I 3.7, where I 0.0 corresponds to the first input. Inputs and outputs may also be accessed by byte, word, or double word—for example IB 8, IW 8, and ID 8 for inputs, and QB 8, QW 8, and QD 8 for outputs. For additional addressing details, refer to *Input Memory* and *Output Memory* in Chapter 1 *Memory and Addressing*, under *S7 Memory Areas*.

Addressing S7-400 Analog I/O

Each analog module inserted in an S7-400 configuration, depending on the number of channels, will reserve sixteen or thirty-two bytes of the peripheral (**P**) memory area. Since each analog input or output channel requires a word (two bytes) to handle its digital representation of the signal, an 8-channel module reserves 16-bytes and a 16-channel module reserves 32-bytes.

Default addressing for S7-400 analog I/O starts with byte-512 of the peripheral memory, for both inputs and outputs. As each module is inserted, its starting byte address is based on the next available byte address following the previous analog input module (for inputs) or previous analog output module (for outputs)—even if a slot is left empty. Like with digital modules in the S7-400, analog input and output modules can have the same starting byte address.

		PIW/PQW	PIW/PQW	PIW/PQW	PIW/PQW	PIW/PQW	
		Slot-4	Slot-5	Slot-6	Slot-7	Slot-8	Slot-9
PS-407	S7-CPU	PIW 512	PIW 544	PIW 576			IM
		PIW 514	PIW 546	PIW 578	PQW 512	PQW 528	
		PIW 516	PIW 548	PIW 580			
		PIW 518	PIW 550	PIW 582	PQW 514	PQW 530	
		PIW 520	PIW 552	PIW 584			
		PIW 522	PIW 554	PIW 586	PQW 516	PQW 532	
		PIW 524	PIW 556	PIW 588			
		PIW 526	PIW 558	PIW 590	PQW 518	PQW 534	
		PIW 528	PIW 560	PIW 592			
		PIW 530	PIW 562	PIW 594	PQW 520	PQW 536	
		PIW 532	PIW 564	PIW 596			
		PIW 534	PIW 566	PIW 598	PQW 522	PQW 538	
		PIW 536	PIW 568	PIW 600			
		PIW 538	PIW 570	PIW 602	PQW 524	PQW 540	
		PIW 540	PIW 572	PIW 604			
		PIW 542	PIW 574	PIW 606	PQW 526	PQW 542	

Figure 27-7. S7-400 default analog addressing, starts at byte-512 for inputs and outputs

In the figure above, 16-channel analog input modules are in slots 4, 5, and 6. Eight-channel analog output modules are installed in slots 7 and 8. Starting with slot-4, bytes 512-to-543 are reserved; in slot-5, bytes 544-to-575 are reserved; and in slot-6, bytes 576-to-607 are reserved. Since analog input and output modules can have the same start address, the output module in slot-7 also uses byte-512 as the start byte address. Slot-8 reserves bytes 528 -to- 543. Each input channel address is prefixed with **PIW** for *peripheral input word*; each output channel address is prefixed with **PQW** for *peripheral output word*. The first analog input channel is PIW 512; the second input is PIW 514. The first analog output channel is PQW 512; the second analog output channel is PQW 514.

Since peripheral memory involves direct access to input and output modules, you must manage analog I/O in your control program. In LAD/FBD, an analog input is read by specifying the input address as the *source address* in a **Move** operation; an analog output is written by specifying the output address as the *destination* address of the **Move** operation. In STL, a *Load* operation reads an analog input by specifying the input address—for example in the statement **L PIW 512**. A value is sent to an analog output with a *Transfer* operation that specifies the output—for example in the statement **T PQW 512**.

408

ASSIGNING SYMBOLIC ADDRESSES

After you have assigned I/O addresses, and perhaps allocated memory elements—for example, timers (**T**), counters (**C**), and bit memory (**M**) in your program, then you should consider assigning symbolic addresses to those allocated absolute addresses. Symbolic addresses may even be assigned to the various code blocks and data blocks that you create in your program. In STEP 7, a symbol name may be created with up to 24-characters. By assigning a meaningful name to each absolute address, you, as well as others, will find the program much easier to understand, interpret, and troubleshoot.

Creating Global Symbols

A *global symbol* is defined using the Symbols Editor. Recall, that the **Symbol Table** is a STEP 7 object located under the **S7 Program** object. Each program has its own symbol table, even in a multi-CPU configuration, where each CPU has its own program, and each program has its own symbol table.

You may assign global symbols to S7 input (**I**), output (**Q**), bit memory (**M**), timer (**T**), counter (**C**), peripheral input (**PI**), and peripheral output (**PQ**) memory areas, as well as to each code block and data block. Symbolic names may be assigned to bit, byte, word, and double word locations. For peripheral memory, only byte, word, and double word locations may be assigned symbol names. As for I/O addresses, you may assign symbols using the Symbols Editor or the Hardware Configuration Tool.

In a S7 program, global symbols are distinguished from local symbols by quote marks that are inserted by the STEP 7 editor—for example 'Motor_1'. Steps for assigning global symbols are shown below.

To assign Global symbols: (See Table 27-4)

1. From the SIMATIC Manager, open the desired project and select the **S7 Program** in which you wish to create or edit existing global symbols.

2. With an **S7 Program** folder selected, from the right pane of the project window, double click on the **Symbols** object, or right click on the object and select **Open Object**.

3. Click in the **Symbol** field and enter up to 24 characters for a symbolic name; click in the **Address** field and enter the absolute address; click in the **Comment** field and enter a descriptive comment of up to 80-characters. The **Data Type** is automatically inserted.

4. When you are done, from the menu select **Symbol Table ➤ Close**.

Creating Local Symbols

A *local symbol* is the **Name** you assign to a variable when it is specified in the declaration table of a code block. As the name implies, use of a local symbol is limited to the block in which it is declared. Local symbols are distinguished from global symbols by a hash mark that is inserted as a prefix by the STEP 7 editor—for example #TEMP_1. In the following example, the local symbol names are assigned to TEMP variables; other variable declaration types may be assigned symbols as well.

To assign Local symbols:

1. From the **S7 Program** open the block in which you wish to create or edit local symbols.

2. Select the **TEMP** object in the left pane of the declaration table.

3. Click in the **Name** field of the declaration row and enter the name for the variable.

4. Click in the **Data Type** field and select the variable's data type from the drop list.

5. Click in the **Comment** field and enter a descriptive comment.

6. Repeat the previous steps to enter additional temporary variables.

Table 27-4. Examples of global symbol Address assignments to S7 memory areas and blocks

Symbol	Address	Type	Comment
STR_PMP	I 40.2	BOOL	EMS START PUMP
STP_PMP	I 40.3	BOOL	EMS STOP PUMP
ALN_BTH	I 32.0	BOOL	CAR ALIGN BOTH
ALN_FWD	I 32.1	BOOL	CAR ALIGN FORWARD
ALN_LFT	I 32.3	BOOL	CAR ALIGNING LEFT
ALN_RHT	I 32.2	BOOL	CAR ALIGNING RIGHT
SPSET_D1	ID 60	DWORD	7-DIGIT BCD SPEED SET DRIVE 1
SPSET_D2	ID 64	DWORD	7-DIGIT BCD SPEED SET DRIVE 2
ARM_TLT_FWD1	Q 108.5	BOOL	ARM TILT FORWARD 1
ARM_TLT_FWD2	Q 108.7	BOOL	ARM TILT FORWARD 2
ARM_TLT_REV1	Q 108.6	BOOL	ARM TILT REVERSE 1
ARM_TLT_REV2	Q 109.0	BOOL	ARM TILT REVERSE 2
SPDSPLY_1	QD 60	DWORD	7-DIGIT BCD SPEED DISPLAY—DRIVE 1
SPDSPLY_2	QD 64	DWORD	7-DIGIT BCD SPEED DISPLAY—DRIVE 2
C_LVL	M 4.6	BOOL	COOLING LEVEL LOW
C_PSL	M 4.7	BOOL	COOLING PRESSURE LOW
CASC_SP1	MW 810	WORD	CASCADE MODE SPRAY 1
CASC_SP3	MW 814	WORD	CASCADE MODE SPRAY 3
SHFT1T1	MD	DWORD	SHIFT 1 TEMPORARY TOTAL RUN1
SHFT1T2	MD	DWORD	SHIFT 1 TEMPORARY TOTAL RUN2
CWFI1	PIW 288	WORD	COOLING WATER FLOW INDICATOR 1
CWFI2	PIW 290	WORD	COOLING WATER FLOW INDICATOR 2
EMSLCC	PIW 274	WORD	EMS LOWER COIL CURRENT
EMSUCC	PIW 272	WORD	EMS UPPER COIL CURRENT
D1SPC	PQW 486	WORD	DRIVE 1 SPEED CONTROL
D2SPC	PQW 498	WORD	DRIVE 2 SPEED CONTROL
L1_CTRL	FC1	BLOCK_FC	CONTROL BLOCK LINE 1
L2_CTRL	FC2	BLOCK_FC	CONTROL BLOCK LINE 2
SPC_FB1	FB 22	BLOCK_FB	SPECIAL CONTROL PROCEDURE 1
SPC_FB2	FB 23	BLOCK_FB	SPECIAL CONTROL PROCEDURE 2
MAINBLK	OB 1	BLOCK_OB	MAIN CYCLICAL PROGRAM BLOCK
PRDDATA	DB 3	BLOCK_DB	PRODUCTION DATA BLOCK STORE
PARMSTR1	DB 22	BLOCK_DB	INSTANCE DATA BLOCK STORE FOR FB22
PARMSTR2	DB 23	BLOCK_DB	INSTANCE DATA BLOCK STORE FOR FB23

Assigning I/O Symbols in the Hardware Configuration

Symbolic addresses are generally entered in the STEP 7 Symbol Table, but with I/O module addresses, you may find that entering symbols is more efficient in the Hardware Configuration tool. As each discrete or analog I/O module is inserted, you may assign a symbolic address to the individual addresses of the module.

When the dialog is opened for each module, the absolute input or output addresses for each module are listed in the **Address** column. You may enter a symbolic address in the **Symbol** column and a description in the **Comment** column. The **Data Type** is automatically inserted by STEP 7. A symbol name may have up to 24 alphanumeric characters.

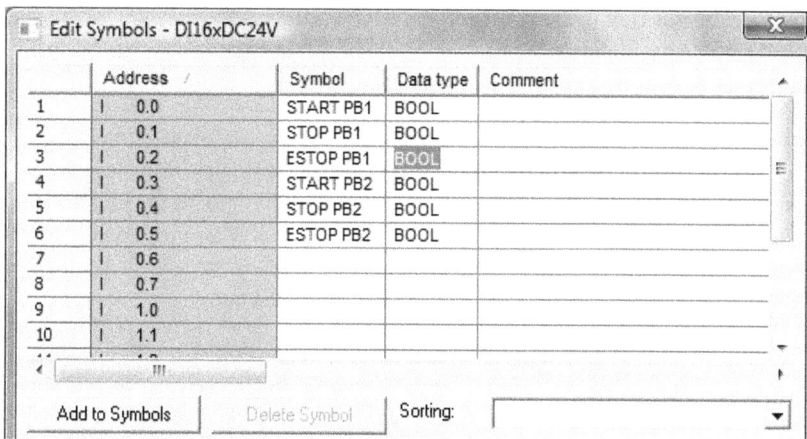

Figure 27-8. Hardware Configuration—Assigning Symbols to a Digital module

To configure I/O symbols in the hardware configuration:

1.   Open the desired station and select the rack containing the modules to which symbolic addresses are to be assigned.

2.   Select a module, right-click and select **Edit Symbolic Addresses**. The symbol table is presented like a spreadsheet, with a column for the absolute **Address, Symbol, Data Type**, and a **Comment**.

3.   In the **Symbol** field enter up to 24-alphanumeric characters for the symbolic address. The **Data Type** of each absolute address is automatically inserted.

4.   In the **Comment** field, you may use up to a maximum of 80-characters to enter a description of the I/O point.

5.   Click the **Add to Symbols** button to insert the absolute address in the Symbol field for any input or output address that you have not yet defined with a symbolic address.

6.   Click in the Sort field, and select a method of sorting the symbol addresses—for example, **Address Ascending, Address Descending**, or **Symbol Ascending**.

7.   Click **Apply** to accept entries and leave the dialog open to resume your work; when done you may confirm your entries and close the dialog with the **OK** button.

CREATING CODE BLOCKS AND DATA BLOCKS

With the task of I/O addressing done, you are now ready to start creating your STEP 7 program. For many, this next step of program development is to create a skeleton outline of the program by simply generating the code blocks and data blocks that will be used—that is in as much as can be known at this stage of development. This process is not yet concerned with writing code, but instead with making a preliminary draft of what blocks are anticipated, and what form the call structure may take.

The Program Call Structure

In this task, you'll determine and generate the blocks that best fit the partitioned structure of your machine or process. Figure 27-2 is an example of such a partitioning. If you have already defined this structure, then from it you can create a rough outline of what blocks will form the program. In the STEP 7 editor, you can print this outline, known as the *Call Structure*. The call structure shows the call order for the blocks of a program, and the block from which each block is called.

In most programs, the first call level involves, what I'll call, *unit manager* blocks—called from OB1. These first level blocks are typically Functions (FCs), and are responsible for managing the overall functionality of a given machine/process unit. Perhaps these blocks would be numbered FC10, FC20, FC40, FC60, FC80, FC100, FC120, and FC140. Unit manager blocks are generally called on each cycle of OB1 in order to ensure that the operations of each unit are processed. The unit manager blocks, in turn, must call the blocks that control the various subtasks of the machine/process unit.

Each *unit subtask* represents an identifiable segment of an overall unit operation that must be controlled. A subtask will generally involve one or more logic networks, and may even require further calls to other blocks (FCs or FBs) that complete the control task. Each unit is likely comprised of multiple subtasks that must be performed to complete the overall unit control.

Depending on the actual number of unit subtasks, and the size of each subtask, you can decide if the operations of each subtask will be controlled in a separate block of its own, or segmented into one or more blocks. Also the specific requirements of the subtask will help determine whether an FC or FB is best suited for the task. By deciding on the call structure and the blocks that will contain the operations of each task, the outline of your program is gradually formed.

Generating Code Blocks and Data Blocks

From the Simatic Manager, you start by generating each of the blocks in which you will eventually write code or insert data. To create a block, you first generate the block object—for example an FC, FB, OB, or DB. The block object, once generated, is essentially a container that is based on a template that differs slightly for each block type. The template, once opened, is the starting place for developing the code. A more in-depth discussion for developing FCs/FB and OBs is presented in Chapter 28 *Developing Functions and Function Blocks*, and Chapter 29 *Developing Organization Blocks*, respectively. Developing and working with data blocks is presented in Chapter 2 *Data Blocks*.

Although your STEP 7 program may eventually consist of several user code blocks and data blocks, the complete program is built incrementally—block-by-block according to specific control and data requirements, and how you have partitioned the functionality of the machine or process. Remember, this step is simply generating the blocks; the next task would be to start writing code or defining the data or data structure for data blocks.

Code blocks are comprised of the *block header, variable declaration area,* and *code section;* data blocks are comprised only of the *block header* and the *variable declaration area.* When you generate a block, you are actually creating the header in which you may define specific attributes of the block. Once generated, you may open the block, use the declaration table to declare any required variables, and, of course, in a code block, you may begin writing your code in the code section of the block.

To generate a new block:

1. From the SIMATIC Manager open the project and select the desired **S7 Program**.

2. Select the offline **Blocks** folder. Offline blocks are listed in the right pane.

3. From the menu, select **Insert ➢ S7 Block**, then choose the block type—for example **Organization Block, Function Block, Function,** or **Data Block**. The block **Properties** dialog is presented.

4. On the **General Part 1** tab, the **Name** field contains the next available block number for the block type you have chosen. Use the default address or enter a new number as the block address.

5. Enter a **Symbol Name** and **Comment** for the symbol address you wish to assign the block.

6. Select the **General Part 2** tab to enter additional properties for the block.

7. Click **OK** to save the block.

Opening, Editing, and Saving Blocks

Once a block object has been generated in STEP 7, you may open it and then enter or edit the code or data as required. A block may be edited offline or online, but in either case it is actually temporarily held in the memory on your PC until saved. When you are done editing a block, you may determine if the block should be stored offline, online, in both places, or simply discarded.

A block that already exists may be opened for editing from the SIMATIC Manager or from the STEP 7 LAD/STL/FBD editor. From the SIMATIC Manager, you will open the offline blocks window of the required project, or to edit online, you will need to open the online blocks window from the toolbar or from the menu. You may click on a block, right click and select **Open Object** to open the block, or double-click on the block.

As a matter of good practice and safety, you should consider making all program changes offline.

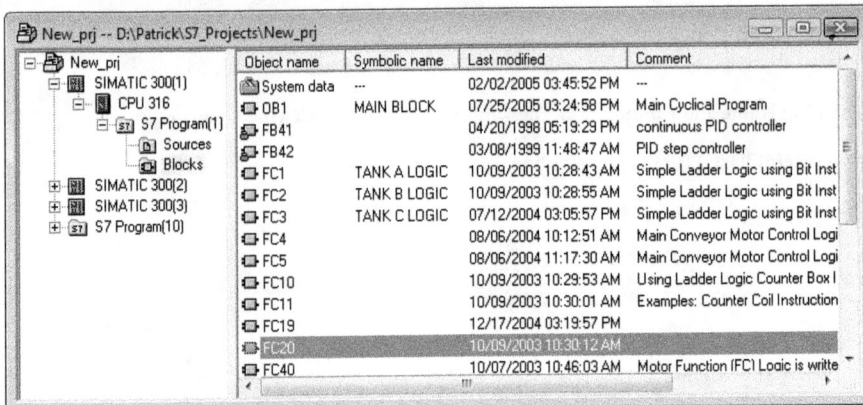

Figure 27-9. The project 'NewProj' is opened with the offline window. The editor will open the block in the offline window, or online window, based on your selection.

To Open, Edit, and Save a STEP 7 code block:

1. From the SIMATIC Manager open the project and select the desired **S7Program**.

2. Select the offline **Blocks** folder. Offline blocks are listed in the right pane.

3. Double-click on a block, or right-click on the desired block, and select **Open ➢ Object**. The LAD/STL/FBD editor opens with the offline block.

4. Select the language of choice, for example select **View ➢ LAD** to edit the block in Ladder format; select **View ➢ FBD** to edit the block in Function Block Diagrams; or select **View ➢ STL** to edit the block in Statement List.

5. Select **View ➢ Overview** if the **Program Elements** tab is not already shown. This operation is not required when programming in Statement List.

6. Find and select the network you wish to edit—for example Network 1.

7. Click on a position in the network, then either insert the basic instructions using toolbar icons, or drag and drop the instruction from the Program Elements browser.

8. To insert a new network following an existing network, first select the network and then select **Insert ➢ Network**, or from the toolbar press the **New Network** icon.

9. To save a block offline select **File ➢ Save**, or from the toolbar click the **Save** icon.

DOCUMENTING THE STEP 7 PROGRAM

A developed data block or code block is not complete until you have added adequate annotations to document your work and the operation or purpose of the block. Each user-written block provides built-in mechanisms for easily documenting your code and data. Using these tools will help to make your program easier to interpret and to troubleshoot. The documentation that you enter for each block is stored in the offline program, but is available when either the offline block, or its online equivalent, is opened from the project.

Documenting Code Blocks

With each block created in LAD, FBD, or STL, you may enter 1) a **Variable Comment** of up to 80-characters, for each declared variable; 2) a **Block Title**, to briefly title the block; 3) a **Block Comment**, to describe the purpose or operation of the entire block; 4) a **Network Title**, to title each logic network; and 5) a **Network Comment** that supports text entry to describe the purpose or operation of each inserted logic network. Blocks written in STL also support entry of a **Line Comment** for each statement. The location of each of these commenting features is illustrated below.

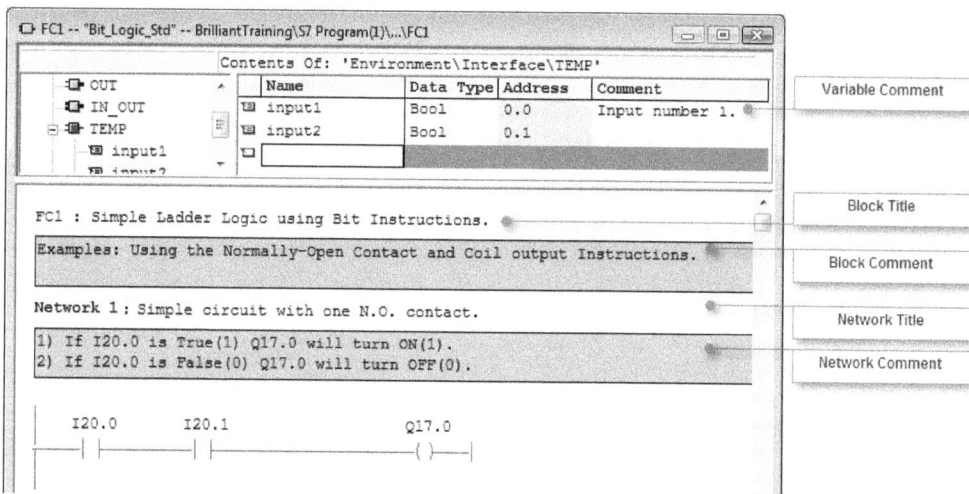

Figure 27-10. Block FC 1 opened in the LAD/STL/FBD editor, with LAD enabled and Display Comments activated

To document a STEP 7 code block:

1. From the menu, select **View ➤ Display with ➤ Comments**. This selection toggles between enabling and disabling comment entry and display.

2. In the **Block Title** field, just to the right of the block name—for example **FC 1**: Title, you may enter a descriptive title of up to 64-characters.

3. In the **Block Comment** field, the gray area above the network, you may enter descriptive commentary of up to 64K-bytes that describe the block operation.

4. Similarly, each network has a **Network Title** of up to 64-characters, and a **Network Comment** of up to 64K bytes. Title the network and enter appropriate comments.

5. If a Symbol Table has been created, symbolic addressing and symbol comments may be viewed by selecting **View ➤ Display with ➤ Symbol Representation** and **View ➤ Display with ➤ Symbol Information** from the menu. This selection may be shown or hidden from display based on specific user preference.

ORGANIZING AND PROCESSING THE PROGRAM

An essential part in STEP 7 program development is determining how each block is called for processing. This step involves knowing what should trigger the call of a block, and how often the block is called. For instance, some blocks are called on every cycle of OB1, while others are called only when certain conditions are enabled. The operational requirements and tasks of each block will determine its call requirements. Before any block can be processed, it must be downloaded to the CPU, and then called from somewhere within the control program. Remember, you do not have to call OBs.

Organizing Block Calls with OBs

How often a block needs processing, and what triggers its call, will depend on the operations that the block performs. In any case, the processing needs of your program may be effectively managed using organization blocks. The task at hand is to determine from which organization block each of the code blocks of your program will be called.

Consider calling blocks that are responsible for initializing or resetting values, at each start-up, from one of the *start-up OBs*—OB 100, OB 101, or OB 102. Call blocks that need processing at fixed periodic intervals—for example every 500 ms, from a *cyclic OB* like OB 30. For calling a service routine that must respond to a module-generated or process interrupt, use a hardware *interrupt OB* like OB40. Code that needs continuous processing is called from OB1—the main organization block. Most blocks, of course, will likely be called from OB1. These are just a few examples of how OBs are used. Organization blocks are covered in detail in Chapter 29 *Developing Organization Blocks*.

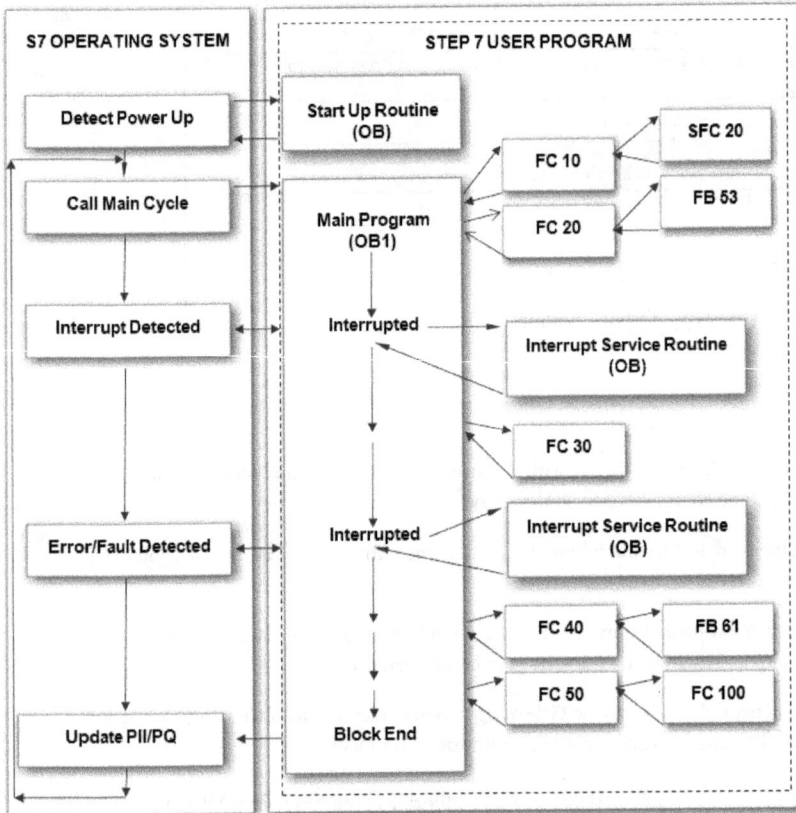

Figure 27-11. Structured S7 Program Processing; and Organization Blocks as Interface between the S7 Operating System and the User Program.

Linear vs. Modular Program Design

In STEP 7, you have two basic choices of how to organize your program. There is the conventional *linear program* in which there is a single contiguous block of code for the entire control program. The second approach involves a *structured program* that uses STEP 7 block resources to create a modular program design where control tasks are handled by different blocks.

As shown in below, if the linear approach is used, the program code could potentially reside entirely in organization block 1 (OB1). You would use as many networks as required. You might also use a single Function or Function Block that is called from OB1. The right part of the illustration shows how a basic call structure of a modular program might look. In this approach, OB1 is the main code block from which other code blocks are called for processing. Each called block may also make calls to other blocks. The linear approach, although possible, is not practical or suitable in large and complex programs, nor does it capitalize on the many advantages of the modular resources provided by STEP 7.

There are many advantages to organizing a structured program—for example each block is a working unit that may be called as needed and perhaps stored in a library for later use. Each block may also be individually tested and started. Code blocks may be copied and reused for the same function, or slightly modified for a similar function. Since each block performs a specific function, the program is simplified and faults in a particular subsystem are more easily pinpointed. Finally, each block may be called for processing on a conditional, or an unconditional, basis. The following discussion examines some basic approaches to developing the call structure of an S7 program.

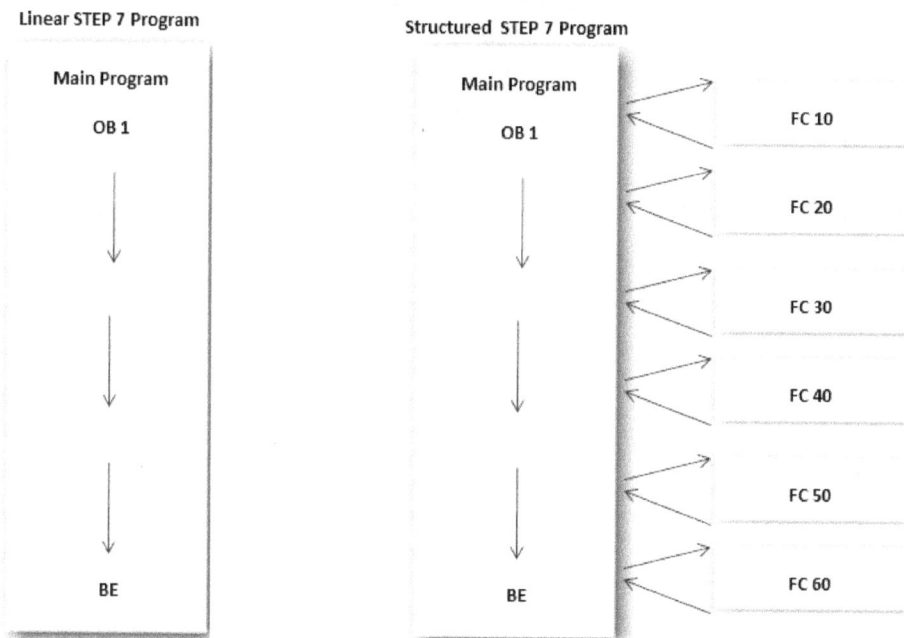

Figure 27-12. Illustration of a Linear S7 Program (left) and a Structured STEP 7 Program (right)

DEVELOPING AN OB 1 CALL STRATEGY

The normal processing of a STEP 7 program is managed by OB 1. Once power is applied, and the CPU is switched to RUN, OB 1 is called cyclically until the CPU is switched to STOP, or until power is removed. Since OB 1 is processed cyclically, blocks called from OB1 are evaluated on each CPU cycle. For this reason, most blocks are called either directly or indirectly from OB1. As we look at structuring calls from OB1, you may want to review the unit structure of the control system in Figure 27-2.

Unconditional Calls from OB1

Some blocks must be called unconditionally—that is regardless of any logic conditions. Because of the tasks that these blocks perform, they simply need to be processed on each processing cycle of OB1. During the normal CPU cycle, OB 1 is processed from the first network to the last, unless jumps or other forms of skips are executed. When calls are made from OB 1 to other blocks, control is returned to OB 1 as each block completes its execution. Using the unit structure of Figure 27-2, the example below illustrates a design in which an unconditional call is made to each of the machine/process unit control blocks. As each block is done, control returns to OB1 and the next block is called.

Example 1: OB1 with unconditional block calls

Network 1:		Calls to unit control blocks
UC	FC 10	// unconditional CALL to FC 10—Manage Unit 1
UC	FC 20	// unconditional CALL to FC 20—Manage Unit 2
UC	FC 30	// unconditional CALL to FC 30—Manage Unit 3
UC	FC 40	// unconditional CALL to FC 40—Manage Unit 4
UC	FC 50	// unconditional CALL to FC 50—Manage Unit 5
UC	FC 60	// unconditional CALL to FC 60—Manage Unit 6
BE	BE	// block end

Conditional Calls from OB1

There will also be blocks that, because of their function, do not need to be processed except for when certain related system conditions are enabled or disabled as the case may be. For these blocks, a conditional call is made based on a condition that is either TRUE (1) or FALSE (0). As in the previous example with unconditional calls, the conditional calls here are using the unit structure of Figure 27-2.

Example 2: OB1 with conditional block calls

Network 1:		Calls to unit control blocks
A	I 4.1	// check if Unit-1 enabled
CC	FC 10	// conditional CALL to FC 10—Manage Unit 1
A	I 4.2	//
CC	FC 20	// conditional CALL to FC 20—Manage Unit 2
A	I 4.3	//
CC	FC 30	// conditional CALL to FC 30—Manage Unit 3
A	I 4.4	//
CC	FC 40	// conditional CALL to FC 40—Manage Unit 4
A	I 4.5	//
CC	FC 50	// conditional CALL to FC 50—Manage Unit 5
A	I 4.6	// check if Unit-6 enabled
CC	FC 60	// conditional CALL to FC 60—Manage Unit 6
BE	BE	// block end

An Alternate Call Approach

Another approach is to only make unconditional calls from OB1 to the subordinate unit control blocks, as shown in the first example on the previous page. Then, from each managing block, determine if processing should continue or not, based on whether certain permissive conditions are met. This approach standardizes the calls from OB1 such that each of the main unit control blocks is always being called. If a fault occurs in a particular subsystem, then a troubleshooter can direct attention to the specific block of the program associated with that subsystem.

In this fashion, the program is designed such that all units are called for processing on every program cycle, but processing only continues inside the unit manager block if the appropriate conditions are satisfied. If the conditions for continuing are not satisfied, control returns to OB1, which in turn calls the next unit manager block for processing. This process of returning to OB1, and processing the next block, is repeated until the last block completes its execution. After the program is ended and the I/O is updated, the program will then return to the beginning of OB1 and repeat the cycle.

This method of allowing each unit manager block to decide whether to continue or terminate execution is illustrated below. In the example, code is shown for FC10—the unit manager block for Unit-1, FC20—the unit manager block for Unit-2, and FC30—the unit manager block for Unit-3.

Example 3: FC10 Unit 1 Manager Block—Called from OB1

Network 1:		Unit 1 Manager
AN	I 4.1	// test permissive to process FC 10
BEC		// terminate FC 10 processing on RLO = 1 (I 4.1 = 0)
		// continue code line for FC 10
		//
		//
		//
BE	BE	// block end FC 10

FC20 Unit 2 Manager Block—Called from OB1

Network 2:		Unit 2 Manager
AN	I 4.2	// test permissive to process FC 20
BEC		// terminate FC 20 processing on RLO = 1 (I 4.2 = 0)
		// continue code line for FC 20
		//
		//
		//
BE	BE	// block end FC 20

FC30 Unit 3 Manager Block—Called from OB1

Network 3:		Unit 3 Manager
AN	I 4.3	// test permissive to process FC 30
BEC		// terminate FC 30 processing on RLO = 1 (I 4.3 = 0)
		// continue code line for FC 30
		//
		//
BE	BE	// block end FC 30

To create OB1:

1. From the SIMATIC Manager, open the project and select the desired **S7 Program**.

2. Select the offline **Blocks** folder. Offline blocks are listed in the right pane.

3. From the menu, select **Insert ➤ S7 Block**, then choose **Organization Block** as the block type. The block **Properties** dialog is presented.

4. Write a brief **Block Title** of OB 1—for example 'Main Cyclical Program Block'.

5. Write a **Block Comment** to describe OB1 purpose—for example 'Call Unit Manager' blocks.

6. To ensure clarity and troubleshooting ease, use OB 1 to make unconditional or conditional calls to code blocks FCs, FBs, SFCs, or SFBs. If possible do not develop control logic in OB1.

7. If appropriate, in your application, use an OB1 call approach as shown in Example 1 on the previous pages. Make unconditional calls to blocks that must be processed on every CPU cycle. You may insert all calls in one network (STL), or you may insert each call in a separate network (LAD, FBD, and STL).

8. If appropriate, make conditional OB1 calls as described in Example 2, on the previous pages. You might insert conditional calls immediately following any unconditional calls; again you may insert all of the conditional calls in a single network (STL), or you may insert each conditional call in a separate network (LAD, FBD, STL). For clarity, insert the entire conditional logic for a call in a single network.

Note: Make conditional calls to blocks that must be processed whenever the driving logic conditions are met. If the RLO =1, the block is processed; if the RLO =0, the block is not processed.

Block Calls and Nesting Depth

In STEP 7, the *nesting depth* refers to the number of block calls that are supported in a horizontal direction, starting with an organization block. For example, if OB1 calls FC10, then the nesting depth for that structure is one level deep; if FC10 calls FB62, then the nesting depth is two levels deep. The nesting depth limit is CPU-dependent, as well as priority-class-(OB)-dependent. A typical nesting depth for a CPU is 8-levels or 16 levels deep. If the block calls of your program exceed the nesting depth, the CPU will stop. The nesting depth is only affected by code blocks (FC, FB, SFC and SFB).

When one processing block calls another block, specific address information is saved in the CPU block stack (b-stack) to allow control to successfully return to the calling block. When a program fault occurs and the CPU is stopped, this information is available in the block stack diagnostic tool as well as in the CPU's diagnostic buffer.

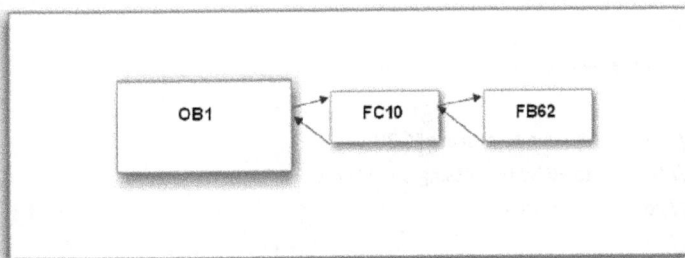

Figure 27-13. Illustration of nesting depth

Chapter 28

Developing Functions and Function Blocks

This chapter covers the application and development of the S7 Function (FC), and Function Block (FB). These two STEP 7 program building blocks are user blocks that you will employ in developing your S7 program code. In this discussion, the S7 System Function (SFC) and System Function Block (SFB) are also introduced. These system blocks, which are integrated in the S7 CPU, support access, from within your program, to built-in routines that may be called to handle various functions such as special controls, communications, and IEC standard operations.

Main Topics

- Introduction
- Declaring Block Variables
- Declaring Block Parameter Variables
- Developing and Calling FCs and FBs
- More on Block Parameters
- Working with System FCs and FBs

INTRODUCTION

The Function and Function Block represent two of three block types (OB, FC, and FB) in which you may write portions of your control program. In the following discussions, we'll look at the application and structure of these two user block types, as well as various methods and techniques for declaring variables, developing code, and calls to these STEP 7 block resources.

Applying Function Blocks

An FB is generally intended for the development of control logic or algorithms where data that is generated or used by the block must be retained from one call of the block to the next. Such a requirement is typical in iterative functions involving calculations, data collection, or where data will be required or generated by the block. To handle this memory requirement, you must assign a data block to each FB that you create.

As introduced previously, a data block associated with an FB is referred to as an *instance DB*. On each instance that the FB is called, the assigned instance DB is opened by STEP 7 for read/write access to the FB. Each call of an FB/DB is therefore referred to as an 'instance'. You can associate an FB with several instance DBs, thereby allowing the FB to perform its function using different sets of data.

Applying Functions

An FC should be considered when there is no memory requirements directly related to the operations that the block performs. The FC differs from the FB in that the FC only supports the use of temporary *local variables*. These temporary variables are not retained after processing of the FC is ended. The fact that the FC only supports temporary local variables makes it more suitable for binary logic applications like relay-equivalent controls and digital operations where retention of variable data is generally not required. In many S7 applications, as much as 70% of the program is comprised of FCs.

If the code in an FC operates on data that needs to be made available to other parts of your program, then you will need to use either S7 bit memory, or a global data block, in which to store and share this data. Bit memory is generally adequate if only a few values must be shared; otherwise a global data block may be more suitable. In the figure below, DB100 provides memory for FB10—DB100 is opened for use whenever FB10 is called. Any block, including FC10 and FC20, has read/write access to DB20.

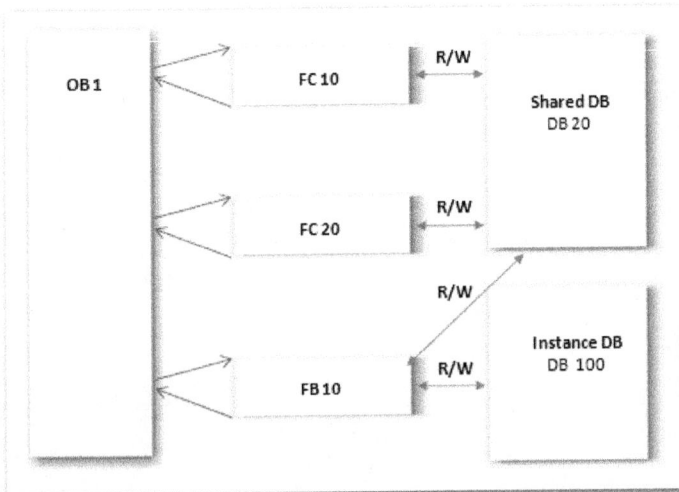

Figure 28-1. FB/FC access to Instance and Shared (Global) Data Blocks

422

The FC Block Template

Development of an FC code block starts with a block window as shown in the figure below—it is the template of every FC as it appears the first time you open an FC in the LAD/STL/FBD editor. The FC is generally developed in a two part process—first there is the step of declaring any variables that may be used by the block; and second there is the entry of the code required for some operation or task.

Each new FC opens in a separate window where the FC number, project name, and program name are displayed in the title bar. The template contains a *declaration section* (top panes), where any variables used in the block are declared; and a *code section* (bottom pane), where the block code and comments are written. The template, as shown below with a new blank network, is how it would appear if the editor is set for LAD. The coding area would look slightly different if the editor was set to FBD or STL.

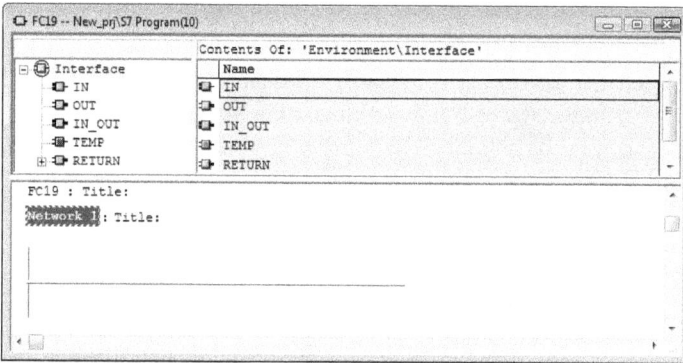

Figure 28-2. FC block template in LAD programming editor

The FB Block Template

Development of an FB code block starts with a block window as shown in the figure below—it is the template of every FB as it appears the first time you open an FB in the LAD/STL/FBD editor. The FB is generally developed in a two part process—first there is the step of declaring any variables that may be used by the block; and second there is the entry of the code required for some operation or task.

Each new FB opens in a separate window where the FB number, project name, and program name are displayed in the title bar. The template contains a variable declaration section (top panes), where you will declare any variables needed by the block; and a code section (bottom pane), where you will write the block code. The template, as shown below with a new blank network, is how it would appear if the editor is set for LAD. The coding area would look slightly different if the editor was set to FBD or STL.

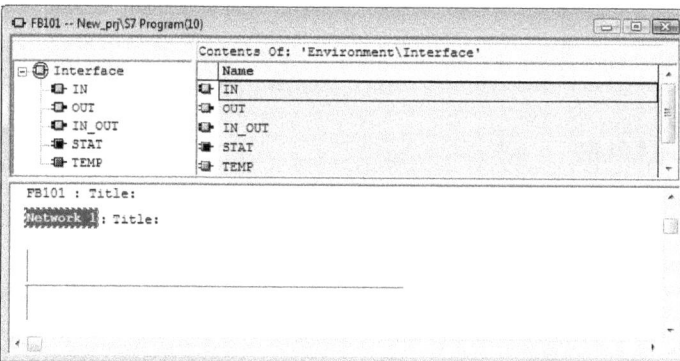

Figure 28-3. FB block template in LAD programming editor

The Declaration Section

The declaration section of a block is where you declare the variables you intend to use in that block. Variables declared in a block, whether an FC, FB, or OB, are restricted to use within that block—in other words, these are local variables. When you declare each variable, you not only define its data characteristics, but more importantly, you define the way in which the variable is used in the block—or more precisely, how the variable interfaces with the block. How a variable is used within a block is based on the declaration type—the declared type also determines where the variable is stored.

The variable *declaration types* are labeled **TEMP, STAT, RETURN, IN, OUT**, and **IN_OUT**. Whereas each declaration type is used to declare a specific use type for a variable— type IN, OUT, and IN_OUT are used to declare what I will refer to as *parameter variables*. For the purposes of our discussion, these variables are described under the heading of 'Declaring Block Parameter Variables'. Declaration types TEMP, STAT, and RETURN are described under the heading of 'Declaring Block Variables'.

As seen in Figure 28-4, the declaration section of a block window is divided into two panes. In the top-left pane, labeled *Interface*, you select the declaration type you wish to define for the variable. Remember, the declaration defines how the variable interfaces with the block. Once variables are defined, a tree structure is formed with the defined objects listed underneath each declaration type.

As you click on a specific declaration type, for example type IN, the variables under that category are displayed in the right pane. A click on the plus sign displays the variables defined under a declaration type; and a click on the minus sign hides those variables. As you click on a specific variable in the tree, its declaration row comes into focus in the right pane, displaying the fields **Name, Data Type, Address, Initial Value**, and **Comment**.

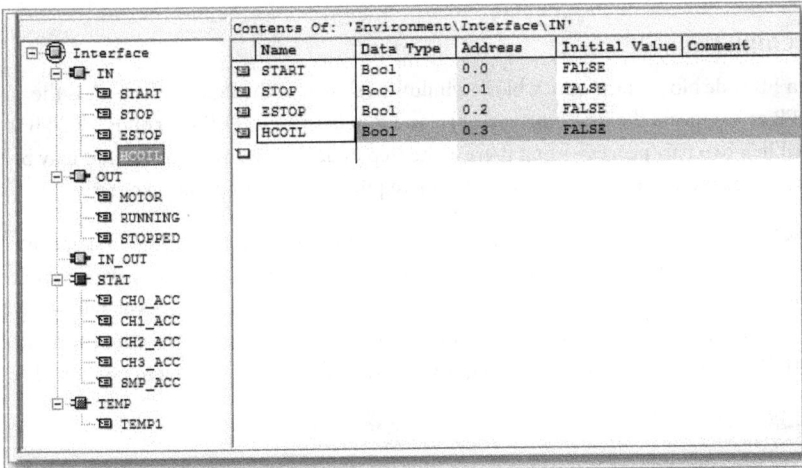

Figure 28-4. Interface panes of the variable declaration section of a Function Block

DECLARING BLOCK VARIABLES

All declared variables are block variables; however here we are referring to those variables that you will declare using declaration types STAT, TEMP, and RETURN. These variables are distinguished from the parameter variables declared using declaration types IN, OUT, IN_OUT. The use of block variables, are important in that they support the development of an FC or an FB that is largely self-contained—that is the block, in as much as possible, does not rely on data outside of itself.

From the block interface pane, you declare a block variable by first designating the declaration type as TEMP, STAT, or RETURN. Declaring these variables is described in the following discussions.

STAT Variables

You declare a variable using the **STAT** declaration type when you want to define a variable that must be retained in memory for use by an FB. Such a variable is referred to as a *static variable*, and may only be declared in an FB. All variables declared as static are automatically stored in the associated instance data block when the DB is generated and associated with an FB. The static variables are also updated during runtime.

When you create a static variable, you may assign it an initial value, since the value will be retained in the instance DB. When declaring static variables, both elementary and complex types may be assigned.

To declare new static variables:

1. Select the **STAT** object in the left pane of the declaration table of the open block.

2. Click in the **Name** field of the declaration row and enter the name for the new variable.

3. Click in the **Data Type** field and from the list select an elementary or complex data type.

4. Click in the **Comment** field and enter a descriptive comment.

5. To insert a new declaration row before or after an existing row, select a row, before or after which you wish to insert a new row, then right click and select **New Declaration Row**.

6. Repeat the previous steps to enter additional **STAT** variables.

TEMP Variables

Block variables that are only needed while the block is processing should be declared using the **TEMP** declaration type. These temporary local variables will reside in the local memory assigned to the block while the block is being processed. Each temporary variable is stored, and addressed, in the order in which it is entered in the block. Typically, local variables are accessed using their assigned name; however, you may access the variable using absolute addressing. With the exception of OBs, where the first 20-bytes are reserved as OB start information, the local variables are accessed starting with relative byte 0 (LB 0).

You may declare TEMP variables in FCs, FBs, and OBs. Since temporary variables reside in the local memory of a block, initial values may not be assigned when you create variables of this type.

To declare new temporary variables:

1. Select the **TEMP** object in the left pane of the declaration table of the open block.

2. Click in the **Name** field of the declaration row, and enter a name for the new variable.

3. Click in the **Data Type** field and from the list select an elementary or complex data type.

4. Click in the **Comment** field and enter a descriptive comment.

5. To insert a new declaration row before or after an existing row, select a row, before or after which you wish to insert a new row; then right click and select **New Declaration Row**.

6. Repeat the previous steps to enter additional **TEMP** variables.

The RETURN Variable

A **RETURN** variable may only be declared in an FC, and only a single RETURN variable may be defined. This variable, which is defined by the name of **RET_VAL**, lets the FC return a value from the block from which it was called. You may specify the data type of the RETURN variable based on the requirement. The RETURN variable can be of any elementary type, as well as complex types, including DATE_AND_TIME, STRING, ANY, POINTER, and UDT. When the FC is terminated, the RETURN value is brought back from the block which the FC was called. If for example, FC10 is called by FC2, then a RETURN variable declared in FC10, can return a value to FC10 from FC2.

To declare new return variables:

1. Select the **RETURN** object in the left pane of the declaration table of the open block. The variable named RET_VAL is displayed in the declaration table in the right pane.

2. Click in the **Data Type** field and, from the list, select an elementary or complex data type.

3. Click in the **Comment** field and enter a descriptive comment of up to 80 characters.

DECLARING BLOCK PARAMETER VARIABLES

Block parameters refer to those variables that may be defined in an FC or FB using declaration types IN, OUT, and IN/OUT. Variables declared as parameters allow input data to be passed to the block, or output results to be passed out from the block. The term 'parameter', as used here, may be compared to those familiar entries that you supply at the inputs or outputs of a LAD/FBD box operation. For example, a multiply operation has two input parameters for the values to be multiplied, and an output parameter for assigning where the result is stored. Here, however, as you develop an FC or FB, you may define parameters as inputs or outputs to act in a way similar to the multiply operation.

Formal Parameters

When you declare an input or output parameter as you develop a reusable FC or FB, the parameter variables are referred to as *formal parameters*. The use of formal parameters support development of an FC or an FB as a black box operation—like a multiplication box instruction, which is used repeatedly throughout a program, and on each use operates on different values. In a similar way, an FC or FB that you develop using formal parameters, may be called as often as needed to perform the identical operation, but on each call use a different set of addresses or values.

As you develop the code for the FC or FB, each formal parameter acts as an address placeholder. Each time you make a call to a parameterized block, in your program, a different set of actual addresses are supplied in place of the formal parameters—these actual addresses are called *actual parameters*, since at runtime, when the block is called, the placeholder addresses are replaced by actual addresses. Entering actual addresses or variable names at the input or output of a block parameter is referred to as '*initializing the parameters*'. See 'More on Initializing Block Parameters', later in this chapter.

Blocks written with formal parameters offer a great advantage in that, while the actual code is written and debugged once, the FB or FC can be called as often as required. Memory use and development time are reduced, and programming errors are minimized. From the block interface pane, you can declare a block parameter by first designating the declaration type as IN, OUT, or IN_OUT. Steps for declaring formal block parameters are described in the following discussions.

IN Parameters

An **IN** parameter represents an input variable to the block— a value to be read. An example might be an analog input signal, or perhaps a digital input that is a START pushbutton to a motor.

When an **IN** parameter is initialized, you may specify a valid address using any of the following S7 memory units, or an appropriate constant or variable. Whereas the specified parameter must match the size and type of the parameter's data type—this is also true if the the input is a constant.

Table 28-1. S7 memory variables (including parameter types) permitted at **IN** block parameters

Declared	Valid S7 Memory Operands
IN	I, IB, IW, ID, M, MB, MW, MD, Q, QB, QW, QD, PIB, PIW, PID, DBX, DBB, DBW, DBD, DIX, DIB, DIW, DID, (T, C, BLOCK_FC, BLOCK_FB, BLOCK_DB, BLOCK_SDB)

OUT Parameters

An **OUT** parameter represents an output variable from the block—a value to be written. An example might be a value sent to an analog output signal, or perhaps a digital output that drives a motor.

When an **OUT** parameter is initialized, an actual address may be specified using any of the following memory units. The specified address must appropriately match the size and type of the data type of the parameter. Parameter types TIMER, COUNTER, and BLOCK_XX may not be specified at an OUT.

Table 28-2. S7 memory variables permitted at **OUT** block parameters

Declared	Valid S7 Memory Operands
OUT	I, IB, IW, ID, M, MB, MW, MD, Q, QB, QW, QD, PQB, PQW, PQD, DBX, DBB, DBW, DBD, DIX, DIB, DIW, DID, (POINTER, ANY—only in FC blocks)

IN/OUT Parameters

As you might guess, the **IN_OUT** declaration type supports definition of a parameter variable that accepts input to the block as well as output from the block—a value that is both read from and written to using the same variable. An example of such a variable might involve an analog output value that drives a speed controller, but also provides feedback input to the block. When an **IN_OUT** parameter variable is initialized, you may specify an actual address using any of the following S7 memory units.

Table 28-3. S7 memory variables permitted at **IN_OUT** block parameters

Declared	Valid S7 Memory Operands
IN_OUT	I, IB, IW, ID, M, MB, MW, MD, Q, QB, QW, QD, DBX, DBB, DBW, DBD, DIX, DIB, DIW, DID, (POINTER, ANY—in FCs and FBs)

To declare a new formal parameter:

1. Select the **IN**, **OUT**, or **IN_OUT** declaration type in the left pane of the declaration table.

2. Click in the **Name** field of the declaration row, and enter a parameter variable name.

3. Click in the **Data Type** field and, from the list, select an elementary or complex data type.

4. Click in the **Comment** field and enter a descriptive comment.

5. To insert a new declaration row before or after an existing row, select a row, before or after which you wish to insert a new row, then right click and select **New Declaration Row**.

6. Repeat the previous steps to enter additional **IN**, **OUT**, or **IN_OUT** parameter variables.

A Use-Case for an FB with Formal Parameters

As the FB introduction noted earlier, this block type is generally used when the solution depends on one or more values being maintained (in memory) from one call of the block to the next. Examples might include an FB that performs analog input filtering, or that tracks production line totals. In the filtering routine, consecutive readings of an analog input are added, and then divided, by the number of samples taken. On each call of a totalizer routine, the total value is either incremented or decremented by adding or subtracting one from the previous total. The filtering could obviously be used for multiple input channels, whereas the totalizer could be applied to multiple production lines—each is surely a case for developing an FB with parameters.

An FB, written with formal parameters to perform analog input filtering on two channels of an input module, might use **CH0_IN, CH1_IN**, and **SAMPLE** as input (IN) parameters; it might use **CH0_AVG, CH1_AVG**, and **DONE** as output (OUT) parameters. The name you use as a formal parameter should always reflect the actual name of the input or output for which it is a placeholder.

When called, in LAD or FBD, an FB with parameters appears as the box shown in Table 28-4—its input parameters are on the left side of the box, its output parameters are on the right side of the box, and the instance DB appears above the box. Parameters of declaration type IN_OUT are not shown. If the FB is called in STL, the block parameters are listed immediately below the **CALL** statement as shown below; the block number and instance DB, separated by a comma, are immediately following the CALL statement. Details for developing this example are provided later in this chapter.

Table 28-4. Required formal parameters for the planned AC Motor FC

Parameter/Variable	Decl.	Name	Type	LAD/FBD Display	STL Display
Channel 0 Input	IN	CH0_IN	INT	DB53 FB53 —EN ENO— —CH0_IN CH0_AVG— —CH1_IN CH1_AVG— —SAMPLE DONE—	**CALL** FB53, DB53 CH0_IN: CH1_IN: SAMPLE: CH0_AVG: CH1_AVG: DONE:
Channel 1 Input	IN	CH1_IN	INT		
Number of Samples	IN	SAMPLE	INT		
Channel 0 Average	OUT	CH0_AVG	INT		
Channel 1 Average	OUT	CH1_AVG	INT		
Done	OUT	DONE	BOOL		
Sample Counter	STAT	SMP_ACC	INT		
Channel 0 Total	STAT	CH0_ACC	INT		
Channel 1 Total	STAT	CH1_ACC	INT		

A Use-Case for an FC with Formal Parameters

Earlier, I noted that an FC is generally used in simple to complex applications that primarily involve binary logic or relay-equivalent operations. An example might include an FC that performs start/stop control of multiple AC motors of a conveyor. The solution does not have memory requirements that depend on data values being maintained (in memory) from one call of the block to the next. The AC motor control routine manages the start/stop circuits of multiple inline motors in which each downstream motor is started after a short time delay. The control of each motor is identical.

The AC motor control function clearly could be used for motors that make up the line, since they are identical in operation. So far, this task looks like a case for an FC with formal parameters.

428

An FC written with formal parameters to control a particular AC motor might use **START**, **STOP**, **ESTOP**, **HCOIL** and **TMRNO** as input parameters; it might use **MOTOR**, **RUNNING**, and **STOPPED** as output parameters. Each name you use as a formal parameter should reflect the actual name of the input or output for which it is a placeholder.

When called, in LAD/FBD, an FC with parameters will appear as the box shown in Table 28-5—its input parameters are on the left side of the box; its output parameters are on the right side; and parameters of declaration type IN_OUT are not shown. If the FC is called in STL, the parameters are listed directly below the **CALL** statement. With the exception of the timer parameter, the details for developing this example are provided later in this chapter. Perhaps, as practice, you can try adding the timer parameter. Hint: The declaration type is IN, and you would be dealing with a parameter type.

Table 28-5. Required Formal Parameters for the Planned AC Motor FC

Parameter/Variable	Decl.	Name	Type	LAD/FBD	STL
Start Pushbutton	IN	START	BOOL	"AC_Motor" / EN ENO	CALL AC_Motor
Stop Pushbutton	IN	STOP	BOOL	??.? — START STOPPED — ??.?	START:
				??.? — STOP RUNNING — ??.?	STOP:
Emergency Stop PB	IN	ESTOP	BOOL	??.? — ESTOP MOTOR — ??.?	ESTOP:
Auxiliary Coil Contact	IN	HCOIL	BOOL	??.? — HCOIL	TIMER:
					HCOIL:
Motor Coil	OUT	MOTOR	BOOL		STOPPED:
					RUNNING:
Motor Running Light	OUT	RUNNING	BOOL		MOTOR:
Motor Stopped Light	OUT	STOPPED	BOOL		

DEVELOPING AND CALLING FCs AND FBs

As you start to develop code blocks, you will very likely begin by asking the question, "Should I use an FC or an FB?" Our earlier introduction on applying FCs and FBs, characterized each block in a way that should help with this decision—there I pointed out that the FC only supports temporary variables, and that it has no memory for storing variables used by, or generated by, the block. An FB, on the other hand, has an associated DB that maintains the data variables that the FB uses or generates. If necessary, you might want to take a quick review of that introduction.

To further illustrate the question to use an FC or FB, we'll continue with the FC and FB use-case examples that were introduced earlier. The difference, however, is that here for both the FC and FB, the first example will develop a block without using formal parameters, and then the second example will be developed using formal parameters. Since all blocks must be called in order to be processed, we'll look at calling each block example after it has been developed.

Develop FC without Formal Parameters

This example of an FC represents one of many like it that would be found in an application that consists largely of discrete logic, and basic operations like timers and counters. If any intermediate data is required by the FC, you would need to declare **TEMP** variables. Recall, that these variables are handled in the local data stack of the block, and are no longer available when the block is done processing. After defining any required **TEMP** variables, you would also need to decide whether a **RETURN** variable needs to be returned from any block that calls this FC.

If neither a **RETURN** variable nor **TEMP** variables are required, as in this example, then your FC will consists only of the code that you write for the block. The code for this FC with no parameters is simply relay-equivalent logic to perform a single function—that is to control multiple motors, of a conveyor, through the use of individual START/STOP control circuits. The logic for each of these motors is identical, except for the addresses used for the required control elements of each device. Each motor has a Start PB, Stop PB, Emergency-Stop PB, and Holding Coil Auxiliary-contacts input for feedback; the outputs include the Motor Coil, an indicator for Running, and an indicator for Stopped.

The control logic for a single motor is shown in the figure below; the control of multiple motors that provide the control for a span of conveyors is entered in the code section of FC30 (See Figure 28-6).

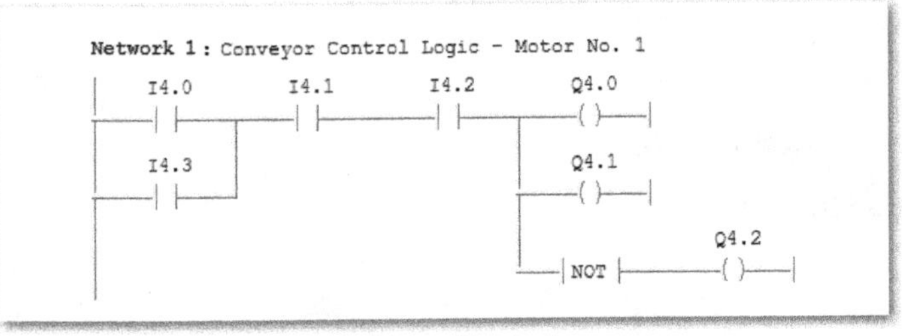

Figure 28-5. AC motor control logic for single motor

To code an FC without formal parameters:

1. From the SIMATIC Manager, create the new FC, in the appropriate program; then double-click on the specific FC to open the block in the LAD/STL/FBD editor.

2. From the menu, select **View ➤ LAD** (FBD/STL) to display the desired program editor.

3. Select View ➤ Overview to display the Program Elements/Call Structure window.

4. **Declare temporary variables**: Select the **TEMP** object in the left pane of the declaration table; then for each temporary variable, enter the **Name, Data Type**, and a descriptive **Comment**. Enter as many **TEMP** variables as required, or skip this step if not required (TEMP variables are not required in this example). See next step.

5. **Insert a temporary variable before an existing variable**: First select the **TEMP** object in left pane of the declaration table to display TEMP records. Then, select a row, before or after which you wish to insert a new temporary variable declaration row; right click and select **New Declaration Row**.

6. **Enter the block code**: Click inside the code section, starting with Network 1, then write the code for each network operation, using absolute or symbolic addresses (if already defined). The code for this example is entered in **FC30**, shown in Figure 28-6.

7. After each network, to start a new network, press the **New Network** button from the toolbar.

8. Select **File ➤ Save** when you are done with entering the last network.

Later we'll look at how you can eliminate the need to write the same code for several devices or operations that have the same control logic. You can write an FC, for example a motor starter, which is written once using formal parameters, yet can be called for multiple motors. Different I/O addresses for the control elements of an individual device (motor in this case) are passed to the FC on each call.

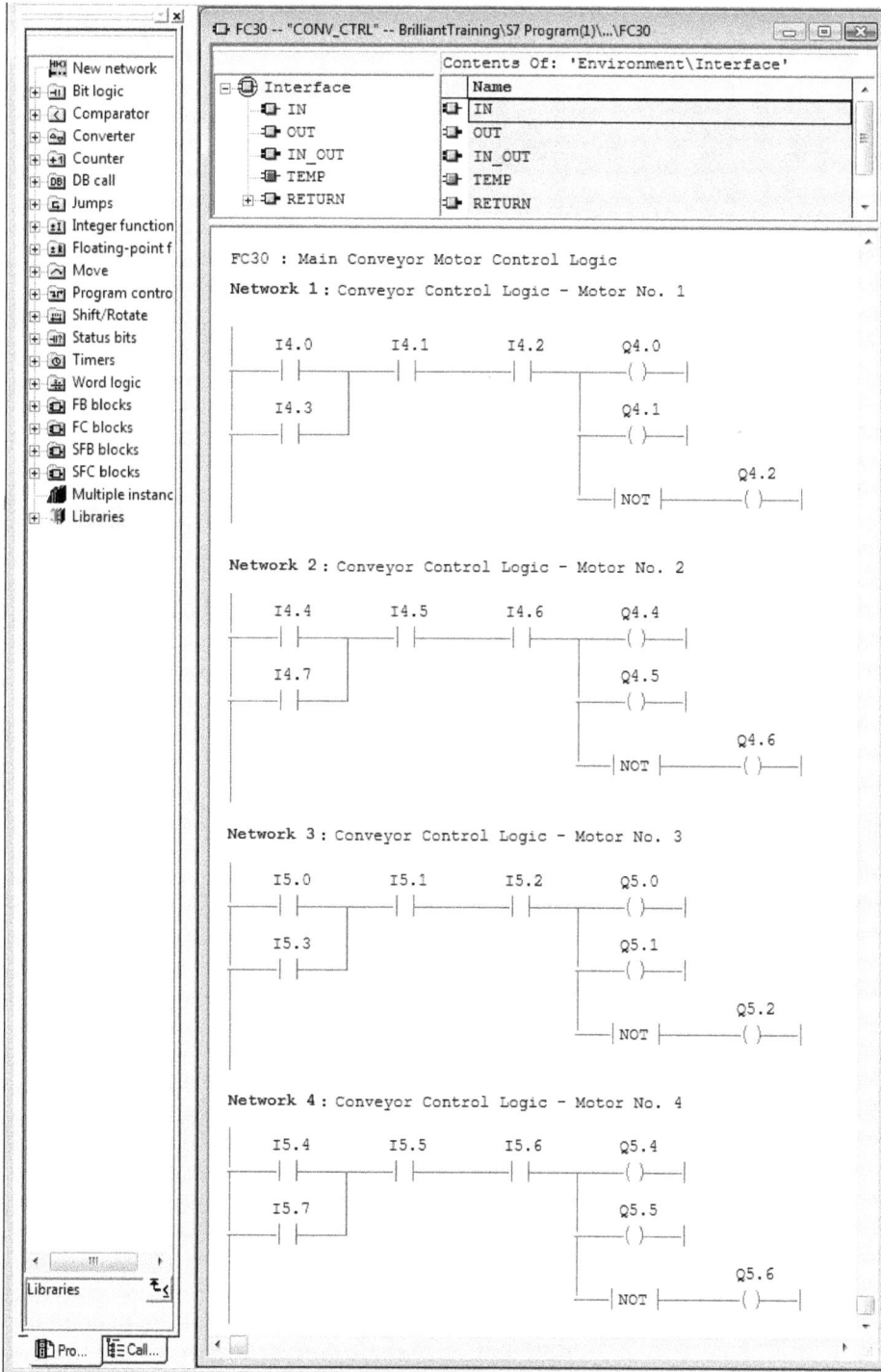

Figure 28-6. A Function (FC30—CONV_CNTRL) written without use of formal parameters

Call to FC without Formal Parameters

Calls to an FC that has no parameters may be conditiona, or unconditional, calls from the main organization block (OB1) or from another block—for example an FC, or another FB. Based on the purpose of an FC, you must determine from which block it is called, how often it is called, and what conditions will trigger the call. For instance, the conveyor control logic of FC30, in our example, might be called on every cycle, from OB 1, whenever the conveyor system is enabled.

A point to remember here is that, while you might call a block for processing on every cycle of OB1, just as a matter of consistency, whether processing continues inside of the block can be determined by checking an enable/disable flag at the beginning of that block.

To call an FC without formal parameters:

1. Open the S7 program and the specific calling block, for example **FC19**, from which you will make a call to an existing FC (**FC30** is our AC Motor example) created without any formal parameters.

2. From the menu, select **View ➤ LAD** (FBD/STL) to display the desired program editor.

3. Select **View ➤ Overview** from the menu to show the instruction browser, if it is hidden from view.

4. From the Program Elements tab, expand the **FC Blocks** folder to display all user-created Functions (FCs) of the currently opened program. The FC you wish to call must have already been written, before if can be called. See previous example.

5. To make the call, click in the code section of the calling block, then from the toolbar press the **New Network** button to enter a new network from which to call the FC.

6. With the network inserted, expand the Program Control instructions folder and double-click on the CALL coil. After the coil is inserted, enter the FC number over the coil, as shown in Figure 28-7, with **FC1** (Network 15) and **FC2** (Network 16); or as an alternate method, see the next step.

7. Locate the desired FC in the FC Blocks folder, for example **FC 30** (CONV_CNTRL), then drag and drop the FC onto the network. When the block is inserted this way, it is presented as a box, as shown in Figure 28-7, with **FC3** (Network 17) and **FC30** (Network 18).

8. **Make conditional call:** Insert control logic in front of the **CALL** coil or the **EN** input to the FC to make a conditional call.

9. **Make unconditional call:** Do not insert any logic conditions in front of the **CALL** coil or the **EN** input to make an unconditional call.

10. Select **File ➤ Save** when you are done with the last network.

In the Figure 28-7 (FC19), FC1 and FC2 are both called using the CALL coil (CALL output in FBD). You may insert the absolute address, or symbolic address, for the FC above the coil. Also notice that the FC may be called with preceding logic (conditional), or without preceding logic (unconditional).

Notice that both FC3, and our example FC30, appear as boxes with the EN/ENO mechanism. These Functions appear in this fashion because of the method by which they were placed in the network. An FC appears as a box operation when you drag and drop it onto the network. Like with the CALL coil, you may insert the box call with or without preceding logic conditions.

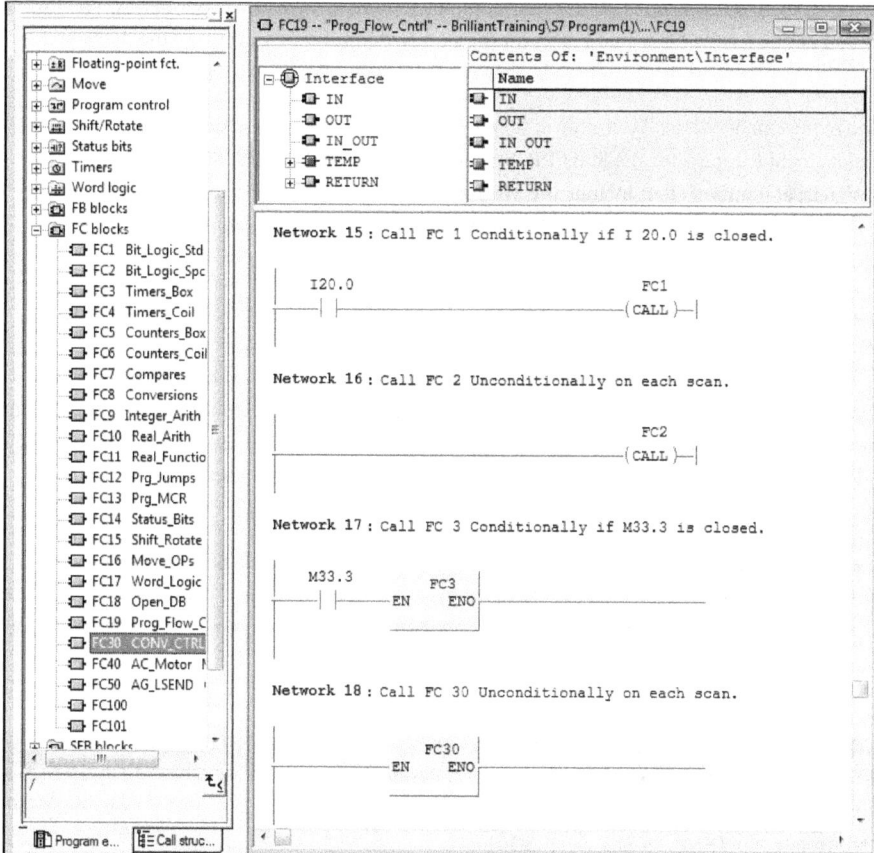

Figure 28-7. Calling an FC without Formal Parameters in LAD, using Call Coil or Call Box

Develop FC with Formal Parameters

In the first FC example, we looked at the development of an FC to control multiple AC motors. The control logic was basically relay-equivalent logic, the same for each motor, without any variables needing retention. For these two reasons, primarily, an FC was sufficient. According to our introductory discussion, a case such as this is one that the code could be written using formal parameters. The formal parameters you define for the FC would represent the standard inputs and outputs of the described AC motor.

Based on the requirements of the AC motor in the example, the parameters needed are shown below. The parameters include inputs for the **Start PB**, **Stop PB**, **Emergency Stop PB**, and **Holding Coil** auxiliary contacts; and outputs that include the **Motor** starter, and status indicators for **Running**, and **Stopped**. Each of these variables is of type BOOL. If temporary variables are required, for example a discrete interlock of type BOOL, you would specify these as **TEMP** declarations.

After defining the required formal parameters and any temporary variables, the code for the block is then written using one or more networks just as you would normally. In this case, the code would be that which is required for our AC motor. An important point here is that the code is functionally the same as what was written for the single motor circuit in Figure 28-5. The difference here is that instead of using the actual input and output addresses, you will specify the formal parameter **NAME**(s), as they have been defined in Table 28-6. See the finished code for the motor FC, in Figure 28-8. Take a few moments to compare this code with the code in Figure 28-5.

I proposed earlier that, as practice, you would later add a delay timer to the block—the timer would provide the delay before starting each downstream motor. That timer variable is considered a parameter type that would allow a unique timer number to be passed on each call of the block. For now, we will not include that timer—just a little something for you to try later.

Table 28-6. Box parameters required for the planned AC Motor Function

Block Parameters	Decl.	Name	Type	LAD/FBD Display	STL Display
Start Pushbutton	IN	START	BOOL	"AC_Motor" EN ENO	**CALL** AC_Motor
Stop Pushbutton	IN	STOP	BOOL	??.?—START STOPPED—??.?	START: STOP:
Emergency Stop PB	IN	ESTOP	BOOL	??.?—STOP RUNNING—??.?	ESTOP:
Holding Coil Contact	IN	HCOIL	BOOL	??.?—ESTOP MOTOR—??.?	HCOIL:
Motor Coil	OUT	MOTOR	BOOL	??.?—HCOIL	STOPPED: RUNNING:
Motor Running Light	OUT	RUNNING	BOOL		MOTOR:
Motor Stopped Light	OUT	STOPPED	BOOL		

To code an FC with formal parameters:

1. From the SIMATIC Manager, create the new FC in the desired S7 program; then from the Blocks folder, double-click on the FC to open the block in the LAD/STL/FBD editor.

2. From the menu, select **View** ➤ **LAD** (FBD/STL) to display the desired program editor.

3. Select **View** ➤ **Overview** to display the Program Elements/Call Structure window.

4. **Declare formal parameters:** In the left pane of the declaration table, select **IN, OUT,** or **IN_OUT** as the declaration type you wish to define. START, STOP, ESTOP, and HCOIL are all **IN** parameters, as they will provide input to the box. STOPPED, RUNNING, and MOTOR are all **OUT** parameters.

5. After selecting a parameter type, for each parameter enter the **Name, Data Type**, and a descriptive **Comment**. Enter as many variables of each parameter type as required.

6. **Declare temporary variables:** Select the **TEMP** object in the left pane of the declaration table; then for each temporary variable enter the **Name, Data Type**, and a **Comment**. Enter as many **TEMP** variables as required. Skip this step if temporary variables are not needed.

7. **Enter the block code:** Click inside the code section, and starting with Network 1, use the formal parameters, as addresses, to write the code for the operation (See Figure 28-8).

8. Select **File** ➤ **Save** when you are done with the last network.

In the top pane of FC40, in Figure 28-8, **IN** is selected as the interface declaration type—therefore, the declaration rows for blocks parameters **START**, **STOP**, **ESTOP**, and **HCOIL** are displayed in the right pane. All of these parameters have a data type of BOOL, since they are discrete signals.

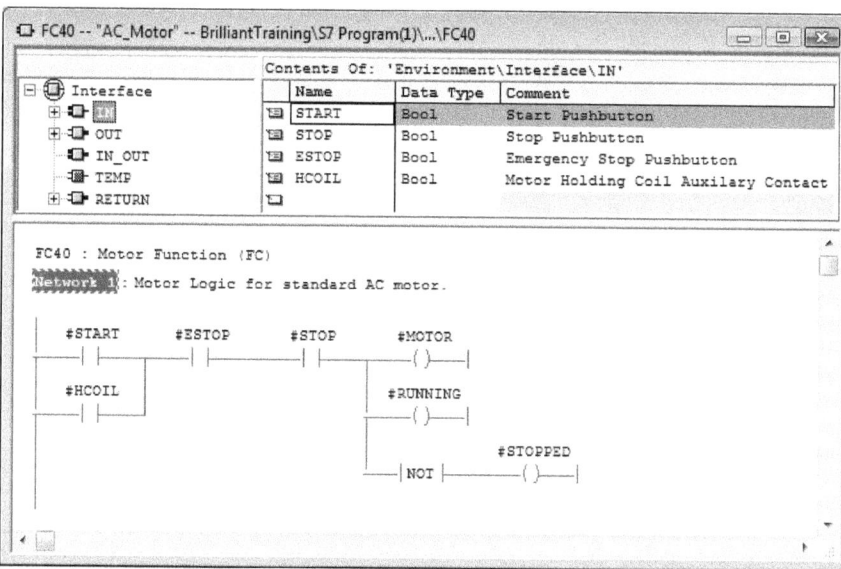

Figure 28-8. FC 40 Motor Control Logic: IN formal parameters shown in declaration table

In Figure 28-9, FC 40 is still displayed, but this time **OUT** is selected as the interface declaration type—therefore, the blocks parameters **STOPPED**, **RUNNING**, and **MOTOR** are displayed in the right pane. These parameters also have a data type of BOOL, since they too are discrete signals.

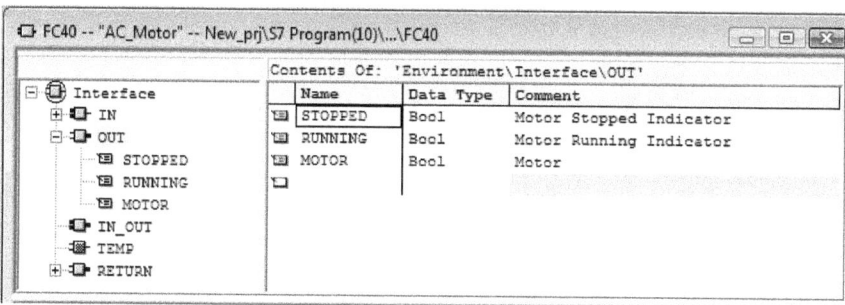

Figure 28-9. FC 40 Motor Control Logic: OUT formal parameters shown in declaration table

Call to FC with Formal Parameters

An FC developed with formal parameters is meant to be called as often as needed throughout your program. Our FC40, written in the previous task, was developed with formal parameters to control AC motors of a specific type. If there are 50 motors of this type, then this block would be called 50 times. Each time you call the *AC Motor FC*, you will need to specify the actual addresses for a different motor. In other words, the placeholder addresses you used when you created the AC Motor FC are replaced with the actual addresses of each motor for each call of the FC (See Figure 28-10).

On each call of the FC, you must supply an absolute or symbolic address at each parameter. As you use an FC with formal parameters, the address that you enter at each parameter must match the data type of the parameter, and must be assigned to the device or operation for which the block is being called. In our AC motor control FC, the parameters included inputs for a Start PB (**START**), Stop PB (**STOP**), Emergency Stop PB (**ESTOP**), auxiliary contacts (**HCOIL**); and outputs for Motor Starter (**MOTOR**), and status indicators for **RUNNING** and **STOPPED**. All of these parameters are of data type BOOL.

As you use an FC with formal parameters, you must ensure that the correct addresses are entered at the appropriate parameters for a unique device or operation, based on the purpose of the block. In this FC, for example, a digital input address is entered for each of the BOOL inputs **START**, **STOP**, **ESTOP**, and **HCOIL**; and a digital output address is entered for each of the BOOL outputs **STOPPED**, **RUNNING**, and **STOPPED**.

To call an FC with formal parameters:

1. Open the S7 program and the specific calling block—for example **FC19**, from which you will make a call to an existing FC (**FC40** is our AC Motor example) created with formal parameters.

2. From the menu, select **View ➤ LAD** (FBD/STL) to display the desired program editor.

3. Select **View ➤ Overview** from the menu to show the instruction browser, if it is hidden from view.

4. From the Program Elements tab, expand the **FC Blocks** folder to display all user-created Functions (FCs) of the currently opened program. The FC you wish to call must have already been written, before if can be called. Find FC40 in the folder.

5. **Make first call to FC with parameters:** Click in the code section of the calling block (FC 19 in this case), then from the toolbar press the **New Network** button to enter the network in which to call the FC. Locate the desired FC number, for example **FC 40** (AC Motor FC), then drag and drop the FC onto the network.

6. For each parameter, enter the actual addresses for the associated device or operation (AC motor in this case). Insert logic at the **EN** input for a conditional call to the FC; insert the FC without logic conditions to make an unconditional call (See Figure 28-10). Depending on each parameter, you may enter a constant, specific address, or variable. In the case of an FC, an actual address must be specified at each parameter.

7. **Make additional calls to FC with parameters:** Create a new network and drag and drop the FC onto the network; then for each additional device or operation, supply the appropriate set of actual addresses (absolute or symbolic). Insert logic at the **EN** input for a conditional call to the FC; insert the FC without logic conditions to make an unconditional call (See Figure 28-10).

8. Select **File ➤ Save** when you are done with the last network.

FC19 -- "Prog_Flow_Cntrl" -- BrilliantTraining\S7 Program(1)\...\FC19

Contents Of: 'Environment\Interface'

Interface
- IN
- OUT
- IN_OUT
- TEMP
- RETURN

Name
- IN
- OUT
- IN_OUT
- TEMP
- RETURN

Move
Program control
Shift/Rotate
Status bits
Timers
Word logic
FB blocks
FC blocks
- FC1 Bit_Logic_St
- FC2 Bit_Logic_Sr
- FC3 Timers_Box
- FC4 Timers_Coil
- FC5 Counters_Br
- FC6 Counters_Cr
- FC7 Compares
- FC8 Conversions
- FC9 Integer_Arith
- FC10 Real_Arith
- FC11 Real_Functi
- FC12 Prg_Jumps
- FC13 Prg_MCR
- FC14 Status_Bits
- FC15 Shift_Rotatr
- FC16 Move_OPs
- FC17 Word_Logi
- FC18 Open_DB
- FC19 Prog_Flow_
- FC30 Conv_Ctrl
- FC40 AC_Motor
- FC50 AG_LSEND
- FC100
- FC101
SFB blocks
SFC blocks
Multiple instances
Libraries

Functions of the project

Network 9: Call Motor Control Function (FC40) for Motor No. 1

```
              FC40
        EN           ENO

I4.0 —START    STOPPED —Q4.0

I4.1 —STOP     RUNNING —Q4.1

I4.2 —ESTOP      MOTOR —Q4.2

I4.3 —HCOIL
```

Network 10: Call Motor Control Function (FC40) for Motor No. 2

```
              FC40
        EN           ENO

I4.4 —START    STOPPED —Q4.3

I4.5 —STOP     RUNNING —Q4.4

I4.6 —ESTOP      MOTOR —Q4.5

I4.7 —HCOIL
```

Network 11: Call Motor Control Function (FC40) for Motor No. 3

```
              FC40
        EN           ENO

I5.0 —START    STOPPED —Q5.0

I5.1 —STOP     RUNNING —Q5.1

I5.2 —ESTOP      MOTOR —Q5.2

I5.3 —HCOIL
```

Network 12: Call Motor Control Function (FC40) for Motor No. 4

```
              FC40
        EN           ENO

I5.4 —START    STOPPED —Q5.3

I5.5 —STOP     RUNNING —Q5.4

I5.6 —ESTOP      MOTOR —Q5.5

I5.7 —HCOIL
```

Program e...   Call stru...

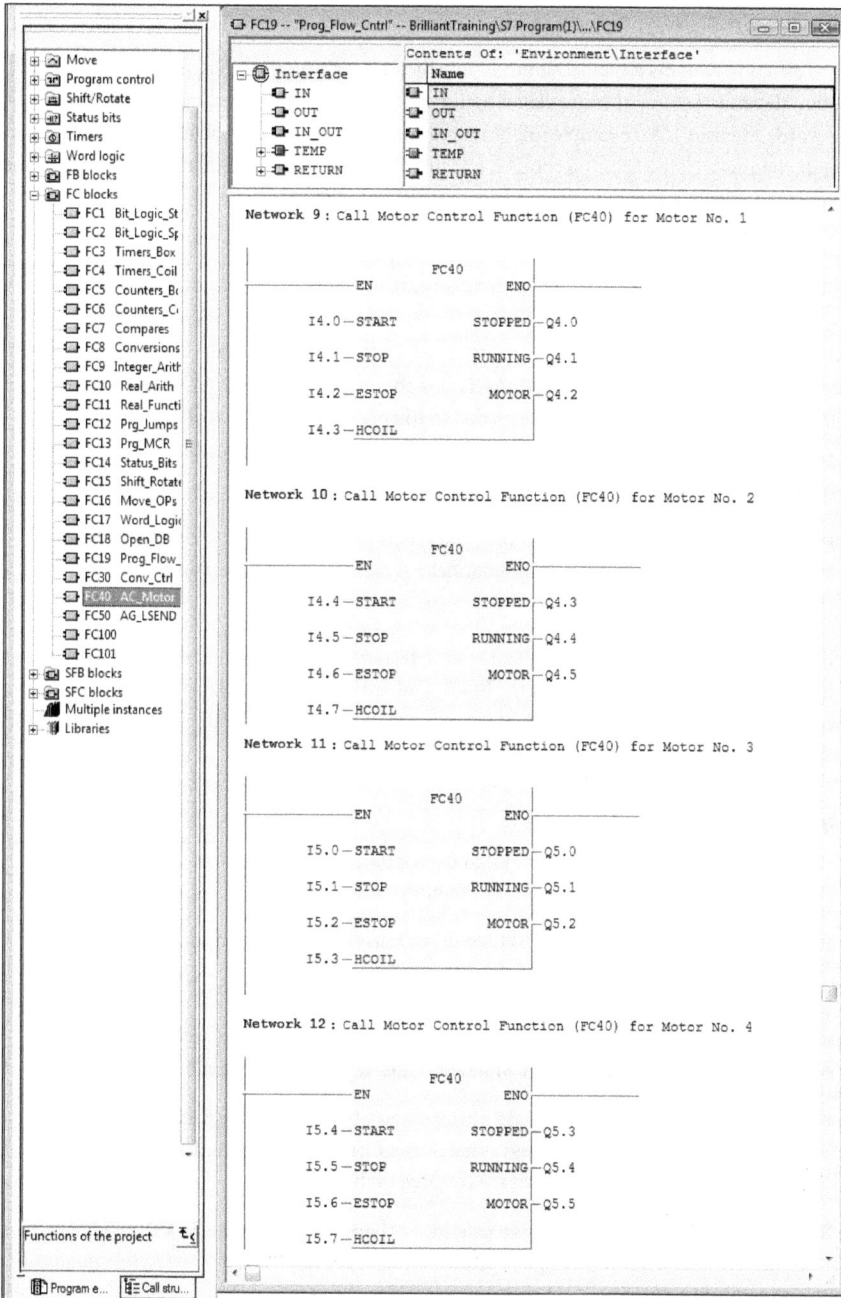

Figure 28-10. Four calls from FC19 to FC40 (AC_Motor); controls four individual motors

Develop FB without Formal Parameters

This example of an FB represents a typical situation in which a Function Block is used primarily because a control solution depends on one or more values being maintained from one call of the block to the next (in other words memory). An example might include an FB that performs analog input filtering of two analog input channels. In the filtering routine, a reading of each channel is taken on each call of the routine; consecutive readings of each analog input are added to the previous reading; and finally the readings total is divided by the total number of sample readings taken.

In the FB example, as described above, formal parameters are not used and declarations of type **IN**, **OUT**, and **IN_OUT** are not required. Variables that must be maintained by an FB must be declared as static (**STAT**) variables. In this example, each analog input channel needs a static variable of type INT as an accumulator. Using the **STAT** variable **CH0_ACC** (for channel 0) and **CH1_ACC** (for channel 1), the new reading of each analog input channel is added to the previous reading. A static variable **SMP_ACC** (sample accumulator) is also used to count the number of sample readings. This single counter is used for both input channels.

After four readings are taken, the sum of each input channel is divided by four, to obtain an average reading. The filtered result for channel 0 and for channel 1 is saved to bit memory locations MW50, and MW52 respectively.

If any intermediate data is required by the FB, you would declare **TEMP** variables. Recall, that these variables are handled in the local data stack of the block, and are no longer available when the block is done. Once you have defined the required variables, you may write the block code using the code section of the block. The 2-channel analog input filtering FB is shown in Figure 28-11.

To code an FB without formal parameters:

1. From the SIMATIC Manager, create the new FB in the appropriate program; then double-click on the specific FB to open the block in the LAD/STL/FBD editor.

2. From the menu, select **View ➤ LAD** (FBD/STL) to display the desired program editor

3. Select View ➤ Overview to display the Program Elements/Call Structure window.

4. **Declare static variables:** Select the **STAT** object in the left pane of the declaration table; then for each variable enter the **Name, Data Type**, and a **Comment**. Enter as many static variables as required.

5. **Declare temporary variables:** Select the **TEMP** object in the left pane of the declaration table; then for each variable enter the **Name, Data Type**, and a descriptive **Comment**. Enter as many temporary variables as required. No temporary variables are required in this example.

6. **Insert a STATIC or TEMP variable before an existing variable:** First select the **STAT** (or **TEMP**) object in left pane of the declaration table, to display records of that type. Then, select a row, before or after which you wish to insert a new row; right click and select **New Declaration Row**.

7. **Enter the block code:** Click inside the code section, starting with Network 1, then write the code for each network operation, using absolute or symbolic addresses (if already defined). The code for this example is entered in **FB51**, shown in Figure 28-11.

8. To start each new network, press the **New Network** button from the toolbar.

9. Select **File ➤ Save** when you are done with the last network.

10. Make sure you create the instance DB and associate it with the correct FB.

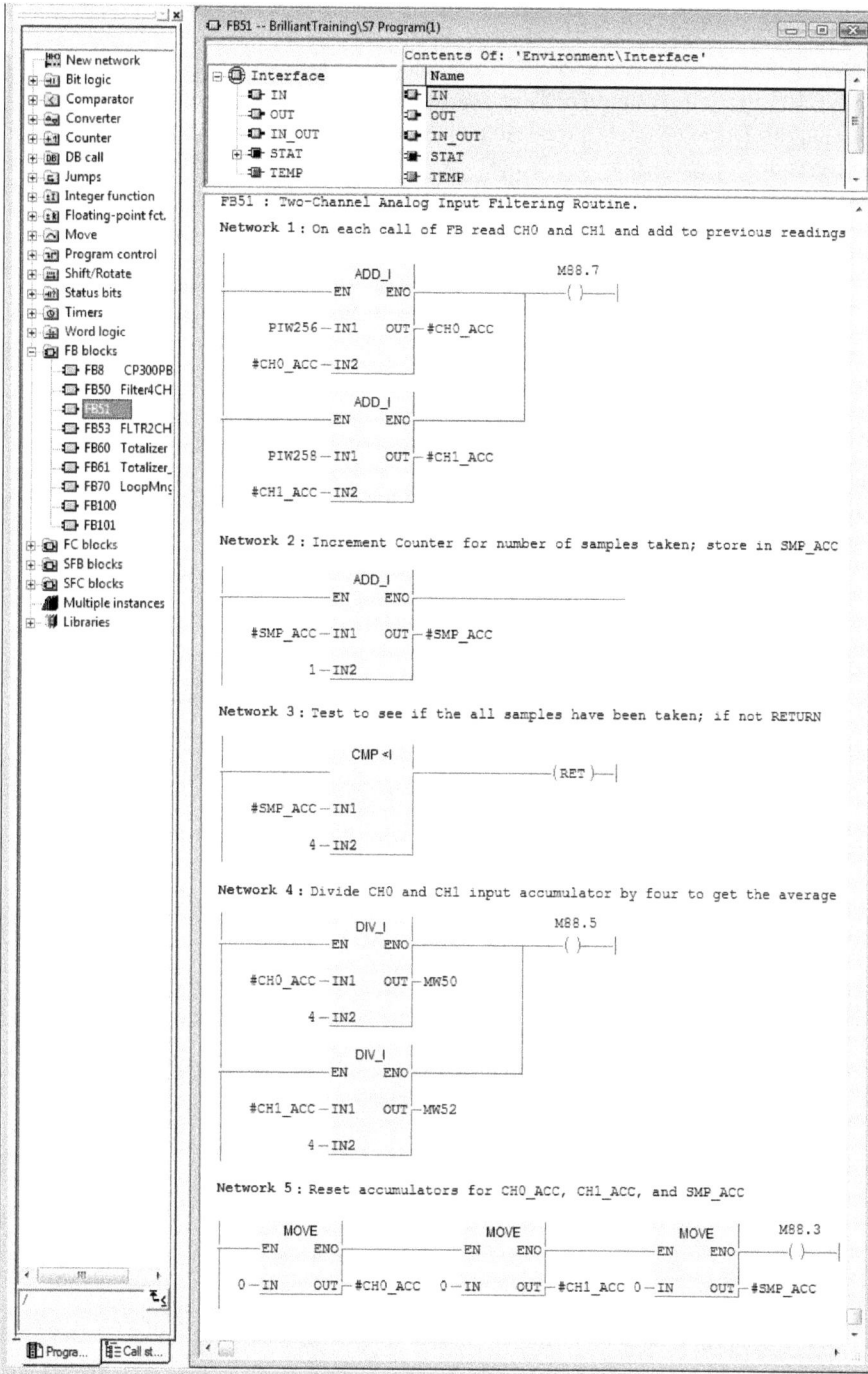

Figure 28-11. An FB without formal parameters, written as a 2-channel Analog Input Filter

Call to FB without Formal Parameters

Calls to an FB with no parameters may be conditional or unconditional, and may be from the main organization block (OB1) or from any other block—including OB, FC, or another FB. Based on the purpose of an FB, you determine where it is called from, how often it is called, and what conditions should trigger the call. Our analog filtering FB (FB61), for instance, might be called from another FB that is responsible for processing analog I/O—for example reading, scaling, and filtering, at periodic intervals. By supplying a different instance DB with each call, different data sets may be used.

In LAD and FBD, an FB with no parameters is called as a **CALL** box with **EN/ENO** lines. All user-created function blocks are located in the FB Blocks folder of the Program Elements tab. To call the FB, you must drag the box from the FB Blocks folder and drop it onto the network, or select the network and then double-click on the FB object in the FB Blocks folder. You must then specify the instance DB, belonging to the FB, above the box. The **EN** input line may call the FB with or without preceding logic. If the call is based on logic conditions, the FB is called when the conditions are satisfied; otherwise, the instruction immediately following the call is processed.

To call an FB without formal parameters:

1. Open the S7 program and the specific calling block—for example **FC19**, from which you will make a call to an existing FC (**FB51** is our 2-channel Analog Input Filter example) created without formal parameters.

2. From the menu, select **View ➢ LAD** (FBD/STL) to display the desired program editor.

3. Select **View ➢ Overview** from the menu to show the instruction browser, if it is hidden from view.

4. From the Program Elements tab, expand the **FB Blocks** folder to display all user-created Function Blocks (FBs) of the currently opened program. The FB you wish to call must have already been written, before if can be called. Find **FB51** in the folder.

5. To make the call, click in the code section of the calling block, then from the toolbar press the **New Network** button to enter a new network from which to call the FB.

6. Locate the desired FB in the **FB Blocks** folder, for example **FB51** (Analog Input Filter FC), then drag and drop the FB onto the network—enter the associated instance DB number FB address above the box.

7. **Make conditional call:** Insert control logic at the **EN** input to the FB to make a conditional call.

8. **Make unconditional call:** Insert the FB without any logic condition to make an unconditional call.

9. Select **File ➢ Save** when you are done with the last network.

In this example, three static variables were declared. Using the **STAT** variable **CH0_ACC** (for channel 0) and **CH1_ACC** (for channel 1), the new reading of each analog input channel is added to the previous reading on each FB call. A static variable **SMP_ACC** was also used to count the number of sample readings. Recall that static variables are maintained in the instance DB associated with the FB. Data block DB51, the instance DB for FB51, is shown below.

	Address	Declaration	Name	Type	Initial value	Actual value	Comment
1	0.0	stat	CH0_ACC	INT	0	0	Accumulator for 1st Analog Input Channel
2	2.0	stat	CH1_ACC	INT	0	0	Accumulator for 2nd Analog Input Channel
3	4.0	stat	SMP_ACC	INT	0	0	Sample Counter

DB51 -- BrilliantTraining\S7 Program(1)

Figure 28-12. Instance DB51. DB51 is opened automatically when FB51 is called

In the Figure 28-13 (FC19), FB51 and FB60 are both called with preceding logic (conditional); FB100 is called without preceding logic (unconditional); FB100 is actually called twice—once with DB100 as the instance DB, and again with DB101 as the instance DB.

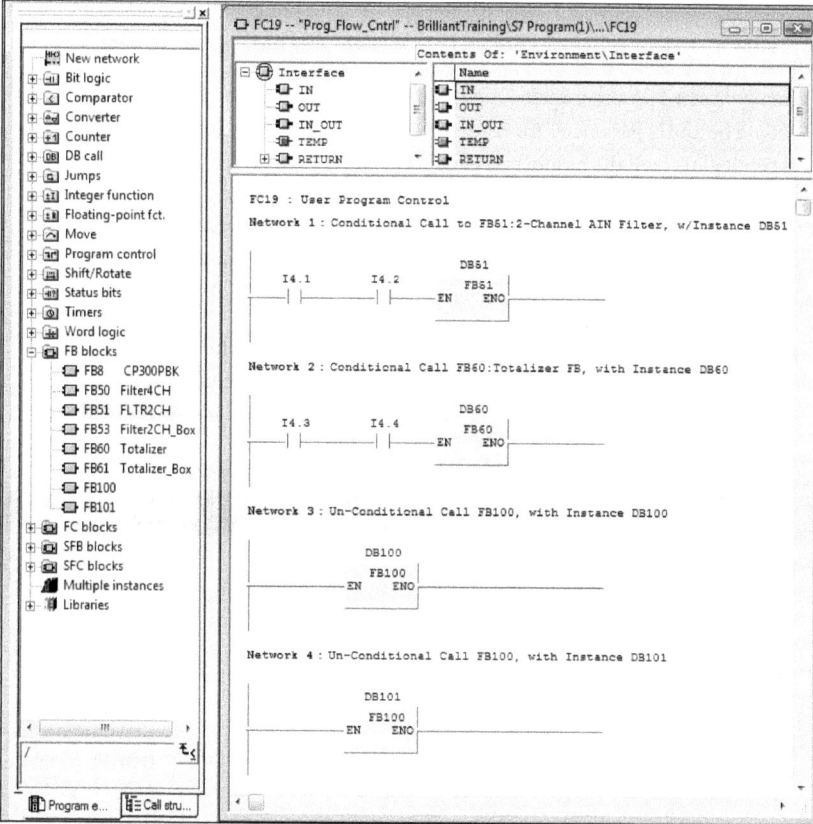

Figure 28-13. Using CALL box in LAD/FBD to call an FB with no formal parameters

Develop FB with Formal Parameters

In the previous FB example, we created an FB that provided filtering of two analog input channels. An FB was used since a number of values were being added that would need to be available from one call of the block to the next. Suppose, however, that there are many of these two-channel input modules that needed filtering. According to the topic on using formal parameters, this would be a case where it would be beneficial to develop an FB written with formal parameters. And, although this example is a 2-channel filter—it could very well be a 4-channel filter, or that 6 samples' readings are needed.

Based on the requirements of a 2-channel analog input filter FB, the parameters needed are shown in Table 28-7. There would be an **IN** parameter for reading each analog input—for example **CH0_IN**, **CH1_IN**; and for accepting input of the number of samples to take—for example, **SAMPLE**. Static integer variables are needed for adding successive readings of each analog input—for example **CH0_ACC** and **CH1_ACC**; and for the number of samples that have been taken—for example **SMP_ACC**. Finally, **OUT** parameters from the FB include the filtered input results, which include **CH0_AVG** and **CH1_AVG**; and the **DONE** parameter to indicate that the filtering is completed.

After defining formal parameters and other required variables, the next step is to write the code using basically the same code as used in the previous example, but instead of using actual input/output addresses, you will enter the **NAME** (s) of the formal parameters and variables as shown in Table 28-7.

Table 28-7. Required Formal Parameters for the Planned 2-channel Analog Input Filtering FB

Parameter/Variable	Decl.	Name	Type	LAD/FBD Display	STL Display
Channel 0 Input	IN	CH0_IN	INT		**CALL**
Channel 1 Input	IN	CH1_IN	INT		FLTR2CH,DB53
Number of Samples	IN	SAMPLE	INT		CH0_IN:
Channel 0 Average	OUT	CH0_AVG	INT		CH1_IN:
Channel 1 Average	OUT	CH1_AVG	INT		SAMPLE:
Done	OUT	DONE	BOOL		CH0_AVG:
Sample Counter	STAT	SMP_ACC	INT		CH1_AVG:
Channel 0 Total	STAT	CH0_ACC	INT		DONE:
Channel 1 Total	STAT	CH1_ACC	INT		

To code an FB with formal parameters:

1. From the SIMATIC Manager, create the new FB, in the appropriate program; then double-click on the FB to open the block in the LAD/STL/FBD editor.

2. From the menu, select **View ➤ LAD** (FBD/STL) to display the desired program editor.

3. **Declare formal parameters:** In the left pane of the declaration table, select **IN**, **OUT**, or **IN_OUT** as the declaration type to define. CH0_IN, CH1_IN, and SAMPLE are all **IN** parameters. CH0_AVG, CH1_AVG, and DONE are all **OUT** parameters. For each parameter enter the **Name, Data Type**, and a descriptive **Comment**.

4. **Declare static variables:** Select the **STAT** object in the left pane of the declaration table; then enter the static variables SMP_ACC, CH0_ACC, and CH1_ACC, inserting the **Name, Data Type**, and a descriptive **Comment** for each variable.

5. **Declare temporary variables:** Skip this step or select the **TEMP** in the left pane of the declaration table, and then enter as many TEMP variables as required. TEMP variables are not required in this example.

6. **Enter code:** Click inside the code section, and starting with Network 1, write the code for the operation using the declared formal parameter and variable names (See Figure 28-14).

In the top pane of FB53, in the figure below, **IN** is selected as the declaration type—the IN parameter variables CH0_IN, CH1_IN, and SAMPLE are displayed. If we were to select STAT, or if we were to select OUT, then the respective parameter variables would be displayed.

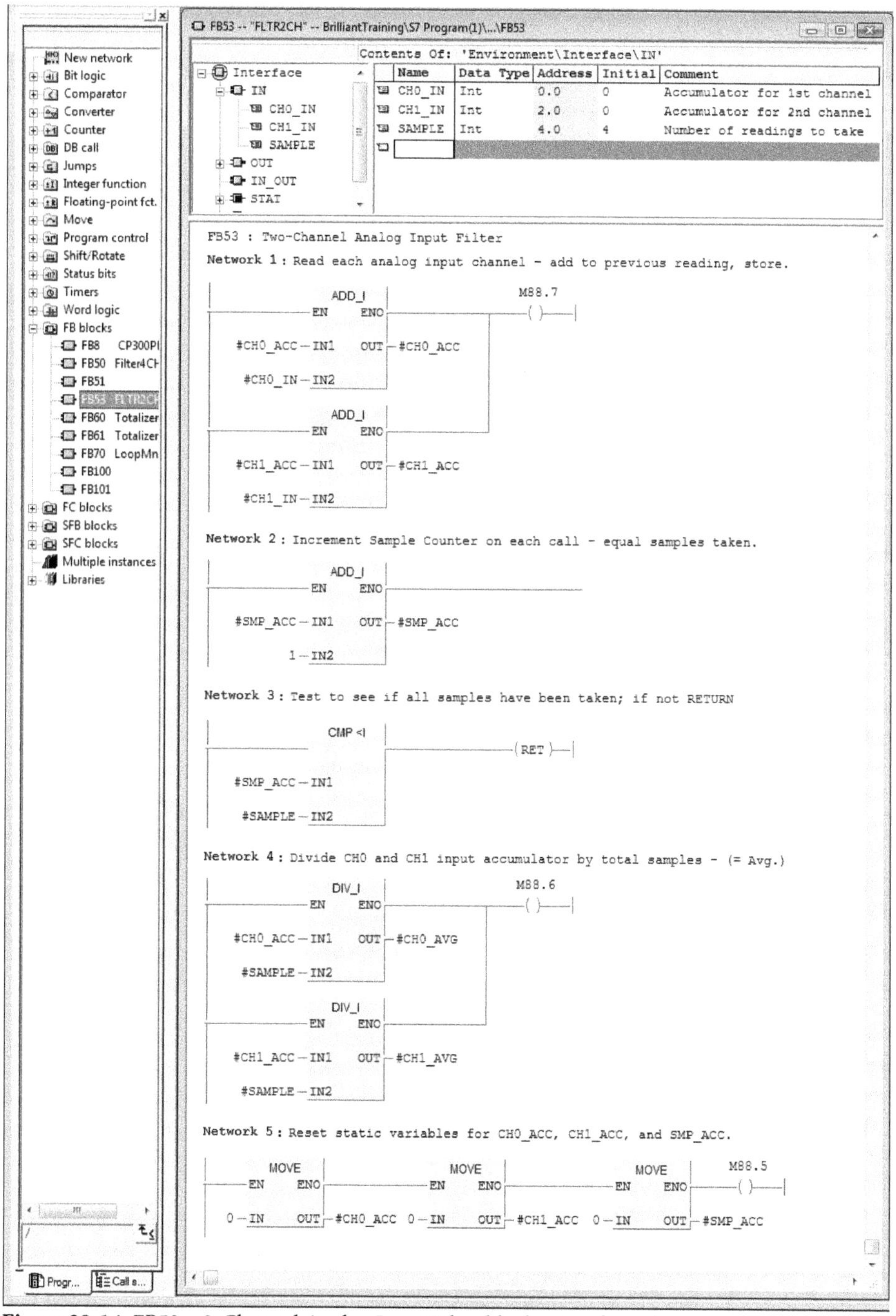

Figure 28-14. FB53—2-Channel Analog Input Filter block using formal parameters

Call to FB with Formal Parameters

An FB developed with *formal parameters* is meant to be called as often as needed throughout your program. Our FB53, written in the previous task, was developed with formal parameters to provide analog input filtering for a 2-channel input module. If there are multiple modules of this type, then this block would be called for as many 2-channel modules as required. Each time you use the *FLTR2CH* FB, you'll specify actual addresses for any two analog input channels. The addresses you specify may be an absolute or symbolic address (See Figure 28-15).

As you use an FB with formal parameters, you must ensure that the correct addresses are entered at the appropriate parameters based on the purpose and requirements of the block. In this FB, for example, a peripheral input address is entered for each analog input (**CH0_IN, CH1_IN**), an integer constant allows for the number of samples (**SAMPLE**) to be specified, and a word-width address or variable may be specified to output each filtered result—**CH0_AVG** and **CH1_AVG**. Finally, to indicate that the operation is completed, a bit memory address or variable may be specified at the parameter **DONE**.

To call an FB with formal parameters:

1. Open the S7 program and the calling block—for example FC19, from which you will make a call to an existing FB created with formal parameters (FB53 is our 2 channel analog filter example).

2. From the menu, select **View ➤ LAD** (FBD/STL) to display the desired program editor.

3. Select **View ➤ Overview** from the menu to show the instruction browser, if it is hidden from view.

4. From the Program Elements tab, expand the **FB Blocks** folder to display all user-created Function Blocks (FBs) of the currently opened program. The FB you wish to call must have already been written, before it can be called. Find FB53 in the folder.

5. **Make first call to FB with parameters:** Click in the code section of the calling block, then from the toolbar press the **New Network** button to enter the network in which to call the FC. Locate the desired FC number, for example **FC40** (AC Motor FC), then drag and drop the FC onto the network.

6. For each parameter, enter the actual addresses for the associated device or operation (AC motor in this case). Insert logic at the **EN** input for a conditional call to the FB; insert the FB without logic conditions to make an unconditional call (See Figure 28-15). Depending on each parameter, you may enter a constant, specific address, or variable.

   Note: In the case of an FB, actual parameters (addresses) do not have to be specified at each block parameter, since these values are stored as static variables in the FB's associated instance DB.

7. **Make additional calls to FB with parameters:** Create a new network and drag and drop the FB onto the network; then for each additional device or operation, supply the appropriate set of actual addresses (absolute or symbolic). Insert logic at the **EN** input for a conditional call to the FB; insert the FB without logic conditions to make an unconditional call (See Figure 28-15).

8. Select **File ➤ Save** when done with the last network.

In our example, as shown in the figure below, **FB53** is called twice—each call is conditional based on a contact closure that generates a one-shot pulse on the positive transition of the contact closure. On the first call, the two analog input channels are peripheral input addresses PIW256 and PIW258; the two filtered results are stored in bit memory locations MW100 and MW102 respectively; the DONE bit is bit address M80.1.

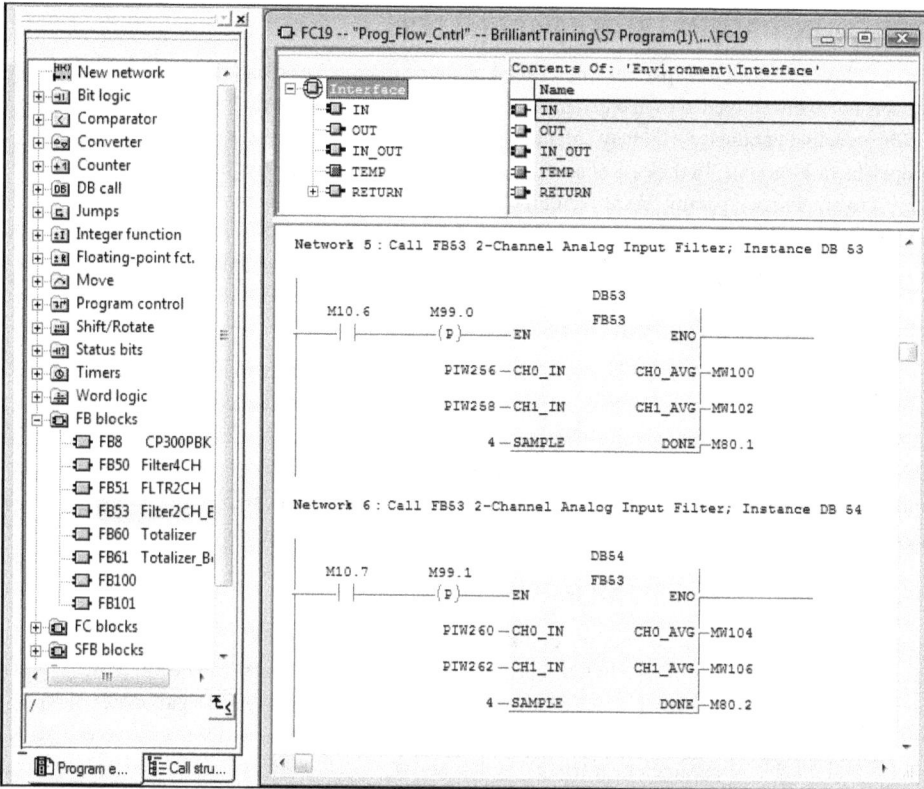

Figure 28-15. Call to FB53 analog filter; two analog input signals are filtered on each call to the FB

The instance data block for **FB53**, shown in the figure below, is **DB53**. Notice that the data elements of the data block include both the static variables and the formal parameters that were declared in our FB53 (FLTR2CH) example. When you create an instance DB, you must associate it with the correct Function Block. When the DB is saved, STEP 7 will automatically update the data block to include the declaration table of the associated FB—**FB53**in this case. Temporary variables would not be stored.

Figure 28-16. Instance DB 53, for FB53. The DB is opened automatically when the FB is called.

Copyright © 2013

MORE ON INITIALIZING BLOCK PARAMETERS

When an FC or FB block with parameters is called in your program, an actual parameter or actual address is normally entered at each block parameter on each call of the block in the program. When entering an actual parameter, you may enter an absolute or symbolic address, a constant where appropriate, or a variable. In the case of an FC, actual parameters must always be specified. While it is common to enter actual parameters at each FB block parameter, it is not required, since block parameters of the FB are stored in the associated instance DB. Table 28-8 shows examples of S7 memory addresses that may be used as actual addresses at block parameters of elementary types.

Initializing with Input/Output Image Addresses

As seen in Table 28-8, both input image (I) and output image (Q) addresses may be supplied at block parameters. As you may note in the table, input addresses may be designated as an IN, OUT, or IN_OUT parameter; the same is true for output addresses—they too may be designated as an IN, OUT, or IN_OUT parameter. Although this practice is possible, it may not always be a wise choice, and not recommended, since these addresses are controlled by the I/O image. In general, input addresses should be designated only at IN parameters and output addresses only at OUT parameters.

Initializing with Bit Memory Addresses

As seen in Table 28-8, bit memory addresses may be used to initialize IN, OUT, and IN_OUT block parameters. For a parameter of type BOOL, specify the bit memory address as Mbyte.bit; if the parameter is of type BYTE/CHAR, specify the bit memory address as M byte; for a parameter of type WORD, INT, S5TIME or DATE, specify the bit memory address as MW n, where n is the start byte number. And finally, for parameters of type DWORD, REAL, DINT, TIME-OF-DAY and TIME, specify the bit memory address as MD n, where n is the start byte number.

Initializing with Peripheral Input/ Output Addresses

As seen in Table 28-8, peripheral input byte (PIB), peripheral input word (PIW), and peripheral input double word (PID) addresses are only specified at IN block parameters. Recall that these addresses access the S7 I/O bus, and therefore only permit access to inputs. On the other hand, peripheral output byte (PQB), peripheral output word (PQW), and peripheral output double word (PQD) addresses are only specified at OUT block parameters, as these addresses access the S7 output bus.

Initializing with Data Addresses

As seen in Table 28-8, all data addresses may be supplied at an IN, OUT, or IN_OUT parameter. For byte width parameters, specify DBn.DBB y (shared DB) or DIn.DIB y (instance DB); for word width parameters specify DBn.DBW y (shared DB) or DIn.DIW y (instance DB). If the data type of the parameter is BOOL, then you may specify the bit address of a shared DB (DB4.DBX 2.0) or the bit address of an instance DB (DI4.DIX 7.2).

When data addresses are used to initialize a block parameter, you may specify a fully- or partially-addressed data location, or the variable name assigned to the data location.

Initializing Parameters of Complex Data Types

A single component of a variable of type ARRAY or STRUCT may be passed to a block parameter where the formal parameter variable of the block is of the same type and structure as the array or structure from which the component element is being passed. The complete array or structure variable itself cannot be passed as the actual parameter—only a component of the variable, specified using the complete component name. Variables of data type STRING can be specified at a block parameter of type

STRING or ANY. The STRING variable, for example, can be copied with the system function SFC20 (BLKMOV). A variable of type DT may be applied at block parameters of type DT (DATE_AND_TIME) or type ANY.

Table 28-8. Memory Operands as Actual Parameters to Block Parameters of Elementary Types

DATA TYPE	Identifier	S7 OPERAND	Example	IN	OUT	IN_OUT
BOOL	I	Input Bit	I 44.3	Y	Y	Y
	Q	Output Bit	Q44.5	Y	Y	Y
	M	Memory Bit	M22.2	Y	Y	Y
	L	Local Bit	L 5.3	Y	Y	Y
	DBX	Shared Data Bit	DB4.DBX 2.0	Y	Y	Y
	DIX	Instance Data Bit	DI4.DIX 4.1	Y	Y	Y
BYTE/CHAR	IB	Input Byte	IB 30	Y	Y	Y
	QB	Output Byte	QB 44	Y	Y	Y
	MB	Memory Byte	MB 23	Y	Y	Y
	LB	Local Byte	LB 2	Y	Y	Y
	DBB	Shared Data Byte	DB4.DBB 6	Y	Y	Y
	DIB	Instance Data Byte	DI4.DIB 6	Y	Y	Y
	PIB	Peripheral Input Byte	PIB 44	Y	N	N
	PQB	Peripheral Output Byte	PQB 44	N	Y	N
WORD/INT S5TIME/ DATE [1]	IW	Input Word	IW 30	Y	Y	Y
	QW	Output Word	QW 44	Y	Y	Y
	MW	Memory Word	MW 24	Y	Y	Y
	LW	Local Word	LW 2	Y	Y	Y
	DBW	Shared Data Word	DB4.DBW 8	Y	Y	Y
	DIW	Instance Data Word	DI4.DIW 8	Y	Y	Y
	PIW	Peripheral Input Word	PIW 66	Y	N	N
	PQW	Peripheral Output Word	PQW 66	N	Y	N
DWORD/REAL/DINT TIME-OF-DAY/ TIME [2]	ID	Input Double Word	ID 30	Y	Y	Y
	QD	Output Double Word	QD 44	Y	Y	Y
	MD	Memory Double Word	MD 8	Y	Y	Y
	LD	Local Double Word	LD 2	Y	Y	Y
	DBD	Shared Data DWord	DB4.DBD 10	Y	Y	Y
	DID	Instance Data DWord	DI4.DID 10	Y	Y	Y
	PID	Peripheral Input DWord	PID 82	Y	N	N
	PQD	Peripheral Output DWord	PQD 82	N	Y	N

1) Constants are not shown in table, but may be used to initialize an IN parameter. **2)** Instance data addresses DIB, DIW, DID, and bit addresses DIX y.x, represent static variables associated with an FB.

WORKING WITH SYSTEM FCs AND FBs

In Parts II and III of this book, you were introduced to STEP 7 LAD/FBD and STL language representations and operations sets. Here in Part IV, you have been introduced to STEP 7 FC, FB, OB, and DB building blocks, as well as an approach to developing an S7 program. Using these operations and resources, you can now build user FCs and FBs to perform various control requirements of your program. Here in this section, you are introduced to system blocks—blocks that are already developed and are available for calling within a program. A list of these blocks is provided in Appendix B.

A system function (SFC) with block parameters is handled in the same manner as a user-developed FC with parameters. The same is true for an SFB—it is handled in the same manner as an FB. To use a parameterized block correctly, the appropriate documentation is needed to know the block's purpose and operation. It is also essential to know the types of memory variables—for example I, Q, M, PI, PQ, and DBDW, as well as the data type—for example BOOL, DINT, INT, and CHAR that are usable at each parameter. Finally, it is important to know if a system block is supported by your CPU. With this understanding, you are ready to make use of available system blocks.

IEC Standard Move Data Functions

In addition to the STEP 7 Move operation that handles a single value at a time, standard System Functions support handling of multiple locations or variables as a contiguous area. These callable System Functions that include SFC20 (BLKMOV—Copy Data Area), SFC21 (FILL—Pre-Fill Data Area), and SFC81 (UBLKMOV—Copy Data Area without Gaps), are available in the STEP 7 library, and integrated in some S7 CPUs. These functions are briefly described here to show some examples of system functions, and how a contiguous area is specified at a parameter of type ANY.

SFC 20, SFC 21, and SFC81 do not support the transfer of timer (T), counter (C), peripheral input (PI), and peripheral output (PQ) memory areas. A maximum of 512 bytes may be transferred using SFC81.

Table 28-9. SFC parameters for SFC 20 (COPY) and SFC 81 (COPY without gaps)

Parameter	Data Type	Description	Valid Areas	SFC/SFB
EN/ENO		Logic Enable /Output Line	I, Q, M, D, L	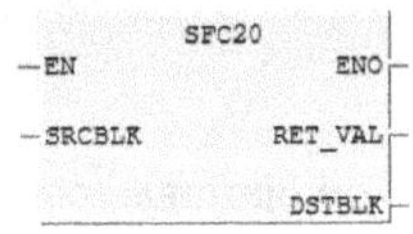
SRCBLK	ANY	Source Data Area	I, Q, M, D, L	
RET_VAL	INT	Return Error Value	I, Q, M, D, L	
DSTBLK	ANY	Destination Data Area	I, Q, M, D, L	

SFC 20 (BLKMOV—Copy Data Area)

System Function SFC20 (BLKMOV) supports the transfer of a contiguous group of memory locations from one area to another area. The area to be moved is called the source area; the area to which the block is moved is called the destination area. For each call of SFC20, you must specify the actual source area at the parameter SRCBLK and the actual destination area at the parameter DSTBLK. Both of these variables are specified as a variable of type ANY.

LAD Example: Call to SFC 20	Example Parameter values:	Comment
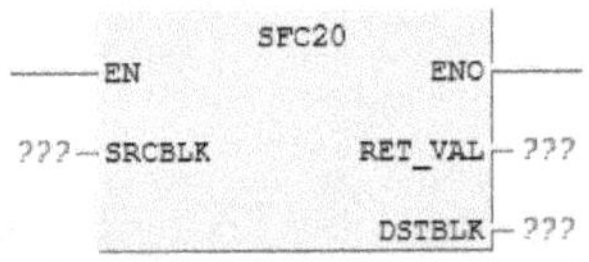	EN/ENO = M 101.1/M 101.7 SRCBLK = P#DB5.DBX0.0 BYTE 8 DSTBLK = P#DB50.DBX0.0 BYTE 8 RET_VAL = MW 200	If RLO =1 at EN > Move Data SRC= 8 bytes in DB 5 from DBB 0 DST= 8 bytes in DB 50 from DBB 0

Figure 28-17. SFC 20 Copy Data Area

SFC81 (UBLKMOV—Copy Data Area without Gaps)

System Function 81 (UBLKMOV) supports the uninterruptible transfer of a contiguous block of memory locations from one area to another area. The area to be moved is called the source block; the area to which the block is moved is called the destination block. For each call of this block, you must specify an actual source area at the parameter SRCBLK, and an actual destination area at the parameter DSTBLK. Both of these variables are specified as a variable of type ANY.

In the following example, the DB 2 destination area of 32 bytes, starting with byte 16 (DBB16) and ending with byte 47 (DBB47), is copied with the source area of 32 bytes from bit memory byte MB 16.

LAD Example: Call to SFC 81	Parameter values:	Comment
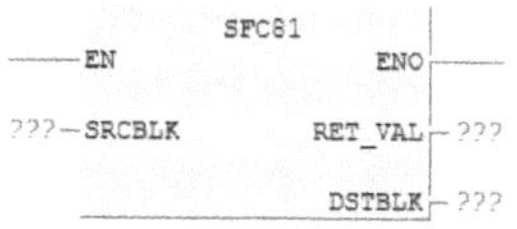 SFC81 —EN ENO— ???—SRCBLK RET_VAL—??? DSTBLK—???	EN/ENO = M 101.2/M 101.6 SRCBLK = P#M 16.0 BYTE 32 DSTBLK = P#DB2.DBX16.0 BYTE 32 RET_VAL = MW 202	If RLO =1 at EN > Copy Data SRC= 32 bytes from MB 16 DST= 32 bytes in DB 2 from DBB 16

Figure 28-18. SFC 81 Copy Data Area without Gaps

SFC 21 (FILL—Pre-Fill Data Area)

System Function SFC21 (FILL) supports the initializing of a contiguous block of memory locations with a specific value or values. For each call of this block, you must specify an actual source area at the parameter BVAL, and an actual destination area at the parameter BLK. Both of these variables are specified as a variable of type ANY. The prefill data area, specified in the parameter BVAL, can contain a single value, for example B#16#00, or several values.

Table 28-10. SFC parameters for SFC 21 (FILL)

Parameter	Data Type	Description	Valid Areas	SFC/SFB
EN/ENO		Logic Enable /Output Line	I, Q, M, D, L	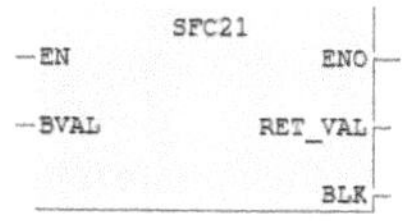
BVAL	ANY	Source Fill Data Area	I, Q, M, D, L	SFC21 —EN ENO—
RET_VAL	INT	Return Error Value	I, Q, M, D, L	—BVAL RET_VAL—
BLK	ANY	Destination Data Area	I, Q, M, D, L	BLK—

If the source or destination pointer is of type BOOL, the specified length of the variable must be divisible by 8; otherwise the block will not be executed.

In the following example the DB 2 area of 32 bytes, starting with DBB16 and ending with DBB31, is prefilled with the value of B#16#00, from the source area MB 16. The value of B#16#00 would have been previously loaded to memory byte MB 16.

LAD Example: Call to SFC21	Parameter values:	Comment
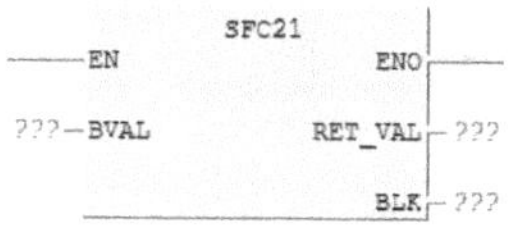 SFC21 —EN ENO— ???—BVAL RET_VAL—??? BLK—???	EN/ENO = M 101.2/M 101.6 BVAL = P#M 16.0 BYTE 1 BLK = P#DB2.DBX16.0 BYTE 16 RET_VAL = MW 202	If RLO =1 at EN > Prefill Destination BVAL = 1 byte from MB 16 BLK = 16 bytes in DB 2 from DBB 16

Figure 28-19. SFC 21 Prefill Data Area

IEC Standard Timer Functions

In addition to standard STEP 7 timers covered in the LAD/FBD and STL chapters on timers, this section presents a brief overview of standard IEC timer functions. These callable System Function Blocks, which include SFB3 (TP—Pulse Generation), SFB4 (TON—Timer On-Delay), and SFB5 (TOF—Timer Off-Delay), are available in the STEP 7 library, and are integrated in some S7 CPUs. STEP 7 also offers a variety of other timer operations as standard FCs and FBs.

IEC standard system block timers SFB3, SFB4, and SFB5 only run in STARTUP and RUN modes, and are enabled when the RLO = 1 at the EN line. By default, any instance of SFB3, SFB4 or SFB5 is reset by the operating system during each cold restart of the CPU. You may also initialize an instance of any of these blocks following a warm restart—to do so you must call the block in OB100. If instances of either of these SFBs are located in any other block, you can reset these instances, as well, by initializing the higher-level block—that is the calling block.

SFB3 (TP—Timer Pulse)

In essence, when triggered by its start input, the output status Q is activated (set to '1') and remains activated for a duration that is equal to the preset time PT. The pulse is generated for the duration, irrespective of changes at the start timer input trigger (hence pulse timer).

Table 28-11. Block parameters for IEC Timer SFB3—PULSE Timer

SFB3	Parameter	Type	Description	Valid Areas
	EN/ENO		Logic Enable /Output Line	I, Q, M, D, L
	IN	BOOL	Start Timer Input	I, Q, M, D, L
	PT	TIME	Preset Time (Pulse Duration)	I, Q, M, D, L
	Q	BOOL	Timer Status Signal	I, Q, M, D, L
	ET	TIME	Elapsed Time Value	I, Q, M, D, L

On a 0-to-1 transition at the start timer input IN, the pulse timer starts timing and continues to time until the preset time PT expires—even if the RLO at the start input IN returns to '0'. When the start timer input IN transitions to '1', the timer status signal Q also transitions to '1' and remains '1' while the timer is timing. The elapsed time value ET reflects the expired time, and will reach a maximum value equal to the timer preset PT. After the timer preset time is reached, the elapsed time value ET is only reset to zero when the RLO = 0 at the start timer input IN.

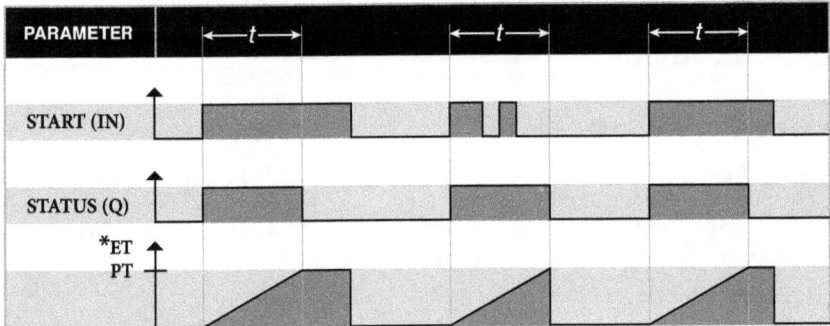

*ELAPSED TIME | PRESET TIME

Figure 28-21. Response diagram for the SFB3 Pulse timer

SFB4 (TON—Timer On-Delay)

In essence, when the on-delay is triggered by its start timer input, the output status Q is activated (set to '1')' after a delay that is equal to that of the preset time PT (hence timer on-delay).

Table 28-12 Block parameters for IEC Timer SFB4—ON-Delay

SFB4	Parameter	Type	Description	Valid Areas
	EN/ENO		Logic Enable /Output Line	I, Q, M, D, L
	IN	BOOL	Start Timer Input	I, Q, M, D, L
	PT	TIME	Preset Time (On Delay)	I, Q, M, D, L
	Q	BOOL	Timer Status Signal	I, Q, M, D, L
	ET	TIME	Elapsed Time Value	I, Q, M, D, L

On a 0-to-1 transition at the start timer input IN, the on-delay timer starts timing and continues to time while the start timer input IN is '1', or until the preset time PT expires. When the preset time expires, the timer status Q is activated (1) and remains activated (1) until the RLO = 0 at the start timer input IN. If the start timer input IN transitions to '0' while the timer is timing, and prior to reaching the preset time PT, the elapsed time ET is reset to zero, and timer output Q remains at '0'.

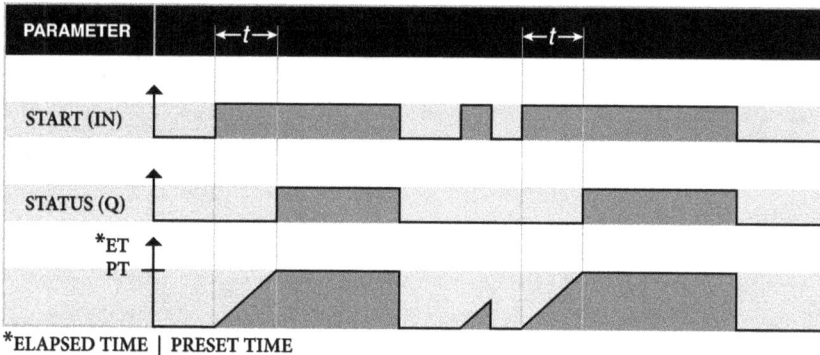

*ELAPSED TIME | PRESET TIME

Figure 28-22. Response diagram for the SFB4 ON-delay timer

SFB5 (TOF—Timer Off-Delay)

In essence, the off-delay timer is enabled by a rising edge of the start timer input, which activates the timer status Q. After a falling edge at the start input, the timer status Q is deactivated (set to '0') after a duration that is equal to the preset time PT (hence timer off-delay).

Table 28-13. Block parameters for IEC Timer SFB5—OFF-Delay

SFB5	Parameter	Type	Description	Valid Areas
	EN/ENO		Logic Enable /Output Line	I, Q, M, D, L
	IN	BOOL	Start Timer Input	I, Q, M, D, L
	PT	TIME	Preset Time (Off Delay)	I, Q, M, D, L
	Q	BOOL	Timer Status Signal	I, Q, M, D, L
	ET	TIME	Elapsed Time Value	I, Q, M, D, L

On a 0-to-1 transition at the start timer input IN, the timer status Q is immediately activated (1). When the start timer input IN transitions from 1-to-0, the timer starts timing and continues to time while the start timer input IN is '0', or until the preset time PT expires. When the preset time expires, the off-delay timer status Q is deactivated. The elapsed time ET value reflects the time expired since the last falling edge at IN. ET is reset when there is a rising edge at the start timer input IN.

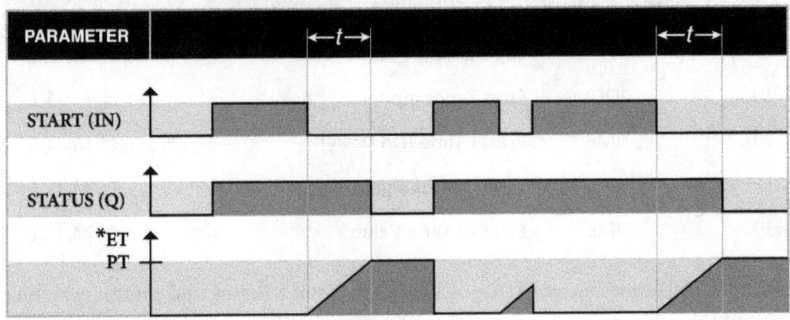

*ELAPSED TIME | PRESET TIME
Figure 28-23. Response diagram for the SFB5 OFF-delay timer

IEC Standard Counter Functions

In addition to the standard STEP 7 counters covered in the LAD/FBD and STL chapters on counters, this section presents a brief overview of standard IEC counter functions. These callable System Function Blocks, which include SFB0 (CTU—Up Counter), SFB1 (CTD—Down Counter), and SFB2 (CTD—Up/Down Counter), are available in the STEP 7 library, and are integrated in some S7 CPUs. STEP 7 also offers a variety of other counter operations as standard FCs and FBs.

SFB0 (CTU—Up Counter)

The SFB0 instance is triggered when the EN line has an RLO = 1. The up counter instance is incremented by +1 on each rising edge at the CU input—when the signal state of CU of the current SFB0 call is compared with the last SFB0 call. The counter discontinues the up count when the count value CV reaches the upper limit of 32767, and no longer responds to the rising edge input at CU. A signal level of '1' at the R input resets the counter value to CV = 0, at any time and regardless of the value at the CU input. The counter status signal Q indicates whether the current counted value is greater or equal to the preset value. Signal Q = 0 for CV < PV; Q = 1 for CV ≥ PV.

By default, any instance of SFB0 is reset by the operating system during each cold restart of the CPU. You may also initialize an instance of SFB0 following a warm restart—to do so, you must call SFB0 in OB100, with the reset input R = 1. If instances of this SFB are located in any other block, you can reset these instances, as well, by initializing the higher-level block—that is the calling block.

Table 28-14. Block parameters for IEC Counter SFB0—Up Counter

SFB0	Parameter	Type	Description	Valid Areas
	EN/ENO		Logic Enable /Output Line	I, Q, M, D, L
	CU	BOOL	Count UP by +1	I, Q, M, D, L
	R	BOOL	Reset Counter Input Signal	I, Q, M, D, L
	PV	WORD	Preset Value	I, Q, M, D, L
	Q	BOOL	Counter Status	I, Q, M, D, L
	CV	WORD	Current Counter Value (0 -to- 32767)	I, Q, M, D, L

SFB1 (CTD—Down Counter)

The SFB1 instance is triggered when the EN line has an RLO = 1. The down counter instance is decremented by +1 on each rising edge at the CD input—when the signal state of CD of the current SFB1 call is compared with the last SFB1 call. The counter discontinues the down count when the counter value (CV) reaches the lower limit of -32768, and no longer responds to the rising edge input at CD. A signal level of '1' at the LOAD input, at any time, sets the counter to the value specified at PV. The counter status signal Q indicates whether the current counter value CV is less than or equal zero. Signal Q = 1 for CV ≤ 0; Q = 0 otherwise.

By default, any instance of SFB1 is reset by the operating system during each cold restart of the CPU. You may also initialize an instance of SFB1 following a warm restart—to do so requires calling SFB1 in OB100, with LOAD = 1 and PV = the desired initial value for CV. If instances of this SFB are in any other block, they can also be reset, by initializing the higher-level block—that is the calling block.

Table 28-15. Block parameters for IEC Counter SFB1—Down Counter

SFB1	Parameter	Type	Description	Valid Areas
??? SFB1 —EN ENO— …—CD Q—… …—LOAD CV—… …—PV	EN/ENO		Logic Enable /Output Line	I, Q, M, D, L
	CD	BOOL	Count Down by +1	I, Q, M, D, L
	LOAD	BOOL	Load Preset Value Input Signal	I, Q, M, D, L
	PV	WORD	Preset Value	I, Q, M, D, L
	Q	BOOL	Counter Status	I, Q, M, D, L
	CV	WORD	Current Counter Value (-32768 -to- 32767)	I, Q, M, D, L

SFB2 (CTD—Up/Down Counter)

System function block SFB2 is used to count both up and down. The counter value CV is incremented by +1 on each rising edge at the CU input; and is decremented by +1 on each rising edge at the CD input. CV is only updated when there is from 0-to-1 transition at either the CU or CD input when the current call of SFB2 is compared with the last call of the block.

The counter discontinues the up count when the count value CV reaches the upper limit of +32767, and discontinues the down count if CV reaches the lower limit of -32768. If either limit is reached, the counter no longer responds to the rising edge input at CU or CD. If a rising edge occurs both at CU and CD during the same cycle, the counter value CV is not changed—the current value is retained.

Table 28-16. Block parameters for IEC Counter SFB2—Up/Down Counter

SFB2	Parameter	Type	Description	Valid Areas
??? SFB2 ——EN ENO—— …—CU QU—… …—CD QD—… …—R CV—… …—LOAD …—PV	EN/ENO		Logic Enable /Output Line	I, Q, M, D, L
	CU	BOOL	Count UP by +1	I, Q, M, D, L
	CD	BOOL	Count Down by +1	I, Q, M, D, L
	R	BOOL	Reset Counter Input Signal	I, Q, M, D, L
	LOAD	BOOL	Load Preset Value Input Signal	I, Q, M, D, L
	PV	WORD	Count Preset Value	I, Q, M, D, L
	QU/QD	BOOL	CU Status/ CD Status	I, Q, M, D, L
	CV	WORD	Current counter value (-32768 -to- 32767)	I, Q, M, D, L

A signal of '1' at the LOAD input, at any time, sets the counter to the value specified at PV—this action is irrespective of the inputs CU and CD. A signal level of '1' at the R input resets the counter value to CV = 0, at any time and regardless of the inputs at CU or CD. The counter status QU indicates whether the current counter value is greater or equal to the preset value (QU = 1 for CV ≥ PV). The counter status QD indicates whether the counter value CV is less than or equal zero (QD = 1 for CV ≤ 0).

By default, any instance of SFB2 is reset by the operating system during each CPU cold restart. You may also initialize an instance of SFB2 following a warm restart—to do so requires calling SFB2 in OB100, with R = 1 when using the block to count up; or with LOAD = 1 and PV = the desired initial value for CV, when using the block to count down. If instances of this SFB are located in any other block, they can also be reset, by initializing the higher-level block—that is the calling block.

Chapter 29

Developing Organization Blocks

This chapter presents Organization Blocks—also called OBs. This category of user blocks provides the STEP 7 developer with a resource for organizing and managing the various processing requirements of the components of the STEP 7 program. Whereas OBs are called by the S7 operating system, each of the blocks of the user program are ultimately called from an organization block.

As indicated in the list of main topics below, organization blocks are grouped into several categories, each of which has a specific event type to which it is associated. Using OBs, you can organize the overall structure of your program based on processing requirements that include startup routines, normal cyclic program processing, interrupt routines, managing system-related errors and faults, and a variety of other processing requirements—all without managing the details of calling these routines.

Although you will determine which OBs must be developed as part of the S7 Program, these blocks become part of the S7 operating system when downloaded to the CPU. And although you will influence what triggers the calls to organization blocks, each OB is called for processing by the CPU when there is an occurrence of the system-related event to which the OB is associated.

Main Topics

- Introduction
- Start-Up Blocks
- The Main Program Block
- Background Processing
- Time-of-Day Interrupts
- Time-Delay Interrupts
- Cyclic Interrupts
- Hardware Interrupts
- Multi-Computing Interrupt
- Asynchronous Errors
- Synchronous Errors

INTRODUCTION

Application Overview

Organization blocks, often called OBs, are an S7 resource for organizing your code to efficiently respond to events and conditions in the controlled process, in your program, and in the PLC itself. As one of three types of blocks in which you may write your program, OBs provide the structured means by which the code blocks (FBs, FCs, SFBs, SFCs) of your program are called for processing during start-up, normal cycle, background, periodic, timed and hardware interrupt, and on error and fault events. The various types of organization blocks and the events that trigger them are listed below.

Table 29-1. Summary of S7 Organization Blocks and their associated triggering events

OB Category	OB No.	Call Event and Application Description
Normal Processing	1	After each CPU start-up and thereafter, cyclically, to process the main user program.
Time-of-Day Interrupts	10-17	On occurrence of any user-configured time-of-day interrupt, to process user code that must be processed at a specific date and time.
Time-Delay Interrupts	20-23	After user-defined time delay expires, to process user code that must be processed after the time delay expires.
Cyclic Interrupts	30-38	Cyclically at precise intervals, for example every 500 ms, to process user code that must be processed regularly—for example PID loops.
Hardware Interrupts	40-47	When any one of the configured process or module-generated interrupts occurs, to process a user-programmed interrupt service.
Multi-Computing Interrupt	60	When an interrupt is generated by any of the CPUs in a multi-CPU configuration. OB60 supports a synchronized CPU response.
Redundancy Errors	70-73	On occurrence of a redundancy error in an S7-400 H-system, to process the user-programmed response.
Asynchronous Errors	80-87	On occurrence of a system-related error—for example power fault, or module failure, to process the user-programmed service routine.
Background Processing	90	When the actual cycle time of the current cycle is less than the user-defined minimum cycle. Background processing, while the CPU is not busy, minimizes CPU wait times for critical processing.
CPU Start-up	100-102	At each CPU start-up event, to process user-defined initialization routines prior to calling OB1.
Synchronous Errors	121-122	At occurrences of program-related runtime errors, to process the user-programmed service routine.

Operating System and User Program Interface

OBs differ from functions (FCs) and function blocks (FBs) in that the operation and processing of each FB and FC, in the program, is completely user-determined. You assign the FC and FB numbers you wish to use, and manage how and when each is called for processing. On the other hand, OB numbers are predefined—and although you must write the code for these blocks, they become part of the S7 operating system when loaded to the CPU. Furthermore, each OB is called by the CPU based on its pre-defined function, and its associated triggering event.

Since OBs are part of your program, yet are called by the CPU, they actually provide an interface between the operating system and your program. As such, with OBs you may determine the CPU's response to errors and faults, interrupt events, as well as how the CPU behaves during start-up and during normal processing. With errors and faults, for example, the CPU will generally transition to STOP if the related OB is not yet programmed and loaded to the CPU. If the appropriate OB is installed, your code can evaluate the condition and determine the reaction—even allow the CPU to continue or to safely trigger a stop.

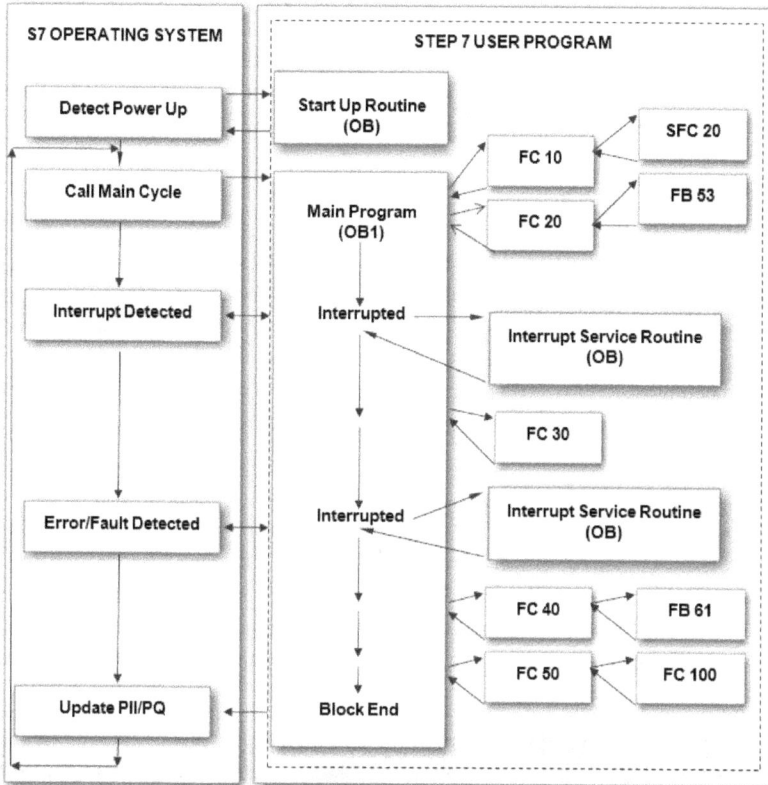

Figure 29-1. Illustration of S7 Program Structured Processing; and Organization Blocks as the Interface between the S7 Operating System and the User Program

OBs and Triggering Events

In the initial definition of organization blocks in Chapter 27 *Designing and Implementing the Step 7 Program*, it was noted that each start-up OB was triggered by a specific CPU start-up type; in a similar fashion, each OB is triggered to process its task based on a specific event to which it is tied. When that event occurs, the S7 operating system responds by calling the correct OB. The actual response is based on code that has been written in the OB. This code is generally referred to as a *service routine*.

By implementing the service routine in the various organization blocks, you determine what code is processed at start-up (OB100, OB101, and OB102), in the event of timed interrupts (OB10-to-OB17) or hardware interrupts (OB40-to-OB47), and based on various system faults (OB80-to-OB87). The code written in OB1 for example is called the main program, and as such it is processed by the CPU on a cyclical basis. In Table 29-1, the general call event is given for each OB category.

Writing Code for Organization Blocks

Generally speaking, you determine which organization blocks are required in your program. Because OB1 is the main program block, from which other blocks are called, it is always required. Whether other specific OBs should be used is dependent on the processing requirements of the program—for example the need to service process loops at precise intervals, the need to respond immediately to process interrupts, or to respond to certain system-related faults—all require specific OBs.

To respond to the events that trigger the call of an OB, you must create the code and transfer the block to the CPU. The code may be written directly in the OB, or in a block called from the OB. Regardless of where the code is written, it should be kept to a minimum. When an OB trigger event occurs, during normal operations, the operating system will interrupt OB1, or whatever block is executing, and call the associated OB. Certain OBs, if not installed when called, cause the CPU to switch to STOP.

Interrupting the Normal CPU Cycle

With the exception of the start-up OBs and the background processing OB, the processing of all other OB tasks require interruption of the normal cycle of OB1. Start-up OBs are called just prior to the normal cycle and I/O update; and background processing shares the normal cycle, after OB1 finishes the minimum cycle requirement. When other events occur, the operating system responds by interrupting OB1 and calling the appropriate OB. When the interrupting OB is done, control returns to OB1. To manage how other OBs may interrupt OB1, as well as one another, the CPU uses a priority scheme. This priority scheme allows OBs of higher priority to interrupt OBs of lower priority.

Priority Classes

The orderly execution of OBs is managed based on the priority class to which each OB is assigned. Priority classes are an established hierarchy of service levels that determine how one OB may interrupt the execution of another. The priority of an OB is based on assigne default values, as shown in Table 29-2. With priorities from 1-to-28—priority 28 is the highest. OB 90 has a special priority of 29, which is interpreted as 0.29, making it the lowest priority—even lower than OB 1, which has a priority of 1.

Table 29-2. Priority Classes of Organization Blocks

OB Type	OB Number (s)	Default Priority	Alt. Priorities
Normal Processing	1	1	Fixed
Time-of-Day Interrupts	10, 11, 12, 13, 14, 15, 16, 17	2	2-24
Time-Delay Interrupts	20, 21, 22, 23	3, 4, 5, 6	2-24
Cyclic Interrupts	30, 31, 32, 33, 34, 35, 36, 37, 38	7, 8, 9, 10, 11, 12, 13, 14, 15	2-24
Hardware Interrupts	40, 41, 42, 43, 44, 45, 46, 47	16, 17, 18, 19, 20, 21, 22, 23	2-24
Multi-Computing OB	60	25	Fixed
Synchronous Cycle Interrupts	61, 62, 63, 64	25	2-26
Technology Synch Interrupt	65	25	Fixed
Redundancy Errors	70, 72/73	25/28	2-26/2-28
Asynchronous Errors	80/81, 82, 83, 84, 85, 86, 87	26 (26 = RUN; 28 = STARTUP)	
Background Processing	90	.29	None
Start-Up OBs	100, 101, 102	27	Fixed
Synchronous Errors	121-122	Priority class of error-causing block	

Priority Defaults and Alternates

The priority is fixed for OB 1, OB 60, OB 90, and for start-up OBs. Time-of-day interrupts OBs all share a default priority of 2—since only a single occurrence is possible at any one time. Time-delay, cyclic, and hardware interrupt OBs are each assigned a unique priority. Time-delay OBs 20, 21, 22, and 23, for example, are assigned priorities of 3, 4, 5, and 6 respectively.

Redundancy error OB 70 has a default priority of 25, and OB 72/73 share a default priority of 28. Asynchronous errors all share priority 26, unless the error occurs during start-up—then priority 28 is assigned. Finally, synchronous error OBs (121 and 122) are called when either a programming or I/O access error occurs while processing the program—the synchronous error OB will execute at the priority class of the block in which the error occurred.

If an OB of higher priority is called while one of lower priority is being processed, then the lower priority OB is interrupted at the completion of the current STEP 7 operation. After the interrupting OB is done, the interrupted OB resumes at the point immediately following the point of interruption. If a number of OBs—for example OB10-OB17, are assigned the same priority, they do not interrupt each other, but instead are processed in sequence. You should avoid assigning the same priority to interrupts, since interrupts can be lost if at least twelve interrupts occur simultaneously.

Start Information

Each OB reserves 20-bytes of pre-defined local memory, as temporary variables, to which the operating system writes during each call. The variable names in each OB are the default names, but they may be modified as required to more suitable names. This *start information* essentially provides a log for data recorded when the OB is called. In each OB, for instance, the first start byte reports the OB start event class; the second byte reports a specific start event. Other reported information includes errors or faults that may have occurred, status and diagnostic information, and the date and time the OB was called. You may view OB start information by opening the block to the declaration area of the OB.

Contents Of: 'Environment\Interface\TEMP'			
Name	Data Type	Address	Comment
OB1_EV_CLASS	Byte	0.0	Bits 0-3 = 1 (Coming event), Bits 4-7 = 1 (Even...
OB1_SCAN_1	Byte	1.0	1 (Cold restart scan 1 of OB 1), 3 (Scan 2-n of...
OB1_PRIORITY	Byte	2.0	1 (Priority of 1 is lowest)
OB1_OB_NUMBR	Byte	3.0	1 (Organization block 1, OB1)
OB1_RESERVED_1	Byte	4.0	Reserved for system
OB1_RESERVED_2	Byte	5.0	Reserved for system
OB1_PREV_CYCLE	Int	6.0	Cycle time of previous OB1 scan (milliseconds)
OB1_MIN_CYCLE	Int	8.0	Minimum cycle time of OB1 (milliseconds)
OB1_MAX_CYCLE	Int	10.0	Maximum cycle time of OB1 (milliseconds)
OB1_DATE_TIME	Date_And_Time	12.0	Date and time OB1 started

Figure 29-2. Structure of the 20-byte OB start information for OB1

Your specific use of an OB will dictate whether you actually need to evaluate the start information. By evaluating the start information, you can determine how to react in your code—for example whether to report specific faults, or determine if a CPU STOP should be initiated. In several of the OBs, the start variable OBxx_EV_CLASS will simply indicate the call of a standard OB; in other OBs, however, this variable may report one of several possible values to indicate the exact nature of the event—for example one value may report '*event entering*' and another reports '*event leaving*'.

While the temporary start information of an OB may be accessed and evaluated while the OB is being executed, you may also evaluate start information from another block, using system function (SFC6 RD_SINFO). When SFC 6 is executed, it reads and returns the start information of the OB that is currently processing (top of the organization block nest).

START-UP BLOCKS

Application Overview

Start-up of an S7 CPU is an operating mode event whereby the CPU is caused to reinitialize itself prior to transitioning to a run state from a stopped state. A start-up may be triggered by manual or automatic means including: *1) CPU power-up, 2) switching the CPU mode from STOP to RUN/RUN-P, 3) a start-up command request from a PG/PC, or from a remote communication partner, 4) a synchronization of CPUs in multi-computing mode, or 5) an initial link establishment in an H-system stand-by unit.* Often start-up OBs are used to reset values or initialize process data values.

Whether manual or automatic, the various start-up events are all start-up requests to the CPU. And, while there are several ways to trigger a start-up, there are only three functional start-up types—*cold restart, complete restart,* and *warm restart.* As shown in the start information of Table 29-4, a start-up may be manual or automatic. Start-up types are differentiated by the initializations that the CPU performs with regard to the I/O update, the point of restart in the program, and the resetting or retaining of specific memory areas. Table 29-3 lists each start-up OB and its start-up request type.

Table 29-3. Start-Up Organization Blocks

OB No.	Priority	Brief Description
100	27	Service Routine—Complete Restart Request (also called hot restart)
101	27	Service Routine—Warm Restart Request
102	27	Service Routine—Cold Restart Request

Typically, you might use the start-up OB code to initialize function modules (FMs) or data values that will be used in the main program or subsequent block calls. For example, preset values or set points may be needed for certain machine or process requirements. Depending on the specific start event, and the application, the start-up code may be slightly different, or the same in all three start-up OBs.

Configuration and Start-up

Unlike most other OBs, start-up OBs do not require configuration to get started. You simply write the OB code and download it to the CPU. There is, however, a CPU dialog that lets you set parameters that affect the start-up behavior. For example, you may wish to set the type that is triggered on each power up. Remember that the startup type affects what the CPU does during the initialization.

If the checkbox box *"Start-up When Expected/Actual Configuration Differ"* is checked, modules in the hardware configuration table are checked against the installed configuration. Each module must exist and must match the station configuration. Any discrepancy would prevent the CPU from starting. To ensure error-free start-up, you can also set monitoring times for the maximum time permitted for the configuration parameters to be transferred to the modules, and for the maximum time permitted for the CPU to receive a 'ready for operation' signal from the modules after power up.

Start Information

Start information for start-up OBs is shown in Table 29-4. When the OB is active, reported by start variable OBxxx_EV_CLASS, the variable OBxxx_STRT_UP reports one of several possible values that indicates the exact startup type—a value of B#16#81 for example, indicates that a manual warm restart was requested. The start variable OBxxx_STOP reports the actual cause of the last CPU stop (also entered in the diagnostic buffer). Additional startup information is reported in the variable OB100_STRT_INFO. Start variable OBnnn_DATE_TIME reports the exact date and time that the start-up OB was called.

Table 29-4. Start Information for Start-Up OBs. Default variable names for OB100 are shown here

Variable Name	Type	Local Variable Value and Description
OB100_EV_CLASS	BYTE	**Event class and identifiers:** Possible values below – B#16#13 = OB call is active
OB100_STRT_UP	BYTE	**Start-up request event:** Possible values below – B#16#81 = manual warm restart request B#16#82 = automatic warm restart request B#16#83 = manual hot restart request B#16#84 = automatic hot restart request B#16#85 = manual cold restart request B#16#86 = automatic cold restart request B#16#87 = master request for manual cold restart B#16#88 = master request for automatic cold restart B#16#8A = master request for manual warm restart B#16#8B = master request for automatic warm restart B#16#8C = standby request for manual restart B#16#8D = standby request for automatic restart
OB100_PRIORITY	BYTE	**Priority class:** 27
OB100_OB_NUMBR	BYTE	**OB number:** 100 (OB100), 101 (OB101), or 102 (OB102)
OB100_RESERVED_1	BYTE	**Reserved:**
OB100_RESERVED_2	BYTE	**Reserved:**
OB100_STOP	WORD	**STOP event:** Number of event that caused the CPU STOP
OB100_STRT_INFO	DWORD	**Start info.:** additional information about current start-up
OB100_DATE_TIME	DT	**Date and Time:** date-time stamp of start-up OB call

Operating Characteristics

When a start-up request occurs, the appropriate OB, if loaded to memory, is called and will execute its code once, and then stop. Although there could be as many as three installed start-up OBs—for example in the S7-400, only one is called at start-up, depending on the request type. OB100 is called after a *complete restart request;* OB101 after a *warm restart request;* and OB102 after a *cold restart request.* If either OB100 or OB102 is not installed when called, then OB1 commences immediately. If OB101 is not installed when called, then OB1 resumes at the point at which it was last interrupted.

Since the CPU cycle monitoring is not active during execution of start-up OBs, there are no restrictions to the code length or the execution time of a start-up OB. The execution of interrupt-driven OBs is not possible while a start-up request is being processed.

MAIN ORGANIZATION BLOCK—OB1

Organization Block 1 (OB1) is the main STEP 7 program block—called cyclically by the S7 operating system. From the time the CPU completes its startup, OB1 is processed continuously until the CPU is stopped or shutdown. After completing each OB1 cycle the CPU performs an I/O update and other system activities, then restarts the OB1 cycle. Since it is processed continuously, most blocks of a STEP 7 program are called for processing either directly or indirectly from OB1. Properly written, OB 1 will facilitate a well-organized program, whose processing tasks are easily identified and followed.

The S7 operating system calls OB1 for cyclical processing, based on either of the following events:

- CPU start-up processing has just completed
- Previous CPU operating cycle is completed

Start Information

Start information for OB1 is shown in Table 29-5. When OB1 is called (OB1_EV_CLASS = B#16#11), you might consider evaluating the variables OB1_SCAN_1, OB1_PREV_CYCLE, OB1_MIN_CYCLE, and OB1_MAX_CYCLE. OB1_SCAN_1 reports one of several possible values that identify the exact event (startup or previous OB1 cycle) that occurred just prior to the first cycle of OB1. A value of B#16#05—for example, reports that the first cycle of OB1 is following a stop of the primary master, and a switchover to the reserve master in an H-system configuration.

The variables OB1_PREV_CYCLE, OB1_MIN_CYCLE, and OB1_MAX_CYCLE report the CPU cycle in milliseconds for the previous cycle, and the minimum and maximum cycles since the last start-up.

Table 29-5. Main Cyclical Program—Organization Block

Variable Name	Type	Local Variable Value and Description
OB1_EV_CLASS	BYTE	**Event class and identifiers:** Possible values below – B#16#11 = OB call is active
OB1_SCAN_1	BYTE	**First OB1 scan after:** Possible values below – B#16#01 = completion of a warm restart B#16#02 = completion of a hot restart B#16#03 = completion of the main cycle B#16#04 = completion of a cold restart B#16#05 = switchover from previous master and start of the new master in a master-reserve configuration
OB1_PRIORITY	BYTE	**Priority class:** 29 (corresponds to priority 0.29)
OB1_OB_NUMBR	BYTE	**OB number:** 1 = OB1
OB1_RESERVED_1	BYTE	**Reserved:**
OB1_RESERVED_2	BYTE	**Reserved:**
OB1_PREV_CYCLE	INT	**Previous cycle:** actual cycle time of the previous OB1 cycle
OB1_MIN_CYCLE	INT	**Minimum cycle:** actual time in milliseconds for the minimum OB1 cycle time since the last startup
OB1_MAX_CYCLE	INT	**Maximum cycle:** actual time in milliseconds for the maximum OB1 cycle time since the last startup
OB1_DATE_TIME	DT	**Date and Time:** date-time stamp of when OB1 was last called

OB1 Operating Characteristics

After power is applied, the CPU is switched to RUN or RUN/P, and after the startup OB is completed, OB 1 is called and is processed cyclically until the CPU is switched to STOP—or power is removed. It is from here in OB1 that the main thread of your program will be called for processing. Here, program processing starts with the first operation and continues until the very last instruction is executed. As blocks such as FCs, FBs, SFCs, and SFBs are called, each is processed until terminated—and upon termination control returns to OB1—until the final OB1 operation is completed. When the current cycle of OB1 has finished its cycle, the operating system starts OB1 again.

Normal Cyclic Duties

After completing an OB1 cycle, and before restarting the next cycle, the operating system performs a series of duties that are in addition to the cyclical program scan of OB1. As with the normal program, these additional duties are carried out on each CPU cycle. Primary duties of the CPU cycle include OB1 execution, update of the process image of inputs and the process image of outputs, management of global I/O data transfer/receipt, and monitoring of the CPU cycle time.

The Maximum OB1 Cycle Time

The maximum cycle time for OB1 has a default preset of 150 ms in the CPU properties dialog. S7 monitors the cycle time to ensure that the maximum preset value is not exceeded. If your program exceeds the maximum cycle time for OB1, the operating system calls OB80 (time error OB); if OB80 is not installed, the CPU will switch to the STOP mode.

By setting a maximum cycle time, you are ensuring a known maximum response time in your program. You may use the set default of 150 ms, or you may modify this value to a longer or shorter cycle as required. In those situations where the cycle time is predictably longer than the set maximum cycle, you may also restart the cycle monitoring using system function SFC43 (RE_TRIGR). Retriggering the cycle effectively restarts the cycle monitoring timer—and thereby avoids exceeding the maximum cycle time, or the need to set a longer preset value as the default.

Minimum OB1 Cycle Time

When a minimum cycle time is set, the operating system delays the start of a new OB1 cycle, if necessary, until the minimum cycle time has been reached. Setting a minimum cycle time lets you ensure that the intervals between successive starts of the OB1 cycle are always the same. Defining a minimum cycle might also be used to eliminate excessive I/O image table updates, and potential timing issues, in programs where there is a very short cycle time.

When assigning the minimum cycle time, it is important to ensure that there is a measureable difference between the minimum cycle and the defined maximum cycle time—this will prevent the cycle time being exceeded unexpectedly in cases where SFC/SFB calls are made from OB90.

OB1 and Background Processing

If the time required for completing a given processing cycle (actual cycle time) is less than the set minimum cycle time, then the CPU will check to see if OB90 is installed. OB90, the background processing OB, is called whenever the actual cycle time of the current OB1 cycle is less than the defined minimum cycle time. If OB90 is not installed, then the CPU will wait until the set minimum cycle time has expired. OB90 is described in the following discussion.

BACKGROUND PROCESSING OB90

Application Overview

When OB1 execution is completed in less than the specified minimum cycle time, including all nested interrupts and system activities, then the operating system responds in one of the following ways:

- Call OB90 to initiate background processing

- If OB90 is not installed, delay the new start of OB1 until the minimum cycle time expires

Background processing using OB 90 is possible in the S7-400 and in some S7300 CPUs. OB90 is called for processing in those time slots in which the minimum cycle time allotted to OB1 is not fully used. With this unused time, you may process a section of your code that is considered non-critical. In such cases, OB90 will execute until finished or interrupted by OB1 at the end of the minimum cycle time. If not completed, OB90 will resume at the point at which it was last interrupted.

Table 29-6. Background Processing Organization Block

OB No.	Priority	Brief Description
90	29	Service Routine—Background Processing Request

Start Information

The OB90 start information is given in Table 29-7. In addition to evaluating byte-1, which indicates whether OB90 is active, you may also consider evaluating the variable OB90_STRT_INF. When OB90 is called, by evaluating byte-2 (OB90_STRT_INF), you can determine if the current start event is one after which OB90 should be started from the first instruction, or where last terminated. B#16#93, for example, reports that OB90 was previously loaded while the CPU was in RUN mode; B#16#92 reports that OB90 was called after a block that was being processed in OB90 was deleted or replaced.

Table 29-7. OB 90 Start Information

Variable Name	Type	Local Variable Value and Description
OB90_EV_CLASS	BYTE	Event class and identifiers: Possible values below – B#16#11 = OB call is active
OB90_STRT_INF	BYTE	Start request for OB90: Possible values below— B#16#91 = after a CPU restart B#16#92 = processed block deleted or replaced B#16#93 = re-installed OB90 on CPU while in RUN mode B#16#95 = after background cycle (OB90) ended and new cycle starts
OB90_PRIORITY	BYTE	Priority class: 29 (corresponds to priority 0.29)
OB90_OB_NUMBR	BYTE	OB number: 90 = OB90
OB90_RESERVED_1	BYTE	Reserved:
OB90_RESERVED_2	BYTE	Reserved:
OB90_RESERVED_3	INT	Reserved:
OB90_RESERVED_4	INT	Reserved:
OB90_RESERVED_5	INT	Reserved:
OB90_DATE_TIME	DT	Date and Time: date-time stamp of when OB90 was called

Operating Characteristics

When a minimum cycle time is set, the operating system will delay the start of a new OB1 cycle, if necessary, until the minimum cycle time is reached. If a given cycle is finished prior to reaching the minimum cycle time, the CPU calls OB90—in other words, OB90 is called whenever the actual cycle time of the current OB1 cycle is less than the defined minimum cycle time.

OB90 has priority class 29, which corresponds to priority 0.29—making it the lowest priority of all. This lowest priority class means that OB90 may be interrupted by any OB. When interrupted, OB90 processes the current operation before passing control to the interrupting block. Normally after an interruption, OB90 resumes processing following the statement at which it was interrupted. Under certain conditions, however, OB90 is processed from the first instruction instead of resuming at the point which it was interrupted—these conditions are indicated by the start variable OB90_STRT_INF.

The following example illustrates a structure for evaluating OB90 start information and processing the OB90 code. The code evaluates the start variable OB90_STRT_INF for two cases—if the comparison for either case is found True (1), then the appropriate code is processed—otherwise processing continues with the normal OB90 code line.

An important note for consideration involves SFCs and SFBs that are called from OB90—in such cases, these system functions are processed under the same priority of OB1, and therefore are not interrupted by OB1. Since the cycle time monitoring does not take OB90 into account, it is essential that there is considerable difference between the set minimum cycle time for OB1 and what you define as the maximum cycle time.

Example: OB90 Service Routine

	L	OB90_STRT_INF	// Load the start variable OB90_STRT_INF
	L	B#16#91	// Load byte constant B#16#91
	==I		// is OB90_STRT_INF = 91
	JC	SI91	// if equal go to label SI91
	L	OB90_STRT_INF	// Load the start variable OB90_STRT_INF
	L	B#16#92	// Load byte constant B#16#92
	==I		// is OB90_STRT_INF = 92
	JC	SI92	// if equal go to label SI92
	JU	NORM	// End background processing
SI91			// Process OB90 after warm restart, cold restart or hot restart
			//
			//
	BEU		// End code for this case
SI92			// Process OB90 after block deleted
			//
			//
	BEU		// End code for this case
NORM			// Normal OB90 program
			//

TIME-OF-DAY INTERRUPT OB

Application Overview

Time-of-day (TOD) interrupt OBs can be configured to interrupt the main program (OB1) and process a section of code once, or at specified intervals, on a specific date and at a specific time. For example, you might need to process code that generates daily messages, collects hourly readings, or outputs monthly production reports. Depending on the specific CPU, up to eight TOD interrupts may be available—as listed in Table 29-8. In most S7-300 CPUs, for example, only OB10 is available.

Table 29-8. Time-of-Day Interrupt Organization Blocks

OB No.	Priority	Alternate Priorities	Brief Description
10	2	2-24	Service Routine—Time-of-day interrupt- 0
11	2	2-24	Service Routine—Time-of-day interrupt- 1
12	2	2-24	Service Routine—Time-of-day interrupt- 2
13	2	2-24	Service Routine—Time-of-day interrupt- 3
14	2	2-24	Service Routine—Time-of-day interrupt- 4
15	2	2-24	Service Routine—Time-of-day interrupt- 5
16	2	2-24	Service Routine—Time-of-day interrupt- 6
17	2	2-24	Service Routine—Time-of-day interrupt- 7

Configuration and Start-up

You can configure the TOD interrupt to execute once, or at periodic intervals, upon reaching the scheduled start time and date. If execution is set for once, the OB is called at the configured date and time, and is processed once and then is terminated; if you set the OB to execute at periodic intervals, you may set an interval of *every minute, hourly, daily, weekly, monthly, end of each month,* or *yearly.*

You may set and activate the TOD parameters using the CPU properties dialog, shown below, or using system functions from within your control program. Once the TOD interrupt OB is configured, you must create the code and transfer the block to the CPU. Remember, that the CPU's system clock must be set accurately in order to correctly process the time-of-day interrupts.

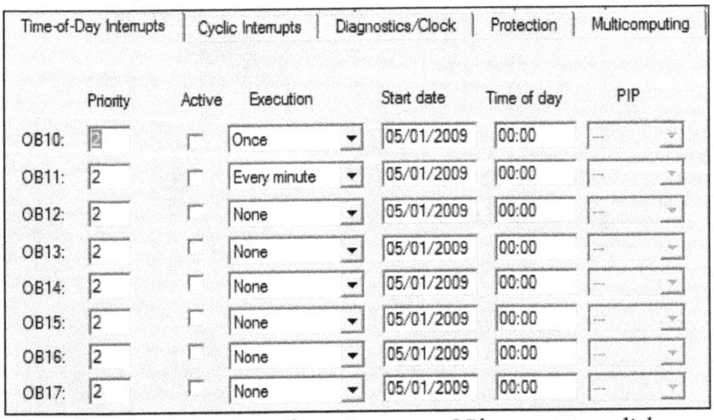

Figure 29-3. STEP 7 Time-of-Day Interrupt OB's parameters dialog

Once downloaded, the TOD interrupt will automatically execute at the schedule date and time if you configured the parameters and activated the OB from the CPU properties dialog. You also have the option of setting the parameters in STEP 7, and then calling SFC30 (ACT-TINT) to activate the OB from your program. A third option is to both set the parameters and activate the OB from the program; this is done by calling SFC28 (SET_TINT), and subsequently calling SFC30 (ACT_TINT).

Start Information

The start information for time-of-day interrupt OBs is given in Table 29-9. When a TOD interrupt event is called (OBxx_EV_CLASS = B#16#11), the byte-2 variable reports a value between 11_{16} and 18_{16}, to indicate a specific OB (OB10 through OB17) for which the TOD start request was made. The following two bytes report the priority class and OB number. The variable OBxx_PERIOD_EXE reports a value that indicates the execution interval—for example, W#16#0000 says the OB will process once and stop; W#16#1001 says the OB will process on a daily interval starting at the specified time.

Table 29-9. Time-of-Day OB Start Information—default variable names for OB10 are shown here

Variable Name	Type	Local Variable Value and Description
OB10_EV_CLASS	BYTE	**Event class and identifiers:** B#16#11 = interrupt is active
OB10_STRT_INFO	BYTE	**Start request info for time-of-day OBs:** Possible values below – B#16#11 = start request for OB10; B#16#15 = start request for OB14 B#16#12 = start request for OB11; B#16#16 = start request for OB15 B#16#13 = start request for OB12; B#16#17 = start request for OB16 B#16#14 = start request for OB13; B#16#18 = start request for OB17
OB10_PRIORITY	BYTE	**Priority class:** default value = 2 (valid alternates = 2-24)
OB10_OB_NUMBR	BYTE	**OB Number:** values = 10 (OB10) to 17 (OB17)
OB10_RESERVED_1	BYTE	**Reserved:**
OB10_RESERVED_2	BYTE	**Reserved:**
OB10_PERIOD_EXE	WORD	**Periodic execution:** Possible values below – W#16#0000 = once and stop; W#16#1201 = once weekly W#16#0201 = once a minute; W#16#1401 = once monthly W#16#0401 = once an hour; W#16#1801 = once yearly W#16#1001 = once a day; W#16#2001 = once end of month
OB10_RESERVED_3	INT	**Reserved:**
OB10_RESERVED_4	INT	**Reserved:**
OB10_DATE_TIME	DT	**Date and Time:** date-time stamp of when the OB was called

Operating Characteristics

When a configured TOD interrupt is triggered, the main cyclical program in OB 1 is interrupted and the time-of-day interrupt OB is called for processing. Based on its configuration, an installed TOD OB will execute once and only once and stop; or if configured to occur periodically, then execution may occur based on the scheduled interval. If a TOD interrupt is set to execute once, the OB instance will not be called if the scheduled date and time are already past when the block is downloaded. If the interrupt is set to execute at intervals, the OB will be called at the next scheduled interval, if the start date and time have already past when the block is downloaded.

Once a TOD interrupt OB is installed, you may manage various aspect of its execution from within your program using system functions (SFCs) SFC39 through SFC42. For instance, by calling SFC39 (DIS_IRT), you may disable a TOD interrupt; by calling SFC40 (EN_IRT), you may enable a disabled time-of-day interrupt; by calling SFC41 (DIS_AIRT), you may delay a TOD interrupt; and finally, by calling SFC42 (EN_AIRT) you may enable a previously delayed TOD interrupt.

TIME-DELAY INTERRUPT OB

Application Overview

A time-delay (TD) interrupt OB is used when parts of your control program must be called for processing based on a time-delay that is not subject to any inaccuracy that may result from the normal program cycle. By calling the appropriate system function (SFC), you can trigger a time-delay interrupt with a specified delay and OB number. After the delay expires, the specified OB is called. Depending on the S7 CPU, up to four time-delay interrupt OBs, numbered OB20 to OB23 are possible. Which OBs are actually available depends on the CPU. In most S7 300 CPUs, only OB20 is available.

Table 29-10. Time-Delay Interrupt Organization Blocks

OB No.	Priority	Alternate Priorities	Brief Description
20	3	2-24	Service Routine—Time-delay interrupt- 0
21	4	2-24	Service Routine—Time-delay interrupt- 1
22	5	2-24	Service Routine—Time-delay interrupt- 2
23	6	2-24	Service Routine—Time-delay interrupt- 3

Configuration and Start-up

Unless the priority of the time-delay OB needs to be modified, there are no other parameters on the properties dialog to be set. The default priorities, 3 through 6, may be substituted for an alternate (2-4) in the S7 400. The S7 300 only supports use of OB 20, for which the priority is fixed at 3.

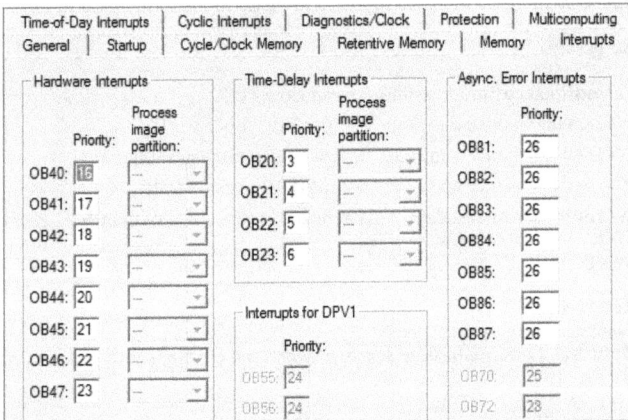

Figure 29-4. STEP 7 dialogue for Time-Delay Interrupt OBs

Once you have written and downloaded the time-delay OB, getting it to execute is simply a matter of calling system function SFC32 (SRT_DINT) to trigger the time-delay interrupt that will call the OB. Although the triggering of this OB is slightly different from how other OBs are called, time-delay OBs are still called by the operating system—you are simply triggering the start of the time delay.

When you enter code to call SFC32, you must specify the time-delay OB number, the delay-time in milliseconds, and a user-defined identifier as input parameters to the function. The identifier is the parameter whose name is 'SIGN'. This same variable is written to the OB start information variable OBnn_SIGN. This variable acts as an identifier in your program when more than one TD interrupt is being used.

Start Information

The time-delay OB start information is shown in Table 29-11. When the OB is active, byte-2 reports a value between 21_{16} and 24_{16}, to indicate which time-delay OB has been requested. The following two bytes report the priority class and OB number. The delay time that was specified in the call to SFC32 is reported in the OBnn_DTIME variable. The variable OBnn_SIGN reports back the identifier that was specified as an input parameter in the call to SFC32. You can evaluate this value to determine which of several time-delay jobs (calls to SFC32) may have been requested in the program.

Table 29-11. Time-Delay Interrupt OB Start Information—default variable names for OB20 are shown

Variable Name	Type	Local Variable Value and Description
OB20_EV_CLASS	BYTE	**Event class and identifiers:** Possible values below – B#16#11 = interrupt is active
OB20_STRT_INFO	BYTE	**Start request info for time-delay OBs:** Possible values below – B#16#21 = start request for OB20; B#16#23 = start request for OB22; B#16#22 = start request for OB21; B#16#24 = start request for OB23
OB20_PRIORITY	BYTE	**Priority class:** Possible values below – default values = 3 (OB20) to 6 (OB23)
OB20_OB_NUMBR	BYTE	**OB number:** values = 20 (OB20) to 23 (OB23)
OB20_RESERVED_1	BYTE	**Reserved:**
OB20_RESERVED_2	BYTE	**Reserved:**
OB20_SIGN	WORD	**User specified ID:** input parameter SIGN from call of SFC32 (SRT_DINT)—a user job identifier
OB20_DTIME	TIME	**Delay time:** elapsed delay time in milliseconds
OB20_DATE_TIME	DT	**Date and Time:** date-time stamp of when the OB was called

Operating Characteristics

During normal operation, an installed time-delay OB is executed only if the CPU is in RUN mode. After the TD interrupt is triggered and the delay expires, the associated OB is called for processing. While a time delay may actually be triggered during the start-up routine, the CPU must be in RUN mode by the time the delay expires. After the delay expires, the OB is called, and, according to priority interrupts, whatever block is being processed. If the time-delay OB, specified with the call of SFC32, is not loaded, the OB priority error OB (OB85) is called. The CPU goes to STOP if OB85 is not loaded.

After a time-delay interrupt OB is installed, you may manage various aspect of its execution from your program using system functions (SFCs) SFC39 through SFC42. For instance, by calling SFC39 (DIS_IRT), you may disable a time-delay interrupt; by calling SFC40 (EN_IRT), you may enable a disabled time-delay interrupt; by calling SFC41 (DIS_AIRT), you may delay a time-delay interrupt; and finally, by calling SFC42 (EN_AIRT) you may enable a previously delayed time-delay interrupt.

Additional system functions available for managing time-delay interrupts include SFC32, SFC33, and SFC 34. For example, a delay time that has expired can be re-started immediately with a call to SFC32 (SRT_DINT). If a triggered time-delay expires and the corresponding OB is still processing, then the timing error OB (OB80) is called. You can call SFC33 (CAN_DINT) to cancel the execution of a time-delay interrupt that has not yet started. Finally, you can query the status of a time-delay interrupt with a call to SFC34 (QRY_DINT).

CYCLIC INTERRUPT OB

Application Overview

A cyclic or 'watchdog' interrupt, is generated by the CPU at fixed periodic intervals. You would use cyclic interrupt OBs to create routines that handle those parts of your program that must be processed regularly, at precise intervals, and unaffected by the normal program cycle. For example, there may be control loops that need to be processed once every 500 milliseconds. That code, perhaps written in a Function Bock (FB), would be called from OB33.

Depending on the CPU, up to nine cyclic interrupt OBs may be available—as listed in Table 29-12. Your CPU will dictate which of these OBs are actually available. In most S7 300 CPUs, only OB35 is available. Each cyclic interrupt OB is listed with its default interval, priority, and alternate priorities.

Table 29-12. Cyclic Interrupt Organization Blocks

OB No.	Priority	Alternate Priorities	Interval	Brief Description
30	7	2-24	5 sec	Service Routine—Periodic cycle of 5 sec
31	8	2-24	2 sec	Service Routine—Periodic cycle of 2 sec
32	9	2-24	1 sec	Service Routine—Periodic cycle of 1 sec
33	10	2-24	500 ms	Service Routine—Periodic cycle of 500 ms
34	11	2-24	200 ms	Service Routine—Periodic cycle of 200 ms
35	12	2-24	100 ms	Service Routine—Periodic cycle of 100 ms
36	13	2-24	50 ms	Service Routine—Periodic cycle of 50 ms
37	14	2-24	20 ms	Service Routine—Periodic cycle of 20 ms
38	15	2-24	10 ms	Service Routine—Periodic cycle of 10 ms

Configuration and Start-up

Configuration and start-up of a cyclic interrupt involves determining which OB will be used, modifying default parameters, if required, creating the code and downloading the OB to your program. The cyclic interrupt OB parameters are set from the CPU Properties dialog.

Where it can be changed, the execution interval is entered in milliseconds. If you configure several cyclic interrupts, you should avoid assigning the same priority to multiple interrupts, and consider entering a phase offset value for each. The phase offset allows the cyclic interrupts to be distributed throughout the CPU cycle. Without a phase offset, cyclic interrupts could be triggered simultaneously, in which case, each is processed in order according to its priority.

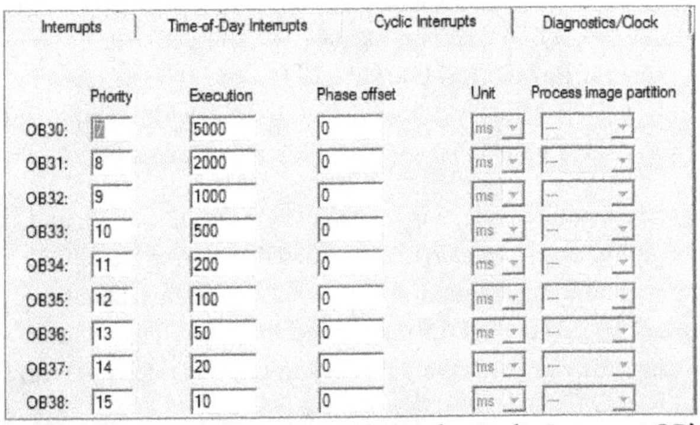

Figure 29-5. STEP 7 Properties dialog for Cyclic Interrupt OB's

470

Start Information

The start information of a cyclic interrupt OB is given in Table 29-13. The event class byte reports a value of 11, to indicate that the OB call is active. Byte-2, reports one of several possible values that indicate the exact cycle type—this information may be helpful in determining what initializations have taken place, or are still required in the cyclic OB. The following two bytes report the priority class and OB number. The phase offset and period interval are both reported in milliseconds.

Table 29-13. Cyclic Interrupt OB Start Information—default variable names for OB35 are shown here

Variable Name	Type	Local Variable Value and Description
OB35_EV_CLASS	BYTE	**Event class and identifiers:** Possible values below – B#16#11 = interrupt is active
OB35_STRT_INF	BYTE	**Start request info for cyclic OBs:** Possible values below – B#16#30 = (special criteria in H-CPUs if explicitly configured) B#16#31 = start request for OB30; B#16#36 = start request for OB35 B#16#32 = start request for OB31; B#16#37 = start request for OB36 B#16#33 = start request for OB32; B#16#38 = start request for OB37 B#16#34 = start request for OB33; B#16#39 = start request for OB38 B#16#35 = start request for OB34; B#16#3A = (see note) **Note:** Start request for cyclic interrupt OB with special criteria (supported by S7-300 only and then only if explicitly configured)
OB35_PRIORITY	BYTE	**Priority class:** Possible values below – default values = 7 (OB30) to 15 (OB38)
OB35_OB_NUMBR	BYTE	**OB number:** values = 30 (OB30) to 38 (OB38)
OB35_RESERVED_1	BYTE	**Reserved:**
OB35_RESERVED_2	BYTE	**Reserved:**
OB35_PHASE_OFFSET	WORD	**Phase offset:** value in milliseconds
OB35_RESERVED_3	INT	**Reserved:**
OB35_EXC_FREQ	INT	**Periodic interval:** value in milliseconds
OB35_DATE_TIME	DT	**Date and Time:** date-time stamp of when the OB was called

Operating Characteristics

When the CPU mode selector is switched from STOP to RUN, the interval clock starts timing for each configured cyclic interrupt OB. Each programmed cyclic OBs is called at the prescribed interval. OB33, for instance will execute once every 500 milliseconds; OB35, every 100 milliseconds. The configured execution intervals and phase offsets will ensure equidistant start times of the cyclic interrupts.

You must ensure that the run time of each cyclic interrupt OB is significantly shorter than its interval, allowing it to complete before starting again. If a cyclic interrupt OB has not been completely executed before it is due to start again, OB80, the time error OB will be triggered. The error-causing cyclic interrupt will be queued to execute later. The CPU will assume the STOP state if OB80 is not installed.

Once a cyclic interrupt OB is installed, you may manage various aspect of its execution from within your program using system functions (SFCs) SFC39 through SFC42. For instance, by calling SFC39 (DIS_IRT), you may disable a cyclic interrupt; by calling SFC40 (EN_IRT), you may enable a disabled cyclic interrupt; by calling SFC41 (DIS_AIRT), you may delay a cyclic interrupt; and finally, by calling SFC42 (EN_AIRT) you may enable a previously delayed cyclic interrupt.

HARDWARE INTERRUPT OB

Application Overview

This group of OBs, in conjunction with hardware interrupts, supports the response to critical events in the machine or process. For example, in response to an over pressure signal, the code can react by sounding an alarm, generating a message, or disabling specific outputs. Hardware interrupts may be generated by a process input to a module, or directly by a module that can generate an interrupt—for example an analog module that reports an over or under limit. Such modules include analog signal modules (SMs), function modules (FMs), and communications processors (CPs).

Up to eight hardware interrupt OBs, as listed below, may be available depending on the CPU—for example, only OB40 is available in most S7 300 CPUs.

Table 29-14. Hardware Interrupt Organization Blocks

OB No.	Priority	Alternate Priorities	Brief Description
40	16	2-24	Service Routine—Hardware interrupt- 0
41	17	2-24	Service Routine—Hardware interrupt- 1
42	18	2-24	Service Routine—Hardware interrupt- 2
43	19	2-24	Service Routine—Hardware interrupt- 3
44	20	2-24	Service Routine—Hardware interrupt- 4
45	21	2-24	Service Routine—Hardware interrupt- 5
46	22	2-24	Service Routine—Hardware interrupt- 6
47	23	2-24	Service Routine—Hardware interrupt- 7

Configuration and Start-up

The first task of configuring hardware interrupts is to define the module interrupt parameters using the hardware configuration tool. If a communications processor (CP) or function module (FM) supports interrupts, then parameters for these modules are defined using the CP/FM configuration tool. For each module, you'll need to enable the module hardware or diagnostic interrupt option, and specify the channel and condition that will trigger the interrupt.

As you enable and configure each interrupt, you'll need to use the module's *Addresses* dialog to specify which OB (OB40-OB47) will respond to the interrupt. Depending on your configuration, a single interrupt OB may be triggered by multiple hardware interrupts.

Start Information

The hardware interrupt OB start information is shown in Table 29-15. When byte-1 reports an active interrupt, the value in byte-2 indicates the line on which the interrupt was generated—in the S7-300 only one interrupt line is available; four lines are available in the S7-400. The following two bytes report the OB number and priority. Byte-6 reports whether the interrupt was generated by an input or output module. The next four bytes report the base address of the module that produced the interrupt. The variable OBnn_POINT_ADDR provides the bit status of up to 32 digital points.

Table 29-15. Hardware Interrupt OB Information—default variable names for OB40 are shown

Variable Name	Type	Local Variable Value and Description
OB40_EV_CLASS	BYTE	**Event class and identifiers:** Possible values below – B#16#11 = interrupt is active
OB40_STRT_INF	BYTE	**Start request info :** Possible values below – B#16#41 = interrupt active on interrupt line 1 (S7300 or S7-400); B#16#42 = interrupt active on interrupt line 2 (S7-400); B#16#43 = interrupt active on interrupt line 3 (S7-400); B#16#44 = interrupt active on interrupt line 4 (S7-400) B#16#45 = interrupt triggered active on PC (WinAC)
OB40_PRIORITY	BYTE	**Priority class:** Possible values below – defaults values = 16 (OB40) to 23 (OB47)
OB40_OB_NUMBR	BYTE	**OB Number:** values = 40 (OB40) to 47 (OB47)
OB40_RESERVED_1	BYTE	**Reserved:**
OB40_IO_FLAG	BYTE	**Module type:** B#16#54 = input module; B#16#55 = output module
OB40_MDL_ADDR	WORD	**Module address:** logical base address of interrupt triggering module
OB40_POINT_ADDR	DWORD	**Point Address for digital modules:** bit status of each point bit-0 to bit-31 (Bit 0 = first module point) **Point Address for analog modules (including CP and FM):** module interrupt status (not relevant)
OB40_DATE_TIME	DT	**Date and Time:** date-time stamp of when the OB was called

Operating Characteristics

Only after entering the RUN mode are hardware interrupts processed. Interrupts triggered during a start-up, prior to RUN, will be lost. When a hardware interrupt occurs, the main cyclical program (OB1) is interrupted, and according to priority, a call is made to the appropriate interrupt OB. If the OB has been installed, it is ready to act as the service routine; if not, OB85 is called; if OB 85 is not installed, then the CPU switches to STOP mode.

On each interrupt, a channel-specific acknowledgement is issued after the OB has executed. If prior to receiving the acknowledgement another interrupt is triggered on the same module/channel, then the new interrupt is lost; if the new interrupt is on the same module, but a different channel, then the interrupt is triggered after the active interrupt is acknowledged. If an OB is active when a new interrupt is triggered on a different module, then the interrupt is serviced only when the interrupt OB is done processing.

Once a hardware interrupt OB is installed, you may manage various aspect of its execution from within your program using system functions (SFCs) SFC39 through SFC42. For instance, by calling SFC39 (DIS_IRT), you may disable a hardware interrupt; by calling SFC40 (EN_IRT), you may enable a disabled hardware interrupt; by calling SFC41 (DIS_AIRT), you may delay a hardware interrupt; and finally, by calling SFC42 (EN_AIRT), you may enable a previously delayed hardware interrupt.

MULTI-COMPUTING INTERRUPT OB60

Application Overview

Using the appropriate CPUs, a multiple-CPU configuration, of up to four S7 CPUs in a rack, is possible in the S7-400. In this type of configuration, the operating modes of the CPUs are always synchronously switched—that is to say that by design, the CPUs are always started together and always stopped together. This switching characteristic makes the multi-computing interrupt an important feature—especially in a critical situation where one of the partners must be stopped.

It is not only essential that participating CPUs be notified in a critical situation, but also that a coordinated response is taken. With the multi-computing interrupt, you can ensure that a critical event in one CPU can result in a synchronized reaction in all of the CPUs. What actions or events may warrant such an action will depend on your configuration and application—one such reason for this kind of system interlocking is described below.

In a multi-CPU arrangement, the complete I/O configuration is generally shared by the participating CPUs. Each installed module (CP, FM, or SM) is assigned to a specific CPU. This arrangement is in contrast with a single-CPU station in which the single CPU governs the entire I/O configuration. In the single-CPU configuration, response to a critical hardware or diagnostic interrupt is left solely to the single CPU—even if the response means shutting down the controller. On the other hand, in a multi-computing configuration whether the controller should be shut down or continue to run will depend on the requirements of all of the CPUs in question. Given the occurrence of a critical condition or event, all of the CPUs must be notified, and a synchronized response is required.

Table 29-16. Multi-Computing Interrupt Organization Block

OB No.	Priority	Brief Description
60	25	Service Routine—CPU interrupt in a Multi-CPU Configuration

Configuration and Start-up

OB60 is the only OB for developing a service routine to handle interrupts generated by the CPUs in a multi-CPU configuration. Unlike with other OBs, there is no explicit configuration required for OB60. As each CPU is installed, it is assigned a CPU number from 1-to- 4. The CPU number corresponds to one of four possible I/O interrupt lines (INT-1, INT-2, INT-3, and INT-4) that can be assigned to a CPU. If the need to initiate a multi-computing interrupt in any one of the CPUs arises, then it initiates the interrupt on its own interrupt line. OB60 is simultaneously triggered in all of the CPUs.

To configure each CPU to respond to an interrupt, generated in itself or in another CPU, you simply need to create the required service routine in OB60, and download the code to each CPU. In order to ensure a synchronized response, OB60 should be the same in each CPU. The multi-computing interrupt is generated when a CPU makes a call to system function SFC35 (MP_ALM). Each CPU, therefore, must incorporate this call somewhere within its program. The call would be triggered by a specific interrupt condition in the local CPU. The result is a synchronized call to OB60 on all of the participating CPUs.

Start Information

The start information for the multi-computing interrupt is given in Table 29-17. When the call to the OB is active, OB60_EV_CLASS = B#16#11, byte-2 will indicate whether the interrupt was generated in the local CPU or one of the other CPUs. This byte-2 value may affect what the local response is in OB60, as well as the response that may be common to up to three other CPUs. The start variable OB60_JOB contains the job ID variable, a value written by the CPU that triggered the interrupt. This value may be used in OB60 to affect what the common response is in all of the CPUs.

Table 29-17. OB 60—Multi-Computing Interrupt Start Information

Variable Name	Type	Local Variable Value and Description
OB60_EV_CLASS	BYTE	**Event class and identifiers:** Possible values below – B#16#11 = Interrupt is active
OB60_STRT_INF	BYTE	**Start request info for multi-computing OB:** Possible values below— B#16#61 = interrupt triggered by own CPU B#16#62 = interrupt triggered by another CPU
OB60_PRIORITY	BYTE	**Priority class:** default = 25
OB60_OB_NUMBR	BYTE	**OB number:** 60 = OB60
OB60_RESERVED_1	BYTE	**Reserved:**
OB60_RESERVED_2	BYTE	**Reserved:**
OB60_JOB	INT	**Job ID:** input variable named JOB from call to SFC35
OB60_RESERVED_3	INT	**Reserved:**
OB60_RESERVED_4	INT	**Reserved:**
OB60_DATE_TIME	DT	**Date and Time:** date-time stamp of when OB60 was called

Operating Characteristics

A multi-computing interrupt is triggered in a CPU when the local program makes a call to SFC35. The triggering of the interrupt is recognized simultaneously in each of the partner CPUs (on same bus segment), by the synchronized starting of OB60. The response is based on the OB60 program in the local, and each of the partner, CPUs.

As part of the call to SFC35, you must supply a job ID using the parameter variable named 'JOB'. This job ID is transferred to all of the CPUs as a TEMP variable in the start information of OB60. The name of the variable, as seen in Table 29-17, is OB60_JOB. This variable may be used in each of the local OB60 service routines to allow each CPU to evaluate and react to a specific event, indicated by the job ID. Up to 16 unique values (0-15) for this variable may be specified when SFC35 is called.

If the code in OB60 is different in the partner CPUs, the result may be different execution times for the OB in each CPU. In such a case, the CPUs will return to the interrupted priority class at different times. If a subsequent multi-computing interrupt is generated by a CPU while OB60 in one of the partners is still executing from the previous multi-computing interrupt, then OB60 is not started in the triggering CPU, nor in any of the partner CPUs. This occurrence is reported in the return function value (RET_VAL) of SFC35. If OB60 is not installed at the time of the call to SFC35, the CPU returns program execution to the priority class that was executing prior to the interrupt.

As with other interrupt OBs, system functions (SFCs) SFC39 through SFC42 may be used to manage various aspects of multi-computing interrupt execution within your program. For instance, by calling SFC39 (DIS_IRT), you may disable a multi-computing interrupt; by calling SFC40 (EN_IRT), you may enable a disabled multi-computing interrupt; by calling SFC41 (DIS_AIRT), you may delay a multi-computing interrupt; and finally, by calling SFC42 (EN_AIRT) you may enable a previously delayed multi-computing interrupt.

SYNCHRONOUS CYCLE INTERRUPT OB

Application Overview

With the use of a synchronous cycle interrupt OB, you have the facility to synchronize the start of a section of your program with the DP cycle or PN send clock. OB61-64 serves as an interface OB to the synchronous cycle interrupt TSAL1. The default priority for synchronous OBs is 25; however you may use the configuration dialog to set the priority from 2 to 26.

When using synchronous cycle OBs, you are cautioned against using the load (L) and transfer (T) operations to access peripheral I/O areas (for example L PIW 4 or T PQW 4) whose process image partition has been allocated to a synchronous OB. This same caution should include the use of system functions SFC14 (DPRD_DAT) and SFC15 (DPWR_DAT).

Table 29-18. Synchronous Cycle Interrupt OBs

OB No.	Priority	Brief Description
61 thru 64	2-26	Service Routine—Synchronizing Program Section with DP cycle

The start information for the synchronous cycle interrupt OBs is shown in Table 29-19—the table specifically lists the default variable names of OB61.

Table 29-19. CPU Hardware Fault OB Start Information—default variable names for OB 84 shown

OB No.	Type	Local Variable Value and Description
OB61_EV_CLASS	BYTE	**Event class and identifiers:** Possible values below – B#16#11 = incoming event, event class 1
OB61_STRT_INFO	BYTE	**Start information ID code:** Possible values below – B#16#64 = Start request for OB61; B#16#65 = Start request for OB62 B#16#66 = Start request for OB63; B#16#67 = Start request for OB64
OB61_PRIORITY	BYTE	**Priority class:** default = 25 = fixed
OB61_OB_NUMBR	BYTE	**OB number:** Possible values below = 61 to 64
OB61_RESERVED_1	BYTE	**Reserved:**
OB61_RESERVED_2	BYTE	**Reserved:**
OB61_GC_VIOL	BOOL	**GC violation:** in PROFIBUS DP
OB61_FIRST	BOOL	**First use of OB:** after STARTUP or STOP mode
OB61_MISSED_EXEC	BYTE	**Missed executions:** Number of failed startup attempts of OB61 since its last execution.
OB61_DP_ID	BYTE	**DP ID:** ID of the DP master system in isochronous mode (1 to 32) or PROFINET IO system ID of the PNIO system in isochronous mode (100 to 115)
OB61_RESERVED_3	BYTE	**Reserved:**
OB61_RESERVED_4	WORD	**Reserved:**
OB61_DATE_TIME	DT	**Date and Time:** date-time stamp of when OB84 was called

TECHNOLOGY SYNCHRONIZATION INTERRUPT OB65

Application Overview

With the use of the technology synchronization interrupt OB (OB65), you have the facility to start a section of your program at the same time the technology blocks are updated. This feature is only applicable to so-called Technology CPUs. OB65 is actually called immediately after the technology blocks are updated. The priority class of OB65 is fixed at priority 25.

Table 29-20. Technology Synchronization Interrupt OB

OB No.	Priority	Brief Description
65	25	Service Routine—Synchronizing Program Section with Technology Blocks Update

The start information for the technology synchronization interrupt OB is shown in Table 29-21.

Table 29-21. CPU Hardware Fault OB Start Information—default variable names for OB 65 shown

OB No.	Type	Local Variable Value and Description
OB65_EV_CLASS	BYTE	**Event class and identifiers:** Possible values below – B#16#11 = incoming event, event class 1
OB65_STRT_INFO	BYTE	**Start Info:** Possible values below – B#16#6A: Start request for OB65
OB65_PRIORITY	BYTE	**Priority class:** default = 25 = fixed
OB65_OB_NUMBR	BYTE	**OB number:** 65 = OB65
OB65_RESERVED_1	BYTE	**Reserved:**
OB65_RESERVED_2	BYTE	**Reserved:**
OB65_RESERVED_3	BOOL	**Reserved:**
OB65_FIRST	BOOL	**First use of OB:** First use of OB65 after startup
OB65_MISSED_EXEC	BYTE	**Mixed executions:** Number of failed startup attempts of OB65 since its last execution.
OB65_RESERVED_4	BYTE	**Reserved:**
OB65_RESERVED_5	BYTE	**Reserved:**
OB65_RESERVED_6	WORD	**Reserved:**
OB65_DATE_TIME	DT	**Date and Time:** date-time stamp of when OB65 was called

ASYNCHRONOUS ERROR OB

Application Overview

Asynchronous errors in S7-300 and S7-400 controllers are system-related errors that can occur independently of the program scan—these faults may occur at any point and at time during program execution. Typical examples of faults that occur in the controller itself include defective modules, rack and power supply faults, and communications link or interface faults.

The occurrence of an asynchronous error will result in the interruption of OB1, and a CPU call to the appropriate error OB. Asynchronous error OBs are listed in Table 29-22, along with the default priority class. Generally, if an asynchronous error OB is not installed when called, the CPU will switch STOP—that is with the exception of OB70, OB72, OB81, and OB 87.

Also counted as asynchronous error OBs, on the CPU properties dialog, are redundancy error OBs. These are faults that may occur in an S7 H-system or F-system. Such faults may occur as a result of loss of CPU redundancy, loss of I/O system redundancy, or loss of redundancy in communications links. Depending on the S7-400 CPU, up to three redundancy error OBs (OB70, 72, and 73) are available.

Table 29-22. Asynchronous Error Organization Blocks

OB No.	Priority	Brief Description
70	25	Service Routine—I/O Redundancy Error (H-System only)
72	28	Service Routine—CPU Redundancy Error (H-System only)
73	0	Service Routine—Communications Redundancy Error (H-System only)
80	26 or 28	Service Routine—Organization Block Timing Error
81	26 or 28	Service Routine—Power Supply Fault
82	26 or 28	Service Routine—Diagnostics
83	26 or 28	Service Routine—Insert/Remove Module
84	26 or 28	Service Routine—CPU Hardware Fault
85	26 or 28	Service Routine—Priority Class Execution Error
86	26 or 28	Service Routine—Rack fault (expansion rack, DP master, distributed I/O station)
87	26 or 28	Service Routine—Communications Error
88	28	Service Routine—Processing Interrupt

Managing Asynchronous Error OBs

As with hardware, cyclic, and TOD interrupts, calls to asynchronous error OBs (exceptions OB70, OB72, OB73) can also be managed using system functions SFC39 to SFC42. With a program call to SFC39 (DIS_IRT), you can disable servicing of all new asynchronous errors; using a call to SFC40 (EN_IRT), you may enable asynchronous errors that were disabled. You may also delay an asynchronous error with a call to SFC41 (DIS_AIRT); and finally, a call toSFC42 (EN_AIRT) is used to enable delayed asynchronous errors.

I/O Redundancy Error—OB70

The operating system of an S7 H-System triggers a call to OB70 when there is an error associated with a critical element in its redundant I/O system. Failure in any of the elements, such as those in the following list, would trigger a call of OB70. OB70 is called both for the event entering, as well as the event leaving.

- Loss of Profibus DP redundancy
- Bus failure in active DP master
- DP slave interface module error

In the event of an I/O redundancy error, and OB70 is not installed, the H-system CPU does not switch to STOP—but instead continues to operate in RUN. If OB70 is installed, and the H system is in the redundant mode, OB70 is triggered on both CPUs. The H-system remains in the redundant mode.

OB 70 Start Information

OB70 start information is shown in Table 29-23. The variable OB70_EV_CLASS reports a value of 72_{16}, indicating that an asynchronous error OB start event is entering; or a value of B#16#73 indicating that a previous event is leaving. The variable OB70_FLT_ID, reports a value of (B#16#A2) when the event is a failure of a DP master, or a DP master system; or a value of (B#16#A3) when a redundancy loss or return of DP slave redundancy occurs. The event is further identified in the variables OB70_INFO_1, OB70_INFO_2, and OB70_INFO_3, based on the specific fault ID reported by OB70_FLT_ID.

Table 29-23. I/O Redundancy Error OB Start Information

Variable Name	Type	Local Variable Value and Description
OB70_EV_CLASS	BYTE	**Event class and identifiers:** **B#16#72** = event incoming; **B#16#73** = event outgoing
OB70_FLT_ID	BYTE	**Fault identification code:** Possible values below – B#16#A2 = Failure of a DP master or a DP master system B#16#A3 = Redundancy Loss/return of DP slave redundancy
OB70_PRIORITY	BYTE	**Priority class:** default = 25
OB70_OB_NUMBR	BYTE	**OB number:** 70 = OB70
OB70_RESERVED_1	WORD	**Reserved**
OB70_INFO_1	WORD	**Based on OB70_FLT_ID as given below:** If OB70_FLT_ID = **B#16#A2**—see following: OB70_INFO_1: = logical base address of the affected DP master If OB70_FLT_ID = **B#16#A3**—see following: OB70_INFO_1: = logical base address of the associated DP master

OB70_INFO_2	WORD	**Based on OB70_FLT_ID as given below:**
		If OB70_FLT_ID = **B#16#A2**—see following:
		OB70_INFO_2: = Reserved
		If OB70_FLT_ID = **B#16#A3**—see following:
		OB70_INFO_2: = affected DP slave as shown below:
		  bits 0-14 = logical base address if S7 slave used, or
		  bits 0-14 = diagnostic address if a DP normal slave is used
		  bits 15 = I/O identification: 1 = input module; 0 = output module

OB70_INFO_3	WORD	This variable is reported based on the fault code ID as shown below: If OB70_FLT_ID = **B#16#A2**—see following: OB70_INFO_3: bits 0-7 = Reserved OB70_INFO_3: bits 8-15 = DP master system ID of affected DP master If OB70_FLT_ID = **B#16#A3**—see following: OB70_INFO_3: bits 0-7 = DP slave station number OB70_INFO_3: bits 8-15 = DP master system ID
OB70_DATE_TIME	DT	**Date and Time:** date-time stamp of when OB70 was called

CPU Redundancy Error—OB72

The operating system of an S7 H-System triggers a call to OB72 whenever there is an error associated with a critical element of its redundant communications system. Failure in any of the elements, such as those listed below, would trigger a call of OB72.

- Loss of CPU Redundancy
- Redundancy Synch Sub-Module Error
- Reserve-Master Switchover Error
- Updating Aborted
- Synchronization Error
- Reserve-Master Memory Comparison Error

OB72 is executed by all CPUs that are in the RUN or STARTUP mode following an appropriate start event. In the event of a CPU redundancy error, and OB72 is not installed, the H-system CPU does not switch to STOP mode—it continues to operate in RUN mode.

OB 72 Start Information

OB72 start information variables are described in Table 29-24. The variable OB72_EV_CLASS reports a value of 72_{16}, to indicate that an OB start event is entering; or reports a value of B#16#73 to indicate that a previous start event is leaving. The start variable OB72_FLT_ID, reports one of several possible values to indicate a specific redundancy error event. The redundancy error event is further identified in the variables OB72_INFO_1, OB72_INFO_2, and OB72_INFO_3, based on the specific fault ID reported by OB72_FLT_ID.

Table 29-24. CPU Redundancy Error OB Start Information

Variable Name	Type	Local Variable Value and Description
OB72_EV_CLASS	BYTE	**Event class and identifiers:** Possible values below – B#16#73; B#16#75; B#16#78 = event outgoing; B#16#79 = event incoming
OB72_FLT_ID	BYTE	**Fault identification code:** Possible values below – B#16#01 = redundancy loss (1 of 2) due to a CPU failure B#16#02 = redundancy loss (1 of 2) due to user-triggered STOP on standby B#16#03 = H-System (1 of 2) changed to redundant mode B#16#20 = error in RAM comparison B#16#21 = error comparing process image output value B#16#22 = error comparing memory bits, timers, or counters B#16#23 = different operating system data recognized B#16#31 = standby-master switchover due to master failure B#16#33 = standby-master switchover due to system change at runtime B#16#34 = standby-master switchover due to sync module connection error B#16#40 = user program synchronization error due to elapsed wait time B#16#41 = user program synch error due to wait at different synch points B#16#42 = synch error in OS due to waiting at different synch. points B#16#43 = synchronization error in OS due to elapsed wait time B#16#44 = synchronization error in OS due to wrong data B#16#50 = No sync module present B#16#51 = modification at sync module without power ON B#16#52 = Sync module removed/synch module inserted (B#16#79/B#16#78) B#16#53 = modification at synch module without reset B#16#54 = Sync module: rack number assigned twice

Variable Name	Type	Local Variable Value and Description
		B#16#55 = Sync module error/module error removed (B#16#79/B#16#78) B#16#56 = illegal rack number set on the sync module B#16#C1 = update attempt aborted; B#16#C2 = update abort due to monitoring time exceeded during n-th attempt (1 <= n <= max. possible number of update attempts after an abort due to the monitoring time being exceeded.)
OB72_PRIORITY	BYTE	**Priority class:** default = 28
OB72_OB_NUMBR	BYTE	**OB number:** 72 = OB72

OB72_RESERVED_1	WORD	Valid for FLT_ID = B#16#03:
		High byte: ID for the content of OB72_INFO_2 and OB72_INFO_3 as follows:
		  B#16#0 = OB72_INFO-2/OB72_INFO_3 = no significance
		  B#16#C4 = Transition to redundant mode after troubleshooting mode was carried out with standby-master switch-over (if OB72_INFO_3=W#16#0001) or without standby-master switch-over (if OB72_INFO_3=W#16#0002).
		
		OB72_INFO_2 is reserved
		B#16#CD = OB72_INFO_2/OB72_INFO_3 contain the actual lock time (in msec) for priority classes > 15
		
		Low byte: reserved

OB72_INFO_1	WORD	**Additional Info- 1:** Valid for FLT_ID = B#16#C2:
		High byte: Identifier values for the exceeded monitoring time as follows:
		  1 = cycle time extension;
		  2 = I/O dead time;
		  4 = communication delay
		
		Low byte = number of the current update attempt (e.g., attempt = 3)

OB72_INFO_2	WORD	**Additional Info- 2:** Valid for FLT_ID = B#16#03: and OB72_RESERVED_1=B#16#CD: High word of the actual lock time (in milliseconds) for priority classes > 15
OB72_INFO_3	WORD	**Additional Info- 3:** Valid for FLT_ID = B#16#03 and OB72_RESERVED_1 = B#16#CD: OB72_RESERVED_1 = B#16#C4: W#16#0001 = Transition to redundant operation after troubleshooting mode was carried out with standby-master switch-over. W#16#0002 = Transition to redundant operation after troubleshooting mode was carried out without standby-master switch-over. OB72_RESERVED_1 = B#16#CD: High word of the actual lock time (in msec) for priority classes > 15
OB72_DATE_TIME	DT	**Date and Time:** date-time stamp of when OB72 was called

Communications Redundancy Error—OB73

In an S7 H-system, the operating system calls OB73 upon the first loss of communications redundancy in a fault-tolerant S7 connection (fault-tolerant connections only support by S7 communications functions). Once a communications redundancy error has occurred, a subsequent loss of redundancy will not trigger a new call to OB73 until all fault tolerant connections are restored. In the event of a communications redundancy error, and OB73 is not installed, the H-system CPU does not switch to STOP mode—instead it continues to operate in RUN mode.

OB 73 Start Information

OB73 start information variables are described in Table 29-25. The variable OB73_EV_CLASS reports a value of 72_{16}, to indicate that a communications redundancy error start event is entering; or reports a value of B#16#73 to indicate that a previous start event is leaving. The start variable OB73_FLT_ID, reports one of several possible values to indicate a specific communications redundancy error event. The variables OB73_INFO_1, OB73_INFO_2, and OB73_INFO_3 are not relevant.

Table 29-25. Communications Redundancy Error OB Start Information

Variable Name	Type	Local Variable Value and Description
OB73_EV_CLASS	BYTE	**Event class and identifiers** B#16#72 = event entering; B#16#73 = event leaving
OB73_FLT_ID	BYTE	**Fault identification code:** Possible values below – B#16#E0 = communications redundancy loss/communications redundancy restored
OB73_PRIORITY	BYTE	**Priority class:** default = 28
OB73_OB_NUMBR	BYTE	**OB number:** 73 = OB73
OB73_RESERVED_1	WORD	**Reserved:**
OB73_INFO_1	WORD	The value in this variable is irrelevant to user
OB73_INFO_2	WORD	The value in this variable is irrelevant to user
OB73_INFO_3	WORD	The value in this variable is irrelevant to user
OB73_DATE_TIME	DT	**Date and Time:** date-time stamp of when OB73 was called

Timing Error—OB80

Timing errors that occur during the execution of any organization block will result in OB80 being called. Conditions that cause timing errors are list below. The condition listed first occurs when the normal program cycle of OB1 has exceeded its set maximum cycle time without acknowledgment.

- The value set as the maximum cycle time is exceeded

- Time-of-day interrupts skipped by moving the CPU time forward

- OB request error (same/other OB still processing) when OB called

In the case of the time-of-day interrupt, a normal call to a TOD interrupt OB would be skipped if you have set the CPU time forward to a time that is past a scheduled call to the OB. An example of the last OB request error would be a call to OB35 (100 ms periodic cycle), while the previous call to the OB35 is still executing. A program execution time longer than the periodic cycle time would result in this error.

If OB80 has not been programmed at the time an OB timing error occurs, then the CPU switches to the STOP mode. Furthermore, if OB80 is called twice in a single program cycle due to the scan time being exceeded, the CPU will also switch to STOP. If within your program there are certain program paths or predictable conditions under which an excessive scan time is expected, then you can retrigger the CPU cycle, at an appropriate point in the program, with a call to SFC43 'RE_TRIGR'.

OB80 Start Information

The OB80 start information variables are described is in Table 29-26. Although not shown below, the BYTE variable OB82_MOD_TYPE, reports as follows: bits 0-3 reports the module class; bit 4 reports that channel information exists; bit 5 reports that user information exists; bit 6 reports a diagnostic interrupt from a configured substitute channel; and bit 7 reports module repairs that are needed.

Table 29-26. Timing Error OB Start Information

Variable Name	Type	Local Variable Value and Description
OB80_EV_CLASS	BYTE	**Event class and identifiers:** B#16#35
OB80_FLT_ID	BYTE	**Fault identification code:** Possible values below – B#16#01 = cycle time exceeded; B#16#02= OB still in execution; B#16#05 = Elapsed TOD interrupt due to moving clock forward B#16#06 = Elapsed TOD interrupt on return to RUN after HOLD B#16#07 = Elapsed TOD interrupt on return to RUN after HOLD. B#16#08, B#16#09, B#16#0A, B#16#0B
OB80_PRIORITY	BYTE	**Priority class:** possible values below - priority = 26—in RUN mode; priority = 28—in start-up mode; priority = 28—in OB request buffer overflow
OB80_OB_NUMBR	BYTE	**OB number:** 80 = OB80
OB80_RESERVED_1	BYTE	**Reserved:**
OB80_RESERVED_2	BYTE	**Reserved:**
OB80_ERROR_INFO	WORD	**Error information:** dependent on fault code
OB80_ERR_EV_CLASS	BYTE	**Error event class:** class of the event causing fault (OB)
OB80_ERR_EV_NUM	BYTE	**Error event no.:** event number of the event causing fault
OB80_OB_PRIORITY	BYTE	**Error information:** value based on priority of the error causing OB
OB80_OB_NUM	BYTE	**Error information:** value based on number of the error causing OB
OB80_DATE_TIME	DT	**Date and Time:** date-time stamp of when OB80 was called

Power Supply Fault—OB81

The power supply fault OB is called upon fault events that occur in the supply voltage or battery backup in a central rack or expansion rack. A power supply fault event is triggered when 1) at least one battery of two is expired; 2) back-up voltage is lost; or 3) 24V power is lost. OB81 is called if the fault is an entering event (incoming), or a leaving event (outgoing). This means that OB81 is also called upon the return of at least one previously expired battery, failed back-up voltage, or lost 24V supply. If there is a power supply fault, and OB81 is not installed, the CPU does not switch to STOP mode.

OB81 Start Information

Start information for OB81 is in Table 29-27. When a fault code ID is reporting an incoming/outgoing power supply fault, the reported error is in byte variable OB81_FLT_ID. Otherwise, faults that occur in expansion racks with the battery backup (B#16#31), back-up voltage (B#16#32), or a 24-V power supply (B#16#33) are reported in reserved variables OB81_RESERVED_4 to OB81_RESERVED_6. These variables, only relevant for fault IDs B#16#31, B#16#32, and B#16#33, indicate which expansion racks are involved. The bits set in these three bytes correspond to specific expansion racks. Start variables OB81_RESERVED_1, OB81_RESERVED_2, and OB81_RESERVED_3 are not shown.

When a power supply fault event occurs, a bit is set for each expansion rack in which at least one battery, back-up voltage, or 24V power supply has failed. The corresponding bit is also set for each expansion rack in which the previous fault is no longer present (event class = B#16#38). When the event is eliminated, and the backup restored on at least one expansion rack, the assigned bits are set.

Table 29-27. Power Supply Fault OB Start Information—default variable names are shown

Variable	Type	Local Variable Value and Description
OB81_EV_CLASS	BYTE	**Event class and identifiers:** Possible values below – B#16#38 = event outgoing; B#16#39 = event incoming
OB81_FLT_ID	BYTE	**Fault identification code:** Fault codes below report faults in either a standard CPU rack (**CR**) or redundant CPU rack (**RCR**) B#16#21 = at least one battery expired/problem removed in CR B#16#22 = back-up voltage loss/problem removed in CR B#16#23 = 24V power supply loss/problem removed in CR B#16#25 = at least one battery expired/problem removed in RCR B#16#26 = back-up voltage loss/problem removed in RCR B#16#27 = 24V power supply loss/problem removed in RCR
OB81_PRIORITY	BYTE	**Priority class:** = 26, in RUN mode; = 28, in START-UP mode
OB81_OB_NUMBR	BYTE	**OB number:** 81 = OB81
OB81_RACK_CPU	WORD	**Bits 0 to 7:** always = B#16#00; **Bits 8 to 15:** as shown below: **- in standard CPU—Bits 8 to 15** = B#16#00 **- in H-system CPU: Bits 8 to 10** = Rack no; **Bit 11** = 0 = Standby CPU, **Bit 11** = 1 = Master CPU; **Bits 12-15** = 1111
OB81_RESERVED_4	BYTE	Relevant for fault codes—B#16#31, B#16#32, and B#16#33 bit-0, bit-1 = Reserved; bit-2 to bit 7 = Expansion Rack 16 to Exp Rack 21
OB81_RESERVED_5	BYTE	Relevant for fault codes—B#16#31, B#16#32, and B#16#33 bit-0 to bit 7 = Expansion Rack 8 to Expansion Rack 15
OB81_RESERVED_6	BYTE	Relevant for fault codes—B#16#31, B#16#32, and B#16#33 bit-0 = Reserved; bit-1 to bit 7 = Expansion Rack 1 to Expansion Rack 7
OB81_DATE_TIME	DT	**Date and Time:** date-time stamp of when the OB81 was called

Diagnostic Interrupt—OB82

A diagnostic interrupt is generated on modules with diagnostic capability, and where the feature is enabled. When enabled, the module will trigger an interrupt and call to OB82—both in the case of an incoming event, and when that event is outgoing. An indication a module component needs repair is an example of incoming event; indication that no component requires repair is the event outgoing. If a diagnostic interrupt occurs, and OB82 is not installed, the CPU switches to STOP.

OB82 Start Information

OB82 start information is given in Table 29-28. The byte contents of the variable OB82_MOD_TYPE, not given in the table, reports as follows: bits 0-3 encodes the module class; bit 4 is '1' if channel information exists; bit 5 is '1' if user information exists; bit 6 reports a diagnostic interrupt from a configured substitute channel; and bit 7 is '1' if module repairs are required.

Table 29-28. Diagnostic Interrupt OB Start Information

Variable Name	Type	Local Variable Value and Description
OB82_EV_CLASS	BYTE	**Event class:** B#16#38 = event outgoing; B#16#39 = event incoming
OB82_FLT_ID	BYTE	**Fault identification code:** B#16#42
OB82_PRIORITY	BYTE	**Priority class:** default = 26—in RUN; default = 28—in START-UP
OB82_OB_NUMBR	BYTE	**OB number:** 82 = OB82
OB82_IO_FLAG	BYTE	**I/O Flag:** B#16#54 = input module; B#16#55 = output module
OB82_MDL_ADDR	WORD	**Module addr.:** logical base address of the module reporting a fault
OB82_MDL_DEFECT	BOOL	**Module defective:** defective = TRUE; not defective = FALSE
OB82_INT_FAULT	BOOL	**Internal fault:** fault = TRUE; no fault = FALSE
OB82_EXT_FAULT	BOOL	**External fault:** fault = TRUE; no fault = FALSE
OB82_PNT_INFO	BOOL	**Point (channel) fault:** fault = TRUE; no fault = FALSE
OB82_EXT_VOLTAGE	BOOL	**External voltage failure:** fault = TRUE; no fault = FALSE
OB82_ FLD_CONNCTR	BOOL	**Front connector:** not plugged-in—Fault = TRUE; No Fault = FALSE
OB82_ NO_CONFIG	BOOL	**No configuration:** Not Configured = TRUE; Configured = FALSE
OB82_ CONFIG_ERR	BOOL	**Config. error:** faulty parameters = TRUE; good parameters =FALSE
OB82_ MDL_TYPE	BYTE	**Module type:** See 'Start Information' above
OB82_ SUB_MOD_ERR	BOOL	**Submodule missing or error:** fault = TRUE; no fault = FALSE
OB82_ COMM_FAULT	BOOL	**Communication fault:** fault = TRUE; no fault = FALSE
OB82_ MDL_STOP	BOOL	**Module operating mode:** STOP = TRUE; RUN = FALSE
OB82_WTCH_DOG_FLT	BOOL	**Watchdog timer fault:** ault = TRUE; no fault = FALSE
OB82_INT_PS_FLT	BOOL	**Internal power supply fault:** fault = TRUE; no fault = FALSE
OB82_PRIM_BATT_FLT	BOOL	**Primary battery expired:** expired = TRUE; no expired = FALSE
OB82_BCKUP_BATT_FLT	BOOL	**Entire backup failed:** failed = TRUE; no failed = FALSE
OB82_RACK_FLT	BOOL	**Rack fault:** fault = TRUE; no fault = FALSE
OB82_PROC_FLT	BOOL	**Processor fault:** fault = TRUE; no fault = FALSE
OB82_EPROM_FLT	BOOL	**EPROM fault:** ault = TRUE; no fault = FALSE
OB82_RAM_FLT	BOOL	**RAM fault:** fault = TRUE; no fault = FALSE

Variable Name	Type	Local Variable Value and Description
OB82_ADU_FLT	BOOL	**A/D conv. or D/A conv.:** fault = TRUE; no fault = FALSE
OB82_FUSE_FLT	BOOL	**Fuse tripped:** tripped = TRUE; not tripped = FALSE
OB82_HW_INTR_FLT	BOOL	**Hardware interrupt lost:** lost = TRUE; not lost = FALSE
OB81_DATE_TIME	DT	**Date and Time:** date-time stamp of when OB82 was called

Insert/Remove Module Interrupt—OB83

In some S7 systems, modules may be inserted or removed during RUN, STOP, and STARTUP modes. Where the insert/removal feature is possible, it applies to all modules except for power supplies, CPUs, adapter modules, and interface modules. With the S7-300, module insertion and removal is only possible with certain CPUs—and then it is not possible with modules in any of the central expansion racks. The feature is only possible in the S7-300 with distributed PROFINET I/O racks.

When a module is either inserted or removed, during RUN, STOP, or STARTUP modes, an entry is made in the CPU's diagnostic buffer, and in the system status list of the affected CPU. If a configured module is removed or inserted, while in the RUN mode, OB83 is started. The CPU will switch to STOP if OB83 is not installed. When you re-insert a module into a configured slot while in RUN, the CPU verifies that the module matches the configuration. If the stored configuration matches the re-inserted module, OB83 is started and the module parameters are assigned.

The operating system monitors module presence at one second intervals. To ensure that module removal and re-insertion is detected, you must wait at least two seconds between removal and re-insertion. Because of this one second monitoring interval, an access error may not be detected until a direct access in the program, or during the I/O image update.

OB83 Start Information

OB83 start variables are described in Table 29-29. The variable OB83_EV_CLASS will report one of several possible values—for example, a value of B#16#32 reports the end of module re-assignment; B#16#38 reports that a module was inserted, and B#16#39 that a module was removed. The start event class identifiers also combine with the variable OB83_FLT_ID to report one of several possible module conditions. These faults are described in Table 29-30. On each call of OB83, the module in question is further identified using the remaining start variables starting with OB83_RESERVED_1, and ending with the variable OB83_MDL_TYPE, based on the specific fault ID reported by OB83_FLT_ID.

Table 29-29. Insert/Remove Module Interrupt OB Start Information—default variable names shown

OB No.	Type	Local Variable Value and Description
OB83_EV_CLASS	BYTE	**Event class and identifiers:** Possible values below – B#16#32 = module parameter re-assignment end (outgoing) B#16#33 = module parameter re-assignment start (incoming) B#16#38 = module inserted; B#16#39 = module removed; module not responding, or end parameter assignment
OB83_FLT_ID	BYTE	**Fault identification code:** Possible values below – B#16#51, B#16#54, B#16#55, B#16#56, B#16#58, B#16#61, B#16#63, B#16#64, B#16#65, B#16#66, B#16#67, B#16#68
OB83_PRIORITY	BYTE	**Priority class:** possible values below – default = 26—in RUN mode; default = 28—in START-UP mode
OB83_OB_NUMBR	BYTE	**OB number:** 83 = OB83

OB No.	Type	Local Variable Value and Description
OB83_RESERVED_1	BYTE	**Identify module or sub module/interface module:** possible values below - B#16#A0 ; B#16#C4; See the variable OB83_RACK_NUM below
OB83_MDL_TD	BYTE	**Module address range:** possible values below – **B#16#54** = peripheral input (PI); **B#16#55** = peripheral output (PQ)
OB83_MDL_ADDR	WORD	**Logical base address of the affected module:** Central/remote Profibus DP: logical base address of the module – for mixed module base address = smallest address on module; – for mixed module and input/output addresses are equal base address = the address that receives the event identifier. – for Distributed Profinet IO: base address of the module/sub module
OB83_RACK_NUM	WORD	**If OB83_RESERVED_1** = B#16#A0: See below for OB83_RACK_NUM: low-byte = sub module no. /interface sub module **If OB83_RESERVED_1** = B#16#C4: OB83_RACK_NUM value as follow: - fault in central rack: value = rack number (S7-400) - fault in Profibus DP: low-byte = station No; high-byte = DP Mast. Sys. ID - fault in Profinet IO: value = physical address as shown below: **bit 15 = 1** = Profinet IO; **bits 11-14** = IO Sys. ID; **bits 0-10** = device no.
OB83_MDL_TYPE	WORD	**Affected module type** in central rack/Profibus DP station (X = irrelevant): W#16#X5XX = analog module (SM); W#16#X8XX = function module W#16#XCXX = comm. processor (CP); W#16#XFXX = digital module **Affected module type** in PROFINET I/O: W#16#8101 = inserted module type = previously removed module type W#16#8102 = inserted module type ≠ previously removed module type
OB83_DATE_TIME	DT	**Date and Time:** date-time stamp of when OB83 was called

The following table shows fault code identification values (OB83_FLT_ID) for when OB83 is triggered. This list includes both events entering (B#16#39), and events leaving (B#16#38). You may evaluate these values in your OB80 service routine.

Table 29-30. Example incoming and outgoing triggering events for OB83

OB83_EV_CLASS	OB83_FLT_ID	Fault ID Description
B#16#39	B#16#51	PROFINET IO module removed
B#16#39	B#16#54	PROFINET IO submodule removed
B#16#38	B#16#54	PROFINET IO submodule inserted—and matches configuration
B#16#38	B#16#55	PROFINET IO submodule inserted –does not match configuration
B#16#38	B#16#56	PROFINET IO submodule inserted—but error with parameters
B#16#38	B#16#58	PROFINET IO submodule—access error corrected
B#16#39	B#16#61	Module removed or not responding **Also see OB83_MDL_TYPE**—it reports the actual module type
B#16#38	B#16#61	Module inserted—module type OK **Also see OB83_MDL_TYPE**—it reports the actual module type
B#16#38	B#16#63	Module inserted—but incorrect module type **Also see OB83_MDL_TYPE**—it reports the actual module type
B#16#38	B#16#64	Module inserted but problem (module ID cannot be read) **Also see OB83_MDL_TYPE**—it reports the configured module type
B#16#38	B#16#65	Module inserted—but error in parameter assignment

OB83_EV_CLASS	OB83_FLT_ID	Fault ID Description
		Also see OB83_MDL_TYPE—it reports the actual module type
B#16#39	B#16#66	Module not responding—load voltage error
B#16#38	B#16#66	Module now responding—load voltage error corrected
B#16#33	B#16#67	Start of module parameter re-assignment
B#16#32	B#16#67	End of module parameter re-assignment
B#16#39	B#16#68	Module re-parameterization terminated with error

CPU Hardware Fault—OB84

The operating system calls OB84 whenever a fault is detected in certain components of the CPU. The following conditions are cases of when OB84 is called—OB84 is called both when the fault condition is an entering event (incoming), and a leaving event (outgoing).

- Memory errors detected/corrected

- Reduced performance detected in redundant link between H-system CPUs

- Operating system error detected in WinAC RTX systems

In the event that a CPU hardware fault occurs, and OB84 is not installed, the CPU does not switch to the STOP mode—instead it continues operating in the RUN mode.

OB84 Start Information

The start information for OB84 is shown in Table 29-31. Note here that the event class and identifiers indicate when the event is incoming (B#16#39) and when the event is outgoing (B#16#38). The actual faults that you may evaluate as incoming or outgoing are indicated by the fault identification code—for example, the fault code is B#16#82 both when the memory error is detected (incoming), and when the memory error is corrected (outgoing). The date and time indicates when OB84 was called.

Table 29-31. CPU Hardware Fault OB Start Information—default variable names for OB 84 shown

OB No.	Type	Local Variable Value and Description
OB84_EV_CLASS	BYTE	**Event class and identifiers:** Possible values below – B#16#35, B#16#39 = event incoming B#16#38 = event outgoing
OB84_FLT_ID	BYTE	**Fault identification code:** Possible values below – B#16#82 = memory error detection and correction B#16#83 = multiple memory error detections and corrections B#16#85 = operation system error detection B#16#86 = negatively affected H-sync link detection B#16#87 = multi-bit memory error detection and correction
OB84_PRIORITY	BYTE	**Priority class:** possible values below – default = 26—in RUN mode; default = 28—in START-UP mode
OB84_OB_NUMBR	BYTE	**OB number:** 84 = OB84
OB84_RESERVED_1	BYTE	**Reserved:**
OB84_RESERVED_2	BYTE	**Reserved:**
OB84_DATE_TIME	DT	**Date and Time:** date-time stamp of when OB84 was called

Note: Start variables OB84_RESERVED_3 and OB84_RESERVED_4 are not shown in the table

Priority Class Execution Error—OB85

The S7 operating system calls the priority class execution error OB if one of the following events occur:

- Start Request Made to Uninstalled OB (Exceptions include OB80, OB81, OB82, OB83, and OB86)

- Error During a Block Call (for example, FB called but instance DB not installed)

- I/O Access Error During Normal I/O Image Update (OB85 call not suppressed in configuration)

If OB85 has not been installed, and a priority class execution interrupt is detected, then the CPU switches to STOP—that is with the above noted exceptions.

OB85 Start Information

The start information for OB85 is shown in Table 29-32. The event class and identifiers indicate when the event is incoming (B#16#39), and when the event is outgoing (B#16#38). The faults that you may evaluate as incoming or outgoing, when OB85 is called, are indicated by the fault identification code.

Table 29-32. Priority Class Error OB Start Information—default variable names of OB85 shown

OB No.	Type	Local Variable Value and Description
OB85_EV_CLASS	BYTE	**Event class and identifiers:** Possible values below – B#16#39 = event incoming; B#16#38 = event outgoing B#16#39 (only with faults B#16#B1, B#16#B2, B#16#B3, B#16#B4) B#16#38 (only with faults B#16#B3 and B#16#B4)
OB85_FLT_ID	BYTE	**Fault identification code:** Possible values below – B#16#A1 B#16#A2 B#16#A3 B#16#A4 B#16#B1 B#16#B2 B#16#B3 B#16#B4
OB85_PRIORITY	BYTE	**Priority class:** default = 26—in RUN mode; default = 28—in START-UP mode
OB85_OB_NUMBR	BYTE	**OB number:** 85 = OB85
OB85_ERR_EV_CLASS	BYTE	**Error event class:** class of the event causing fault
OB85_ERR_EV_NUM	BYTE	**Error event no.:** event number of the event causing fault
OB85_OB_PRIOR	BYTE	**OB priority:** Priority of the active OB on error occurrence
OB85_OB_NUM	BYTE	**OB number:** Number of the active OB on error occurrence
OB85_DATE_TIME	DT	**Date and Time:** date-time stamp of when OB85 was called

Rack Failure Interrupt—OB86

The S7 operating system calls OB86 whenever there is a failure detected in critical elements of the rack system components. These critical conditions, listed below, trigger a call of OB86 both when the fault condition is an entering event (incoming), or a leaving event (outgoing). If OB86 has not been installed, and a rack failure interrupt is detected, then the CPU switches to the STOP mode.

- Failure in S7-400 central expansion rack (power supply, communication link, faulted IM)

- Failure in a DP master system subnet

- Failure in a DP master station

- Failure in a PROFIBUS DP/PROFINET distributed I/O station (including S7-300/S7-400)

OB86 Start Information

The start information for OB86 is shown in Table 29-33. Note that the event class and identifiers indicate when an event is incoming (B#16#39), and when that same event is outgoing (B#16#38). The actual faults that you may evaluate are indicated by the fault identification code.

As it relates to the call of OB86, in some cases the coming and leaving of a fault is indicated by the same fault ID, such as with B#16#C1—in other cases, the coming and leaving of a specific fault is indicated by different fault IDs—as with B#16#C3 and B#16#C4. Still yet, there are those cases where an additional fault ID is used to indicate that a specific situation is operational, but specific problems still exist—as in the case of B#16#C2. Since multiple slave racks may be faulted, OB86 is also called with the return of every DP slave station.

Table 29-33. Rack Failure OB Start Information—default variable names of OB86 shown

OB No.	Type	Local Variable Value and Description
OB86_EV_CLASS	BYTE	**Event class and identifiers:** Possible values below – B#16#39 = event incoming; B#16#38 = event outgoing
OB86_FLT_ID	BYTE	**Fault identification code:** Possible values below – B#16#C1 = expansion rack fault/expan. rack operational—B#16#39/38 B#16#C2 = expan. rack operational –but expected/actual config differ B#16#C3 = DP master system fault—only with B#16#39 B#16#C4 = DP master system fault cleared—only with B#16#38 B#16#C4 = DP slave station fault—B#16#39/#38 B#16#C5 = return of DP slave station—but slave faulty; B#16#39/38 B#16#C6 = expan. rack operational—module still has parameter errors B#16#C7 = return of a DP-rack—module parameter assignment error B#16#C8 = return of a DP-rack—module config/actual config. differ (B#16#39) B#16#CA = PROFINET IO rack failure—only with B#16#39 B#16#CB = PROFINET IO rack failure/station return—B#16#39/38 B#16#CC = PROFINET IO rack return—but with problem—B#16#39/38 B#16#CD = PROFINET IO rack return—expected/actual config. differ B#16#CE = PROFINET IO rack return—but error in module config.
OB86_PRIORITY	BYTE	**Priority class:** default = 26 in RUN mode; default = 28 in START-UP
OB86_OB_NUMBR	BYTE	**OB number:** 86 = OB86
OB86_MDL_ADDR	WORD	Depends on error code
OB86_RACKS_FLTD	ARRAY	Depends on error code
OB86_DATE_TIME	DT	**Date and Time:** date-time stamp of when OB86 was called

Note: Start variables OB86_RESERVED_1 and OB86_RESERVED_2 are not shown in the table

Communication Error Interrupt—OB87

OB87 is called by the operating system whenever a communications error event occurs. These error events are listed in Table 29-34 as fault codes. If one of these events occurs in the S7-300, and OB87 has not been installed, the CPU switches to the STOP mode; in S7-400 controllers, the CPU does not switch to STOP—instead it continues in RUN mode.

OB87 Start Information

Start information for OB87 is in Table 29-34 with each variable described. Note that the variables whose names include 'reserved' are only relevant when the variable OB87_FLT_ID is reporting certain values. When the fault identification, for example, is reporting the values B#16#E1, E3, and E4—the variable OB87_RESERVED_3 reports a value that represents the interface ID as either K-bus or MPI.

Table 29-34. Communication Error OB Start Information—default variable names of OB87 shown

Variable	Type	Local Variable Value and Description
OB87_EV_CLASS	BYTE	**Event class and identifiers:** Possible values below – B#16#35; B#16#38 (only with OB87_FLT_ID—**fault ID** = B#16#DA) B#16#39 (only with OB87_FLT_ID—**fault ID** = B#16#DA)
OB87_FLT_ID	BYTE	**Fault identification code:** Possible values below – B#16#D2 = transmission of diagnostic entries currently not possible B#16#D3 = synchronization messages cannot be transmitted (master). B#16#D4 = illegal time-of-day jump due to clock synchronization. B#16#D5 = error when receiving synchronization time (slave). B#16#E1 = incorrect frame ID during global data communication. B#16#E2 = GD packet status cannot be entered in the DB B#16#E3 = frame length error during global data communication. B#16#E4 = illegal GD packet number received B#16#E5 = DB access error during data exchange via communication FB B#16#E6 = GD group status cannot be entered in DB
OB87_PRIORITY	BYTE	**Priority:** default = 26—in RUN mode; default = 28—in START-UP
OB87_OB_NUMBR	BYTE	**OB number:** 87 = OB87
OB87_RESERVED_1	BYTE	On block errors, the block type may be read as the following values: B#16#88 = **OB**; B#16#8A = **DB**; B#16#8C = **FC**; B#16#8E = **FB**
OB87_RESERVED_3	WORD	**Depends on OB87_FLT_ID:** possible values below – FLT_ID = B#16#D2,D3,D4,D5: **variable** = No further information FLT_ID = B#16#E1,E3,E4: **variable** = Interface ID: 0 = K-bus; 1 = MPI FLT_ID = B#16#E2: **variable** = No further information FLT_ID = B#16#E5: Then **variable** = Reserved for CPU use FLT_ID = B#16#E6: Then **variable** = No further information
OB87_RESERVED_4	DWORD	**Depends on OB87_FLT_ID:** possible values below – FLT_ID = B#16#D2,D3,D4,D5: **variable** = No further information FLT_ID = B#16#E1, E3, E4: **variable** = High-byte = GD circuit No.; Low-byte = No further information; FLT_ID = B#16#E2: **variable** = High word = No further information Low word = GD packet No. (low-byte); GD Circle No. (High byte) FLT_ID = B#16#E5: **variable** = Low word = Rel. Addr. of error-causing MC7 command; High word = Block Number of error-causing block FLT_ID = B#16#E6: Then **variable** = No further information
OB87_DATE_TIME	DT	**Date and Time:** date-time stamp of when OB87 was called

Processing Interrupt—OB88

When it comes to errors involving block processing, OB88 is generally called—for example, a block call is made to an FB whose instance DB is not installed; or a call is made to an FC that is not installed; even an OB that is triggered by an error event, but the OB is not installed. OB88 is also triggered by errors involving block calls—but these are slightly different. The S7 operating system calls OB88 after the execution of a block has been aborted. The following examples are cases for which OB88 is called:

- The nesting depth of synchronous errors is exceeded

- The nesting depth of block calls (I-Stack) is exceeded

- There is an runtime error during allocation of local data

If OB88 is not installed when it is triggered by the call of a block execution that is aborted, the CPU will switch to STOP mode. If execution of a block is aborted in priority class 28, the CPU switches to STOP mode. The start information variables for OB88 are described in Table 29-35.

Table 29-35. Diagnostic Interrupt OB Start Information

Variable Name	Type	Local Variable Value and Description
OB88_EV_CLASS	BYTE	**Event class and identifiers:** B#16#35
OB88_SW_FLT	BYTE	**Software fault identification code:** Possible values below – B#16#71: the nesting depth of the nesting stack is exceeded B#16#72: the nesting depth of the master control relay is exceeded B#16#73: the nesting depth of synchronous errors is exceeded B#16#74: the nesting depth of block calls (I stack) is exceeded B#16#75: the nesting depth of block calls (B stack) is exceeded B#16#76: Error during allocation of local data B#16#78: Unknown instruction B#16#7A: Jump instruction with destination outside of current block
OB88_PRIORITY	BYTE	**Priority class:** default = 28
OB88_OB_NUMBR	BYTE	OB number: 82 = OB88
OB88_BLK_TYPE	BYTE	**Block type:** type of block at which the error occurred; possible values below – B#16#88 = **OB** B#16#8C = **FC** B#16#8E = **FB** B#16##00 = Unable to determine the interrupt source
OB88_RESERVED_1	BYTE	**Reserved:**
OB88_FLT_PRIORITY	BYTE	**Fault priority:** priority class of the OB that has caused the error
OB88_FLT_OB_NUMBR	BYTE	**Fault OB number:** number of the OB that has caused the error
OB88_BLK_NUM	WORD	**Block number:** number of the block containing the instruction that has caused the error
OB88_PRG_ADDR	WORD	**Program address:** relative address of the error causing instruction
OB88_DATE_TIME	DT	**Date and Time:** date-time stamp of when OB87 was called

SYNCHRONOUS ERROR OB

Application Overview

In the S7-300 and S7-400, faults encountered while processing the user program are classified as synchronous errors. These errors are called synchronous because they occur as a direct result of processing the program—more specifically, an instruction. Synchronous errors are placed in two groups—*programming errors* are those faults that are related to processing a specific operation; and *I/O access errors* are those faults that are related to failed attempts to access an I/O module.

If a programming error occurs during execution, then OB 121 is called; if an I/O access error occurs, then OB 122 is called.

Synchronous error OBs are listed in the following table.

Table 29-36. Synchronous Error Interrupt Organization Blocks

OB No.	Priority	Brief Description
121	2	Service routine for programming error
122	2	Service routine for I/O access error

Managing Synchronous Errors

Synchronous errors occur during the execution of a particular instruction. When such errors occur, the operating system makes an entry in the I-stack and then starts the appropriate synchronous error OB. To respond to these errors, you must create the code in the appropriate organization block (OB121 or OB122) and transfer the block to the CPU. If the error occurs during normal operations, the operating system will interrupt whatever block is executing and call the associated organization block. A typical response may be to generate an operator message, or to ensure that certain process conditions are brought to an orderly and safe shutdown.

Each synchronous error OB is executed according to the same priority class as the block that was being executed when the error was detected. OB121 and OB122 can therefore access the values in the accumulators and other system registers as they were at the time when the error occurred. These values can be used to react to the error condition, and then to return to normal program processing. For example, if an access error occurs on an analog input module, a substitute value can be specified in OB122 using SFC44 RPL_VAL.

The following system functions are available for managing response to OB121 and OB122:

- SFC36 (MSK_FLT)—to masks specific error codes
- SFC37 (DMSK_FLT)—to unmasks error codes masked by SFC36
- SFC38 (READ_ERR)—to read the contents of the error register

Programming Error Interrupt—OB121

OB121 is called whenever a program-related processing error occurs—for example, if your program tries to access a byte memory address, using an incorrectly specified pointer; or a fully-addressed data location is accessed, but the data block is not installed. When OB121 is called, it is executed in the same priority class of the block in which the error occurs. If OB121 has not been installed, when called, then the CPU switches from RUN to STOP mode.

OB121 Start Information

OB121 start information variables are described in Table 29-37, with additional information described in Table 29-38. The start variable OBSW_FLT reports one of several possible values to indicate a specific type of programming access error event. The specific fault is further identified in the variable OB121_FLT_REG, based on the fault ID reported by OB121SW_FLT. Additional info is also provided in start variable OB121_RESERVED_1.

Table 29-37. Programming Error Interrupt Organization Block—default variable names for OB121

Variable	Type	Local Variable Value and Description
OB121_EV_CLASS	BYTE	**Event class and identifiers:** B#16#25—call programming error OB121
OB121_SW_FLT	BYTE	**Software fault code:** Possible values below – B#16#21, B#16#22, B#16#23, B#16#24, B#16#25, B#16#26, B#16#27, B#16#28, B#16#29, B#16#30, B#16#31, B#16#32, B#16#33, B#16#34, B#16#35, B#16#3A, B#16#3C, B#16#3D, B#16#3E, B#16#3F
OB121_PRIORITY	BYTE	**Priority:** priority class = priority of OB in which error occurred (that is the OB from which faulted block was called).
OB121_OB_NUMBR	BYTE	**OB number:** 121 = OB121
OB121_BLK_TYP	BYTE	**Block type:** in which error occurred—possible values below – B#16#88 = **OB**; B#16#8A = **DB**; B#16#8C = **FC**; B#16#8E = **FB** **Note:** No valid value is reported for the S7-300
OB121_RESERVED_1	BYTE	**For OB121_SW_FLT fault codes** = B#16#22, B#16#23, B#16#28, and B#16#29, OB121_RESERVED_1 reports the incorrect byte address, data area and access type in question. **OB121_RESERVED_1 is encoded as follows:** **Bits 7-4 = memory access type during error:** **value = 0** = bit access; **value = 1** = byte access; **value = 2** = word access; **value = 3** = double word access **Bits 3-0 = memory area:** **value = 0** = I/O; **value = 1** = input image; **value = 2** = output image **value = 3** = bit memory; **value = 4** = global DB; **value = 5** = instance DB **value = 6** = own local data; **value = 7** = local data of calling block
OB121_FLT_REG	WORD	**Fault register:** Exact fault source depends on value in OB121_SW_FLT: See Table 29-33
OB121_BLK_NUM	WORD	**Block number:** block number of block where error occurred (S7-400)
OB121_PRG_ADDR	WORD	**Program address:** relative address where the error occurred (S7-400)
OB121_DATE_TIME	DT	**Date and Time:** date-time stamp of when error was detected

The following table describes the meaning of the fault indication values that are reported in the fault register variable OB121_FLT_REG, based on the software fault code that is currently being reported by the variable OB121_SW_FLT. Generally speaking, the programming error will fall into one of the following fault categories, which you may evaluate in your OB121 service routine. When each of these errors occurs, an essential value regarding the error source is found in the variable OB121_FLT_REG.

- Register where BCD conversion error occurred
- Incorrect address during read/write access
- Incorrect area identification in the case of area access error
- Incorrect number of a referenced block, timer, or counter

Table 29-38. Programming errors that trigger OB121; OB121_SWFLT = Fault type

OB121_SW_FLT	Programming Fault Description	OB121_FLT_REG Contains
B#16#21	BCD Conversion error	= ID of concerned register W#16#0000: ACCU1
B#16#22	Area length error when reading	See Table 29-37 variable OB121_RESERVED_1.
B#16#23	Area length error when writing	
B#16#24	Range error when reading	Low byte = ID of memory area where range error occurred. For example, B#16#86 = in own local data area (See Table 29-37, OB121_RESERVED_1).
B#16#25	Range error when writing	
B#16#26	Error for timer number	= invalid timer number
B#16#27	Error for counter number	= invalid counter number
B#16#28	Read access error to byte, word, or double word – where the pointer bit address is not '0'	See Table 29-37 variable OB121_RESERVED_1.
B#16#29	Write access error to byte, word, or double word— where the pointer bit address is not '0'	
B#16#30	Write access error to write-protected global DB	= the protected DB number
B#16#31	Write access error to write-protected instance DB	= the protected DB number
B#16#32	DB number error accessing a global DB	= the invalid DB number
B#16#33	DB number error accessing an instance DB	= the invalid DB number
B#16#34	FC number error in call to FC	= missing FC number
B#16#35	FB number error in call to FB	= missing FB number
B#16#3A	Access to a DB that is not installed— the DB number is in the permitted range	= missing DB number
B#16#3C	Access to an FC that is not installed— the FC number is in the permitted range	= missing FC number
B#16#3D	Access to an SFC that is not installed— the SFC number is in the permitted range	= missing SFC number
B#16#3E	Access to an FB that is not installed— the FB number is in the permitted range	= missing FB number
B#16#3F	Access to an SFB that is not installed— the SFB number is in the permitted range	= missing SFB number

I/O Access Error Interrupt—OB122

The S7 operating system calls OB122 whenever an error occurs while directly accessing an I/O module to either read module data, or to write module data. When OB122 is called, it is executed in the same priority class of the block in which the error occurs. If OB122 has not been installed, when called, then the CPU switches from RUN to STOP mode.

OB122 Start Information

OB122 start information variables are described in Table 29-39. The start variable OBSW_FLT reports one of several possible values to indicate a specific type of programming access error event. The variable OB122_BLK_TYP tells whether the error occurred in an OB, FC, or FB. From the variable OB122_MEM_AREA, it reports the memory area (I/O, Input Image, Output Image), and whether the error involved bit, byte, word, or double word access. The remaining variables MEM_ADDR, BLK_NUM, and PRG_ADDR are described in the Table 29-39.

Table 29-39. I/O Access Error Interrupt Organization Block—default variable names for OB122

Variable	Type	Local Variable Value and Description
OB122_EV_CLASS	BYTE	**Event class and identifiers:** B#16#29
OB122_SW_FLT	BYTE	**Software fault code:** Possible values below: B#16#42 = I/O read access fault B#16#43 = I/O write access fault
OB122_PRIORITY	BYTE	**Priority:** priority class = priority of OB in which error occurred (that is the OB from which faulted block was called).
OB122_OB_NUMBR	BYTE	**OB number:** 122 = OB122
OB122_BLK_TYP	BYTE	**Block type:** in which error occurred—possible values below – B#16#88 = OB; B#16#8C = FC; B#16#8E = FB **Note:** No valid value is reported for the S7-300
OB122_MEM_AREA	BYTE	**Memory area**—Memory area and access type where error occurred **bits 7 to 4 = access type:** 0 = bit; 1 = byte; 2 = word; 3 = double word **bits 3 to 0 = memory area:** 0 = I/O; 1 = input image; 2 = output image
OB122_MEM_ADDR	WORD	**Memory address:** address where the access error occurred
OB122_BLK_NUM	WORD	**Block number:** block number of block where error occurred (S7-400)
OB122_PRG_ADDR	WORD	**Program address:** relative address where error occurred (S7-400)
OB122_DATE_TIME	DT	**Date and Time:** date-time stamp of when error was detected

Appendices

APPENDIX A: ORGANIZATION BLOCKS

As described earlier in this book, you will be required to write the code for Organization Blocks (OBs) you decide to include in your program. All OBs, while not supported by all CPUs, are included in the Standard Library provided in the STEP 7 Basic package. The OBs are only provided as templates for the declaration area of each of the OBs. To include an OB in your program, you must insert the block as you would with any other block. Remember, not all OBs are available in all CPUs—you must open *Module Information* for your CPU to determine which OBs are supported in the CPU you are using.

Table A-1. Organization Blocks (OBs) list. Provided in STEP 7 Standard Library

OB	Default Priority	Description
		Cyclical Processing
OB 1	1	Main Program Block
		Time-of-Day Interrupt Processing
OB 10	2	Time-of-Day Interrupt—0
OB 11	2	Time-of-Day Interrupt—1
OB 12	2	Time-of-Day Interrupt—2
OB 13	2	Time-of-Day Interrupt—3
OB 14	2	Time-of-Day Interrupt—4
OB 15	2	Time-of-Day Interrupt—5
OB 16	2	Time-of-Day Interrupt—6
OB 17	2	Time-of-Day Interrupt—7
		Time-Delay Interrupt Processing
OB 20	3	Time-Delay Interrupt—0
OB 21	4	Time-Delay Interrupt—1
OB 22	5	Time-Delay Interrupt—2
OB 23	6	Time-Delay Interrupt—3
		Cyclic Interval (Watchdog) Interrupt Processing
OB 30	7	Cyclic Interval Interrupt 0—Default Interval = 5 sec
OB 31	8	Cyclic Interval Interrupt 1—Default Interval = 2 sec
OB 32	9	Cyclic Interval Interrupt 2—Default Interval = 1 sec
OB 33	10	Cyclic Interval Interrupt 3—Default Interval = 500 msec.
OB 34	11	Cyclic Interval Interrupt 4—Default Interval = 200 msec.
OB 35	12	Cyclic Interval Interrupt 5—Default Interval = 100 msec.
OB 36	13	Cyclic Interval Interrupt 6—Default Interval = 50 msec.
OB 37	14	Cyclic Interval Interrupt 7—Default Interval = 20 msec.
OB 38	15	Cyclic Interval Interrupt 8—Default Interval = 10 msec.

APPENDIX A: ORGANIZATION BLOCKS

Table A-1. Organization blocks (OBs) list—continued

OB	Default Priority	Description
		Hardware Interrupt Processing
OB 40	16	Hardware Interrupt 0
OB 41	17	Hardware Interrupt 1
OB 42	18	Hardware Interrupt 2
OB 43	19	Hardware Interrupt 3
OB 44	20	Hardware Interrupt 4
OB 45	21	Hardware Interrupt 5
OB 46	22	Hardware Interrupt 6
OB 47	23	Hardware Interrupt 7
		Specialty Interrupt Processing
OB 60	25 (fixed)	Multi-Computing Interrupt
OB61- OB64	25	Synchronous Cycle Interrupt OBs
OB65	25 (fixed)	Technology Synchronization Interrupt
		Asynchronous Error Processing
OB 80	26, 28	Timing Error
OB 81	26, 28	Power Supply Fault
OB 82	26, 28	Diagnostic Interrupt
OB 83	26, 28	Insert/Remove Module Interrupt
OB 84	26, 28	CPU Hardware Fault
OB 85	26, 28	Priority Class Execution Error
OB 86	26, 28	DP Error
OB 87	26, 28	Communications Error
OB 88	26, 28	Processing Interrupt
		Background Processing
OB 90	29	Background Processing
		Start-up Processing
OB 100	27	Warm Restart
OB 101	27	Hot Restart (S7 400 only)
OB 102	27	Cold Restart
		Synchronous Error Processing
OB 121	27	Programming Error
OB 122	27	I/O Access Error

APPENDIX B: SYSTEM BLOCKS

System blocks are another resource available to the STEP 7 program developer. These blocks, which are integrated in the S7 operating system, may be called in your program. As part of the CPU, the actual code for these blocks is built into the operating system and does not use any of the user memory. To call these blocks, however, you will need the calling interface associated with each SFC or SFB. These interfaces are provided offline in the Standard Library supplied with STEP 7 under the program name "*System Function Blocks.*" The block will appear in the *Blocks* folder of your program, once used.

Calling an SFB or SFC in your program is the same as for calling a user-created FB or FC. Simply insert a new network, or open an existing network, and call the system block as required. If the block requires that actual parameters be supplied, you will need proper documentation that defines each parameter, and the correct operand requirements. Remember that SFBs are like other FBs, and require an instance DB as its memory. You must create and download the instance DB to your program.

Table B-1. System blocks, including SFBs and SFCs. Provided in STEP 7 Standard Library

SFB/SFC	Abbreviation	Description
IEC Counters and Timers		
SFB 0	CTU	UP Counter
SFB 1	CTD	Down Counter
SFB 2	CTUD	UP/Down Counter
SFB 3	TP	Pulse Timer
SFB 4	TON	On Delay Timer
SFB 5	TOF	Off Delay Timer
S7 Communications via Configured S7 Connections		
SFB 8	USEND	Uncoordinated Send
SFB 9	URCV	Uncoordinated Receive
SFB 12	BSEND	Block-Oriented Send
SFB 13	BRCV	Block-Oriented Receive
SFB 14	GET	Read (Fetch) Data from Partner
SFB 15	PUT	Write Data to Partner
SFB 16	PRINT	Write Data to Printer
SFB 19	START	Initiate Complete Restart in Communications Partner
SFB 20	STOP	Initiate STOP in Set Communications Partner
SFB 21	RESUME	Initiate Restart in the Communications Partner
SFB 22	STATUS	Check Status of Communications Partner
SFB 23	USTATUS	Receive Status of Communications Partner
SFC 62	CONTROL	Check Communications status

APPENDIX B: SYSTEM BLOCKS

Table B-1. System blocks, including SFBs and SFCs—continued

SFB/SFC	Abbreviation	Description
		Handling Integrated I/O Functions (CPUs with Integrated I/O)
SFB 29	HS_COUNT	High-Speed Counter
SFB 30	FREQ_MES	Frequency Meter
SFB 38	HSC_A_B	Control 'Counter A/B' (On CPU)
SFB 39	POS	Control 'Positioning' (On CPU)
SFB 41	CONT_C	Continuous-Action Closed-Loop Control (On CPU)
SFB 42	CONT_S	Step-Action Closed-Loop Control (On CPU)
SFB 43	PULSEGEN	Generate Pulse (On CPU)
SFC 63 *	AB_CALL	Call Assembler Code Block
		Drum Sequencer
SFB 32	DRUM	Drum
		Generating Block-Related Messages
SFB 33	ALARM	Messages with Acknowledgement Display
SFB 34	ALARM_8	Messages without Embedded Values
SFB 35	ALARM_8P	Messages with Embedded Values
SFB 36	NOTIFY	Messages without Acknowledgement Display
SFB 37	AR_SEND	Send Archive Data
SFC 9	EN_MSG	Enable Messages (Block-Related, Symbol-Related, Group Status)
SFC 10	DIS_MSG	Disable Messages (Block-Related, Symbol-Related, Group Status)
SFC 17	ALARM_SQ	Generate Block-Related Message with ACK
SFC 18	ALARM_S	Generate Block-Related Message with Permanent ACK
SFC 19	ALARM_SC	Check ACK Status of Last ALARM_SQ Entering State Message
		CPU Clock/Run-Time Meter Operations
SFC 0	SET_CLK	Set System Clock
SFC 1	READ_CLK	Read System Clock
SFC 2	SET_RTM	Set Run-Time Meter
SFC 3	CTRL_RTM	Control Run-Time Meter Start/Stop
SFC 4	READ_RTM	Read Run-Time Meter
SFC 48	SNC_RTCB	Synchronize Slave Clocks
SFC 64	TIME_TCK	Read System Time
		H-System Operations
SFC 90	H_CTRL	Control Operation in H Systems

APPENDIX B: SYSTEM BLOCKS

Table B-1. System blocks, including SFBs and SFCs—continued

SFB/SFC	Abbreviation	Description
		Copy Data Area/Block Operations
SFC 20	BLKMOV	Copy Variables
SFC 21	FILL	Initialize a Memory Area
SFC 22	CREAT_DB	Create a Data Block
SFC 23	DEL_DB	Delete a Data Block
SFC 24	TEST_DB	Test a Data Block
SFC 25	COMPRESS	Compress User Memory in CPU
SFC 44	REPL_VAL	Transfer Substitute Value to ACCU1
SFC 81	UBLKMOV	Un-Interruptible Block Move
		Program Control Operations
SFC 43	RE_TRIGR	Retrigger CPU Cycle Time Monitoring
SFC 46	STP	Switch the CPU to STOP Mode
SFC 47	WAIT	Delay User Program Execution
		Handling Interrupt Events
SFC 28	SET_TINT	Set a Time-of-Day Interrupt
SFC 29	CAN_TINT	Cancel a Time-of-Day Interrupt
SFC 30	ACT_TINT	Activate a Time-of-Day Interrupt
SFC 31	QRY_TINT	Query a Time-of-Day Interrupt
SFC 32	SRT_DINT	Start a Time-Delay Interrupt
SFC 33	CAN_DINT	Cancel a Time-Delay Interrupt
SFC 34	QRY_DINT	Query a Time-Delay Interrupt
SFC 35	MP_ALM	Trigger a Multiprocessor Interrupt
SFC 36	MSK_FLT	Mask Synchronous Faults
SFC 37	DMSK_FLT	Unmask Synchronous Faults
SFC 38	READ_ERR	Read Error Register
SFC 39	DIS_IRT	Disable New Interrupts and Asynchronous Errors
SFC 40	EN_IRT	Enable Interrupts and Asynchronous Errors
SFC 41	DIS_AIRT	Disable (or Delay) New Interrupts and Asynchronous Errors
SFC 42	EN_AIRT	Enable (previously disabled) Interrupts and Asynchronous Errors
		System Diagnostic Operations
SFC 6	RD_SINFO	Read OB Start Information
SFC 51	RDSYSST	Read System Status List/Partial List
SFC 52	WR_USMSG	Write User-Defined Diagnostic Event to Diagnostic Buffer

APPENDIX B: SYSTEM BLOCKS

Table B-1. System blocks, including SFBs and SFCs—continued

SFB/SFC	Abbreviation	Description
		Distributed I/O Operations
SFC 7	DP_PRAL	Trigger DP-master Hardware Interrupt
SFC 11	DPSYC_FR	Synchronize DP-slave Groups
SFC 12	D_ACT_DP	Deactivate/Activate DP-slaves
SFC 13	DPNRM_DG	Read DP-slave Diagnostic Data
SFC 14	DPRD_DAT	Read DP-slave Consistent Data
SFC 15	DPWR_DAT	Write DP-slave Consistent Data
		Process Image Update Operations
SFC 26	UPDAT_PI	Update Process Image of Inputs
SFC 27	UPDAT_PO	Update Process Image of Outputs
SFC 79	SET	Set a Range of Outputs
SFC 80	RSET	Reset a Range of Outputs
		Address/Access Module Operations
SFC 5	GADR_LGC	Get Logical Address of a Channel
SFC 49	LGC_GADR	Check Module Slot of a Specific Logical Address
SFC 50	RD_LGADR	Check Module Logical Addresses
		Data Record Transfer
SFC 54	RD_PARM	Read Defined Parameters
SFC 55	WR_PARM	Write Dynamic Parameters
SFC 56	WR_DPARM	Write Default Parameters
SFC 57	PARM_MOD	Assign Parameters to a Module
SFC 58	WR_REC	Write a Data Record
SFC 59	RD_REC	Read a Data Record
		Use with S7 Global Data Communications
SFC 60	GD_SND	Send a Global Data
SFC 61	GD_RCV	Accept Received Global Data
		Use with S7 Communications via Un-Configured Connections
SFC 65	X_SEND	Send Data to External Partner
SFC 66	X_RCV	Receive Data from Ext. Partner
SFC 67	X_GET	Read Data from External Partner
SFC 68	X_PUT	Write Data to External Partner
SFC 69	X_ABORT	Abort Connection to Ext. Partner
SFC 72	I_GET	Read Data from Internal Partner
SFC 73	I_PUT	Write Data to Internal Partner
SFC 74	I_ABORT	Abort Connection to Int. Partner

APPENDIX C: MODIFY/FORCE I/O AND MEMORY VARIABLES

The entries of Table C-1 represent typical address entries when using the STEP 7 Monitor/Modify Variables utility. The table presents bit, byte, word, and double word addresses from S7 memory areas including Inputs (I), Outputs (Q), Bit Memory (M), Peripheral Inputs (PI), Peripheral Outputs (PQ), Data Blocks (DB), Timers (T), and Counters (C). The Monitor/Modify utility facilitates the online/runtime viewing and modifying, as well as the forcing, of I/O and memory variables.

When a variable is modified, it will remain as modified unless overwritten by the user program; a forced variable will remain modified (forced) until the force status is removed.

Table C-1. Typical I/O and Memory Addresses with Valid Modify/Force Values

Possible Bit Addresses (I, Q, M, DBX)		
Address	**Force/Modify Value**	**Description**
I4.5	false	BOOL variable; FALSE also valid
M4.2	true	BOOL variable; TRUE also valid
Q11.6	0	Bit variable without binary identifier
DB5.DBX1.0	1	Bit variable without binary identifier
I4.4	2#1	Binary variable with binary identifier
M5.5	2#0	Binary variable with binary identifier
Possible Byte Addresses (IB, QB, MB, PQB, DBn.DBBn)		
Address	**Force/Modify Value**	**Description**
IB 10	2#11011001	BYTE variable as binary constant with binary identifier
MB 22	b#16#FF	BYTE variable as hex constant with hex identifier
MB 24	FF	BYTE variable as hex constant w/o hex identifier
QB 10	'y'	BYTE variable as CHAR constant
DB5.DBB 2	10	BYTE variable as positive INT constant
PQB 4	-12	BYTE variable as negative INT constant
Possible Word Addresses (IW, QW, MW, PQW, DBn.DBWn)		
Address	**Force/Modify Value**	**Description**
IW 3	2#0000111100001111	WORD variable as binary constant with binary identifier
MW 42	w#16#ABYN	WORD variable as hex constant with hex identifier
MW 44	ABYN	WORD variable as hex constant w/o hex identifier
QW 12	b#(20,44)	WORD variable as 2-byte unsigned decimal values
DB5.DBW 2	'yn'	WORD variable as CHAR constant
PQW 6	-10500	WORD variable as negative INT constant (-32,768, to 32,767)
MW 2	10500	WORD variable as positive INT constant (-32,768, to 32,767)
MW 4	S5T#10m30s	WORD variable as S5TIME value with S5TIME identifier
MW 8	c#175	WORD variable as COUNT value with Count identifier
MW 10	d#2005-07-04	WORD variable as DATE value w/o DATE identifier

Note: You may specify the format in which you wish to view the monitored variable.

APPENDIX C: MODIFY/FORCE I/O AND MEMORY VARIABLES

Table C-1. Typical I/O and Memory Addresses with Valid Modify/Force Values—Continued

Possible Double Word Addresses (ID, QD, MD, PQD, DBn.DBDn)		
Address	**Force/Modify Value**	**Description**
ID 4	00011000111001010111010111010111	DWORD variable as binary constant w/o identifier
MD 2	10 E3	DWORD variable as REAL constant
MD 6	9	DWORD variable as hex value w/o identifier
QD 20	DW#16#abcdef10	DWORD variable as hex value with identifier
QD 24	ABCDEF10	DWORD variable as hex value w/o identifier
DB5.DBD 6	B#(12,34,56,78)	DWORD variable as 4-unsigned decimal numbers
PQD 4	'abcd'	DWORD variable as STRING constant
MD 20	L# -12	DWORD variable as DINT value with identifier
MD 24	L#12	DWORD variable as DINT value with identifier
MD 28	-123456789	DWORD variable as DINT value w/o identifier
MD 32	123456789	DWORD variable as DINT value w/o identifier
MD 36	T#30s250ms	DWORD variable as TIME value
MD 40	TOD#1:2:34.567	DWORD variable as Time-of-Day value
MD 44	p#e0.0	DWORD variable as POINTER value
Timer Memory Addresses (Tn)		
Address	**Force/Modify Value**	**Description**
T 10	0	Conversion to milliseconds (ms)
T 11	20	Conversion to ms
T 12	12345	Conversion to ms
T 13	S5TIME#10m30s	Timer value with S5TIME (or S5T) identifier
Counter Memory Addresses (Cn)		
Address	**Force/Modify Value**	**Description**
C 15	0	WORD variable as COUNT value w/o identifier
C 16	35	WORD variable as COUNT value w/o identifier
C 17	c#150	WORD variable as COUNT value with identifier

APPENDIX D: ANALOG I/O DIGITAL REPRESENTATION

Table D-1 presents an overview of standard analog I/O signal ranges for S7 modules with bipolar operating ranges. Bipolar ranges are represented, along with intermediate values within the normal operating range, as well as the values at over-range and absolute overflow, or underflow conditions.

Table D-1. Analog Input/Output Integer Representation for common Bipolar Measurement Ranges

Range	± 500 mV	± 1 V	± 2.5 V	± 5 V	± 10 V	± 20 mA	
+ 32,767	≥ 587.96	≥ 1.1760	≥ 2.9398	≥ 5.8796	≥ 11.759	≥ 23.5160	OVERFLOW
+ 32,511	+ 587.94	+ 1.1750	+ 2.9397	+ 5.8794	11.7589	23.5150	OVER-RANGE
.	.	.	.	.	.	.	
.	.	.	.	.	.	.	
+ 27,649	+ 500.02	+ 1.0004	+ 2.5001	+ 5.0002	10.0004	20.0007	
+ 27,648	+ 500.00 mV	1.0000 V	+2.5000 V	5.000 V	+10.00 V	+20.00 mA	
+ 24,192	+437.50	+0.8750	+2.1875	+4.375	+8.75	+17.50	
+ 20,736	+375.00	+0.7500	+1.8750	+3.750	+7.50	+15.00	
+ 17,280	+312.50	+0.6250	+1.5625	+3.125	+6.25	+ 2.50	
+ 13,824	+250.00	+0.5000	+1.2500	+2.500	+5.00	+10.00	
+ 10,368	+187.50	-0.3750	+0.9375	+1.875	+3.75	+7.50	
+ 6,912	+125.00	+0.2500	+0.6250	+1.250	+2.50	+5.00	
+ 3,456	+62.50	+0.1250	+0.3125	+0.625	+1.25	+2.50	
0	0.00	0.0000	0.0000	0.000	0.00	0.00	NORMAL
- 3,456	-62.50	-0.1250	-0.3125	-0.625	-1.25	-2.50	
- 6,912	-125.00	-0.2500	-0.6250	-1.250	-2.50	-5.00	
- 10,368	-187.50	-0.3750	-0.9375	-1.875	-3.75	-7.50	
- 13,824	-250.00	-0.5000	-1.2500	-2.500	-5.00	-10.00	
- 17,280	-312.50	-0.6250	-1.5625	-3.125	-6.25	-12.50	
- 20,736	-375.00	-0.7500	-1.8750	-3.750	-7.50	-15.00	
- 24,192	-437.50	-0.8750	-2.1875	-4.375	-8.75	-17.50	
- 27,648	-500.00 V	-1.000 V	-2.5000 V	-5.00 V	-10.00 V	-20.00 mA	
- 27,649	-500.02	-1.0004	-2.5001	-5.0002	-10.0004	-20.0007	OVER-RANGE
			.	.	.	.	
	-587.96	-1.1750	-2.9398	-5.8796	-11.7590	-23.5160	
- 32,768	≤ -588.98	≤ -1.176	≤ -2.935	≤ -5.880	≤ -11.76	≤ -23.517	UNDERFLOW

APPENDIX E: S7/STEP 7 ABBREVIATIONS AND ACRONYMS

Table E-1 presents a list of some of the common abbreviations and acronyms used when working with S7 PLCs and the STEP 7 programming environment.

Table E-1. Common Abbreviations and Acronyms

Abbreviation	Description
AI	Analog Input
AO	Analog Output
CFC	Continuous Function Chart
CP	Communications Processor Module
CPU	Central Processing Unit Module
DB	Data Block
DI	Digital Input
DO	Digital Output
DP	Distributed Periphery (or Distributed I/O)
EN	Enable Input Line
ENO	Enable Output Line
FB	Function Block
FBD	Function Block Diagram language
FC	Function
FM	Function Module
IM	Interface Module
LAD	Ladder Diagram Language
MPI	Multi-Point Interface
OB	Organization Block
OP	Operator Panel
PG/PC	Programming Device/PC Programming Device
PII	Process Image of Inputs
PIQ	Process Image of Outputs
RLO	Result of Logic Operation
SCL	Structured Control Language
SDB	System Data Block
SFB	System Function Block
SFC	System Function
SM	Signal Module
STL	Statement List Language
UDT	User Defined Type
VAT	Variable Table

APPENDIX F: SYMBOL TABLE ADDRESSES AND DATA TYPES

Table F-1 presents a list of IEC address operand prefixes for S7 memory areas and block objects that may be given symbolic labels. The table provides a description of each memory area or object, its valid data types, and the range of addresses that may be assigned a symbolic address. The selection of the IEC (English), or SIMATIC (German), mnemonics (address prefixes) that will be used throughout the entire program is selected and switched from the SIMATIC Manager.

Table F-1. Addresses and Data Types permitted in symbol tables

IEC	Description	Valid Data Types	Address Range
I	Input bit	BOOL	0.0 to 65535.7
IB	Input byte	BYTE, CHAR	0 to 65535
IW	Input word	WORD, INT, S5TIME, DATE	0 to 65534
ID	Input double word	DWORD, DINT, REAL, TOD, TIME	0 to 65532
Q	Output bit	BOOL	0.0 to 65535.7
QB	Output byte	BYTE, CHAR	0 to 65535
QW	Output word	WORD, INT, S5TIME, DATE	0 to 65534
QD	Output double word	DWORD, DINT, REAL, TOD, TIME	0 to 65532
M	Memory bit	BOOL	0.0 to 65535.7
MB	Memory byte	BYTE, CHAR	0 to 65535
MW	Memory word	WORD, INT, S5TIME, DATE	0 to 65534
MD	Memory double word	DWORD, DINT, REAL, TOD, TIME	0 to 65532
PIB	Peripheral input byte	BYTE, CHAR	0 to 65535
PQB	Peripheral output byte	BYTE, CHAR	0 to 65535
PIW	Peripheral input word	WORD, INT, S5TIME, DATE	0 to 65534
PQW	Peripheral output word	WORD, INT, S5TIME, DATE	0 to 65534
PID	Peripheral input double word	DWORD, DINT, REAL, TOD, TIME	0 to 65532
PQD	Peripheral output double word	DWORD, DINT, REAL, TOD, TIME	0 to 65532
T	Timer	TIMER	0 to 65535
C	Counter	COUNTER	0 to 65535
FB	Function block	FB	0 to 65535
OB	Organization block	OB	0 to 65535
DB	Data block	DB, FB, SFB, UDT	0 to 65535
FC	Function	FC	0 to 65535
SFB	System function block	SFB	0 to 65535
SFC	System function	SFC	0 to 65535
VAT	Variable table	-	0 to 65535
UDT	User-defined data type	UDT	0 to 65535

Glossary

- A -

absolute address: A direct reference to a memory location, represented by a character identifier and a number. Q 6.7 and M 44.3 are absolute addresses. See symbolic address.

absolute addressing: Use of the direct references of memory locations like I 5.6, instead of substitute or name references such as 'STOP_PB' to reference memory locations. See absolute address.

accumulator (ACCU): Memory registers in the CPU that serve as buffers for load and transfer operations, as well as for comparisons, math, and conversion operations.

actual parameter: The address that replaces a formal parameter when a function block (FB) or function (FC) is called. For example, the formal parameter 'STOP' may be replaced by the actual parameter 'I 60.2'. Also see block parameters; formal parameter.

address: 1) A reference number that identifies a unique memory bit, byte, word, or double word location. 2) An identifying number designating a unique system entity or object.

address identifier: A single or double letter designation that precedes an address, and indicates the specific memory area and size of the memory unit being referenced. The first letter indicates the area, and is always a bit if only one letter is present; a second letter will indicate the unit size—'B' for byte, 'W' for word, and 'D' for double word.

asynchronous error OBs: The group of organization blocks called when one of the asynchronous errors occurs.

asynchronous errors: Run-time errors that are not associated with, or resultant from, any particular aspect of the user program (e.g., power supply error, module inserted/removed). See asynchronous error OB.

- B -

background processing: A method of processing non-critical code (in OB90) in the slice of time that starts with the end of the program scan time and continues until the expiration of the most set minimum CPU scan. If the actual execution cycle time is less than the user-specified minimum cycle time (S7 400), OB90 is processed for the remaining duration of the minimum cycle time. Then, the OB1 cycle is re-started.

bit memory: The area of S7 system memory containing what may be referred to as the CPU's internal storage bits or software relays. The letter 'M' precedes each bit memory address. For example, M10.0 is a bit, MB 10 is a byte, MW 10 is a word, and MD 10 a double word.

blocks: Blocks make up the STEP 7 user program, and are distinguishable by function, structure, or purpose. STEP 7 blocks may be grouped as logic code blocks of types FB, FC, OB, SFB, and SFC; data blocks of types DB, and SDB, and user-defined data types (UDT).

block parameters: Refers to the named input/output channels that allow data to be passed into (input parameters), or output from (output parameters), a block function.

- C -

CC0 status bit: Condition code status bit of the CPU status word.

CC1 status bit: Condition code status bit of the CPU status word.

code block: S7 program resource that allows program construction in modular sections of code. These sections are referred to as blocks. Logic block types include Organization blocks (OBs), Function blocks (FBs), and Functions (FCs). System Functions (SFCs) and System Function Blocks (SFBs) are code blocks that are part of the CPU operating system.

cold restart: A restart in which the CPU reinitializes itself (including system memory and retentive memory), both process image tables (input image/output image), and I/O module memory. Upon cold restart, the program always calls OB 102 and upon completion starts at the beginning of OB1

complete restart: A restart in which the CPU reinitializes itself and and I/O modules, and clears the non-retentive memory, and the process images (inputs/outputs). Upon cold restart, the program always calls OB 100 and upon completion starts at the beginning of OB1.

control logic: 1) The combination of conditions that must be satisfied to control a particular device or perform a particular function. 2) Refers to the entire program logic or relay logic that will control a given machine or process.

counter memory (C): The area of the system memory containing locations used for software counters. Each counter word, when set, contains the user-defined preset value as a BCD value. The letter 'C' precedes each S7 counter address, as in the counters C0, C1, C2, and C3.

counter, down: An S7 software counter instruction that decrements its count value by one for each off-to-on transition of its count-down input.

counter, up: S7 software counter instruction that increments its count value by one for each off-to-on transition of its count-up input.

counter, up-down: A STEP 7 software counter instruction that combines the actions of both the up-counter and down-counter.

cyclic interrupt: S7 300 and S7 400 systems provide cyclic interrupt OBs to support periodic processing of certain portions of the user program, independent of the normal cyclic processing time of the CPU. Cyclic interrupt are available at intervals of 10 milliseconds to 5 seconds.

cyclic interrupt OBs: OB 30 to OB 38. Organization blocks provided for servicing cyclic interrupts.

- D -

Data Block (DB): A STEP 7 block that allows users to store data associated with the program and the process. There are two types of data blocks: the shared data block, whose data is accessible by all logic code blocks, and the instance data block, whose data is associated with a specific function block (FB).

Data Block, instance: See data block.

Data Block, shared: See data block.

data type: Defines the characteristics of a data element with respect to its size in bits, its STEP 7 representation, and its permissible number range. Some example data types include BOOL, INT, REAL, ARRAY, and STRING. In STEP 7, there are two categories of data types—elementary and complex.

data type, complex: Category of data types that reserve memory for variables that are comprised of two or more data elements of elementary data type. Complex data types include DATE-AND-TIME, STRING, ARRAY, STRUCT, and UDT.

data type, elementary: Category of data types that reserves a single bit, byte, word, or double word of memory. Elementary data types include BOOL, BYTE, WORD, DWORD, INT, DINT, REAL, DATE, TIME, S5TIME, TIME-OF-DAY, and CHAR.

data type, user defined: A user-defined structure of any combination of elementary or complex data types, to be used as a template for creating data blocks of the same data structure, or for declaring a frequently used structure as a data type.

data type, parameter: Group of data types that represents variables that allow timers and counters, as well as blocks, to be passed as block parameters in an FC or FB. Parameter data types include TIMER, COUNTER, BLOCK_FC, BLOCK_FB, BLOCK_DB, BLOCK_SDB, POINTER, and ANY.

declaration section: The area of a STEP 7 code block, or data block, in which variables may be defined.

direct addressing: A method of addressing in which the memory location to be accessed or operated upon, is the address referenced by the instruction. The address can be absolute (e.g., I4.7) or symbolic (e.g., E-STOP).

direct I/O read: An instruction operation that uses the peripheral input (PI) address area to access an input module's data directly from the module or I/O bus, as opposed to normal reading from the process image of inputs (PII).

direct I/O write: An instruction operation that uses the peripheral output (PQ) address area to access an output module's data directly via the module or I/O bus, as opposed to normal writing to the process image of outputs (PIQ).

- E -

error, asynchronous: Run-time errors that are associated with, or resulting from, some aspect of the user program (e.g., addressing error, I/O module access).

error, synchronous: Run-time errors that are not associated with, or resulting from, the user program (e.g., power supply fault).

error OBs: Organization blocks that allow users to write code to respond to system-related (asynchronous), or program-related (synchronous), errors. There is an associated OB for each error type. If the OB is loaded in the CPU as part of the user program, it will be called by the operating system if the associated error occurs.

extended pulse timer: See Pulse Timer Extended.

- F -

FBD: Abbreviation for Function Block Diagram—one of the three language representations of the basic STEP 7 programming package. Control logic programming based on Boolean gate-logic diagrams. Also, see LAD, STL.

formal parameter: A placeholder for an actual parameter, used when creating a logic block that can be passed parameters. In FBs and FCs, formal

parameters are defined by the user; in SFBs and SFCs, formal parameters are predefined. Also, see actual parameter.

Function (FC): According to the IEC 1131-3 standard, functions are logic blocks that do not reference an instance data block, meaning they do not have a 'memory'.

Function Block (FB): According to the IEC 1131-3 standard, function blocks are logic blocks that reference an instance data block for use as memory.

Function Block Diagram: See FBD.

- G -

global symbol: A symbolic address known to the entire user program (i.e., all blocks), as opposed to a local symbol, defined in a block, and which is known only to the specific block. See symbolic address.

- H -

hardware interrupt: An ability of some S7 modules to interrupt the CPU when certain process, or module, generated events occur.

- I -

image table: See process image.

indirect addressing: An addressing method in which the desired absolute address is not used directly in an instruction operation; instead the address is contained in another memory location, used as a pointer to the actual address.

input image table: See input memory.

Input (I) memory: Memory area that stores snapshot status of connected digital inputs. Each connected input has a bit in the input memory that corresponds to the configured byte address for the module. The address identifier 'I' indicates an input memory address when used in the program (e.g., I10.0 is a bit, IB 10 is a byte, IW 10 is a word, and ID 10 a double word. Also, see process image inputs.

instance: An 'instance' is the call of a function block. If a Function is called four times in a STEP 7 program, then there are four instances. An instance data block is assigned to each call. See data block.

instance Data Block: See data block.

interrupt: An efficient method of requesting the immediate attention of a central processor to gain some type of service. Interrupts are usually classified as hardware (e.g., process signal, I/O module with interrupt lines), or software (e.g., timed interrupts).

I/O image table: See process image.

- L -

LAD: Abbreviation for Ladder—one of the thrêe language representations of the basic STEP 7 programming package. Control logic programming based on relay-like instructions. Also, see FBD, STL.

Ladder: See LAD.

ladder network: See ladder rung.

ladder rung: A single Ladder diagram network, part of a complete program or block, that performs the desired control logic in part, or in whole, for a single device or function.

library: A STEP 7 object that contains reusable programs or program components. Users may create libraries from the Simatic Manager, and certain standard libraries are supplied with STEP 7 basic and optional packages

library, standard: A STEP 7 library that is provided with the basic package.

linear program: Refers to a STEP 7 user program designed such that the entire main program is coded in organization block 1 (OB1). See partitioned program; structured program.

local data: Temporary data stored in the local (L) stack and accessed exclusively by the executing block. The local data area is reserved in a block when temporary (TEMP) variables are defined. Also, see local stack.

local data stack: See local stack.

local data: Temporary data stored in the local (L) stack memory, and used exclusively by the executing block for temporary results. The local variables defined in the block header determine the size requirements of the local data. See local stack.

local memory (L): See local stack.

local stack: Temporary memory locations made available by the CPU to each code block as it is called. The so-called L-stack is released when a block terminates, and is available for use by the next called

block. The block header reserves the L-stack length, however, the maximum length is CPU dependent.

local symbol: A name assigned to a block variable and known only to the block in which it is defined, as opposed to global symbols assigned in the Symbol editor, and that are known to the entire user program (i.e., all blocks). See symbolic address.

- M -

maximum cycle time: A CPU parameter that allows setting of a maximum time for processing the main program. If the actual processing time of the main program (OB1) is more than this specified maximum cycle time, then a call for OB1 is retriggered, and program processing starts from the beginning.

minimum cycle time: A CPU parameter that allows setting of a minimum time for processing the main program. If the actual processing time of the main program (OB1) is less than the specified minimum cycle time, then the CPU will wait until the specified minimum cycle time expires before starting the next cycle of OB1.

multi-computing interrupt: An S7 400 feature that allows any of four possible CPUs operating in a multiple processor configuration, to generate an interrupt to all of the other participants such that a synchronized response is possible. See multi-computing interrupt OB.

multi-computing interrupt OB: OB 60 is provided in S7 400 for servicing user-configured multiprocessing. See multiprocessing interrupt; multiprocessing.

- N -

nesting depth: Refers to the number of block calls that can be made in a horizontal direction, starting for example, from the main organization block (OB1). Nesting-depth is CPU-dependent. The first OB represents nesting depth of one, hence six more block calls in a horizontal direction would be a nesting depth of seven.

network: Divides Ladder Logic (LAD) and Function Block Diagram (FBD) blocks into logic rungs or Statement List code, into smaller segments. A network normally consists of a single string of logic.

- O -

OB 1: Organization Block 1; the main code block of a STEP 7 program. Once downloaded to the CPU, OB1 processes cyclically. All other blocks are called directly or indirectly from OB1, or from another organization block.

object: Refers to an item in a STEP 7 folder that can be opened and edited, automatically starting the appropriate application in the software; for example, block, source file, or station.

off-delay timer (S_OFFDT): A timer whose output is activated when the enabling signal goes TRUE, and starts timing when the enable goes FALSE. After the programmed timed delay expires, the timer coil de-energizes.

on-delay timer (S_ODT): A timer that starts timing when the enable signal goes TRUE and continues to time unless the enable goes false, or the timer is reset. The output is activated after the preset time has elapsed, and stays energized until the enable signal goes FALSE, or the timer is reset.

Organization Block (OB): A special category of STEP 7 code block that provides the user program with an interface to the CPU's operating system. Organization blocks allow processing of the main user program, as well as organized program response to various categories of system related conditions. What OBs are available is CPU dependent.

output image table: See output memory.

Output (Q) memory: System memory area for storing the status of connected digital outputs. Each connected output has a bit in the output memory that corresponds exactly to the terminal to which the output is connected. The letter 'Q' precedes each output memory address when used in the program (e.g., Q10.0 is a bit, QB 10 is a byte, QW 10 is a word, and QD 10 a double word). Also, see process image outputs.

- P -

parameter: 1) A variable input to or output from an S7 code block. 2) A variable that can be set on an S7 module to determine one or more aspects of the module's behavior.

partitioned program: A STEP 7 user program that, in essence, is a linear program; however the program is subdivided into blocks of code. The blocks are then called in sequence. See linear program; structured program.

Peripheral inputs (PI): An address area that allows direct read access to the data of an input module via the I/O bus. In other words, use of a PI address allows a direct or immediate input read. See direct I/O read.

peripheral outputs (PQ): An address area that allows direct write access to the data area of an output module via the I/O bus. In other words, use of a PQ address allows a direct or immediate output write to a module. See direct I/O write.

priority: A method of assigning the order in which portions of code are executed based on level of importance. A program of higher priority can interrupt one of lower priority.

priority class: A program hierarchy level into which a STEP 7 organization block can be placed. There are 28 groupings, or priority classes, to which OBs are assigned. Priority classes determine how OBs may interrupt one another. OBs assigned the same priority class do not interrupt one another, but are processed in sequence.

process image: In the S7 world, that area of system memory that contains a snap-shot image of the status of digital inputs and digital outputs—the image is updated on each CPU scan. Also, see process image inputs; process image outputs.

process image inputs (PII): The system memory area that contains a snap-shot image of the input status bits (ON = 1, OFF = 0) as read from the connected digital input modules, at the beginning of each CPU cycle.

process image outputs (PIQ): The system memory area that contains an image of the output status bits as set by the control logic (ON = 1, OFF = 0), and are transferred to the connected digital output modules at the end of each CPU cycle.

pulse timer (S_PULSE): A timer that starts to time when the trigger input (RLO) goes from low to high, and continues to time for the programmed duration, or until the trigger signal goes from high to low. The timer is reset when the reset input goes high while the timer is timing.

pulse timer extended (S_PEXT): An S7 timer that starts to time when the trigger input transitions from low to high, and continues to time until the programmed duration expires; regardless of any change at the trigger input before the timer times out.

- R -

restart, complete: See complete restart.

restart, cold: See cold restart.

restart, warm: See warm restart.

result of logic operation (RLO): The current result of a series of logic operations in the user program, which is then used to further process signals digitally. Certain instructions are either executed, or not, depending on the last RLO.

retentive bit memory: Part of the bit memory area designated to retain its contents or signal states even under off-circuit conditions. Retentive bit memory, by default, starts at byte 0 of the (M) area, and ends according to user designation.

retentive on-delay timer (ODTS): A timer that starts timing when the enable signal goes TRUE and continues to time even if the enable goes false. If the enable signal changes back to '1' before the timer expires, the timer will restart. The output is activated after the preset time has elapsed, and remains activated until the timer is reset.

rising edge: The rising edge of a signal—that is the OFF-to-ON or 0-to-1 transition.

rung: See network.

run-time meter: A CPU function that counts hours. Run-time meters are used for tasks such as determining total CPU run-time, or the running time of certain devices connected to the CPU.

- S-

STL: Abbreviation for Statement List language— one of the three language representations of the basic STEP 7 programming package.

standard blocks: Turnkey S7 blocks provided as part of a library. Standard blocks are delivered as a library part of the STEP 7 basic or optional software.

start information: Every organization block reserves 20-bytes of local data that the operating system supplies as start information when the OB is started. The start information reports the OB start event, the date and time the OB started, errors that have occurred, and diagnostic events.

Start-up OBs: Organization Blocks that are called on one of the three types of S7 start-up modes: warm restart (OB 100); hot restart (OB 101); and cold restart (OB 102). Start-up OBs are called just prior to starting the cyclical program (OB1).

Statement List: See STL.

static parameter: An S7 module parameter that can only be set or modified using the STEP 7 software, contrasted to dynamic parameters that may be set via the STEP 7 user program. An example static is the input delay on a digital input module.

static variable: One of the variable types that can be defined in an S7 function block (FB). Static variables are variables that must be maintained from one call of the FB to the next. These variables are saved in the associated instance data block when the FB is terminated.

status bits: Binary indicator flags used by the CPU during binary logic operations, and set by the CPU during digital operations like conversions, arithmetic, and word logic. Status bits are available as bit instructions that may be combined with other bit instructions.

structure: A composite grouping of variables with different data types. An example where a structure may be required is a recipe, using data types REAL, INT, TIME, and DATE. It could also contain an array.

structured program: Refers to a STEP 7 user program designed to take advantage of STEP 7 resources that allow programming of modular block functions that can be reused throughout the program to minimize code writing and redundant code development.

symbol: See symbolic address. Also, see global symbol, local symbol.

symbolic addressing: Use of substitute or name references like 'STOP_PB' to reference memory locations, instead of absolute addresses, like I 6.7. See absolute address.

symbols table: A STEP 7 program component that supports the definition of symbol names that represent absolute addresses. Symbol definition involves specifying the name (symbol), absolute address, data type, and an associated comment.

Symbol table definitions are considered global. See global symbol.

synchronous error OBs: Those OBs called when one of several possible synchronous errors occur.

synchronous errors: Run-time errors that are associated with, or resultant from, a particular aspect of the user program (e.g., addressing error, error accessing I/O module). See synchronous error OBs.

system blocks: S7 blocks that are actually components of the operating system. Such blocks include system functions (SFCs), system function blocks (SFBs), and system data blocks (SDBs).

system data object: The compiled configuration data that is loaded to the CPU. The data contains configured hardware, networks, and connection s.

System Function (SFC): An S7 Function, integrated in the operating system of an S7 CPU. SFCs are called from the user program, just as any other Function.

System Function Block (SFB): An S7 Function Block, integrated in the operating system of an S7 CPU. SFBs are called from the user program, just as any other Function Block.

- T -

temporary local data: Variables defined using the (temp) designation in the declaration table of a block. Variables designated as such are only available to the local block, and only while the block is being processed. See local stack.

time base: A unit of time generated by the system clock and used by software timer instructions. Normal time bases are 0.01, 0.1, and 1.0 second.

time-delay interrupt: An S7 interrupt capability that causes an interrupt to the CPU after a specific amount of time has expired (e.g., every 100 ms). S7 300 and S7 400 systems provide time-delay-interrupt OBs to allow creation of service routines to respond to user-configured time-delay interrupts. See time delay interrupt OBs.

time-delay interrupt OBs: OB 20 to OB 23. Organization blocks provided for servicing time-delay interrupts. The code blocks called from these OBs will support timed regulated processing of certain portions of the user program.

time-of-day interrupt: S7 300 and S7 400 systems provide time-of-day interrupts to support processing of certain portions of the user program at a specific time of day, either once only, or periodically (e.g., hourly, daily, weekly, monthly, yearly, or every minute). See time-of-day interrupt OBs.

time-of-day interrupt OBs: OB 10 to OB 17. Organization blocks provided for servicing user-configured time-of-day interrupts. See time-of-day interrupt.

timer memory (T): The area of the system memory containing locations used for software timers. Each timer word, when set, contains the timer preset value in BCD, and the user-defined time-base of the timer. The letter 'T' precedes each S7 timer address (e.g., T0, T1, T2, and T3 each is a unique timer.).

timer preset value: A programmed value that determines the number of time-base intervals to be counted in a software timer, and, subsequently, the programmed time duration.

- U -

UDT: User defined data type. A composite of user-defined elements of arbitrary data types that, once defined, may be used to define local variables of a block or data block; or as a template when defining an entire data block.

User Defined Type (UDT): See UDT.

- V -

variable: A factor that can be altered, measured, or controlled; quantity that can change in value.

variable declaration: The act of defining a variable for use in the user program. Declaration requires naming the variable, and defining its data type (e.g., BOOL, INT, etc.).

variable declaration table: The part of a block in which its local variables are defined. In the case of code blocks, depending on the block type, these are the block parameters, as well as the temporary or static data. In the case of data blocks, the declared variables are the data addresses.

variable table (VAT): User created table containing a group of variables to be displayed, monitored and perhaps modified or forced. Up to 255 variable tables, from VAT 1 to VAT 255, may be created as resources for monitoring and debugging the control program.

VAT: See variable table.

- W -

warm restart: Only possible in S7-400 CPUs. Upon warm restart, the program always resumes or continues from the point at which the program was interrupted, and finishes to completion. The warm restart can occur when no changes took place in the program while the CPU was in STOP.

Upon warm restart, the program always calls OB 101 and thencompletes the remaining portion of the OB1 cycle.

Index

Copyright © 2013